Open Distributed Processing

IFIP – The International Federation for Information Processing

IFIP was founded in 1960 under the auspices of UNESCO, following the First World Computer Congress held in Paris the previous year. An umbrella organization for societies working in information processing, IFIP's aim is two-fold: to support information processing within its member countries and to encourage technology transfer to developing nations. As its mission statement clearly states,

> IFIP's mission is to be the leading, truly international, apolitical organization which encourages and assists in the development, exploitation and application of information technology for the benefit of all people.

IFIP is a non-profitmaking organization, run almost solely by 2500 volunteers. It operates through a number of technical committees, which organize events and publications. IFIP's events range from an international congress to local seminars, but the most important are:

- the IFIP World Computer Congress, held every second year;
- open conferences;
- working conferences.

The flagship event is the IFIP World Computer Congress, at which both invited and contributed papers are presented. Contributed papers are rigorously refereed and the rejection rate is high.

As with the Congress, participation in the open conferences is open to all and papers may be invited or submitted. Again, submitted papers are stringently refereed.

The working conferences are structured differently. They are usually run by a working group and attendance is small and by invitation only. Their purpose is to create an atmosphere conducive to innovation and development. Refereeing is less rigorous and papers are subjected to extensive group discussion.

Publications arising from IFIP events vary. The papers presented at the IFIP World Computer Congress and at open conferences are published as conference proceedings, while the results of the working conferences are often published as collections of selected and edited papers.

Any national society whose primary activity is in information may apply to become a full member of IFIP, although full membership is restricted to one society per country. Full members are entitled to vote at the annual General Assembly, National societies preferring a less committed involvement may apply for associate or corresponding membership. Associate members enjoy the same benefits as full members, but without voting rights. Corresponding members are not represented in IFIP bodies. Affiliated membership is open to non-national societies, and individual and honorary membership schemes are also offered.

Open Distributed Processing
Experiences with distributed environments

Proceedings of the third
IFIP TC 6/WG 6.1 international
conference on open distributed
processing, 1994

Edited by

Kerry Raymond and Liz Armstrong
CRC for Distributed Systems Technology
Brisbane
Australia

Published by Chapman & Hall on behalf of the
International Federation for Information Processing (IFIP)

 CHAPMAN & HALL
London · Glasgow · Weinheim · New York · Tokyo · Melbourne · Madras

Published by Chapman & Hall, 2–6 Boundary Row, London SE1 8HN, UK

Chapman & Hall, 2–6 Boundary Row, London SE1 8HN, UK

Blackie Academic & Professional, Wester Cleddens Road, Bishopbriggs, Glasgow G64 2NZ, UK

Chapman & Hall GmbH, Pappelallee 3, 69469 Weinheim, Germany

Chapman & Hall USA, 115 Fifth Avenue, New York, NY 10003, USA

Chapman & Hall Japan, ITP-Japan, Kyowa Building, 3F, 2-2-1 Hirakawacho, Chiyoda-ku, Tokyo 102, Japan

Chapman & Hall Australia, 102 Dodds Street, South Melbourne, Victoria 3205, Australia

Chapman & Hall India, R. Seshadri, 32 Second Main Road, CIT East, Madras 600 035, India

First edition 1995

© 1995 IFIP

Printed in Great Britain by TJ Press, Padstow, Cornwall

ISBN 0 412 71150 8

A catalogue record for this book is available from the British Library

CONTENTS

Preface   ix

**3rd International IFIP TC 6 Conference on Open
Distributed Processing**   xiii

List of Referees   xiv

PART ONE   Invited Presentations   1

1 Reference model of open distributed processing (RM-ODP): Introduction
  K. Raymond   3

2 RM-ODP: the architecture
  P. Linington   15

3 New ways of learning through the global information infrastructure
  J. Slonim and M. Bauer   34

PART TWO   Reviewed Papers   51

Session on Architecture   53

4 The A1✓ architecture model
  A. Berry and K. Raymond   55

5 Interoperability of distributed platforms: a compatibility perspective
  W. Brookes, J. Indulska, A. Bond and Z. Yang   67

6 A general resource discovery system for open distributed processing
  Q. Kong and A. Berry   79

Session on Management - a Telecommunications Perspective   91

7 Management service design: from TMN interface specifications to ODP
  computational objects
  A. Wittmann, T. Magedanz and T. Eckardt   93

8 Analysis and design of a management application using RM-ODP and OMT
  E. Colban and F. Dupuy   105

9 Distributing public network management systems using CORBA
  B. Kinane   117

Session on Trading   131

10 Designing an ODP trader implementation using X.500
   A. Waugh and M. Bearman   133

11 An evaluation scheme for trader user interfaces
    A. Goodchild   145

12 AI-based trading in open distributed environments
    A. Puder, S. Markwitz, F. Gudermann and K. Geihs   157

Session on Interworking Traders   171

13 A model for a federative trader
    L.A. de Paula Lima Jr. and E.R.M. Madeira   173

14 Enabling interworking of traders
    A. Vogel, M. Bearman and A. Beitz   185

15 An explorative model for federated trading in distributed computing
    environments
    O.-K. Lee and S. Benford   197

16 Cooperation policies for traders
    C. Burger   208

Session on Realizing ODP Systems   219

17 Charging for information services in ODP systems
    M. Warner   221

18 Intercessory objects within channels
    B. Kitson   233

19 A fault-tolerant remote procedure call system for open distributed
    processing
    W. Zhou   245

Session on Managing Distributed Applications   257

20 Towards a comprehensive distributed systems management
    T. Kochi and B. Krämer   259

21 Flexible management of ANSAware applications
    B. Meyer and C. Popien   271

Session on Experience with Distributed Environments   283

22 DDTK project: analysis, design and implementation of an ODP
    application
    S. Arsenis, N. Simoni and P. Virieux   285

23 Experiences with groupware development under CORBA
    T. Horstmann and M. Wasserschaff   297

24 Performance analysis of distributed applications with ANSAmon
    B. Meyer, M. Heineken and C. Popien   309

Session on Quality of Service   321

25 Class of service in the high performance storage system
    S. Louis and D. Teaff   323

26 An approach to quality of service management for distributed
multimedia applications
A. Hafid and G. v. Bochman 335

27 Integration of performance measurement and modeling for open
distributed processing
R. Friedrich, J. Martinka, T. Sienknecht and S. Saunders 347

28 A quality of service abstraction tool for advanced distributed
applications
A. Schill, C. Mittasch, T. Hutschenreuther and F. Wildenhain 359

29 Quality-of-service directed targeting based on the ODP engineering model
G. Raeder and S. Mazaher 372

30 Quality of service management in distributed systems using
dynamic routation
L.J.N. Franken, P. Janssen, B.R.H.M. Haverkort and G. v. Liempd 384

Session on Using Formal Semantics 397

31 Some results on cross viewpoint consistency checking
H. Bowman, J. Derrick and M. Steen 399

32 Maintaining cross viewpoint consistency using Z
J. Derrick, H. Bowman and M. Steen 413

33 ODP types and their management: an object-Z specification
W. Brookes and J. Indulska 425

Session on Integrating Databases in Distributed Systems 437

34 Multiware database: a distributed object database system for
multimedia support
C.M. Tobar and I.L.M. Ricarte 439

35 ObjectMap: integrating high performance resources into a distributed
object-oriented environment
M. Sharrott, S. Hungerford and J. Lilleyman 451

Session on DCE Experiences 463

36 Experiences with the OSF distributed computing environment
J. Dilley 465

37 The TRADEr: Integrating trading into DCE
K. Müller-Jones, M. Merz and W. Lamersdorf 476

38 Experiences using DCE and CORBA to build tools for creating
highly-available distributed systems
E.N. Elnozahy, V. Ratan and M.E. Segal 488

PART THREE   Position Statements and Panel Reports 501

39 The open systems industry and the lack of open distributed management
L. Travis 503

40 Murky transparencies: clarity using performance engineering
   J. Martinka, R. Friedrich and T. Sienknecht 507

41 Quality of service workshop
   J. de Meer and A. Vogel 511

Index of contributors 513

Keyword index 515

Preface

Advances in computer networking have allowed computer systems across the world to be interconnected. Despite this, the heterogeneity in interaction models prevents interworking between systems. Open Distributed Processing (ODP) is an emerging technology which is attempting to solve the software interaction problem by proposing a common framework for distributed systems. ODP systems are those that support heterogeneous distributed processing both within and between autonomous organisations.

The third IFIP International Conference on Open Distributed Processing (ICODP'95) was held in Brisbane, Australia, in February 1995, hosted by the Cooperative Research Centre for Distributed Systems Technology. The Honourable Wayne Goss, MLA, Premier of Queensland, officially opened the conference with an address that outlined the growth in the information technology and telecommunications industries within the Queensland and Australian economies.

The theme for ICODP'95 was "Achieving Open Distributed Systems - Experiences with the Environments" which attracted a number of case study papers documenting the largely successful development of systems using such middleware technologies as DCE, CORBA and ANSAware. These papers illustrate how earlier research in open distributed processing is now being deployed and sends a strong positive message that current and future research must continue to provide the methodologies and technologies to meet the growing demand within organisations throughout the world.

A workshop on Quality of Service was held within ICODP'95. The focus was on extending research and technology for quality of service in communications towards a framework for quality of service in distributed applications.

A total of 94 papers and 11 position statements were submitted to ICODP'95, an increase of approximately 50% from the previous ICODP, demonstrating that open distributed processing is an increasingly active research and development topic. Papers were submitted from all over the world, from every continent except Antarctica! The Programme Committee observed that the overall quality of submitted papers is increasing with each ICODP, making it a difficult task to select 35 papers and 2 position statements for presentation at the conference. Unfortunately, it was impossible to include all papers recommended for acceptance by the referees.

The selected papers were grouped into a number of themed sessions: Architecture, Management - a Telecommunications Perspective, Trading, Interworking Traders, Realising ODP Systems, Managing Distributed Applications, Experience with Distributed Environments, Using Formal Semantics, Integrating Databases in Distributed Systems, DCE Experiences, and the Quality of Service Workshop.

The selected papers were complemented by a set of invited talks. Jan de Meer (GMD-FOKUS, Germany) presented a talk on "Quality of Service in Open Distributed Processing" as an opening address for the Quality of Service Workshop, while the other invited speakers gave presentations directed at the main conference theme of applying open distributed processing to both current and future applications. Jacob Slonim (IBM, Canada) envisaged future education being delivered through open distributed processing in his talk "New Ways of Learning through the Global Information Infrastructure", while Peter Richardson (Telecom Research Laboratories, Australia) addressed the more immediate needs of the telecommunications providers in his talk "TINA - A Telecommunications Initiative in ODP". Art Gaylord (University of Massachusetts at Amherst, USA) summarised his experiences with current technology in his talk "Developing an enterprise-wide computing infrastructure using OSF DCE".

ICODP'95 included three panel sessions: "Future Trends in Open Distributed Processing" chaired by Melfyn Lloyd (CRC for Distributed Systems Technology, Australia), "Adding Quality of Service to Middleware" chaired by Jan de Meer, and "DCE for Fun and Profit" chaired by Jim Curtin (OSF, Japan). The panels commenced with panellists expressing their views on issues raised by the panel chair and by the audience, followed by a period of open discussion involving panellists and audience.

A day of tutorials provided an introduction to ICODP'95. There were two parallel strands, one on the Reference Model for Open Distributed Processing (RM-ODP) and the other on Distributed Environments.The Reference Model strand commenced with Kerry Raymond's (CRC for Distributed Systems Technology, Australia) "RM-ODP: Introduction" followed by Peter Linington (University of Kent at Canterbury, UK) with his tutorial "RM-ODP: The Architecture". Mirion Bearman (CRC for Distributed Systems Technology, Australia) is the international rapporteur for ISO's ODP Trader standard and she presented a tutorial on "Trading in Open Distributed Environments" followed by Andrew Berry (CRC for Distributed Systems Technology, Australia) with his tutorial on "RM-ODP: Modelling and Specification". The Distributed Environments strand consisted of two long tutorials, "OSF DCE" by Andy Bond (CRC for Distributed Systems Technology, Australia) and "OMG CORBA" by Richard Soley (OMG, USA).

In terms of attendance, ICODP goes from strength to strength, and ICODP'95 was no exception. At ICODP'95 and its tutorials, there were 195 participants from 17 nations, an increase of 75% from the previous ICODP. Approximately half the participants came from Australia, reflecting the strong local interest in open distributed processing. Due to the high cost of travel, many Australian researchers and practitioners cannot attend events outside Australia, and therefore took full advantage of the opportunity to attend ICODP'95. The other participants represented nations around the globe: Brazil, Canada, Denmark, Finland, France, Germany, Ireland, Japan, Korea, Netherlands, New Zealand, Norway, Sweden, Switzerland, UK, and USA.

ICODP'95 has seen a broadening of the participant base with greater involvement by industry, through participation of organisations that provide or use open distributed processing in the course of their business. ICODP'95 participants span a broad spectrum from pure academic researchers through applied researchers to technology providers and technology users. Using organisational affiliation as a coarse approximation, 45% of ICODP participants were from universities and research establishments, while 55% were from industry (technology providers or technology users). This industrial participation is very high compared with previous ICODPs, probably reflecting the very large Australian industry presence.

ICODP'95 represents the start of a new direction for ICODP. Only a few weeks before ICODP'95, the major technical parts of the Reference Model of Open Distributed Processing, a joint standard of ISO (International Organisation for Standardisation) and ITU (International Telecommunications Union), was voted to International Standard status. Although ICODP is not specifically focussed on the Reference Model, there has been considerable overlap in both research topics and researchers between ICODP and RM-ODP, leading to a healthy cross-fertilisation of ideas in both forums. However, the number of submitted papers and the extensive participation in ICODP'95 illustrates the enormous interest in open distributed processing, above and beyond RM-ODP itself. Therefore, future ICODPs must focus on populating the framework established by RM-ODP, developing conformant technologies, and using the principles of RM-ODP to develop applications that will increase the efficiency of industry and betterment of society.

With so much interest in open distributed processing, it begs the question of whether ICODP should become an annual event rather than a biennial event. There is considerable support for this idea within the ICODP community. Unfortunately, the preparation time required for a well-organised ICODP makes it impossible to hold an ICODP in 1996, and so the next ICODP will be held in 1997 in Toronto, Canada, in adherence to the current schedule. However, to meet the demands of particular research areas within open distributed processing, the ICODP programme committee is proposing to organise some smaller special interest workshops in 1996. There is already a firm proposal to hold a workshop on the use of formal methods in open distributed processing in France, and there are tentative proposals in the areas of quality of service and DCE. During 1995, the ICODP Programme Committee will establish a long-term plan for ICODP with a view to holding ICODP annually from 1997 onwards.

A successful international conference does not happen by waving a magic wand. Therefore, I would like to thank the following people:

- the participants and authors of submitted papers for choosing ICODP as the forum to express their interest in open distributed processing

- the Programme Committee and the team of paper reviewers that tackled the large and difficult task of assessing and selecting papers

- the tutorial presenters, invited speakers, session chairs, and panellists, for informing and entertaining the participants

- the staff of the ParkRoyal Hotel and other suppliers to ICODP

- IFIP for providing a coordinating framework for conferences and the publication of their proceedings

- our major sponsors, Aspect Computing, Bay Technologies, IBM, Open Environment Corporation, Open Software Foundation, and supporters, DSTO and Fujitsu, for their financial assistance

- the Board and Research Executive of the Cooperative Research Centre for Distributed Systems Technology (DSTC) for their moral support in winning and organising this conference

- DSTC administration for secretarial, accounting, and systems administration services

- DSTC Architecture Unit for being prepared to lend a hand at any time for any task

- ICODP Organising Committee, all from DSTC, for their planning

- Levare Quick, DSTC, for handling registrations

- Jenny MacKay, DSTC, for administering the papers and the reviewers' reports

- Liz Armstrong, DSTC, for her excellent planning and coordination of all local matters; ICODP'95 could not have happened without her!

Finally, I extend my best wishes to Jacob Slonim *(jslonim@vnet.ibm.com)* of IBM Canada and Jerry Rolia *(jar@sce.carleton.ca)* of Carleton University in Ottawa, as they prepare to host the next ICODP in Toronto, Canada, in 1997.

Brisbane, March 1995

Kerry Raymond
kerry@dstc.edu.au
ICODP'95 Programme Chair

3rd International IFIP TC6 Conference on Open Distributed Processing

Organised by:

Cooperative Research Centre for Distributed Systems Technology (DSTC)

Programme Chair:

Kerry Raymond, CRC for Distributed Systems Technology, Australia
Jan de Meer, GMD-FOKUS, Berlin
Jacob Slonim, IBM, Canada

Programme Committee:

Gregor von Bochmann, University of Montreal, Canada
Eng Chew, DEC, Australia
Vladimir Gorodetzki, AOS St. Petersburg, Russia
Andrzej Goscinski, Deakin University, Australia
Andrew Herbert, APM, United Kingdom
Jan Kikuts, ESTI Riga, Latvia
Laszlo Kovacs, Academy of Sciences, Hungary
Hidehito Kubo, NEC, Japan
Guy Leduc, University of Liege, Belgium
Peter Linington, University of Kent at Canterbury, United Kingdom
Elie Najm, ENST Paris, France
Peter Richardson, Telecom Research Laboratories, Australia
Tom Rutt, AT&T, USA
Gerd Schuermann, GMD-FOKUS, Germany
Martin van Sinderen, University of Twente, The Netherlands
Otto Spaniol, RWTH Aachen, Germany
Jean-Bernard Stefani, CNET Paris, France

Local Organisation Committee:

Elizabeth Armstrong (Chair), CRC for Distributed Systems Technology
David Barbagallo, CRC for Distributed Systems Technology
Ashley Beitz, CRC for Distributed Systems Technology
Lois Fordham, CRC for Distributed Systems Technology
Melfyn Lloyd, CRC for Distributed Systems Technology
Jenny MacKay, CRC for Distributed Systems Technology
Andreas Vogel, CRC for Distributed Systems Technology

Sponsors:

Aspect Computing Pty Ltd *(Major Sponsor)*
Bay Technologies *(Major Sponsor)*
IBM Australia Limited *(Major Sponsor)*
Open Environment Corporation *(Major Sponsor)*
Open Software Foundation *(Major Sponsor)*
Defence Science and Technology Organisation *(Supporter)*
Fujitsu Australia *(Supporter)*

List of Referees

Gopi Attaluri
Michael Bauer
Olivier Bonaventure
Bob Brownell
Andre Danthine
Damian De Paoli
Juergen Dittrich
M. Feeley
Patrick Finnigan
Yuri A. Golovin
Ahmed Guetari
Lex Heerink
Michael Hobbs
Robert Huis in't Veld
David Jackson
Toshihiko Kato
Jeff Keegan
Janis Kikuts
Eckhart Koerner
Harro Kremer
Thomas Kunz
R. Langerak
Guy Leduc
Ian Lewis
Hanan Lutfiyya
Laurent Mathy
Eddie Michiels
Shoichiro Nakai
Ying Ni
Charles Pecheur
Claudia Popien
Axel Rennoch
Rudolf Roth
Gerd Schuerman
Siva Sivasuramanian
Y.A. Solntsev
Jean-Bernard Stefani
Akira Tanaka
George Tseung
Marten van Sinderen
V.I. Vorobiev
Geoff Wheeler
Zhonghua Yang

Yves Baguette
Ashley Beitz
Martin Brachwitz
Joerg Burmeister
Peter Davids
Mark de Weger
Joubine Dustzadeh
Arnaud Fevrier
L.J.N. Franken
V.I. Gorodetsky
A. Hafid
Andrew Herbert
James W. Hong
Mike Husband
Philip Joyce
Joost-Pieter Katoen
Rudolf K. Keller
Barry Kitson
I.V. Kotenko
Olaf Kubitz
Dean Kuo
Michael Lawley
Luc Leonard
Dave Liddy
Kazutoshi Maeno
T.H. Merrett
Rokia Missaoui
Shin Nakajima
Raj Mohan Panadiwal
Nathalie Perdigues
Dick Quartel
Peter Richardson
Tom Rutt
Henning Schulzrinne
Jacob Slonim
Otto Spaniol
Tomoyoshi Sugawara
Marijana Tomic
Mitsukazu Uchiyama
Rolf Velthuys
Alexei Vovenko
Greg Wickham
Stefan Zlatintsis

Kees Bakker
J.P. Black
Dexter Bradshaw
Eng Chew
Jan de Meer
Petre Dini
Klaus-Peter Eckert
Colin Fidge
Jan Gecsei
Andrzej M. Goscinski
Boudewijn R. Haverkort
Oliver Hermanns
Hirokazu Horiguchi
Hidehiro Ishii
Yuri G. Karpov
Brian Keck
Pr. Brigitte Kerherve
Kazutomo Kobayashi
Laszlo Kovacs
Hidehito Kubo
Akihisa Kurashima
Stephen Leask
M. Levy
Peter Linington
P. Martin
Bernd Meyer
Elie Najm
Tetsuo Nakakawaji
Ajeet Parhar
Luis Ferreira Pires
Kerry Raymond
Jerry Rolia
Peter Schoo
V. Shiishkin
Michael I. Smirnov
Andrew Speirs-Bridge
Yosuke Takano
Volker Tschammer
Gregor von Bochmann
Andreas Vogel
Nigel Ward
Ing Widya

PART ONE

Invited Presentations

1

Reference Model of Open Distributed Processing (RM-ODP): Introduction

Kerry Raymond

kerry@dstc.edu.au
CRC for Distributed Systems Technology
Centre for Information Technology Research
University of Queensland
Brisbane 4072 Australia

Abstract

The Reference Model of Open Distributed Processing (RM-ODP) was a joint effort by the international standards bodies ISO and ITU-T to develop a coordinating framework for the standardisation of open distributed processing (ODP). The model describes an architecture within which support of distribution, interworking, interoperability and portability can be integrated. The RM-ODP framework defines ODP concerns using five "viewpoints" (abstractions), namely enterprise, information, computational, engineering, and technology. This tutorial introduces the reference model, describing the viewpoints and some of the ODP functions and transparencies.

Keyword Codes: C.2.4
Keywords: Computer-Communication Networks, Distributed Systems

1. WHAT IS RM-ODP?

Advances in computer networking have allowed computer systems across the world to be interconnected. Despite this, heterogeneity in interaction models prevents interworking between systems. Open distributed processing (ODP) describes systems that support heterogeneous distributed processing both within and between organisations through the use of a common interaction model.

ISO and ITU-T (formerly CCITT) have developed a Reference Model of Open Distributed Processing (RM-ODP) to provide a coordinating framework for the standardisation of ODP by creating an architecture which supports distribution, interworking, interoperability and portability.

1.1. The Goals and Deliverables of RM-ODP

RM-ODP aims to achieve:

- portability of applications across heterogeneous platforms
- interworking between ODP systems, i.e. meaningful exchange of information and convenient use of functionality throughout the distributed system
- distribution transparency, i.e. hide the consequences of distribution from both the applications programmer and user

The reference model provides a "big picture" that organises the pieces of an ODP system into a coherent whole. It does not try to standardise the components of the system nor to unnecessarily influence the choice of technology.

There are many challenges in developing a reference model. RM-ODP must be adequate to describe most "reasonable" distributed systems available both today and in the future, so RM-ODP is abstract, but not vague. RM-ODP carefully describes its components without prescribing an implementation.

1.2. Structure of RM-ODP

The RM-ODP standard is known as both ISO International Standard 10746 and ITU-T X.900 Series of Recommendations and will consist of four parts:

- Part 1: Overview and Guide to Use (ISO 10746-1/ITU-T X.901) [1]
- Part 2: Descriptive Model (ISO 10746-2/ITU-T X.902) [2]
- Part 3: Prescriptive Model (ISO 10746-3/ITU-T X.903) [3]
- Part 4: Architectural Semantics (ISO 10746-4/ITU-T X.904) [4]

Part 1 contains a motivational overview of ODP and explains the key concepts of the RM-ODP architecture. Part 2 gives precise definitions of the concepts required to specify distributed processing systems. Part 3 prescribes a framework of concepts, structures, rules, and functions required for open distributed processing. Part 4 describes how the modelling concepts of Part 2 can be represented in a number of formal description techniques.

This tutorial focuses on the framework established in Part 3.

1.3. Status of RM-ODP

In January 1995, Parts 2 and 3 of RM-ODP were successfully balloted to become International Standards; official copies of the standard will be soon be available, after the necessary administrative processing.

Parts 1 and 4 are based on Parts 2 and 3. Therefore, the standardisation of Parts 1 and 4 follows the standardisation of Parts 2 and 3. Parts 1 and 4 are currently Committee Drafts and expected to become Draft International Standards in April 1995 and International Standards in early 1996.

2. VIEWPOINTS

Part 3 of RM-ODP prescribes a framework using *viewpoints* from which to abstract or view ODP systems. A set of concepts, structures, and rules is given for each of the viewpoints, providing a "language" for specifying ODP systems in that viewpoint.

RM-ODP defines the following five viewpoints:

- Enterprise Viewpoint (purpose, scope and policies)
- Information Viewpoint (semantics of information and information processing)
- Computational Viewpoint (functional decomposition)
- Engineering Viewpoint (infrastructure required to support distribution)
- Technology Viewpoint (choices of technology for implementation)

Specifying an ODP system using each of the viewpoint languages allows an otherwise large and complex specification of an ODP system to be separated into manageable pieces, each focused on the issues relevant to different members of the development team. For example, the information analyst works with the information specification while the systems programmer is concerned with the engineering viewpoint. Figure 1 shows how the RM-ODP viewpoints can be related to the software engineering process.

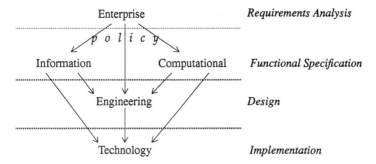

Figure 1: RM-ODP Viewpoints and Software Engineering

3. ENTERPRISE VIEWPOINT

The enterprise viewpoint is used to organisational requirements and structure. In the enterprise viewpoint, social and organizational policies can be defined in terms of:

- objects — both "active" objects, e.g. bank managers, tellers, customers, and "passive" objects, e.g. bank accounts, money
- communities — groupings of objects intended to achieve some purpose, e.g. a bank branch consists of a bank manager, some tellers, and some bank accounts; the branch provides banking services to a geographical area
- roles of the objects within communities, expressed in terms of policies:
  * permission — what can be done, e.g. money can be deposited into an open account
  * prohibition — what must not be done, e.g. customers must not withdraw more than $500 per day
  * obligations — what must be done, e.g. the bank manager must advise customers when the interest rate changes

The enterprise language is specifically concerned with *performative actions* that change policy, such as creating an obligation or revoking permission. In a bank, the changing of interest rates is a performative action as it creates obligations on the bank manager to inform the customers. However, obtaining an account balance is not a performative action as obligations, permissions, and prohibitions are not affected. Thus, an enterprise specification of a bank need not include the obtaining of account balances; such functionality will be identified in the computational specification.

By preparing an enterprise specification of an ODP application, policies are determined by the organisation rather than imposed on the organisation by technology (implementation) choices. For example, a customer should not be limited to having only one bank account, simply because it was more convenient for the programmer.

4. INFORMATION VIEWPOINT

The information viewpoint is used to describe the information required by an ODP application through the use of schemas, which describe the state and structure of an object; e.g., a bank account consists a balance and the "amount withdrawn today".

A *static schema* captures the state and structure of a object at some particular instance; e.g., at midnight, the amount-withdrawn-today is $0.

An *invariant schema* restricts the state and structure of an object at all times; e.g., the amount-withdrawn-today is less than or equal to $500.

A *dynamic schema* defines a permitted change in the state and structure of an object; e.g. a withdrawal of $X from an account decreases the balance by $X and increases the amount-with-drawn-today by $X. A dynamic schema is always constrained by the invariant schemas. Thus, $400 could be withdrawn in the morning but an additional $200 could not be withdrawn in the afternoon as the amount-withdrawn-today cannot exceed $500.

Schemas can also be used to describe relationships or associations between objects; e.g., the static schema "owns account" could associate each account with a customer.

A schema can be composed from other schemas to describe complex or composite objects; e.g., a bank branch consists of a set of customers, a set of accounts, and the "owns account" relationships.

The information specification of an ODP application could be expressed using a variety of methods, e.g., entity-relationships models, conceptual schemas, and the **Z** formal description technique.

5. COMPUTATIONAL VIEWPOINT

The computational viewpoint is used to specify the functionality of an ODP application in a distribution-transparent manner. RM-ODP's computational viewpoint is object-based, that is:

- objects encapsulate data and processing (i.e. behaviour)
- objects offer interfaces for interaction with other objects
- objects can offer multiple interfaces.

A computational specification defines the objects within an ODP system, the activities within those objects, and the interactions that occur among objects. Most objects in a computational specification describe application functionality, and these objects are linked by bindings through which interactions occur. Binding objects are used to describe complex interaction between objects.

Objects in a computational specification can be application objects (e.g. a bank branch) or ODP infrastructure objects (e.g. a type repository or a trader, see Section 8.3.1 and Section 8.3.2). Figure 2 illustrates a bank branch object providing a bank teller interface and a bank manager interface. Both interfaces can be used to deposit and withdraw money, but accounts can be created only through the bank manager interface. Each of the bank branch object's interfaces is bound to a customer object.

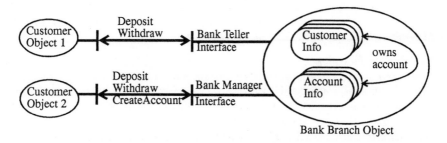

Figure 2: Bank Branch Object with Bank Manager and Bank Teller Interfaces

5.1. Computational Interaction

RM-ODP provides three forms of interaction between objects: operational, stream-oriented, and signal-oriented.

Operational interfaces provide a client-server model for distributed computing—client objects invoke operations at the interfaces of server objects (i.e. the remote procedure call paradigm). Operational interfaces consist of named operations with parameters, terminations, and results. Operations in RM-ODP can be either interrogations (which return a termination) or announcements (which do not return a termination).

For example, a bank branch object offers a number of BankTeller operational interfaces, whose signature is defined as:

```
BankTeller - Interface Type {

  operation Deposit (c: Customer, a: Account, d: Dollars)
    returns OK (new_balance: Dollars)
    returns Error (reason: Text);

  operation Withdraw (c: Customer, a: Account, d: Dollars)
    returns OK (new_balance: Dollars)
    returns NotToday (today: Dollars, daily_limit: Dollars)
    returns Error (reason: Text);
}
```

Note that the notation used in the example above is merely illustrative. RM-ODP does not prescribe any particular notation for defining operational interface types.

Stream interfaces provide (logically) continuous streams of information flowing between producer and consumer objects. Consumer objects connect to the stream interfaces of producer objects or vice-versa, and several streams can be grouped in a single interface, e.g., an audio stream and a video stream. Stream interfaces have been included in RM-ODP to cater for multi-media and telecommunications applications.

Underlying both operational interfaces and stream interfaces are signal interfaces which provide very low-level communications actions. The OSI service primitives (REQUEST, INDICATE, RESPONSE, and CONFIRM) are examples of signals.

5.1.1. Interface Subtyping

The concept of interface type is particularly important in RM-ODP. Interfaces in the computational model are strongly typed and inheritance of an interface type (usually) creates a subtype relationship. Subtypes of an interface type are substitutable for the parent type (or any supertype).

Figure 3 illustrates interface subtyping.The BankManager and LoansOfficer interface types are subtypes of the BankTeller interface (super-)type; either can substitute for a BankTeller as they can perform the Deposit and Withdraw operations expected of a BankTeller. Neither a Bank-Teller nor a LoansOfficer can replace a BankManager, as neither can provide the CreateAccount operation.

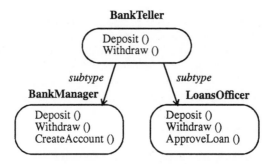

Figure 3: Example of Interface Subtyping

5.2. Computational Activity

The computational viewpoint also defines the actions that are possible within a computational object. These are:

- creating and destroying an object
- creating and destroying an interface
- trading for a interface (see Section 8.3.2)
- binding to an interface
- reading and writing the state of the object
- invoking an operation at an operational interface
- producing/consuming a flow at a stream interface
- initiating or responding to a signal at a signal interface.

These basic actions can be composed in sequence or in parallel. If composed in parallel, the parallel activities can be dependent (the activity is forked and must subsequently join at a synchronisation point) or independent (the activity is spawned and cannot join).

5.3. Environment Contracts

The refinement of a computational object and its interfaces might require the specification of requirements on the realization of that object or its interfaces (and, hence, of the objects with which it interacts). For example, a bank must protect the customer's money and must ensure that interaction is secure against a variety of fraudulent activities, e.g. capturing and replaying operations. Therefore, the actual interactions must either be communicated over a secure network or employ end-to-end security checks.

Ideally, environment contracts will be expressed in high-level quality-of-service terms rather than, e.g., specifying a particular network or a particular encryption scheme (either of which presupposes the environment in which the ODP system will operate).

Currently, the state of the art falls short of this ideal. However, it is important that RM-ODP be "future-proof", capable of incorporating both current and expected future technologies.

6. ENGINEERING VIEWPOINT

The engineering viewpoint is used to describe the design of distribution-oriented aspects of an ODP system; it defines a model for distributed systems infrastructure. The engineering viewpoint is not concerned with the semantics of the ODP application, except to determine its requirements for distribution and distribution transparency.

The fundamental entities described in the engineering viewpoint are objects and channels. Objects in the engineering viewpoint can be divided into two categories—basic engineering objects (corresponding to objects in the computational specification) and infrastructure objects (e.g., a protocol object — see below). A channel corresponds to a binding or binding object in the computational specification.

6.1. Channels

A channel provides the communication mechanism and contains or controls the transparency functions required by the basic engineering objects, as specified in the environment contracts in the computational specification. Figure 4 illustrates the channel between a Customer Object and the Bank Branch object in Figure 2. The shaded area is the channel, composed of stubs, binders, and protocol objects. Stubs and binders are used to provide various distribution transparencies.

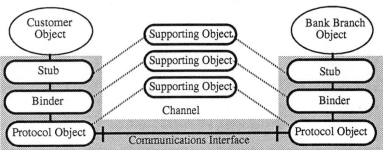

Figure 4: Structure of a Channel

Stubs are used when the transparency involves some knowledge of the application semantics, e.g., maintaining a log of operations for an audit trail.

Binders are used when application semantics are not required; they merely transport the messages (bit streams). Binders are responsible for managing the binding between the basic engineering objects; e.g., binders could use sequence numbers to foil capture-and-replay attempts.

Protocol objects interact via a communications interface; this models networking.

Outside of the channel, supporting objects assist the stub, binder, and protocol objects within the channel. Typically, supporting objects are repositories of information required by the stubs, binders, and protocol objects. For example, binders register and retrieve interface locations via a supporting object known as the relocator (see Section 8.3.3) in order to achieve location transparency.

6.2. Engineering Structures

The RM-ODP engineering viewpoint prescribes the structure of an ODP system. The basic units of structure are:

- cluster — a set of related basic engineering objects that will always be co-located
- capsule — a set of clusters, a cluster manager for each cluster, a capsule manager, and the parts of the channels which connect to their interfaces
- nucleus object — an (extended) operating system supporting ODP
- node — a computer system

Figure 5 illustrates the structure of a node.

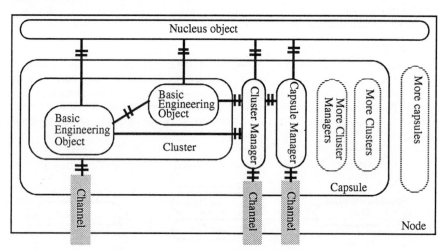

Figure 5: Structure of a Node

Given these definitions, the following structuring rules are defined:

- a node has a nucleus object
- a nucleus object can support many capsules
- a capsule can contain many clusters
- a cluster can contain many basic engineering objects
- a basic engineering object can contain many activities
- all inter-cluster communication is via channels

An implementation of an ODP system can choose to constrain the structuring, for example, by allowing:

- only one object per cluster
- only one cluster per capsule

7. TECHNOLOGY VIEWPOINT

A technology specification of an ODP system describes the implementation of that system and the information required for testing. RM-ODP has very few rules applicable to technology specifications.

8. ODP FUNCTIONS

The ODP functions are a collection of functions expected to be required in ODP systems to support the needs of the computational language (e.g. the trading function) and the engineering language (e.g. the relocator). The following subsections outline the major function groups in RM-ODP; a few of the functions are discussed in more detail to illustrate the scope of RM-ODP.

8.1. Management Functions

RM-ODP defines a number of functions to manage the engineering structures, including:

- node management function (provided by the nucleus) for creating capsules and channels
- capsule management function (provided by the capsule manager) for instantiating clusters and checkpointing and deactivating clusters in a capsule
- cluster management function (provided by the cluster manager) for checkpointing, deactivating and migrating clusters
- object management function (provided by the basic engineering object) for checkpointing and deleting basic engineering objects

8.2. Coordination Functions

RM-ODP defines a number of functions aimed at coordinating the actions of a number of objects, clusters, or capsules in order to produce some consistent overall effect. These include:

- checkpoint and recovery
- deactivation and reactivation
- event notification
- groups and replication
- migration
- transactions

8.2.1. Transaction Function

In the information viewpoint, state change appears to happen as a single indivisible action. However, in a computational and engineering viewpoints, this state might be distributed throughout the ODP system and be concurrently accessed by many parallel activities. In order to develop reliable ODP systems, it will be necessary to coordinate the behaviour of objects to achieve the desired degrees of:

* visibility — the degree to which the intermediate effects of an operation (or other interaction) are visible to other operations
* recoverability — the state after the failure of the operation (which of its effects are undone?)
* permanence — the consequences of the failure of the operation on completed operations (are their effects altered?)

RM-ODP defines a very generalised transaction function; this is another example of "future-proofing" in RM-ODP. Realistically, the ACID transaction model will be the only style of transaction mechanism supported by most ODP systems for a number of years. Consequently, RM-ODP defines an ACID transaction function as specialisation of its generalised transaction function.

8.3. Repository Functions

In addition to a general storage function and a general relationship repository, RM-ODP defines a number of specific repository functions, concerned with maintaining a database of specialised classes of information.

8.3.1. Type Repository

In most computer systems, type definitions are not explicitly maintained within the system. Instead, types are documented in manuals or defined according to some local conventions (e.g. use of mnemonic file names). ODP systems must make type information available through the ODP system itself; the primary need is to support type checking during trading and interface binding.

In RM-ODP, the type repository is a registry for type definitions, particularly for interface types.The type registry maintains a type hierarchy (subtype relationships) and other relationships between types.

8.3.2. Trader

The ODP Trader provides "a dating service for objects"; its purpose is to support dynamic binding by allowing services to be discovered at run-time. The trader is a repository of service advertisements.

Server objects advertise their services through a trader; the service advertisement specifies the interface type and service attributes. Servers manipulate their service advertisements by using the *export* operations provided by the trader. Clients choose services by specifying the required type and attributes in *import* operations.

The ODP Trader is also the subject of standardisation, separate from RM-ODP. An introduction to this standard can be found in [5].

9.2. Relocation Transparency

Relocation transparency frees a basic engineering object (and the programmer of the object) from needing to know if an interacting object is relocated.

Relocation transparency can be achieved by configuring the channel with binders, which:

- inform the relocator (see Section 8.3.3) of the location of the interface it supports
- obtain from the relocator the location(s) of the other interface(s) connected to the channel

Binders will typically cache location information. If the location of an interface changes, the use of the old location will cause an error. With relocation transparency, the binder will automatically obtain the new location from the relocator, reconnect the channel, and replay the interaction. The basic engineering object should remain unaware of the change in location.

9.3. Transaction Transparency

Unlike access and relocation transparency which are achieved through configuring engineering channels with clever components, transaction transparency cannot be achieved by this mechanism alone.

The correct operation of the transaction function requires the reporting of the execution (or undo-ing) of certain "actions of interest" (e.g. reading or writing a piece of transaction-managed data). These events occur internal to the objects and are not visible to a stub or binder configured in the channel. Therefore, transaction transparency must involve the refinement of a transaction-transparent specification into a specification which reports the execution of these actions of interest to the transaction function.

10. SUMMARY

RM-ODP is a reference model, not an implementation standard; it defines a framework for the standardisation of open distributed processing. The RM-ODP model defines five viewpoints which decompose the specification of ODP applications by focusing on separate concerns.

The enterprise viewpoint defines a model for policy analysis while the information viewpoint provides a model for information analysis. The computational viewpoint defines a model for distributed programming languages; the run-time support for these languages is provided by the distributed systems infrastructure based on the engineering viewpoint model and the ODP infrastructure functions. The technology viewpoint is used to describe implemented systems.

ACKNOWLEDGEMENTS

The author thanks Andrew Berry for his assistance in preparing this paper and all of the participants in the Australian and international RM-ODP standards groups for the many hours of lively discussions.

The participation of the author in the standardisation of RM-ODP has been supported by:

- Telecom (Australia) Research Laboratories through the Centre of Expertise in Distributed Information Systems (CEDIS)
- the Cooperative Research Centre for Distributed Systems Technology through the Cooperative Research Centres Program of the Department of the Prime Minister and Cabinet of the Commonwealth Government of Australia
- Standards Australia through their travel assistance scheme.

REFERENCES

[1] ISO/IEC CD 10746-1, "Basic Reference Model of Open Distributed Processing - Part 1: Overview and Guide to Use", July 1994.

[2] ISO/IEC DIS 10746-2, "Basic Reference Model of Open Distributed Processing - Part 2: Descriptive Model", February 1994.

[3] ISO/IEC DIS 10746-3, "Basic Reference Model of Open Distributed Processing - Part 3: Prescriptive Model", February 1994.

[4] ISO/IEC CD 10746-4, "Basic Reference Model of Open Distributed Processing - Part 4: Architectural Semantics", July 1994.

[5] M.Y. Bearman, "ODP-Trader", International Conference on Open Distributed Processing, Berlin, September 1993.

[6] ISO/IEC CD 10181, "Security Frameworks in Open Systems".

2

RM-ODP: The Architecture

P.F. Linington

Computing Laboratory, University of Kent,
Canterbury, Kent CT2 7NF, United Kingdom

The Reference Model for Open Distributed Processing is a joint ISO/ITU Standard which provides a framework for the specification of large scale, heterogeneous distributed systems. It defines a set of five viewpoints concentrating on different parts of the distribution problem and a set of functions and transparency mechanisms which support distribution. The resulting framework is being populated by more detailed standards dealing with specific aspects of the construction and operation of distributed systems.

Keyword Codes: C.2.4
Keywords: Distributed Systems

1. INTRODUCTION

The Reference Model for Open Distributed Processing is a standard produced jointly by the International Organization for Standardization (ISO) and the International Telecommunications Union (ITU). Experts from these two organizations have been working together on this framework for Open Distributed Processing (ODP) for some seven years, and the resulting architecture has recently been approved for publication by ISO; ITU approval is expected at a meeting later in 1995.

This work is based on recent research and best practice; it has drawn upon a wide range of experience in the implementation of distributed systems and the formulation of a general architecture. Input has been taken from advanced industrial research, such as that in the ANSA consortium, from various ESPRIT and RACE activities, and from practical experience with platforms such as the OSF-DCE. In the standards world, ideas from OSI management and the security frameworks have been incorporated. Major input from the ITU has introduced the requirements from TMN, INA and TINA. More recently, the work has benefited from a strong, two-way liaison with the Object Management Group.

The resulting standard is a framework, which documents key decisions, relates components and sets the technical agenda for future, more detailed, standardization. The Reference Model for Open Distributed Processing (RM-ODP) is made up of four parts [1-4].

These are

Part 1:	**Overview**: this contains a motivational overview of ODP, giving scoping, justification and explanation of key concepts, and an outline of the ODP architecture. It contains explanatory material on how the RM-ODP is to be interpreted and applied by its users, who may include standards writers and architects of ODP systems. It also includes a categorization of required areas of standardization, expressed in terms of the reference points for conformance identified in Part 3.

Part 2:	**Foundations**: this contains the definition of the concepts and analytical framework for the description of arbitrary distributed processing systems. It introduces the principles of conformance to ODP standards and the way in which they are applied.

Part 3:	**Architecture**: this contains the specification of the required characteristics that qualify distributed processing as open. These are the constraints to which ODP standards must conform.

Part 4:	**Architectural semantics**: this contains a formalization of the ODP basic modelling concepts defined in Part 2. The formalization is achieved by interpreting each concept in terms of constructs of the different standardized formal description techniques.

The current paper concentrates on Part Three of the RM-ODP — the architecture which makes a distributed system be specifically an ODP system.

2. MODELLING FOUNDATIONS

The architecture is supported by a set of modelling concepts which provide the foundation for expressing it. These concepts are object based, and are very general; they can be applied in many different areas of the architecture.

The most basic concepts defined are those of object, action and interaction; an object encapsulates its state, and this state can only be modified by interaction with other objects or by the internal actions of the object. The interactions between objects take place at interfaces; an object can have any number of interfaces. Interfaces can be located at particular points in space. These interfaces are the basis for the description of configurations of objects and the transfer of information about the availability of objects.

A second set of concepts supports the structuring of specifications, introducing ideas of composition and refinement, type and class and of the instantiation of objects or interfaces from templates which describe them. Following this, a variety of more abstract organizational concepts, such as domain, contract and liaison are introduced to express relationships between objects.

Finally, there is a basic framework for the definition of conformance to the ODP specifications, and for the statement of where such conformance applies. The conformance of implementations to ODP standards can be tested in a number of ways, depending on whether the requirement is for interworking, software portability, correct generation of interchange media (such as removable discs), or correct interaction with human users and the outside world in general.

3. ARCHITECTURAL FRAMEWORK

Distributed systems can be very large and complex, and the many different considerations which influence their design can result in a substantial body of specification, which needs to be given structure if it is to be managed successfully. A good framework should allow different parts of the design to be worked on separately if they are independent, but should identify clearly those places where different aspects of the design constrain one another. There are two main structuring approaches used in the ODP architecture: the definition of viewpoints and the definition of transparencies.

3.1. Viewpoints

The RM-ODP defines five viewpoints. A viewpoint is a subdivision of the specification of a complete system, established to bring together those particular pieces of information relevant to some particular area of concern during the design of the system. The viewpoints are not completely independent; key items in each are identified as related to items in the other viewpoints. However, the viewpoints are sufficiently independent to simplify reasoning about the complete specification.

Each of the viewpoints in the set can be related to all the others. They do not form a fixed sequence like a set of protocol layers, nor are they created in a fixed order according to some design methodology. The architecture is expressed in terms of the complete set of related viewpoints, without laying down how this complete specification is to be constructed.

The five viewpoints defined are:

a)   the enterprise viewpoint: a viewpoint on the system and its environment that focuses on the purpose, scope and policies for the system.

b)   the information viewpoint: a viewpoint on the system and its environment that focuses on the semantics of the information and information processing performed.

c)   the computational viewpoint: a viewpoint on the system and its environment that enables distribution through functional decomposition of the system into objects which interact at interfaces.

d)   the engineering viewpoint: a viewpoint on the system and its environment that focuses on the mechanisms and functions required to support distributed interaction between objects in the system.

e)   the technology viewpoint: a viewpoint on the system and its environment that focuses on the choice of technology in that system.

In each viewpoint, key terminology is established and constraints expressed which characterize the architecture, making the systems described distinctively ODP systems, not just any distributed systems. These terms and constraints are expressed as the initial terms and grammar of a set of abstract viewpoint languages. Specifications expressed in these languages conform to their grammar, and so the systems designed are ODP systems, at least from an architectural point of view.

The various viewpoint languages differ in the strengths of the constraints their use implies. Those concerned with organizing distribution and providing common solutions to its problems (the computational and engineering viewpoints) place a significant number of constraints that must be observed, and in so doing give guarantees of interworking between and portability of components. Those which express requirements for the system as a whole (the enterprise and information viewpoints) place fewer constraints. Constraints on the complete system effectively limit its scope, and so make the architecture less general-purpose. Therefore few language constraints have been defined in these viewpoints in order to support as wide a range of applications as possible.

It should be noted that the expression of the architectural constraints in the form of an abstract language does not imply any particular syntax or notation; there may be many notations consistent with the architecture, derived from a variety of programming practices or development methods, and embodying choices not expressed in the ODP architecture itself.

3.2. Transparencies

The second structuring approach taken is to identify a number of transparencies. When contemplating a distributed system, a number of problems become apparent which are a direct result of the distribution: the systems components are heterogeneous, they can fail independently, they are at different and, possibly, varying locations, and so on. These problems can either be solved directly as part of the application design, or standard solutions can be selected, based on best practice.

If standard mechanisms are chosen, the application designer works in a world which is transparent to that particular problem; the standard mechanism is said to provide a transparency. Application designers simply select which transparencies they wish to assume, and where in the design they are to apply.

The transparency approach can lead directly to software reuse. Selection of transparencies in the system specification can lead to the automatic incorporation of well-established implementations of the standard solutions by the system building tools in use, such as compilers, linkers and configuration managers. The designer expresses system requirements in the form of a simplified statement of the interactions required and the transparency properties that they should possess.

The transparencies defines in the RM-ODP are

a) **access transparency,** which masks differences in data representation and invocation mechanisms to enable interworking between objects. This transparency solves many of the problems of interworking between heterogeneous systems, and will generally be provided by default.

b) **failure transparency,** which masks from an object the failure and possible recovery of other objects (or itself), to enable fault tolerance. When this transparency is provided, the designer can work in an idealized world in which the corresponding class of failures does not occur.

c) **location transparency,** which masks the use of information about location in space when identifying and binding to interfaces. This transparency provides a logical view of naming, independent of actual physical location.

d)   **migration transparency**, which masks from an object the ability of a system to change the location of that object. Migration is often used to achieve load balancing and reduce latency.

e)   **relocation transparency**, which masks relocation of an interface from other interfaces bound to it. Relocation allows system operation to continue even when migration or replacement of some objects creates temporary inconsistencies in the view seen by their users.

f)   **replication transparency**, which masks the use of a group of mutually behaviourally compatible objects to support an interface. Replication is often used to enhance performance and availability.

g)   **persistence transparency**, which masks from an object the deactivation and reactivation of other objects (or itself). Deactivation and reactivation are often used to maintain the persistence of an object when the system is unable to provide it with processing, storage and communication functions continuously.

h)   **transaction transparency**, which masks coordination of activities amongst a configuration of objects to achieve consistency.

In each case, the definition of the transparency involves both a set of requirements and a solution that satisfies it. The set of requirements states where the transparency is needed (i.e. which interactions it affects). This may simply be a statement that it applies throughout a system, or may be a more selective statement involving specific interfaces, defining, for example, the interactions which make up a transaction or selecting the objects and interfaces to be supported by replication. The solution takes the form of specific rules for the transformation from a specification in which the transparency is requested to a more detailed one which expands selected interaction or objects so as to include mechanisms which provide a means of interaction with the requested properties.

4. THE ENTERPRISE LANGUAGE

The aim of an enterprise specification is to express the objectives and policy constraints on the system of interest. This involves the identification of the main roles involved in the system. These roles represent, for example, the users, owners and providers of information processed by the system. Creating a separate viewpoint to convey this information decouples the objectives set for the system from the way it is to be realized.

One of the key ideas in the enterprise language is that of a contract, linking the performers of the various roles and expressing their mutual obligations. A contract can express the common goals and responsibilities which distinguish roles in a community, such as a business and its customers or an government organization and its clients, as being related in particular ways in a single activity or enterprise.

A federation is one particular kind of community; a federation is a coming-together of a number of groups answering to different authorities (and thus representable as distinct domains) in order that they may jointly cooperate to achieve some objective. Since the evolution of distributed systems will repeatedly result in the merging of existing, separately managed sub-systems to share information or support commercial interests, the creation of federations and the expression of the rules which are to govern them forms an important part of system specification in the enterprise viewpoint.

Where appropriate, an enterprise specification will also express aspects of ownership of resources and responsibility for payment for goods and services in order to identify, for example, constraints on accounting and security mechanisms within the infrastructure which supports the system.

Different notations for enterprise specification can be expected to support specific organizational structures and business practices, but architecturally, ODP is neutral, requiring only that an appropriate specification be generated; few constraints are placed on the form that organizations should take.

5. THE INFORMATION LANGUAGE

The individual components of a distributed system must share a common understanding of the information they communicate when they interact, or the system will not behave as expected. Some of these items of information are handled, in one way or another, by many of the objects in the system. To ensure that the interpretation of these items is consistent, the architecture identifies the information viewpoint to specify the information to be handled, independently of the way the information processing functions themselves are to be distributed.

The information specification consists of a set of related schemata, just as in familiar data modelling activities. In the RM-ODP, a distinction is made between invariant, static and dynamic schemata. An invariant schema expresses relationships between information objects which must always be true, for all valid behaviour of the system. A static schema expresses assertions which must be true at a single point in time, and a dynamic schema specifies how the information can evolve as the system operates.

These schemata may apply to the whole system, or they may apply to particular domains within it. Particularly in large and rapidly evolving systems, the reconciliation and federation of separate information domains will be one of the major tasks to be undertaken in order to manage information.

Different information specification notations model the properties of information in different ways. Emphasis may be placed on classification and reclassification of information types, or on the states and behaviour of information objects. The approach to be taken will depend on the modelling technique and notation being used.

Since both the information and enterprise viewpoints consider the system as a whole, conformance to them must be assessed as a whole. Sets of observations at any of the points where system components are defined to interact should all be consistent with the requirements expressed in these viewpoints. If a set of observations is not consistent with the requirements of, for example, an invariant information schema, the implementation of the system does not conform to that part of its specification.

6. THE COMPUTATIONAL LANGUAGE

In distinction to the two viewpoints described so far, which consider the distributed system as a whole, the computational viewpoint is directly concerned with distribution. It does not address interaction mechanisms, but it does decompose the system into objects performing individual functions and interacting at well-defined interfaces. The computational specification thus provides the basis for decisions on how to distribute the jobs to be done, because objects can be located independently and communications mechanisms can be defined to support the behaviour at their interfaces.

The heart of the computational language is the object model which defines the form of interface an object can have, the way that interfaces can be bound and the forms of interaction which can take place at them. The computational language also defines the actions an object can perform, so that new objects and interfaces can be created and bindings established. This model provides the basis for specification languages, programming languages and communication mechanisms all to perform in a consistent way, thus allowing open interworking and portability of components.

6.1. Object interaction

There are two basic kinds of computational interface: operation interfaces, which support discrete interactions, and stream interfaces, which support continuous flows. Often, these interactions can be considered as indivisible, and nothing further needs to be said about how the interactions take place.

If, however, details of the progress of the interaction need to be expressed, for example, to define time delays or other aspects of quality of service, then the interactions can be structured into a sequence of signals, each of which has a precise location and time of occurrence. Timing requirements can then be stated in terms of the delay between particular pairs of signals or by more complex timing statements, as necessary. This use of signals as a common representation may in future be used to define new interaction types.

For the present, however, the kinds of interaction supported are strictly limited to those for which efficient communication mechanisms are known to exist. Thus there are only two kinds of operation interaction: interrogations, in which there is a request followed by a reply, and announcements, in which there is a single unidirectional transfer of information. More complex forms, such as rendezvous, which are expensive to provide in practice, are not supported.

Stream interfaces can be made up of a number of independent flows, each with a specified direction, so that a stream might consist of linked audio and video flows, or paired audio flows in opposite directions. Streams were introduced into the architecture primarily to support multimedia interactions, but they can also be used to represent other forms of unstructured information transfer, such as sequences of announcements giving regular updates from some sensor, if the exact repetition rate of the readings is not of major concern in the application design.

6.2. Binding

Before two objects can interact, the interfaces to be involved must be associated by creating a binding. In simple cases, one of the objects involved can perform a **primitive binding** action, linking it to another object. This is adequate for the expression of, for example, a straightforward client-server binding. However, in other situations, it may be necessary to model the binding process in more detail. This requirement may arise, for example, if there is a need to manage the quality of service of the expected interactions, or if the binding is to link more than two objects, supporting some sort of group or multicast interaction. In particular, stream bindings are often set up by third parties, and generally require some kind of explicit control during their lifetime.

To give the necessary control, the notion of a **compound binding** is introduced. In this form of binding (which can be expressed in terms of some piece of object behaviour involving primitive bindings), the originator of the binding instantiates a binding object, which, in turn, is bound to the set of objects which are to interact. The required rules associated with the interactions to be performed are then expressed via the behaviour of the binding object. Whenever a binding object is created, knowledge of a control interface is returned to its creator, so that changes in behaviour or configuration can be requested, or the binding object asked to terminate itself. Apart from the way its creation is parameterized by the set of interfaces to be bound, and its special status in encapsulating and controlling part of the communication function, a binding object is just an ordinary computational object.

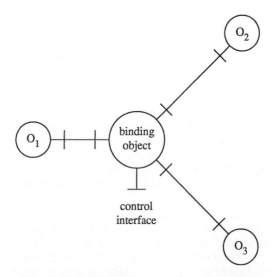

Figure 1 - An example of a compound binding

The behaviour required from a binding object can be quite complex. For example, a number of full duplex interfaces may be linked by a binding object which encapsulates the rules of a conference system for allowing the audio flow from a selected producer (the current talker) to be delivered to all consumers. Varying degrees of application control might be exercised via the binding control interface to provide explicit floor control. Simpler binding objects might represent full duplex audio links or synchronized audio and video flows.

However, in the simple client-server case, using a default quality of service, the visibility of the binding adds little to, and may complicate, the specification. In such cases, the designer of a computational notation may opt to conceal the whole binding process, providing **implicit binding**. If so, the notation specifies with each interaction being invoked the identity of the other interface involved; it is then left up to the supporting infrastructure to create any necessary binding and either to discard it after the interaction completes or to retain it, for a while, for possible reuse. Some notations may support a mixture of implicit and explicit binding for different interfaces.

6.3. Interface types and subtyping

Every interface has an associated type which characterizes it. One important component of this type is the interface signature, which expresses the static aspects of the interface — the operations available and their associated parameter types (or, for a stream, its flow types), together with the interface's role in the cause and effect sequence of interaction.

The computational language does not require that interfaces participating in a binding be of identical types. To do so would create a barrier to system evolution by making the phased introduction of new services and functions more difficult. Instead, a set of subtyping rules is defined, and the constraints on interface binding expressed in terms of them.

The computational language specifies appropriate type matching rules for each of the kinds of interface defined. The rules are in terms of signatures because they are easy to check during the interaction; behavioural aspects of the interface type may require much more knowledge of the previous history than is available. For operation and signal interfaces, a "no surprises" rule is applied. Interfaces can be bound only if none of the interactions involved introduce unexpected behaviour or fail to supply expected information. However, not all features of the responding party in the interaction need be exercised. Thus, in a client-server interaction, for example, the server may support operations unknown to the client, and thus not used, but the client must not invoke operations which are unknown to the server.

The situation for stream bindings is more complex, because there is an element of application choice in the selection of the binding rules. The computational language requires that flows be coupled only if they are compatible, but admits the possibility of applications in which some flows may be left without a matching counterpart. Thus, for example, one might choose to allow participation in a video conference from a mobile (non-video) 'phone.

6.4. Support for portability

Further definitions are provided so that the computational language stipulates which actions an object can perform (effectively defining an object based virtual machine) and enumerating the possible failure modes of these actions.

A set of portability rules, using the actions defined, identifies the requirements on a computational notation which is to be used to support the portability of objects between different environments. Notations may be referred to as basic or complete, depending on the sets of actions they support.

7. THE ENGINEERING LANGUAGE

The engineering language focuses on the way object interaction is achieved and with the resources needed to do so. Thus the computational viewpoint was concerned with when and why objects interact, but the engineering viewpoint is concerned with how they interact. In the engineering language, the main concern is with the support of the individual interactions between computational objects. It is here that one of the most direct links between viewpoints is found; computational objects are visible in the engineering viewpoint as **basic engineering objects** and primitive computational bindings are visible as channels or local bindings.

7.1. Clusters, capsules and nodes

The engineering language deals with the basic engineering objects and with various other engineering objects which support them. It relates these objects to the available system resources by identifying a nested series of groupings.

At the outer level, objects are physically located and associated with processing resources by grouping them into **nodes**, which can be thought of as representing independently managed computing systems. A node can be anything which has a strongly integrated view of resources, as long as the system designer can consider it as a whole. Thus a tightly coupled parallel processing system can be considered a node, so long as it has one scheduling and allocation policy — one operating system.

The node is under the control of a **nucleus** which is responsible for initialization, for creating groups of objects, for making communications facilities available, and for providing basic services like timing and the source of unique identifiers.

Within a node, there may be a number of **capsules**. A capsule owns storage and a share of the node's processing resources. It can be thought of in terms of a traditional protected process, with its own address space. A capsule is thus the unit of protection and is generally the smallest unit of independent failure supported by the operating system. There is a special object, called the capsule manager, associated with each capsule, and for descriptive purposes, a capsule is controlled by interactions with this manager.

A capsule will typically contain many objects; the grouping of objects into capsules is done to reduce the cost of object interaction. This is because communication between traditional processes is slow and expensive, because of the checks which need to be performed; however, the compiling tools that build capsules can be trusted to validate and structure the interactions between closely related objects to a sufficient extent to let them share resources. Resources within a capsule will be controlled by some kind of language-specific run-time system.

The smallest grouping of objects is into a set of **clusters** within a capsule. The objects in a cluster are grouped together in order to reduce the cost of manipulating them. The objects in a cluster can be checkpointed together, transferred to persistent storage, reactivated or moved to another node altogether. This manipulation of complete clusters as a single operation opens the way to the management of very fine-grain object-based systems at reasonable cost. For example, a geographical information system might consider data about individual points on a map to be objects, but could not sustain the cost of giving each of these objects a completely separate existence. Communication between objects in a cluster can be highly optimized, since the objects are created together, in the same language, and are expected to stay together.

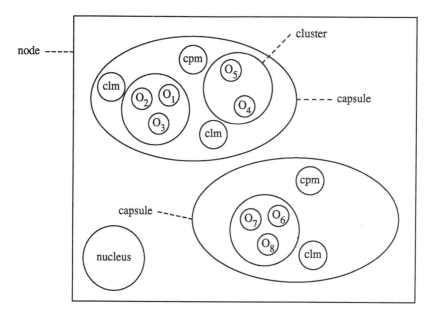

Figure 2 - Clusters, capsules and nodes

Interaction within a cluster might therefore be supported by a simple local method invocation or equivalent.

Clusters are controlled and actions on them initiated by interaction with an associated cluster manager object.

7.2. Channels

When objects in different clusters interact, there is a need for a good deal of supporting mechanism. Even if the objects are currently within the same capsule or node, mechanisms are needed to cope with the possibility of one or other of them terminating, failing or moving elsewhere. The set of mechanisms needed to do this constitute a channel, which is made up of a number of interacting engineering objects.

The objects within a channel can be divided into three types, based on the job that they do. **Stubs** are concerned with the information conveyed in an interaction, **binders** are concerned with maintaining the association between the set of basic engineering objects linked by the channel, and **protocol objects** manage the actual communication.

Stubs interact directly with the basic engineering objects they support, and perform functions such as the marshalling and unmarshalling of parameter, or the logging of information about the interaction being performed. Thus the stubs need access to information about the type of the interaction, or, more generally, the type of the interface that is being supported. This distinguishes them from binders and protocol objects, which transfer complete messages without concern for their internal structure.

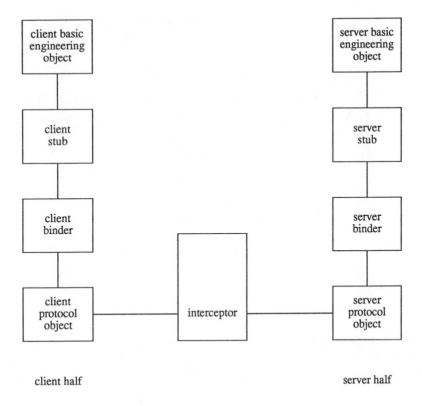

Figure 3 - An example of a client-server channel

Depending on the design of the system, a stub may be directly associated with a particular basic engineering object, or it may be shared between a number of such objects. Sharing will generally imply the need to transfer some additional information to identify, and thus distinguish between, the objects being supported.

Binders need to solve many of the problems of distribution. They are responsible for maintaining the end-to-end integrity of the channel, and so have to handle changes of configuration and communication or object failures. The binder has to establish the binding when the channel is created, and has to keep track of the other endpoints if objects move or fail and are replaced; this is the process of object relocation. The binders are thus involved in the provision of many of the distribution transparencies.

The protocol objects provide for communication of sufficient quality and reliability between the binders they serve. In addition to handling whatever peer protocols are in use, they will provide access to supporting services, such as directory services for translating addresses, where necessary.

Any of these three kinds of engineering object may itself need to communicate with other parts of the system, in order to obtain the information it needs to do its job, or to supply management information to other objects. Such communication may itself need the various distribution transparencies, and so the communication from these objects to elsewhere is by means of a channel; from this point of view, the objects within one channel can play the role of basic engineering objects in another. Similarly, any of these objects can support control interfaces, via which they can be managed. For example, a protocol object may provide a control interface through which the target quality of service for the channel can be adjusted.

In cases where the channel crosses some technical or organizational boundary, there may be a need for additional checks or transformations to match the requirements on the two sides. These functions are performed by **interceptors**, which form part of the channel. They may need to perform format or protocol conversion, or may provide accounting or access control checks. An interceptor may be built up from protocol objects, binders and stubs, depending on the nature of the job it has to do.

For simplicity, channels have been described here as linking two basic engineering objects. However, channels with many endpoints can be defined, supporting various forms of group communication or multicast. In such channels, the binders are responsible for coordinating communication, but the multicast mechanisms may be provided by either binder or protocol, depending on the technology available. Multi-endpoint channels are used to support replication transparency.

7.3. Interface references

When an interface is created, an interface reference for it is generated. The nucleus is involved in this process, so as to make the reference unambiguous, and sufficient resources are allocated and initialized for the objects in that node to participate in bindings if asked to do so.

The interface reference is the key for access to a large amount of information. Given such a reference, it is possible to discover the type of the interface, a communications address at which binding to it can be initiated, and other information about the expected behaviour of stubs, binders and protocol objects within the channel, which is needed for a subsequent binding to succeed. It is also the starting point for calling upon the functions needed to handle errors; knowledge of an interface reference makes it possible to contact an appropriate relocator.

This does not imply, however, that the information is all encoded as part of the interface reference; to do so might make it a very big item to manipulate. The architectural requirement is that there should be some prescription for obtaining the necessary information, starting from the interface reference, but the exact prescription, in terms of decoding and enquiry from other objects, can be chosen differently in different system designs.

In addition to these design variations, there will also be variations arising from the existence of multiple naming domains and the allocation of references with respect to these domains. For both these reasons, it will be necessary for interceptors, or other objects in the channel, to transform interface references when they are passed across domain boundaries.

7.4. Binding

There are two kinds of engineering binding. Within a cluster, or between the objects which cooperate within a node to provide a channel, there are **local bindings**, which are provided by system-specific mechanisms. Such bindings are regarded as primitive in the architecture. On the other hand, the bindings supported by channels provide appropriate distribution transparencies; these are called **distributed bindings**, and creating them will generally involve some interaction between a number of nodes to establish the channel.

7.5. Conformance

The structuring of the engineering specification into clusters, capsules and nodes, and the support of interaction by structured channels gives rise to a large number of interfaces, any of which can be selected as a conformance point, allowing for observation and conformance testing.

The various interfaces can be used to provide the different kinds of conformance. The interface between protocol objects is an interworking conformance point, providing for familiar methods, like OSI testing, based on observation of the communication behaviour. Most of the other interfaces are internal to a node, and represent boundaries between software modules; they are programmatic reference points and allow testing for software compatibility and portability. Some of the interfaces to basic engineering objects may allow other forms of conformance testing, for interchange or perceptual conformance (correct interaction with the real world).

8. THE TECHNOLOGY LANGUAGE

The technology viewpoint provides a link between the set of viewpoint specifications and the real implementation, by listing the standards used to provide the necessary basic operations in the other languages. The aim of the technology language is thus to provide the extra information needed for implementation and testing by selecting standard solutions for basic components and communication mechanisms. Such a selection is necessary to complete the system specification, but is largely divorced from the rest of the design process.

There are consequences of the technology selection, however. One area in which the selections in the technology viewpoint feed back to other aspects of the system design is in the provision of a specific quality of service. The selections in the technology viewpoint determine the performance costs of interactions and thus, indirectly, the quality of service which can be achieved by the behaviour defined in other viewpoints.

The technology language plays a major role in the conformance testing process. It supplies the information needed to interpret the observations a tester can make in terms of the vocabulary and concepts used in the other viewpoints of the system specifications. For example, it allows valid interactions to be recognized, so that their appropriateness can be checked against some specified object behaviour.

9. CONSISTENCY BETWEEN VIEWPOINTS

The five viewpoint specifications constructed must be linked by defining the relations between key terms in them. It is these statements of the relationships between viewpoints that make them specify a single system, rather than being completely independent documents. See [5] for some examples of demonstration of viewpoint consistency.

Many of the links needed will be provided implicitly by the notations used, resulting from correspondences between names. However, some of the key constraints need to be stated explicitly. In the architecture, constraints are placed on the relations between terms in the viewpoint languages themselves, establishing some limits on the mappings which can be established. Most of the constraints placed are between terms in the computational and engineering languages, and are defined so as to create consistent interpretations when system components, such as those supporting the ODP functions, are specified separately.

Clear mappings between viewpoints are necessary if the processes of identifying interfaces and of providing transparencies are to be supported automatically by development tools. For example, a computational object may be realized as a set of linked engineering objects, but a single engineering object cannot represent multiple computational object; a computational interface cannot be divided into separate engineering interfaces supported by unconnected channel structures; computational interfaces can always be identified unambiguously by engineering identifiers. These kinds of constraint help to ensure that common engineering mechanisms will be able to support the full range of possible computational behaviours.

10. ODP FUNCTIONS

In addition to the five viewpoint languages, the RM-ODP gives brief definitions of a number of common functions. Most of these are either introduced in the engineering language to provide support needed for its structures, or form convenient building blocks for the provision of transparencies. Functions are provided by objects, although it is generally left for more detailed standards or individual implementors to decide whether each function is provided by a single object, or several functions by one object, or a function provided by a set of interacting objects.

The specification of how one of these functions is to be provided may be complex and may itself needs to structured. It amounts to the design of a small, special purpose distributed system, in which the interactions between the object providing the function and objects having roles which relate to it are defined. It is natural, therefore, to structure the specifications by using the RM-ODP framework, considering the provision of the function as a small enterprise with its own information and computational models.

10.1. Management functions

Management functions are needed to control the lifecycle of objects and of the various groupings of objects identified in the engineering language. For each of the management functions, there will be a corresponding management interface type, the details of which will depend on the kind of grouping being managed. Management functions exist to control individual objects, clusters, capsules and nodes.

For individual objects, a management function can request checkpointing of the object's internal state or deletion of the object. Object checkpoints are combined into cluster checkpoints, which are essentially templates for recreating the cluster in the state it had when the checkpoint was taken, if necessary. Cluster management is primarily concerned with using this information to deactivate, move, reactivate or recover clusters. These activities form the basis of a number of the transparency mechanisms.

The capsule management functions perform a similar job, but at a coarser level of granularity, controlling the resources of the cluster as a whole and instructing the capsule managers to create or remove their clusters.

At the coarsest level, the node manager controls the basic resources of the node, and makes allocations from them on request. It handles execution resources such as threads and timers, communication resources and naming responsibilities. It creates new capsules when necessary.

10.2. Coordination functions

The second class of functions is concerned with the coordination of distributed activities and the management of distributed groups of objects. These functions support various kinds of consistency and information dissemination mechanisms.

The first coordination function deals with event notification — the establishment of objects which maintain historical records of which events have happened and take responsibility for informing other objects of when events occur. The event notification function thus makes it possible for a coordinated group of objects to maintain an interest in the occurrence of selected events in a consistent way.

The next set of functions is concerned with the coordinated maintenance of checkpoint records of the state of the clusters which an application depends on. The architecture distinguishes checkpointing and recovery functions, which are concerned with keeping and using records, from deactivation and reactivation functions, which control the activity of interest directly.

The group function coordinates some number of objects which are participants in a multi-party binding; it is concerned with group membership, distributed management of group interactions, and combination of results to ensure a consistent outcome. It has a specialized form supporting exact replication of a number of objects, coordinating their interactions such that they remain replicas of each other at all times.

Built on these is the migration function, which supports the movement of a cluster from one capsule to another. Migration can be achieved in one of two ways, depending on the performance requirements. It can be achieved by deactivation followed by reactivation in the new location, or it can be achieved by replication of a new copy in the desired location, followed by deletion of the original copy. The first involves less communication, but the second provides a more continuous service.

Sequences of actions can be coordinated to achieve a consistent result by using the transaction functions. Both a general transaction function and a specialization of it to provide ACID properties are defined.

Finally, an interface reference tracking function is defined to maintain records of what interface references exist and where copies of references are held, and to support whatever garbage collection policy the system designer selects.

10.3. Repository functions

The repository functions are all concerned with persistent storage. There is a general storage function and then a number of specializations of it, supporting different types of repository. The basic storage function just allows an object using it to make any data item persistent, that is, to have a longer lifetime than the object itself.

The information organization function stores information about the various objects and interfaces in the system, and supports structured queries on the information stored. It can be used to maintain information about relationships between and attributes of objects.

The relocation function provides a specialized store of information about interface references, which can be used to update an interface reference if the object concerned moves, of fails and is restarted. For this mechanism to work, it is necessary for each of the mechanisms which might be involved in altering the interface reference data to record the appropriate information with the relocator.

The type repository function provides a source of information about the various type definitions supporting the system, particularly interface types, and can record type identifiers, type definitions and assertions of subtyping relationships between them.

In contrast, the trader stores information about interface instances and properties associated with them. These properties can give information both about the expected behaviour at an interface and less tangible properties of the service available from it. The trader can be queried for services by type and by properties, allowing suitable instances of the service to be discovered. Service offers are placed in the trader's records by a service exporter and queries performed by an importer, which intends to use or pass on knowledge of the service. The trader is one of the most important ODP functions, because it allows the dynamic configuration and evolution of distributed systems. Objects using it can seek out the services they need. Because of its importance, it is the first of the ODP functions to be the subject of formal standardization.

10.4. Security functions

The RM-ODP identifies a full range of security functions, largely by reference to the established standard security frameworks. It identifies functions for access control, security audit, authentication, integrity, confidentiality, non-repudiation and key management.

11. ODP TRANSPARENCIES

The ODP transparencies are defined by giving a prescription for translating from a set of system specifications in the computational, information and enterprise viewpoints to an engineering specification which incorporates the various ODP functions needed; the functions are used in a coordinated way so as to provide the necessary transparency. The transformation may result in changes to the original object's behaviour and interface signatures, in order to incorporate any control interactions and information needed to guarantee the transparency. Thus, for example, the transaction transparency may require additional commit interactions and the addition of transaction identifiers to the parameters carried by existing interactions.

11.1. Access and location transparencies

Access transparency is normally provided as part of the basic function of the engineering stub object, and so the transformation to be performed to provide it is a straightforward refinement to introduce the channel structure.

In a similar way, location transparency will be provided by some combination of the stubs and protocol objects.

11.2. Failure transparency

Failure transparency is requested in terms of the kinds of failure that should not be allowed to disrupt the application. It can be provided in a number of ways. Firstly, the objects involved can be placed in an environment which is inherently sufficiently reliable, such as a non-stop system. Secondly, the checkpointing and recovery mechanisms can be used to overcome faults when the occur. Thirdly, the replication mechanisms can be used to make faults non-damaging to the application.

Which of these approaches is to be taken will depend on the relative cost and performance objectives to be met, particularly whether the system has to give real-time guarantees. These choices will be made on the basis of the enterprise policies which have been established.

11.3. Migration transparency

The migration transparency is expected to minimize the effect of movement of objects. One part of the information needed to determine the support needed for migration is the set enterprise policies which constrain object mobility, since if the objects do not move, there is no problem.

The decision as to whether objects should be moved will have to take into account issues of resource management, performance targets and security. Once the mobility constraints are established, suitable migration strategies can be determined and the mechanisms needed to support them incorporated.

11.4. Persistence and relocation transparencies

Both the persistence and relocation transparencies will involve the use of the relocation function. This is because they are both involved in changes which may invalidate current interface references, causing subsequent attempts to create a binding to fail.

In both cases, the relocation function will need to be used in a way which is coordinated with the resource management and recovery activities, to ensure that sufficient information is provided for the relocator to do its job.

11.5. Replication transparency

The provision of replication transparency is potentially complex because of the need to consider both the client and server roles of any object being replicated, to ensure that the behaviour of the system as a whole remains consistent.

The transformations involved to support replication are therefore potentially less localized than in some of the other transparencies. The cluster managers supporting the replica copies, at least, will need to be involved in the coordination.

Similar concerns also apply to transaction transparency.

12. CONCLUSIONS

The reference model described in this paper provides a firm basis for the construction of families of open distributed processing systems, capable of supporting a wide range of applications.

ISO and the ITU now plan to populate this framework, drawing on the work of the more forward looking of the industry consortia, where appropriate, to speed the process. The framework is now stable, and has been defined with sufficient flexibility and with a broad enough scope to satisfy the needs of distributed system builders for many years to come.

REFERENCES

[1] ITU Recommendation X.901 | ISO/IEC CD 10746-1, Open Distributed Processing - Reference Model - Part 1: Overview (1994).

[2] ITU Recommendation X.902 | ISO/IEC 10746-2: 1995, Open Distributed Processing - Reference Model - Part 2: Overview.

[3] ITU Recommendation X.903 | ISO/IEC 10746-3: 1995, Open Distributed Processing - Reference Model - Part 3: Overview.

[4] ITU Recommendation X.904 | ISO/IEC CD 10746-4, Open Distributed Processing - Reference Model - Part 4: Overview (1994).

[5] Bowman, H., Derrick, J., Steen, M., "Some Results on Cross Viewpoint Consistency Checking", Proc. ICODP'95, Brisbane, Australia, February 1995

3

New Ways of Learning through the Global Information Infrastructure

Jacob Slonim [a] and Michael Bauer [b]

With increased access to computer-based information sources and the growth of national information infrastructures, society is shifting to one that is information-based. The result of this shift is new types of workers and new applications; both will become the driving forces behind the evolution of systems, particularly open systems. What is required, then, is a broader view of an "open system".

In this paper, we first examine some of the technological and social transitions underway as society becomes more information-based. We then look at a particular emerging application area, life-long learning, and its technological and social-economic requirements. We broaden the notion of an open system and discuss the challenges faced by developers of open systems prompted by the application demands.

Keyword Codes: C.2.4., H.4.3, K.3.1
Keywords: open systems, distributed applications, information infrastructure,
     life-long learning

1 INTRODUCTION

The growing accessibility of computing resources, communication infrastructures and digitized information is causing a major shift in society's behaviour. As computer-based information sources become more accessible and as the need for timely information in many different organizations grows, our technical-based society is shifting to become an information-based one. In looking at this shift from a broad perspective, one can see three fundamental dimensions. First, there is technology: the rapid evolution and deployment of computers and communications technology. Second, a growing consumer population is demanding more access to information, more varieties of information and new applications (see Figure 1). Much of this demand is driven by home entertainment. Third, emerging application domains offer opportunities for new applications made possible by the technology and made viable by the potential consumers.

Consumers and application domains are the driving forces behind the technology. In particular, we see a shift from a technological-centric evolution of systems to a solution-centric one driven by users and applications. Applications in new domains, such as the

---

[a] Centre for Advanced Studies, IBM Canada Laboratory, 844 Don Mills Road, North York, Ontario, Canada M3C 1V7 (e-mail: jslonim@vnet.ibm.com)

[b] Department of Computer Science, Middlesex College Building, The University of Western Ontario, London, Ontario, Canada N6A 5B7 (e-mail: bauer@csd.uwo.ca)

environment, health, and education, become possible because of the new technology. Solutions to problems in these areas require further advances in technology and, we believe, that they require a broader definition of "open systems".

Until now, the concept of an "open system" has been driven by technology experts. If we look at the broader view entailing applications and consumers, open systems become more "real". Moreover, this implies a significant shift in the "perception" of open systems: the concrete reality of applications means that openness is no longer an "if", but a "when". Openness is a key to providing solutions to problems in these domains, and the key aspect of openness is the *degree of transparency*. The mechanisms involved in how one achieves such transparency, such as CORBA[1] or DCE[2] or ODP[3], become "details". "Proprietary" systems will not disappear; rather, they too will become transparent, especially to the information consumers.

In this paper, we look at the implications of this view on open systems, particularly the challenges that must be addressed in the overall "solution". In other papers [4, 5, 6] we have looked at some of the technical challenges central to achieving open systems. This paper takes the perspective of the end-user. We look at a particular domain, education, to explore this perspective. The educational domain is of particular interest for two reasons. First, the rapid advances in technology mean that it is very difficult for those in the information-technology field to keep current and maintain and develop new skills. Second, there is an enormous demand for ongoing, life-long learning, in general. Traditional ways of learning, classrooms, universities, etc., are no longer satisfactory. New and emerging technologies will make alternatives possible, and at the same time will raise many issues and challenges for those working on open systems. Some of these will be technological, but many will be economic and societal.

The following section, explores some of the transitions, both technological and societal, already underway. Section examines the requirements and issues arising in life-long learning, including social issues. Section discusses the implications of a broader view of open systems and the challenges facing those developing such systems. The final section provides a brief summary and conclusions.

2 TRANSITIONS

The shift toward an information-centric society is being driven by rapid advances in computer and communication technology, increased availability of personal computing systems and software and growth in online, digital information. Trends in some of these areas are illustrated in Figure 2. They are creating changes i.e., transitions[7, 8], in many different aspects of society; global, social and technological.

2.1 Global transitions

The emergence of national and global information infrastructures and global computing are changing the way business is done and the ensuing economic issues.

National borders are disappearing. Organizations are becoming increasingly international; different organizations are finding it mutually beneficial to cooperate. Competitiveness is not constrained to or protected by national borders. Advances in communications mean that physical location is no longer a key factor in competitiveness. Work schedules are no longer restricted to 9-5 or east vs. west coast. Information and information technology services are provided as easily by an organization across the street or on the other side of the world.

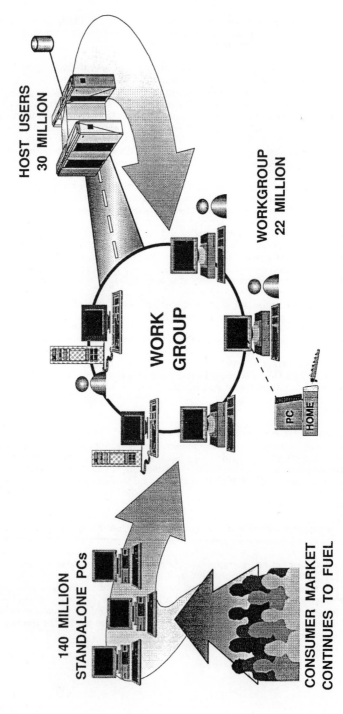

Figure 1: Trends in Technology

As marketing and sales methods change, new services are emerging. Pizza Hut's recent advertising and accessibility via the internet represent an initial effort to explore these new avenues. In many respects, the services parallel those available via newspapers: advertising, job posting/hunting, current information, etc. The convergence of computer communications, cable television and satellites raises many opportunities in home shopping, advertising and sales services. Access to timely and diverse discussion groups, such as the 60,000 or so available over the internet, means that information exchanges can be dynamic and user-motivated. The provision of these services is not limited to local geographical regions.

As borders become transparent, governments face new challenges. Global competitiveness pushes countries to group together to foster growth, remove restrictions on multinational cooperation, and address regulatory issues. Workers with limited skills are being replaced by knowledge workers. Countries are faced with significant changes in their work forces; appropriate education, will enable them to compete globally. Many of these societal issues are more challenging than the technological issues facing the developers of open systems.

2.2 Technological transitions

Coming from technology organizations, most of us are familiar with the transitions within the computer and communication field.

A number of general trends rely on advances in hardware and software. There is a push to provide the means to integrate components, whether they be hardware or software, within an open system framework. We are seeing a shift from centralized computing environments to client-server environments; we believe that this shift will continue toward peer-to-peer computing environments.

The emergence of a consumer population for information and software is forcing the information technology industry to shift from being computer-centric to network-centric and, eventually, to being human-centric. As open networks become reality, the role of the individual workstation will evolve. (For example, see [7]). In the past, there has been a great deal of focus on the individual workstation and its capabilities. With open systems, these workstations become part of a much larger computing environment and are no longer the prime focus. The network will provide access to and will enable use of resources for an individual from any of his or her computers. (An individual may have access to several.) A good analogy is the telephone. The individual telephone has a limited number of well-defined functions and features. A multitude of services can be provided via the telephone if it is connected to the intelligent switches now in operation. The switches provide the additional, value-added services, and they are most critical in communications. As a harbinger for developers of open systems, these switches may involve 20 million lines of code. Competitive advantages go to those who can make changes rapidly without introducing problems.

Further, as consumers make greater use of information technology and place greater demands on what is provided, they will become the driving force behind the technology. This demand will necessitate a further shift toward a human-centric view of information technology. As more users with a limited knowledge of technology gain access to information technology in their jobs, there will be a greater need to focus on user interaction. New devices and modes of communication, for example, voice input and output, will be required. The underlying systems will inevitably change. Interfaces will have to become "standardized", and systems will have to change to comply with these "standards". Again, by way of analogy, consider the telephone; even with some variation, the interface is standard.

Interestingly, the shift toward network-centric and human-centric is happening concurrently. As technology drives toward openness, there is a push toward a network-centric view. Simultaneously, as the population of consumers of information technology grows, there is pressure to move toward their view. This additional pressure has motivated our concern for a broader view of "openness".

Some of the trends in hardware have already been described. (See Figure 2). Processors continue to double in speed and capacity every eighteen months. The speed of system-to-system communications is increasing at an incredible rate. Advances in secondary storage technology continues to decrease the cost of a megabyte, whether in memory, on secondary storage, or on optical disks. Hardware components are fast becoming commodities with well-defined and variable characteristics, for example, a processor with a specific clock speed or performance level, a hard disk with a certain access speed or storage capacity. Other device-specific details are often irrelevant to the choice. Other factors are important to the consumer, such as price, availability, and reliability.

The new application domains made plausible through advances in the above technologies and the growth of a consumer population will also require transitions in software technology. These new applications will require domain-specific skills and knowledge and the use of various building blocks, such as objects or processes. It is likely that many of the future applications will be "composed" by domain-specialists from these components or niche applications, such as word processors or spreadsheets. (See Figure 3.) These new applications may also require new languages – a shift from the procedural languages of today to ones which are declarative or functional.

These applications will also demand new forms of data. Structured data will be augmented with unstructured data, such as text, voice, video. New approaches for storing, accessing, and searching such data will be required. No single data model will suffice for all forms of data or applications; multiple models will have to be integrated. Online data sources will grow; terabyte databases will become commonplace and grow to petabyte or exabyte sizes.

We also see a shift from data to information to "knowledge"[9]. Not only will there be growing sources of data, basic digital representations of numbers, characters, audio, video, etc., but there will be an even greater growth in information, that is, a capturing of the relationships among the data. Moreover, the "information" is very likely to grow as it is used by more people and as it is digested and new relationships drawn. Finally, it will become necessary to synthesize this information, that is, to extract "knowledge" from it. "Intelligent agents" may be needed to help users extract knowledge pertinent to their needs.

Mobile computing means that applications and information sources will be accessible from just about anywhere in the world. There is already some movement of jobs from the office to homes. With the prevalence of mobile computing, the "office" will move with the user.

Further, new applications will present much greater challenges in terms of the integration of components. New needs will arise that can only be addressed by a response to the multiple facets of both hardware and software technologies.

Just as there is a shift from mainframe computing environments to distributed ones, there will be a shift from managing centralized computing environments to managing

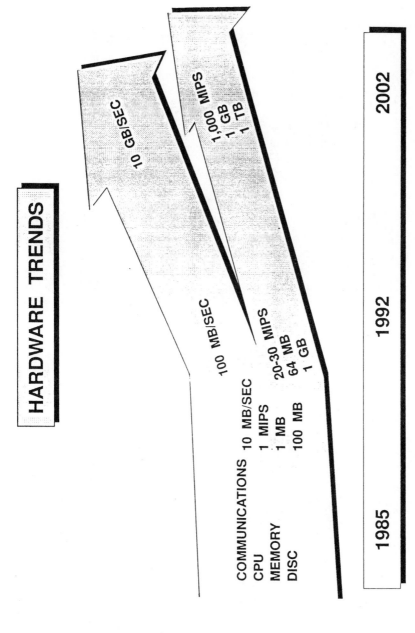

Figure 2: Trends in Technology

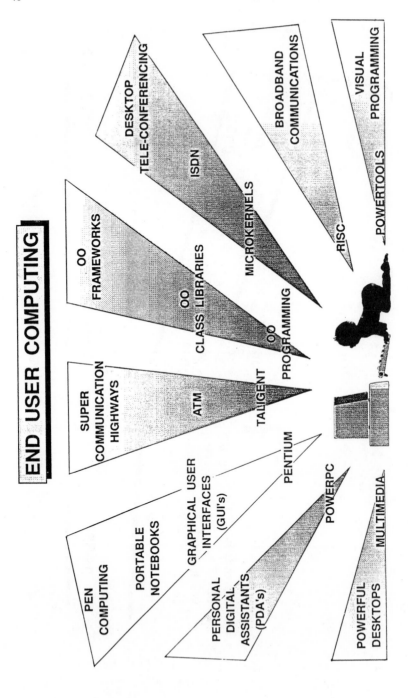

Figure 3: End User Computing

large, distributed computing ones. This shift also presents a number of significant challenges. They include a heterogeneous computing environment, one which is not "static" because the comings and goings of individual workstations and mobile computing devices; the dynamic reallocation of software resources, such as servers; and quality of service requirements for certain applications. All of these will require bandwidth on demand, placing additional complexity on the management of the entire network. Configuration management will become more important as organizations cope with a dynamic, diverse computing environment.

Management will be further complicated as devices and systems are embedded with more "intelligence" of their own. Home computers, televisions, radios, even refrigerators, will be, at least potentially, "manageable". It will be possible to monitor, perhaps even control, millions of devices. Information technology may fundamentally change the home. Delivery of a variety of different types of information, multiple household users of technology, its pervasiveness in many different devices may result in the "home" being simply another extension of the network. As suggested in a recent newspaper article about Bill Gates' home, future homes may be like his and have a wall or walls of one or more screens for communication with others, for entertainment, for learning, etc.

2.3 Social transitions

Societal changes are inevitable as we move into an information-based society and the consumers of information technology increase. Information technology will have to become more human-centric; systems and services will have to be designed around users and their needs. There is already a shift from a non-computerized generation to one that is computer literate or at least computer "aware" and an increase in the number of nontechnical users requiring access and use of information technology[10]. A generation of knowledge workers is emerging.

As economies become based on information technology and knowledge-dependent jobs, the common notion of the "work place" will change. Organizations are now experimenting with work at home. As noted, mobile computing will mean that an individual can take the office along. Will this change lead to a generation of individuals unable to interact in social settings?

Distributed work, the pervasiveness of information technology, and many other aspects of an information-based society, threatens the privacy and security of the individual. This concern will have to be addressed through advances in technology and in changes and additions to international regulations and processes.

The growth in users of information technology will have other affects as well. Many jobs requiring minimal skills will be eliminated and replaced with new ones, as yet unknown, requiring knowledge skills. The creation of new applications, for example in the application domains, and new information sources require knowledge workers to acquire new skills on an ongoing basis. Moreover, the individual will have to be more responsible for his or her own education, specifically within the context of an individual job and its needs. Education will have to be timely, accommodate greater numbers of nontechnical users, and have to become personalized. Learning will be life-long.[11, 8]. Individuals will not be able to return to long-term classroom education; education must be continuous. Those providing information technology and open systems as well as those in the education systems must face the challenges of education. Subsequent sections discuss these concerns.

Given the current levels of job-related pressure and stress, how will individuals be able to take responsibility for their own ongoing education and spend the time being educated? We cannot continue at the current pace and still expect individuals to remain up-to-date. Education will fail if individuals have no time to relax and to spend time with family and friends. We may have to change our work scenarios so that employment consists of three days of work and two days of self-directed education. Such an employee would still receive remuneration at current levels, but have two days of rest and relaxation in order to maintain effectiveness.

How will those currently unable to afford an education be trained? On one hand, affordable technology may mean that access to educational resources available throughout the network can make education available to many more individuals and at an overall lower cost per student. On the other hand, will such resources be freely available[10]? If there are costs, who will pay? How is continuous education to be funded? Responding to these questions will significantly affect educational systems.

Finally, we are already seeing the beginnings of a merger of information technology with entertainment. What will be the boundaries between entertainment and information, and who will own the content? Will there be boundaries? Should there be? As with the current entertainment industry, how will censorship be handled, especially with the delivery of products across national boundaries, and who will assess the accuracy, etc. of the information?

3 LIFE-LONG LEARNING

As information technology will play a greater role in helping to provide ongoing education. The following focuses on ongoing education, commonly called *continuing education*, and the requirements placed on information technology in general. The implications for open systems, in particular, are discussed in the section . We do not address the totality of the educational environment. Some of the technological advances implied in the following may also be of great use in other aspects of education, such as in K-12.

3.1 Educational technology requirements

Advances in information technology, and, in particular, open systems, will make this new education possible[8]. The use of such technology in providing education to the worker does not preclude its use of traditional educational settings. We believe that advances in educational technology for the workplace can augment and expand and, perhaps in some instances, even replace how students are taught in traditional settings[12]. The following points identify some of the requirements of life-long learning:

- the need for timely, accurate information: Information on specific topics or skills must be provided when an individual needs it. Needs may arise during "normal" working hours, but also at other times as individuals take on the responsibility of their own education.

- Need to filter information: The multitude of information sources and enormous amounts of detail mean that a user needs tools to acquire relevant information at the level of detail the user requires. Intelligent agents can help an individual locate sources of information and filter their contents.

- Multiple delivery mechanisms: Similar information might exist in different formats, for example, an audio/video form and a textual one[13]. Different delivery mechanisms will be required to satisfy the needs of different individuals. For example, a textual version may be needed by someone wanting to study a topic in great detail, whereas a straightforward audio/video presentation may just provide an overview.

- Geographical transparency: The individual will not be bound to "courses" provided by local institutions, instructors or information sources. Global education resources and expertise will be available. Current efforts at distance education are already moving in this direction.

- Self-paced education: Different individuals will have different needs in the kinds and level of detail and will also learn at different rates. While self-paced instruction may be adequate for updating knowledge, gaining some basic introductory understanding of an area, etc., will it create an expert? How will curricula have to change to accommodate this need? New educational approaches will be required.

- User-oriented delivery: Education must become personalized. Individual needs and preferences must be accommodated. Individuals will study at different times, require different amounts of detail, move at different pace, etc. Software must be provided that can be customized to a user's preferences or that facilitates the user's own selection. Further, physically challenged individuals may require specialized interface devices, software, etc.

- Dynamic and scalable curriculum: In some instances, extended periods of study will be required. An individual may need to study material over weeks, or months, or even longer. Curricula would be required to lead the individual through his or her studies. Such curricula would have to be tailored and customizable based on a user's needs, preferences, capabilities. (See next point.)

- Multidisciplinary: The needs of an individual may require knowledge from several different disciplines. Material from different sources and domains may have to be brought together, perhaps in ways not anticipated by designees.

- Life-long plan for education: As an individual changes jobs and careers, educational needs will change as well. Education will become part of planning a career.

- Need to locate experts: Part of an individual's learning will involve studying bodies of material, but part will also involve consulting with experts. Such consultation will require advanced technology and will require tools to help individuals contact experts.

- Tools for interaction: One advantage of classroom or seminar settings is the ability to learn from other individuals and to debate and discuss ideas. Advances in technology are making online, real-time, multiparty communication and interaction a possibility[14]. Group collaboration is important, such collaboration would also have to be available. The development of interfaces, devices and appropriate curricula will be required to facilitate such interaction.

- Multi-lingual communications: Given the global information community, there may be sources of information or experts anywhere in the world; "courses" may involve individuals from different countries. Simply providing for the exchange of messages will not be enough; agents to help translate or to find experts to translate will be needed.

- Multi-modal interactions: There will also be a need for expanded interaction mechanisms. Keyboards may be adequate for many purposes and users, but others will want to use a routine pen-like device or touch screens to access information and run applications. Better displays and audio will be required for multiple interactions, videos, etc. Applications will have to accommodate different combinations of input and output devices depending on the needs and desires of the individual. Interactions may also come via less familiar devices. A user's stereo or television may draw upon digital sources for delivery, diagnosis of device problems, such as a malfunctioning washing machine or automobile, may be handled from remote sources and therefore require connection to the network.

- Convey human emotion: A greater challenge exists in providing the means to convey human emotions. A lecturer's excitement about a particular topic cannot be adequately conveyed via strict textual exchanges; the vehemence of a debate is not effectively conveyed in a news group.

- Entertaining: As the entertainment and information technology industries become more closely aligned, it may be possible to make education fun, exciting as well as rewarding. Expertise from both industries may be needed to effectively deliver educational material to an individual or group of individuals.

- Packaging: Just as curriculum design will need to become more flexible and more individual oriented, there will be a need to combine technologies with sources of information, with services and delivery mechanisms. Information is of little use if the content is limited, if the technology is inadequate to display or deliver the information or if the service is unreliable.

- Ownership, copyrights and plagiarism: There are questions about ownership of online material. In an educational setting, incentives are needed to ensure that the best educators are encouraged to prepare material and interact with students and knowledge workers. How to ensure the intellectual property of on-line information must be addressed. Because digital information makes reproduction of material easy, there must be assurances that an individual's work is his or her own.

- Affordable: Even with great technological advances in hardware and software, the use of information technology for education will not become a reality unless it is affordable. On the other hand, vendors and providers of information must be able to survive in an increasingly competitive world.

- Digital libraries: As more information becomes digitized, we will see a shift from the traditional library to an online one. Intelligent search agents will help locate reference material, perhaps at distant sites. Changes taking place within the digital library framework are discussed in greater detail elsewhere[15].

While these requirements are not necessarily complete, they raise many challenges, especially for those working on open systems. In section , we explore some of these challenges.

3.2 Socio-economic requirements

The future learning and education environment represents many challenging technological problems. A number of social (see [16] for a discussion of some) and economic requirements must ultimately be addressed.

First, as suggested above, there are many questions surrounding ownership and copyrights. These will have to be addressed in the broader context of governments and international regulations.

A second issue arises out of the question of cost. An educational system must be affordable, yet service providers, information providers, integrators, experts, etc., will be unwilling to participate unless they can do so profitably. How will educational "services" be paid for? Should they be free to the individual, funded by governments? Should they be on a pay-per-use basis? How is public education separated from work place education? Should it be?

Security and privacy will continue to be issues, even in the context of education. Providers of information sources will want to ensure that only authorized access to their databases is permitted. In a competitive environment, much can be gleaned by knowing what "research" individuals in one organization are doing. There is a need to ensure that some of the educational activities pursued by an individual are kept personal.

Transparency itself complicates the above issues. Being able to access certain courses or information sources regardless of where they are or who offers them, but there may be cost implications that the user may need to know about. Similarly, not know who is accessing a particular educational provider's material may not be desirable.

Finally, although providing an individually oriented educational environment via technology is an exciting possibility, the overall management of the entire educational infrastructure represents a great challenge. How is access provided for all individuals within a single country? How is online, distance education integrated with local classroom education? How are such services, such as public libraries integrated? How are these, in turn, integrated with digital libraries? How will future teachers and educators themselves be trained (or retrained)? Many of the existing governmental educational structures will have to change to accommodate radical changes in education.

4 Implications and challenges for open systems

As we have suggested, there is a need to broaden our view of open systems, where an open system is composed of pieces that can fit together. In the last few years, the view of open systems has been vendor-centric: how to facilitate the cooperation of different vendor systems composed of different hardware, software, communications components. To a great extent, this need has been driven by the rapid advances in hardware and communication technologies. More recently, advances in the development of midware have begun to address system and service interactions.

We believe that there is a need to go further, to consider openness from the perspective of the consumer or from the needs of different application domains. In the following, we discuss some of the implications of these requirements on the notion of openness. Then we briefly discuss the challenges facing the developers of open systems and their constituents.

4.1 Implications The broader view of openness we discuss is primarily motivated by our consideration of education and life-long learning. However, many of the aspects discussed are common to many of the other application domains mentioned in the introduction. Where differences are likely to appear of course, is in the specific individual applications and solutions in those domains.

- User-centric environments: Open systems should be driven by the needs of the users. Users will include traditional students and teachers at all levels of education, as well as knowledge workers seeking to upgrade their skills and expand their knowledge. Users may be individuals at home simply seeking to enhance their quality of life, for example, increasing their gardening skills.

- Technological transparency: The hardware, communications, and software should be transparent to the end user and should accommodate the user's choice or needs.

- Component independence: Objects, at least those within the domain of relevance to a user, should be independent of underlying systems and platforms; it should be possible to easily move an object from one system to another. More advanced objects may capture data, procedures and models of a user. For example, a "lesson" may have the relevant data, procedures for delivery, testing, etc., as well as model of the user, for example, beginner, expert, preferences, etc., to accommodate delivery and pacing of the lesson itself.

- Representational independence: The user should not know that there are different kinds of data, of different types, stored in different types of database systems. Data should be easily accessible from multiple locations.

- Scalability: Open systems should be scalable to accommodate new systems, applications, devices without visible changes or effects on users. They must also scale to accommodate components of different sizes - systems and data could be potentially unbounded. Building in size limits becomes restrictive, and alternative approaches must be found.

- Peer-to-peer: Open systems will have to evolve from client-server views to peer-to-peer; the need for scalability, user-based objects and demands of end users will require the greater flexibility of peer-to-peer.

- Quality of service: The demands of technology throughout the workplace and its role in the knowledge worker's domain will mean that service quality will have to include many aspects of information technology, for example, data access and delivery of information. Open systems will have to be designed to accommodate such service demands.

- Reliability and dependability: Related to quality of service, our view of open systems must be reliable and dependable across multivendor platforms and systems.

- Open, yet proprietary: Openness does not imply that platforms and systems are no longer proprietary. Solutions can still be proprietary, but the interfaces must be open. What is in the box is irrelevant, providing that the interface to the box is open and that it provides what the interface specifies.

- Manageability: As systems become increasingly decentralized and interoperate, management becomes increasingly more complex. Open Systems must be manageable if they are to be successful.

- Charging and costing: Openness implies interoperability and transparency of different vendor systems. For open systems to be successful, it must be possible to recoup costs, even if access and delivery of services cross various geographical boundaries. Certain aspects of charging and costing must be open to enable cooperation, yet there must be sufficient flexibility to permit competition in providing unique, complete or quality information sources and to charge for them.

- Ownership: As information sources become widely accessible, distributed and replicated, and as new generations of applications begin to emerge, such as those in education, questions of ownership and protection become critical. Openness implies accessibility, but, without financial rewards and protection of "information assets", it will not become a reality.

- Regulatory and business policies: Certain policies, for example, charging, will have to be altered to accommodate open systems. Changes in regulatory policies of governments and specific business policies will be required.

Although many technological problems remain to be addressed before open systems become a reality, transitions already in place suggest that they have become of a question of "when" rather than "if". An expanded view of open systems is required for the understanding and subsequently addressing of technological challenges. Some of these challenges are discussed in the following.

4.2 Challenges

Given the requirements of a life-long learning environment, one can identify a number of significant technological challenges. Many of these challenges are also faced in other application domains.

Not surprisingly, the interoperability among systems, services and applications remains a significant challenge. We have seen progress in addressing these issues, but validation of approaches and their scalability remain.

A key to interoperability, since underlying proprietary systems are unlikely to disappear, is the specification and development of interfaces to a variety of components in an open system. What interfaces are required and how are they specified? How do these interfaces evolve as the nature and demands of new applications emerge? How is the consistency of different interfaces ensured, especially in ensuring privacy and security?

Significant challenges exist in ensuring that open systems can provide the appropriate mechanisms for privacy and security. How does one provide openness and interoperability, yet ensure that the openness does not provide unauthorized access to restricted information sources or systems? How can the individual identity of an individual or the actions of an individual be protected? What are appropriate interfaces for ensuring privacy and security concerns between interoperating components?

Open systems will have to accommodate multiple, heterogeneous databases and different types of information. The access, storage and creation of such information must be possible throughout an open system. There may be specialized servers and systems that can provide support for certain types of data, for example, video servers. How is access provided in an open environment? Security and accessibility are critical for vendors supplying such servers.

The multiplicity of information types and sources presents challenges in the delivery infrastructure of open systems. A user at a poor quality device or a mobile device may have to be constrained to a subset of information or even information from a different source. Timely delivery of information across communication channels is only one aspect; quality of service encompasses more.

Open systems will have to encompass peer-to-peer interactions, as argued earlier and elsewhere[5, 6]. Current work does not address many of the issues in peer-to-peer computing and the composition of components.

Interoperability and the dynamic computing environment likely within an open system mean that the naming and location of resources and systems will be key. It is unlikely that a single naming system will suffice, and even if one is adopted by technology providers, it may be insufficient to support user needs. Directories and name services to accommodate multiple naming domains arising in different proprietary platforms, different databases, etc., remain a challenge.

Even with advances in hardware and communication technology, performance will be a primary concern in operational open system environments. Developers need to keep this in mind in designing interfaces for components and in identifying required embedded services. Performance is important in providing quality of service for end-user applications.

Evident in the information technology environment to support life-long learning is the need for agents to operate on behalf of a user. There are challenges in how these agents are developed, their characteristics, behaviour, etc. There are also questions of how they interact, and how security and privacy are ensured, especially since some agents may act on behalf of a single individual, while others may interact with several different individuals.

The need to determine costs means that appropriate accounting mechanisms must be put into place. The open system must provide accounting hooks for multiple vendors, for information providers and for service providers, etc. With access to information from multiple countries and regions, currency conversions, accounting, etc. must be accommodated.

Finally, there is the challenge in creating standards to support open systems. Interfaces between components must be standardized. The focus should not be on standardizing how things are to be done.

5 SUMMARY AND CONCLUSIONS

There is major progress in the last five years to build infrastructure, hardware and software, so that new applications can become possible. We need to expand our definition of "openness" to include much more. In particular, we need to move from the technological-driven view of openness to one based on the needs of consumers and applications.

There is a shift in the development environment toward openness integration into the overall networking structure. To address the new domains, developers must address more dynamic issues: configuration management, and systems and application management. In time there must be a move toward total transparency of the technologies themselves. Much more emphasis should be placed on the human-centric aspects of open systems.

As with anything new, requirements and evolving requirements will shift research and development efforts. To find solutions for life-long learning there will be a shift to the digital library; time and place become less important. There will be increased requirements on devices and their connectivity and management. A nontechnical user needs easy-to-use applications. Life-long learning will evolve into a broader discussion of education in general, since many of the issues and requirements are also relevant in public school, high school and universities. Open interfaces are the fuse key to use and acceptance.

At the same time, security, ownership, cost, etc., become magnified in an open environment and must be addressed by researchers. Finally, one should be aware of and prepare for the social issues that open systems raise.

REFERENCES

[1] R.M. Soley. *Object Management Architecture Guide.* Object Management Group, November 1990.

[2] OSF. *The OSF Distributed Computing Environment Rationale.* Open Software Foundation, Cambridge MA, May 1990.

[3] ITU-TS. *Basic Reference Model of Open Distributed Processing Part 1: Overview and Guide to the Use of the Reference Model.* ITU-TS Rec X.901, ISO/IEC 10746-1, July 1992.

[4] Michael Bauer, Neil Coburn, Doreen Erickson, Patrick Finnigan, James Hong, Paul Larson, Jan Pachl, Jacob Slonim, David Taylor, and Toby Teorey. A Distributed System Architecture for a Distriubted Application Environment. *IBM Systems Journal,* 33(3):399–425, September 1994.

[5] Jacob Slonim, Michael Bauer, Patrick Finnigan, Paul Larson, Toby Teorey, Alberto Mendelzon, Richard McBride, Shaula Yemini, and Yechiam Yemini. Distributed Programming Environment: Challenges. *Proc. of the International Conference on Open Distributed Processing,* pages 379–394, Berlin, Germany, September 1993.

[6] Jacob Slonim, James Hong, Patrick Finnigan, Doreen Erickson, Neil Coburn, and Michael Bauer. Does Midware Provide an Adequate Distributed Application Environment. *Proc. of the International Conference on Open Distributed Processing,* pages 34–46, Berlin, Germany, September 1993.

[7] Rick Boucher. The Challenge of Transition. *EDUCOM Review,* 27(5):30–36, September/October 1992.

[8] Information Infrastructure Task Force Committee on Applications and Technology. *A transformation of learning: Use of the NII for education and lifelong learning.* U.S. Government Printing Office, 1994.

[9] Richard Lucier. Towards Knowledge Management Environment: A Strategic Framework. *EDUCOM Review,* 27(6):24–31, November/December 1992.

[10] Marilyn Van Bergen. Electronic Citizenship and Social Responsibility. *EDUCOM Review,* 28(3):45–47, May/June 1993.

[11] Robert Cavalier. Shifting Paradigms in Higher Education and Educational Computing. *EDUCOM Review,* 27(3):32–35, May/June 1992.

[12] Robert Cavalier. Course processing and the Electronic *Agora*: Redesigning the Classroom. *EDUCOM Review,* 27(2):32–37, March/April 1992.

[13] EDUCOM. Multimedia. *EDUCOM Review,* 27(1):*entire issue,* January/February 1992.

[14] Elliot Soloway, Mark Guzdial, and Kenneth Hay. Reading and Writing in the 21st Century. *EDUCOM Review,* 28(1):26–29, January/February 1993.

[15] J.Slonim. Networked Information Systems as Digital Libraries. Technical Report to appear, IBM Academy, January 1995.

[16] EDUCOM. Legal, Social and Ethical Issues: Operating in the '90's. *EDUCOM Review,* 27(4):*entire issue,* July/August 1992.

PART TWO

Reviewed Papers

SESSION ON

Architecture

4

The A1√ Architecture Model

Andrew Berry and Kerry Raymond

CRC for Distributed Systems Technology, The University of Queensland, Australia

The A1√ (the "√" is silent in speech) model provides the basis for development of distributed applications and distributed infrastructure within the CRC for Distributed Systems Technology. The model has been developed to overcome some of the deficiencies in the RM-ODP standard, in particular, to address the diverse requirements of participants in the CRC. This paper introduces the major concepts of the A1√ model, and discusses both its relationship with RM-ODP and how it overcomes deficiencies identified in RM-ODP.

Keyword Codes: C.2.4; D.2.10
Keywords: Distributed Systems; Design

1. INTRODUCTION

The A1√ model has been created as a basis for development of distributed applications and development of an infrastructure to support these applications. The A1√ model underpins the work of the CRC for Distributed Systems Technology (the DSTC), a consortium of research, industrial and end-user organisations in Australia. The model defines a minimal set of concepts and rules for distributed systems.

The involvement of the DSTC in the ISO/ITU-T RM-ODP standardisation effort [2],[3] has a significant influence on the A1√ model, noting that the DSTC also has a substantial influence on the RM-ODP standard. Although there are similarities between our work and RM-ODP, the A1√ model addresses some of the deficiencies in RM-ODP that become evident in an organisation with such diverse interests as the DSTC. In particular:

- the RM-ODP viewpoints are effective for describing systems, but less effective for software engineering;
- the current RM-ODP computational model is inelegant when applied to complex applications. RM-ODP was originally created with RPC and client-server applications in mind, and is not flexible enough to accommodate legacy systems, which is an essential requirement within the DSTC.
- the computational type rules of RM-ODP are too prescriptive;
- the engineering language of RM-ODP prescribes unnecessary structuring detail;
- RM-ODP does not distinguish fundamental infrastructure functionality from other, merely desirable, functionality;

This paper introduces the major concepts of the A1√ model, and discusses how it overcomes the deficiencies identified in RM-ODP. The complete A1√ model is described in [1].

Section 2 identifies the important goals of the architecture. Section 3 of this paper describes our development model for distributed systems, identifying two component models, a specification model and an infrastructure model. Sections 4 and 5 describe the specification and infrastructure models respectively. Section 6 describes a number of approaches to the transformation of a specification to an implementation. Section 7 discusses the relationship of our work to RM-ODP and future plans for the A1✓ model. Section 8 concludes the paper.

2. PHILOSOPHY AND GOALS

The A1✓ model considers only the aspects of distributed systems related to distribution. For openness, no model for data, objects or execution should be imposed, provided that model can co-exist with the constraints required to support distribution of applications across heterogeneous platforms. Therefore, the model primarily discusses the concepts needed to describe the creation, deletion and interaction between entities of a distributed system.

2.1. Openness

The A1✓ model can describe both open or closed (proprietary) implementations. Support for open implementations will, in general, impose some constraints on the model. Since the term "openness" has many interpretations, we see openness as implying the following properties:

- interworking

  The model must allow interworking between heterogeneous software systems. However, we cannot guarantee interworking, since it is only possible to achieve interworking by conformance with interworking standards (both *de-facto* and *de-jure*), or by supporting conversions between systems. The aim is to facilitate interworking, but there is no commitment to total interworking,

- portability

  It should be possible to build applications that are portable across implementations supporting the model. However, portability requires that standards (for example, application programming interface standards) are adhered to, hence some applications might not be portable. Again, the aim is to facilitate portability, but there is no commitment to total portability.

- technology independence

  The model should be independent of any vendor-specific technology, meaning that adopters of the model are not unduly constrained in their choice of vendor.

- autonomy of administration and ownership

  The model must be capable of modelling systems with multiple autonomous administrations and diverse ownership of resources.

Note that these goals are often conflicting, for example, a way to ensure portability might be to buy from a single vendor. At some stage, it will be necessary for an adopter of the model to make a decision regarding the relative importance of these attributes. It is not always possible to satisfy all of these goals in a particular implementation or use of the A1✓ model.

2.2. Evolution, reconfiguration and legacy systems

A legacy system is one that interacts with another system outside the universe of discourse of the original system specification. That is, if a system has to interact with some system that was not visible or known in the original specification, then a system is a legacy system. All systems will become legacy systems at some point.

To support interoperability with legacy systems, and to provide flexible, robust and configurable distributed systems, implementations supporting the A1✓ model should be dynamic in nature. In the model, reconfiguration and evolution of systems and software is treated as the normal rather than the exceptional case. The concepts of dynamic binding and flexible type description are fundamental to this goal.

Our intention is that implementations supporting the model are capable of interworking with legacy applications. While perpetual backward compatibility with legacy systems is possible, most implementations are expected to sacrifice some backward compatibility in order to achieve progress or meet new challenges.

3. THE "BIG PICTURE"

In order to build a useful distributed system, a real-world problem undergoes a series of transformations to an implemented solution. In the A1√ model, these transformations are based around two component models:

- a specification model for applications, which defines the fundamental abstractions used for design of distributed applications;
- a distributed infrastructure model, which defines the fundamental services provided by the distributed system.

Based on these two models, there are three major transformations necessary to build a distributed system:

- a transformation of the real-world problem to a specification of an application;
  This transformation is necessary to exclude aspects of the problem that cannot or should not be solved by a computer system. The power and flexibility of the specification model is crucial to this transformation step. This step is similar to requirements specification in software engineering, and will be based on a requirements analysis.
- a transformation of the specification into an application program that is executable over the basic services provided by the infrastructure;
  Transforming a specification into an executable program is application programming. Note that application programming can be assisted by a variety of tools, for example, IDL compilers or application frameworks.
- a transformation of local operating systems and hardware into an environment that supports the infrastructure model.

The concepts and rules of A1√ model are positioned in terms of these two models and three transformations. Our "Development Model for Distributed Systems" is the combination of the models and transformations and is illustrated in figure 1.

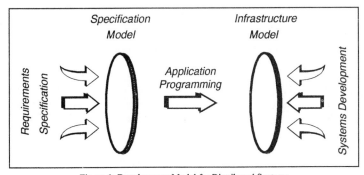

Figure 1: Development Model for Distributed Systems

4. SPECIFICATION MODEL

Distributed applications in the A1√ model are specified in terms of objects and their behaviour, including interactions. The goal of the specification model is to describe the externally significant behaviour of objects, without regard for how those objects will be physically positioned within the distributed system. That is, the implemented objects could be split across many nodes, or could be combined with others on a single node without violating the specification.

Specifications within this model are intended to describe the primary functionality of the application space, and should not consider the supporting infrastructure. Since the specification model only considers externally significant behaviour, it does not address issues local to objects, for example, concurrency within objects (threads, processes, spawning etc.), but equally does not constrain them. Note that some non-visible behaviour of objects is externally significant, for example, deadlock.

4.1. Objects

An object is an entity that encapsulates state, behaviour and activity. An object has a unique identity, and at any point in time can be associated with a non-empty set of types. An object can evolve over time, changing the set of types satisfied by the object. Objects can only interact by exchanging strongly typed messages. Objects are not passed as parameters to interactions, that is, only identifiers of objects can be passed as parameters (note that this derives from the fact that objects encapsulate activity— activity cannot be transmitted). Objects are autonomous, that is, they are capable of independent action.

Object granularity is the responsibility of the application designer, so the designer can specify the distributed application in terms of individual integer objects if desired, or might equally specify the application in terms of database table objects. The model is object-based—we avoid the use of the term "object-oriented" since inheritance and polymorphism are not explicitly specified.

4.2. Bindings

A binding is a association between a set of objects that allows the objects to interact. Bindings are strongly typed—a binding type defines the roles of objects in a binding and the interaction that can occur between objects fulfilling those roles. In order to provide a general model for distributed applications, we do not prescribe any specific binding types—if desired, it is possible to define binding types specific to each application. The behaviour associated with a binding can be infinite. Bindings are explicitly identified, meaning multiple bindings can exist for any given set of objects.

Roles of a binding are filled by the objects participating in the binding. For example, a binding to support RPCs must have a "client" role and a "server" role. Each role of a binding specifies an interface type that must be satisfied by objects fulfilling that role—objects participating in the binding must therefore instantiate an interface that is a compatible with their role in the binding. Taking our RPC example, the client role specifies that the client object requires a set of operations to be implemented by the server in the binding, whereas the server role specifies that the server object must implement the operations required by the client.

Note that the actual interface types offered by objects for a binding need only be compatible with the roles specified in a binding. In general, compatibility is defined by "substitutability", that is, the ability of one object to substitute for another object that offers exactly the requested interface type. For example, a file object offering "open-read-write-close" operations is able to be used for a binding which only requires "open-read-close".

Although our model permits arbitrary binding semantics, it is expected that a number of common binding types will be used regularly. To simplify the use of such binding types, parameterised, pre-defined types can be used, for example:

- DCE RPC binding
- Unix pipe binding
- ftp binding

Note that to fully describe a binding, these generic types must be parameterised with the interactions and data types (e.g. as defined in DCE IDL) for a specific binding. Higher level binding types can also be defined, capturing the common semantics of several types of binding, for example generic RPC binding. The specification of sub-type relationships between the higher level binding types and more specific binding types provide the basis for determining and supporting interoperability between similar systems.

The fundamental unit of interaction in a binding is a single, strongly typed message, although interactions can be considerably more complex. For example, an RPC interaction consists of a pair of messages between two objects, a request and a response, with the response occurring after the request and transmitted in the opposite direction. For ease of specification, generic, widely used interaction types are defined by the model, including operations (RPC), multicast and strongly-typed flows of data.

4.3. Interfaces

An interface is instantiated to fulfil the role of an object in a particular binding. Interfaces are strongly typed—an interface type describes the possible the structure and semantics for interactions of an object during a binding. In other words, an interface type describes the structure of messages and the object behaviour associated with messages sent and received by that object during a binding. An interface type can also describe constraints on interactions—this might constrain the types of bindings in which the object can participate. By definition, interfaces must exist during a binding and hence must be created either prior to or during the creation of the binding.

In the case of interface creation prior to binding, the type satisfied by the interface is static and not subject to negotiation or transformation during binding. Usually, objects are expected to offer interface types that can be bound, but for maximum flexibility, these types will often define additional interactions not required for all bindings. When a binding is created, an object should instantiate an interface satisfying the role assigned to that object in the binding—this interface need not necessarily implement all of the offered interface type. For example, a file object might offer an interface type describing the operations "open-read-write-close". A particular binding might only require "open-read-close", so the interface instantiated for that binding could take advantage of the reduced functionality required by the binding to optimise, for example, concurrency control or caching.

4.4. Types and relationships

A type is a predicate describing a set of entities. In the specification model, types are used specifically to classify and describe objects, bindings, interfaces and relationships. In the specification model, it is assumed that type information is dynamically available—objects can dynamically find, interpret and create types. Interactions between objects must satisfy typing constraints specified by the participating objects and the binding template. Type information can, however, be hidden to satisfy security or other constraints.

The model also prescribes the maintenance of information about relationships between types, including subtype relationships defining the substitutability of types. An explicit definition of subtyping is not given by the model—since there are many different approaches to subtyping, we simply identify the relation and describe how it can be used to support flexible bindings.

The ability to define general relationships is a key aspect of the model. Relationships are simply associations between entities that satisfy some predicate (i.e. type). As described above, information about relationships between types is maintained, but relationships are also necessary for many other purposes, for example, finding an executable program that offers a particular interface type, or finding the object file from which an object was created.

Methods for description, storage and usage of types and relationships is a area of research being developed within the DSTC. Further detail of this research can be found in [5].

4.5. Fundamental activities

A number of fundamental activities of objects have been identified in the specification model, specifically:

- instantiation and deletion of objects;
- creation and termination of bindings;
- interaction between objects;
- object composition.

These are discussed in more detail in [1]. Of particular interest, however, is the creation of bindings.

Objects can create bindings to other objects. In order to do so, an object must provide a template for the binding, identify the objects that are expected to participate in the binding and their roles in that binding. A binding can only be valid if the identified objects are capable of fulfilling the roles specified for them. Objects participating in a binding can determine that such a binding has been created and have access to an identifier for the binding. Third-party binding is permitted, allowing objects to create bindings without actually participating in those bindings.

For example, an RPC session between a client and server in a DCE application is a binding. The binding template used to establish this binding is specified by the IDL (interface definition language) file. "Client" and "server" are the roles in this binding. The client is implicitly identified since it initiates the binding. The server is explicitly identified when creating the binding. If the server cannot fulfil the role of server as specified in the IDL, then the binding fails. If the client cannot fulfil the role of client in the binding (e.g. if it does not have stubs for the RPCs defined in the IDL) then the binding fails.

4.6. Illustrating the concepts

To illustrate the concepts of the specification model, the following diagrams show the three stages associated with establishment of a binding. The illustration shows the creation of a binding between a bank, a customer, an employer, and a credit provider. The binding is intended to support the banking activities of a customer, including automatic withdrawals (e.g. American Express withdrawing a monthly account payment) and automatic payments (e.g. employer pay-ing salary). We model this activity as a multi-party binding involving all of these objects to capture the dependencies between the interactions.

1. Prior to binding, each of the objects assert their ability to support a particular interface type. This is a capability—there is no interface, just an assertion that the object is capable of instantiating such an interface. In this case, we have Clancy asserting the ability to be a bank customer, the employer asserting the ability to be a depositor, and so on.

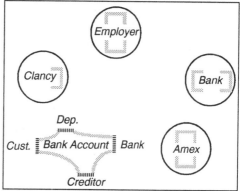

2. Clancy locates a suitable binding type to support the multi-party banking activity. The bank account binding type has roles (indicated by flat surfaces) for each of the parties. For creation of a binding, the interfaces of the participants must match (subtyping) the roles in the binding type. The binding type specifies the interactions that occur in response to activity at an interface, for example, the pay deposit by the employer might lead to the depositing of the pay at the bank and a notification being sent to Clancy indicating the amount deposited.

3. Once an appropriate binding type has been found, Clancy can initiate the binding. Each participant must agree to their role in the binding, and instantiate an interface that satisfies their role. Note that this need not be exactly the interface type offered—it simply needs to satisfy the associated role. The infrastructure instantiates the binding, connecting each of the interfaces. In this example, Clancy has initiated the binding. In general, however, any object can initiate a binding between any set of objects, including third-party binding (for example, by a trader).

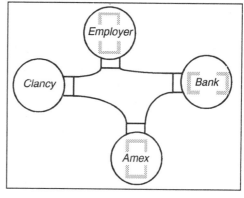

5. INFRASTRUCTURE MODEL

This section describes the high-level design of an infrastructure to support the concepts of the functional specification model. The infrastructure model is intended to be independent of any particular implementation or platform, but aware of the existence of independent nodes and communications between nodes, hence the distribution of objects. The design is expressed as a set of definitions and concepts. The mapping of the specification model concepts to infrastructure model concepts is also described.

The infrastructure provides an abstract machine for execution of object-based program specifications. It enforces the interaction types in the specification, and provides the functionality required to support the abstractions of the specification model.

5.1. Classifying objects

Within the infrastructure model, an object is a unit of distribution (i.e. an identifiable entity that exists on a single node) for distributed applications and services. Note that objects defined in the specification model are not constrained in this manner. Objects are capable of communication through communication networks. The following subsections outline a set of categories for objects that are explicitly identified in the infrastructure model. The definition of the categories is based on the activities or functionality of the objects concerned. More detail for these definitions is given in [1].

5.1.1. Fundamental objects

Fundamental objects are those that are essential to the operation of the infrastructure. That is, they must exist to support the functional specification model. Fundamental objects provide access to the resources offered by nodes and communication networks, for example, communication protocol stacks and operating system kernels.

The activities supported by fundamental objects include instantiation and deletion of entities, that is, objects, relationships, types, interfaces and connections. In addition, naming, encapsulation sufficient for self defence (security), and interaction must be supported by fundamental objects. These activities must be supported by all implementations of the model. Many other activities in distributed systems are important (for example, maintaining security), but are not fundamental since they can be constructed from the fundamental activities. In other words, the fundamental activities are the basis for construction of all other distributed system activities.

5.1.2. Utility objects

Utility objects are objects considered necessary for practical use of the infrastructure. In general, utility objects provide persistent, shared services that will be used by most distributed applications, but are not essential to the functioning of the infrastructure. That is, distributed applications can be built without using utility objects, but it is not generally effective to do so. Some examples of utility objects are:

- a trader object (for service trading)
- a key distribution object (supplying keys for encryption)
- a notification object (providing a notification service for objects)

5.1.3. Auxiliary objects

An auxiliary object supports a particular application or part thereof, but does not directly implement application functionality, and is not essential to the operation of the infrastructure. Auxiliary objects provide programming abstractions or transparencies, and are typically derived from binding types or from requirements described in the non-functional specification of an application.

Auxiliary objects are obtained either by configuration of "off-the-shelf" components or are application-specific and generated during programming. Some examples of auxiliary objects are:

- RPC stubs which provide marshalling of parameters for RPCs (application specific)
- objects that guard the interactions of an object to provide security (off-the-shelf)

Auxiliary objects differ from utility objects in that they are not shared or persistent, existing only for the lifetime of their application.

5.1.4. Derived objects

A derived object is one that implements part of the application described in a functional specification. The mapping between specification objects and derived objects is not necessarily one-to-one, allowing for the possibility that a specification object is distributed across nodes or that several specification objects are combined into a single implemented object. However, programming tools will support common mappings, for example one-to-one mappings or replicated objects, between specification and derived objects. Derived objects will often include additional functionality required to realise specification-level concepts as infrastructure-level primitives, for example:

- establishing connections to support the bindings described in a functional specification;
- implement interactions in the specification as a set of communication primitives;
- maintaining the consistency of replicas.

5.2. Connections

A connection is a configuration of objects and communication paths that implement a binding (as defined in the functional specification model). A single connection might involve multiple interaction protocols, for example, DCE RPC and FTP.

Connection establishment can involve negotiation between the local infrastructures supporting the interacting objects. A set of connection constraints and parameters compatible with the needs of all objects involved in the interaction are generated. This might include, for example, type matching, data representation or security policy.

Where the objects participating in an interaction are co-located, some optimisation of the connection process might be possible, but it will still exist as a logical step. The complexity and duration of the connection are a reflection of the interactions possible over the binding. For example, asynchronous interactions are typically very short and might be implemented by packaging the connection establishment and interaction data into a single datagram.

Connections are explicitly terminated. However, connections can also fail due to changes in components of the connection that invalidate the configuration. A connection can be re-established after a failure, and this is the mechanism typically used to cope with, for example, migration or replication of an object during interaction. The termination and re-establishment of a connection does not necessarily terminate the binding implemented by the connection. This allows failure and migration to be transparent in a specification. It also allows the infrastructure to establish and terminate a connection implementing a particular binding on-demand, to optimise resource usage.

5.3. Illustrating the concepts

In order to illustrate the concepts of the infrastructure model and the relationship to specification concepts, the following figures and text describe the creation of the binding illustrated in 4.6. The major difference between the illustrations is that nodes and networks are now explicitly visible, and must be dealt with.

1. The assertion of interface type must be made to the local supporting infrastructure (fundamental objects). This information is needed for establishment of bindings, and is independent of any advertisement of the interface types used for location or resource discovery. We assert that derived objects have pre-established connections with local fundamental objects.

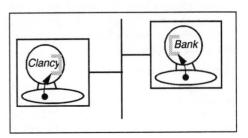

2. Clancy must find each of the other objects to participate in the binding. We assume that Clancy is already aware of the employer and American Express. However, Clancy chooses a bank through interaction with the trader. Note that use of a trader is not prescribed—it is simply used for this illustration (and we do not show the establishment of a connection to the trader).

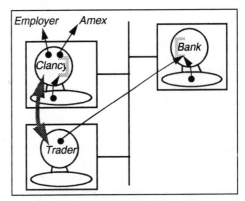

3. Clancy then asks the infrastructure to create a connection to support a binding type (supplied by Clancy). The connection is created, including auxiliary objects (e.g. security "black boxes" as shown in the figure) and utility objects (e.g. a notary to support non-repudiation) required to support any requirements specified in the binding type. Note that auxiliary objects need not necessarily be co-located with the derived objects and that the connection may require multiple protocols and networks.

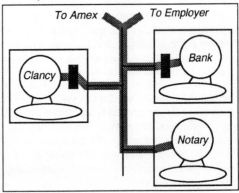

6. PROGRAMMING TRANSFORMATION

The aim of programming is to transform a specification into an implementation. Within our model, this means mapping the set of specification objects and bindings onto implemented objects and connections.

Ideally, once the objects, bindings and interactions of an application are specified, the programmer can simply create code for the objects without considering the distribution of those objects in the target system or even considering the fact that the objects are distributed. Realistically, the programmer still needs to consider some aspects of distribution. Transparencies and tools assist (rather than replace) the programmer.

6.1. Transparencies

The infrastructure can be hidden from a programmer through use of transparencies. Transparencies can be achieved through one or more of:

- transformation of specifications (to specifications for programs);
- encapsulation of objects, adding additional functionality;
- using complex bindings to hide, for example, replication.

Transparencies can be implemented through pre-processing of programs, generation of "stubs, or through run-time configuration. The infrastructure can instantiate auxiliary objects to support the transparency requirements of a binding and its participating objects.

7. A1√ VERSUS RM-ODP

In comparing the A1√ model to RM-ODP, this section describes how the A1√ model addresses the deficiencies in RM-ODP identified in the introduction to the paper.

7.1. Viewpoints, models and software development

The A1√ model introduces two models and three transformations. As pointed out in section 3, the transformations map directly onto software development processes, and the models map onto specification languages and distributed systems infrastructure respectively. The RM-ODP approach of defining five viewpoints is far less clear in this regard.

There is a relatively straightforward mapping between our components models of A1√ and the viewpoints of RM-ODP. The A1√ specification model corresponds to the information and computational viewpoints and the infrastructure model corresponds to the engineering viewpoint. Equivalent concepts for the enterprise and technology viewpoints do not feature in the A1√ model—the development of a distributed application will absorb enterprise and technology modelling into the process (i.e. the transformations).

7.2. Computational language issues

The concept of bindings in A1√ gives us a simple yet powerful basis to describe both existing and new, complex interaction patterns amongst objects in a distributed application—all interactions are described in terms of strongly-typed messages, with more complex interactions built from this single mechanism. The similar RM-ODP concepts of signals, signal interfaces, binding objects and control interfaces are cumbersome, and are not integrated with the higher-level operational and stream interaction mechanisms and their associated interfaces and types. The plethora of concepts used to describe the three distinct interaction mechanisms complicates the computational language. For legacy systems, the ability to simply and effectively describe the interactions supported by a legacy object is paramount.

The concept of bindings also provides the basis for determining compatibility and subtyping rules. A1✓ allows alternative type rules to co-exist, and this can be captured in a binding type. For example, an ANSAware binding requires ANSAware type rules to be satisfied, whereas DCE binding requires the satisfaction of DCE type rules.

7.3. Engineering and function issues

A1✓ relies on object composition to provide a flexible, generic structuring mechanism in the infrastructure model. The RM-ODP concepts of capsule and cluster provide no additional expressive capability, and the confusion and prescription associated with cluster and capsule managers is avoided.

The clear distinction between fundamental and optional functionality in A1✓ removes the need to prescribe the optional functionality. This makes the A1✓ model minimal yet relatively complete, where RM-ODP describes a number of functions that could well be left for standardisation outside the core model. The optional functionality for A1✓ is being described separately—the DSTC A2 [4], A3 [5] and A4 [6] projects are currently addressing particular aspects of the optional functionality.

8. CONCLUSIONS

The A1✓ model provides a sound basis for the development of distributed applications and infrastructure, and overcomes some specific deficiencies in RM-ODP. These deficiencies have been identified because of the diverse requirements of participants in the DSTC. This paper has described the major concepts of the A1✓ model and discussed the relationship of A1✓ to RM-ODP, in particular, how the A1✓ model addresses the identified deficiencies in RM-ODP.

ACKNOWLEDEMENTS

The authors would like to thank numerous staff of the DSTC, and in particular the Architecture Unit, for lively debate and fruitful discussion. This work is funded in part by the Cooperative Research Centres Program through the department of the Prime Minister and Cabinet of the Commonwealth Government of Australia. This research was also supported by Telecom (Australia) Research Laboratories via the Centre of Expertise in Distributed Information Systems.

BIBLIOGRAPHY

1. A. Berry and K. Raymond (eds), *The DSTC Architecture Model*, DSTC Technical Report, June 1994.

2. ISO/IEC JTC1/SC21, *Draft Recommendation X.902: Basic Reference Model of Open Distributed Processing—Part 2: Descriptive Model* (ISO/IEC DIS 10746-2), February, 1994.

3. ISO/IEC JTC1/SC21, *Draft Recommendation X.903: Basic Reference Model of Open Distributed Processing—Part 3: Prescriptive Model* (ISO/IEC DIS 10746-3), February, 1994.

4. A. Bond and D. Arnold, *Open Distributed Environments—Misadventure or Masterpiece?*, submitted to the International Conference of Open Distributed Processing, Brisbane, Australia, February 1995.

5. W. Brookes and J. Indulska, *A Type Management System for Open Distributed Processing*, TR 285, Department of Computer Science, The University of Queensland, January 1994.

6. A. Beitz and M. Bearman, *An ODP Trading Service for DCE*, Proceedings of the First International Workshop on Services in Distributed and Networked Environments (SDNE), June 1994, Prague, IEEE, p 42-49.

5

Interoperability of Distributed Platforms: a Compatibility Perspective

W. Brookes[a], J. Indulska[a], A. Bond[b], Z. Yang[b]

[a]Department of Computer Science, CRC for Distributed Systems Technology,
The University of Queensland, Brisbane 4072, Australia

[b]CRC for Distributed Systems Technology, Level 7, Gehrmann Laboratories,
The University of Queensland, Brisbane 4072, Australia

Interoperability of heterogeneous distributed platforms is affected by differences in their type systems, including both differences in syntax and semantics of types and approaches to inclusion polymorphism. This paper examines the differences of well known distributed platforms (DCE, ANSAware, CORBA) as well as models for distributed and object oriented systems. The examination is illustrated by a scenario of their cooperation. The scenario is based on resolving differences during the federation process. The federation involves extension of persistent type repositories for each domain to incorporate mappings among domains. The scenario focuses on differences in compatibility relationships.

Keyword Codes: C.2.4; D.2.6
Keywords: Distributed Systems, Programming Environments

1. INTRODUCTION

Current trends in distributed computing are towards creating open distributed systems, where co-operation is necessary in spite of the heterogeneity and autonomy of systems present in an open environment. The open, heterogeneous nature of the environment introduces a number of difficult problems which are not encountered in centralised systems, or in closed homogeneous systems. The open nature of the system means that objects are created and destroyed frequently, and more importantly, new types of objects are frequently created and objects with new types are required to interwork with existing components. Late binding is assumed as a mechanism to allow existing objects to work with newly introduced components.

A number of different distributed computing platforms attempt to address these issues (to different extents) including DCE [21, 22], CORBA (and OMA) [17, 18], ANSAware [2] and

---

The work reported in this paper has been funded in part by the Cooperative Research Centres Program through the Department of the Prime Minister and Cabinet of the Commonwealth Government of Australia. It was also partially supported by an Australian Government Postgraduate Research Scholarship (APRA).

COMANDOS [7]. Also, the standard under development, the ISO Reference Model of Open Distributed Processing (RM-ODP) [12], attempts to create a framework for interoperability of open distributed systems. Moreover, research on type-safe interoperability of object-oriented applications [1, 13, 14] adds additional insight into the problem.

This paper examines type-safety issues related to interoperability of heterogeneous and autonomous platforms. We analyse the requirements for type-safe interoperability and examine some existing platforms for distributed computing against these requirements showing the differences in type models which make interoperability difficult. The examination is supported by both a scenario of interoperability for these platforms and a proposal for possible solutions. The scenario focuses on differences in compatibility relationships defined for the platforms.

For the analysis, we assume an object-based model of distributed systems, similar to that described in RM-ODP. Using this model, an open distributed system consists of a large number of heterogeneous communicating objects managed by many independent, autonomous authorities. Meaningful interoperability requires that objects can be discovered, bound to, dynamically invoked and also replaced by compatible objects at run time. Therefore, if the objects are from different type models, either a common framework which will allow mapping between the objects is required [15], or one-to-one mappings have to be provided. The framework or one-to-one mappings must capture basic differences in the syntax of types, but also some semantic differences including differences in compatibility relationships.

A large amount of research has been already carried out on a unifying type model supporting interoperability [5, 6, 7, 9, 12, 15, 16, 19, 20]. As this work approaches the problem from different backgrounds, a unification of the models is still necessary. In the meantime, cooperation between distributed platforms can be achieved by a one-to-one mapping approach. In this paper we analyse such a mapping between type systems of distributed domains. The goal of the analysis is twofold: to show the extent of interworking between platforms if one-to-one mapping is assumed and to facilitate work on a common framework by highlighting the differences which should be captured in the common model.

The remainder of the paper is organised as follows. Section 2 discusses requirements for type-safe interoperability of distributed systems. Section 3 presents approaches of existing platforms considering the mapping of types and relationships. Section 4 illustrates the problems involved using a case study of existing distributed system models and showing an approach to mapping relationships between these models.

2. REQUIREMENTS FOR DISTRIBUTED PLATFORM INTEROPERABILITY

Interoperability requires that resource type selection is possible during resource discovery and the same for type matching during binding and invocation. Both selection and matching can cross platform boundaries (therefore also type system boundaries) and should also work for evolving systems. In spite of crossing boundaries between different type systems, the operations (resource selection, binding and invocation) should be type safe, *i.e.* free from mismatched types. Many object-oriented languages and platforms for distributed computing already provide descriptions of objects, their interfaces and also a definition of compatibility which differs for different platforms [19]. Therefore, the features of the type models of existing languages which are visible during interoperability, *i.e.* interaction types (operations, streams, notifications, *etc.*), interface types, binding types and relationship types have to be captured in a common framework which will allow mapping between models.

Supporting such mapping implies that the framework must include a means for the description of some elements of objects' semantics and some semantics of relationships. When type matching is performed in an open system to search for compatible types, as much as possible of this matching should be performed automatically and invisibly to users of the system. In the general case, it is not always possible to perform transparent, automatic type matching when type semantics are involved, but the type model should be designed with support for all kinds of matching: automatic, semi-automatic or manual. Even in the latter case, when matching is performed manually by a user it should be supported by a description of elements of semantics which will facilitate the user's task of determining matches.

In addition to substitutability, the framework should allow the creation of generic objects. This implies that the model should capture two kinds of polymorphism which support genericity and substitutability, *i.e.*

- parametric polymorphism allowing definition of generic operations/objects,

- inclusion polymorphism defining substitutability (subtyping/compatibility) rules, especially substitutability of interface types and relationship types which may involve semantics.

To create such a framework, the model should be able to describe both the structure and some semantics of data types, action types (interaction types, interfaces, bindings between objects) and relationship types. Existing Interface Definition Languages [2, 3, 18, 21, 22] and their type models aim to provide a description of types visible during interoperability. They support polymorphism to a certain extent, however the models do not address all the requirements and there is still a need for a richer and unifying framework. The type model encapsulating these features visible at platform boundaries is necessary if interworking of heterogeneous platforms is to be supported.

In addition the model should be extensible to encompass new domains and their type models, and therefore the ability to dynamically learn about new object types or new relationship types is necessary.

In the next section, we compare the syntax and semantics of types of data, actions and relationships in type models of existing distributed and object-oriented platforms. Type safety of their type matching operations is also considered.

3. EXISTING APPROACHES

Existing platforms for distributed computing have adopted their own type models to support application interoperability. The approaches that are compared here include DCE [21, 22], OMG's CORBA (and OMA) [17, 18], ANSAware [2], RM-ODP-based systems [12] and work on type-safe interoperability of object-oriented applications including Liskov & Wing's work [13, 14]. and the POOL language [1].

All of these models provide features for supporting interoperability to different degrees. The three distributed environments (DCE, CORBA and ANSAware) and the POOL language all provide a standard set of basic data types and constructors. DCE has a richer set than the other systems, but all provide a set which is satisfactory for supporting application interoperability.

However, neither set is 'canonical' in the sense of being more general than other models. RM-ODP does not specify data types important for interoperability.

In terms of action types, all models at least support the concept of operations (methods). All models also support either the concept of interfaces (logical groupings of operations) or objects or both. DCE, CORBA, ANSAware and RM-ODP also support notifications, and RM-ODP supports signals (one way typed messages). Of the five models, only America (POOL language) and Liskov & Wing provide any description of semantics of action types, where the semantics are those of the action itself, not the semantics of the language in which the actions are described. Liskov & Wing use invariants and pre- and post-conditions to describe the semantics of operations. The POOL language uses properties to capture semantics of operations and objects, however the designers note the need for more expressive semantic description [1].

In terms of supporting relationships between types, only the OMG approach supports the notion of multiple, user-defined relationships, through the use of the Relationships Service. The other models use only built-in relationships (usually for inclusion polymorphism, namely subtyping or inheritance).

DCE supports interface subtyping through the use of versions. If one interface has the same major version and a larger minor version number, then it is a subtype. Subtyping is not part of the DCE IDL language, and is only an abstract concept defined on operation signatures. DCE uses interface subtyping by extension, namely that one interface is a subtype of another if the subtype has at least the same operations as the supertype, and possibly more. Additionally, the operations of the subtype must be in the same order as the supertype. Arguments and results of the operations must be of exactly the same type. It does not include any semantics of the subtyping relationship.

In addition to the user-defined relationships in the Relationships Service, CORBA defines interface inheritance based on type syntax (not semantics) supporting interface extension by the addition of operations. OMA, the OMG architecture [17] at a higher level than CORBA, recognises the need for a subtyping relationship. OMA also notes that subtyping and inheritance are different concepts, although inheritance often implies subtyping. OMA defines subtyping as follows:

> Formally, if S is declared to be a subtype of T (and conversely, T is a supertype of S), then for each operation $o_i \in Ops(T)$ there exists a corresponding operation $o_j \in Ops(S)$ such that the following conditions hold:
>
> 1. the name of the operations match
>
> 2. the number and types of the parameters are the same
>
> 3. the number and types of the results are the same.
>
> Thus, for every operation in T there must be a corresponding operation in S, though there may be more operations in $Ops(S)$ than $Ops(T)$.

ANSAware supports interface inheritance through an *is_compatible_with* relationship. This information is not used in ANSAware, apart from for reuse. ANSAware also defines a subtyping relationship in the Type Manager (type conformance) which it assumes is related to the interface inheritance. The ANSAware Type Manager stores only names in the subtyping graph (based on information supplied by a user), and type descriptions are not taken into account when determining subtyping. Again the relationship is only based on syntax. It is very similar to subtyping in DCE in that it only supports subtyping by extension, assumes that argument and result types are identical and requires the same order of operations.

RM-ODP also defines interface subtyping. Here we will only consider RM-ODP's operational interfaces. The difference between the RM-ODP subtyping definition and that of DCE, OMG and ANSAware is the addition of subtyping of operation arguments and results, in addition to subtyping by extension (allowing extra operations). Parameter subtyping follows the "contravariant arguments, covariant results" rule [19]. RM-ODP [12] informally defines its subtyping rules as follows (a more rigorous definition of the subtyping rules are also included in RM-ODP):

Operational interface X is a subtype of interface Y if the conditions below are met:

- for every operation signature in Y, there is an operation signature in X (the corresponding signature in X) which defines an operation with the same name;

- for each signature in Y, the corresponding signature in X has the same number and names of arguments;

- for each signature in Y, every argument type is a subtype of the corresponding argument type in the corresponding signature in X;

- the set of termination names of an operation signature in Y contains the set of termination names of the corresponding signature in X;

- for each operation signature in Y a given termination in the corresponding signature in X has the same number and names of results in the termination of the same name in the signature in Y;

- for each operation signature in Y, every result type associated with a given termination in the corresponding signature in X is a subtype of the result type with the same name in the same termination name, in the signature in Y.

Liskov & Wing's major contribution is their definition of an interface subtyping relationship which includes subtyping of operations based on described semantics as well as syntax. In addition, it allows for the renaming of operations by a subtype. Semantics are described as pre- and post-conditions in a language based around the Larch specification languages [10], and the subtyping is based on the weaker precondition and stronger postcondition rule [13, 14]. It also includes syntax-based subtyping based on contravariant arguments and covariant results. POOL is a language which incorporates the theory of behavioural subtyping, but in the implementation restricts the behaviour description to the use of properties to tag behaviour [1].

Both approaches to behavioural subtyping describe operation behaviour using pre- and post-conditions specified as state predicates. To formally relate a subtype to a supertype, Liskov & Wing define a *correspondence mapping* consisting of an abstraction function for relating values of the subtype to values of the supertype, a renaming function to relate subtype operations to supertype operations and an extension map to explain the effects of any additional operations provided by the subtype. If such a correspondence mapping can be found between two types, subject to both structural and semantic subtyping rules, then one type is a behavioural subtype of another. Using an informal notation, the semantic subtyping defined by Liskov & Wing is given by the following two rules:

1. pre (super) ⇒ pre (sub)
2. (pre (sub) ⇒ post (sub)) ⇒
      (pre (super) ⇒ post (super))

where pre (super) is the precondition of the supertype operation, post (super) is the post-condition of the supertype operation, pre (sub) is the precondition of the subtype operation and post (sub) is the postcondition of the subtype operation. The first rule known by Liskov & Wing as the pre-condition rule and the second as the predicate rule. POOL uses a slightly different (and stronger) formulation, but is very similar in approach.

POOL is the only model that supports parametric polymorphism, through its use of 'bounded genericity'. CORBA supports universal polymorphism through introducing the type ANY, however this is a weaker form of polymorphism than true parametric polymorphism where it is possible to create type templates which can be parameterised to create actual types [8].

As discussed earlier, there are three basic points at which type matching can occur in a distributed environment: type selection, binding and invocation of interactions. DCE supports matching (based on versions) during selection and binding only, and not during invocation. CORBA assumes matching at all three points. ANSAware supports matching during selection, but not during binding or invocation. RM-ODP assumes matching is supported at all three points. Both Liskov & Wing's approach and the POOL language are models for object-oriented languages, and thus it is irrelevant to consider when matching occurs (as models do not address concrete issues such as selection, binding and invocation).

Of all the approaches listed, they can be clearly divided according to the definition of the subtyping relationship they use. DCE, ANSAware and CORBA all only provide subtyping based on extension, that is, the addition of operations to a subtype. RM-ODP takes this one step further and also supports subtyping of operation arguments and results. Both Liskov & Wing's approach and the POOL language support behaviour description, fully in theory, and limited in practical implementation. In addition, Liskov & Wing's subtyping definition supports renaming.

The differences in subtyping relationships are summarised in Table 1.

	subtyping definition				type safety		
	interface extension	parameter subtyping	behaviour description	component renaming	selection	binding	invocation
DCE	✓				✓	✓	
OMA/CORBA	✓				✓	✓	✓
ANSAware	✓				✓		
RM-ODP	✓	✓			✓	✓	✓
Liskov & Wing	✓	✓	✓	✓	N/A	N/A	N/A
POOL	✓	✓	✓		N/A	N/A	N/A

Table 1: Differences in subtyping relationships

It is clear from the preceding description that the type models presented can express distributed system concepts to varying degrees. Difficulties arise when applications from different domains (with different type models) must interwork. There are two basic models of interoperability which can be applied to achieve interworking:

- creating a common framework for types including relationship types (in particular subtyping and inheritance relationships),

- supporting cooperation based on 'known differences', *i.e.* define one-to-one mappings between type models (which may incur losses and gains in expressiveness).

When presenting a scenario of cooperation between the classes of distributed systems defined above, we apply the 'known differences' approach. This has the disadvantage of incurring an *n(n-1)* mapping problem, but illustrates the differences in subtyping relationships and shows a possible solution. The 'known differences' approach could be easily mapped to a common framework, provided that the latter is given.

4. INTERWORKING SCENARIO

In this section we describe an interworking scenario for the platforms (DCE, CORBA, ANSAware) and models (RM-ODP, Liskov & Wing's approach) compared earlier. The models are also included, because they forecast future changes in building distributed systems and some of them already have partial implementations. The scenario highlights the differences in type systems and proposes a working solution based on the 'known differences' approach to ensure type safe cooperation between the described systems. For the scenario the following assumptions were made:

- the scenario is restricted to compatibility relationships in heterogeneous distributed systems — other type system differences must be considered, but are not fully presented here,

- the scenario considers only interface extension, component renaming, parameter subtyping, and parameter conversion.

The basis for cooperation is provided by persistent type information in each domain. We call the objects managing this type information "type managers". They are responsible for resolving type differences among heterogeneous systems. For the purpose of this scenario we assume that type managers for the domains are centralised. We focus on information about type system differences only, not on the management of this information.

4.1. Federation of type managers

To cope with the differences between type systems of various distributed domains it is essential that the type managers federate. The federation can be limited to subdomains. The federation process includes learning about the differences in type systems and possible losses which may be incurred by translation from one type system to another. In our scenario only differences and losses in inclusion polymorphism are emphasised. We assume that starting points for querying type lattices in type managers are defined prior to resolution of subtyping differences. When the federation process is accomplished, the federated type managers are able to perform inter-domain type matching automatically with the limitations of the matching already known to the domains. A cut-down view of an example federation is shown in Figure 1.

4.2. The banking scenario

To create a scenario of interworking between the domains we consider a family of banking interfaces and we examine how previously described type domains are able to interpret relationships among the interfaces. All the approaches to inclusion polymorphism described earlier

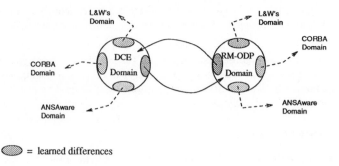

◯ = learned differences

Figure 1: An example federation

are illustrated in this scenario. For each model we analyse the representation of its subtyping relationship and the mapping of the relationships which can be used for interworking. Each type manager must know the subtyping relationships of other federated domains and be able to map its own relationship onto the others. In other words, during federation the relationships between subtyping relationships are established.

The remainder of this section analyses the basis for establishing relationships among subtyping relationships while illustrating it using real life examples of substitutability for banking applications. Semantic subtyping which is present only in Liskov & Wing's (L&W) model has not been fully addressed in the scenario because the result of mapping is obvious. No subtype relationship in any of the domains can be accepted as a subtype in the L&W domain without a loss. The result of a query to other domains can only give a 'hint' about possible subtyping relationship among the returned types, but not the exact answer. Querying the L&W domain from any other domain is different. Everything allowed by the structural subtyping described below can be accepted if the semantics are added.

We start with a simple banking interface:

```
INTERFACE Bank : [
    Deposit (acct:acct_t, amount:US_dollar_t);
    Withdraw (acct:acct_t, amount:US_dollar_t);
];
```

4.2.1. Interface extension

Let us assume that there also exists a bank with the following interface:

```
INTERFACE StatementBank : [
    Deposit (acct:acct_t, amount:US_dollar_t);
    Withdraw (acct:acct_t, amount:US_dollar_t);
    Statement (acct:acct_t);
];
```

StatementBank is a subtype of Bank because it extends the interface by the addition of an operation. The analysis of the five models leads to the following observations:

- DCE uses sequential minor version numbers to represent the compatibility relationship ... Bank v1.0, StatementBank v1.1. Interface operation order must be identical.

- ANSAware uses type conformance to specify that StatementBank is a subtype of Bank.

- CORBA uses interface inheritance. StatementBank inherits from Bank.

- RM-ODP uses the definition of subtyping relationship to recognise that StatementBank is a subtype of Bank (by comparing the syntax of the two).

- L&W uses the definition of the subtyping relationship to check relationships between common operations and defines an extension mapping between Bank and StatementBank. However, these two interfaces are in a subtype relationship if pre- and post-conditions are also in a subtype relationship as defined by the model.

As can be seen, the mapping of an extension-based subtyping relationship between type models is straightforward in all cases and the translation between domains will not incur any loss of information (assuming that the semantic losses for the L&W domain are not considered).

4.2.2. Component renaming

Let us assume, that another bank is available, which provides an equivalent service, but names of the interface, operations and parameters and types differ:

> INTERFACE BanqueFrancaise : [
>   Deposer (compte:compte_t, somme:US_dollar_t);
>   Retirer (compte:compte_t, somme:US_dollar_t);
>];

The analysis of the models leads to the following results:

- DCE cannot legitimately represent the relationship between Bank and BanqueFrancaise. The simple versioning subtyping scheme could be used as long as the types are mapped into basic types, because interactions between clients and servers involves no names, only numbered operations, parameters, and base type references. However this is an exploitation of implementation-specific details of current DCE systems.

- ANSAware uses type conformance to specify the relationship between Bank and BanqueFrancaise, however again this relies on the fact that the ANSAware implementation does not check names during invocation.

- CORBA (OMA) interface subtyping only stipulates that the operation names and parameter types must be equivalent, but that parameter names can change. As the operation names are different, the relationship between types is not recognised.

- RM-ODP will not recognise any relationship between Bank and BanqueFrancaise.

- L&W uses its subtyping definition which allows renaming. It will recognise the relationship provided that a renaming map is defined, and that any pre- and post-conditions are also in a subtype relationship.

As the evaluation shows, the mapping between relationships can incur some losses. For example, when a query about types matching Bank will be forwarded from CORBA (or an RM-ODP-like system) to the L&W domain, the result of the query will have to be filtered in the CORBA type manager to remove all subtypes with renamed operations or parameter types. DCE and ANSAware would be able to cope with renamed parameters in some cases (parameter names), but in others the result would have to be filtered. L&W's model is able to take the result of the query from any other domain (again semantic differences are ignored).

4.2.3. Parameter subtyping

Let us assume, that there exists a banking interface which accepts only small transactions:

>   INTERFACE SmallAccount : [
>   Deposit (acct:acct_t, amount:US_dollar_t [1 .. 1000]);
>   Withdraw (acct:acct_t, amount:US_dollar_t [1 .. 1000]);
>   ];

If we analyse the subtyping relationships of our five models, the conclusions are:

* DCE, ANSAware and CORBA have no means to represent subtyping on subranges, and will not be able to determine that Bank is a subtype of SmallAccount,

* RM-ODP and L&W apply their respective subtyping definitions to recognise the relationship between SmallAccount and Bank.

Translation from RM-ODP or L&W's model to DCE or ANSAware or CORBA will lose all parameter subtyping information.

4.2.4. Parameter type conversion

The last stage of the scenario is not directly related to subtyping relationships. It tries to illustrate that in real heterogeneous distributed systems, a more general meaning for compatibility should be assumed than subtyping provides. Let us take into account the following interface:

>   INTERFACE BritishBank : [
>   Deposit (acct:acct_t, amount:brit_pounds_t);
>   Withdraw (acct:acct_t, amount:brit_pounds_t);
>   ];

If we assume that *brit_pound_t* can be converted to *US_dollar_t* then Bank and BritishBank are compatible (provided that the conversion is made). Some of the compared models may incorrectly believe that Bank and BritishBank are in a subtype relationship (in both directions, *i.e.* they are equivalent). None of the models has sufficient expressive power to show the constraints of this compatibility (conversion required). Liskov & Wing's mapping rules associated with relationships are most applicable to this problem, but does not provide the full solution. To provide an 'open' compatibility, a general type relationship mechanism must be introduced which will allow the definition of relationship types and mapping among them including various kinds of equivalence, subtyping and conversion relationships [11].

4.2.5. Summary

The scenario presents examples of banking interfaces and applies to them various kinds of subtyping on interfaces. It also suggests possible mappings between subtyping relationships when type managers of the various domains federate provided that the type managers can understand relationship types and are able to relate them (*e.g.* the type manager presented in [4]). Table 2 summarises the possible relationship translations including interface extension, renaming, parameter subtyping, and parameter conversion. The information about these translations has to be introduced to the type managers during federation and used for implementing 'filters' for the results of type queries directed to other domains. The left column in the table shows the model to be translated "from" and the top the translated "to" models. Those marked with a dagger incur some loss of information in the translation as noted in the previous sections.

	DCE	ANSAware	CORBA	RM-ODP	L&W
DCE	*	extension, renaming	extension, renaming†	extension	extension, renaming
ANSAware	extension, renaming	*	extension, renaming†	extension	extension, renaming
CORBA	extension, renaming†	extension, renaming†	*	extension	extension, renaming†
RM-ODP	extension	extension	extension	*	extension, parameter subtyping
L&W	extension, renaming	extension, renaming	extension, renaming†	extension, parameter subtyping	*

Table 2: Possible relationship translations

5. CONCLUSION

The paper has examined interoperability features of well known platforms for distributed computing (DCE, ANSAware, CORBA) and some models built for distributed and object-oriented systems (including RM-ODP). The purpose of the examination was to highlight the differences in their type systems which make type safety of interoperability among heterogeneous distributed domains difficult. Requirements for interoperability have been discussed. An examination was carried out showing differences in the definition of data types, action types and relationship types, and in particular types of compatibility relationships. The latter were also illustrated by a cooperation scenario based on one-to-one mappings between various kinds of compatibility relationships. These mappings can be a basis for creating a federation of the heterogeneous domains. Through highlighting the differences, the comparison also facilitates work on a future common interoperability framework to capture the differences among the type systems.

REFERENCES

[1] P. America. "Designing an Object-Oriented Programming Language with Behavioural Subtyping". In *Foundations of Object-Oriented Languages,* number 489 in Lecture Notes in Computer Science, pages 60–90, REX School/Workshop, Noordwijkerhout, The Netherlands, 1991. Springer-Verlag.

[2] APM Ltd, Cambridge UK. *ANSAware 4.1 Application Programmer's Manual*, Mar. 1992. Document RM.102.00.

[3] J. Auerbach, A. Goldberg, G. Goldszmidt, A. Gopal, M. Kennedy, J. Russell, and S. Yemini. "Concert/C Specification: Definition of a Language for Distributed C Programming". Technical Report RC18994, IBM T. J. Watson Research Center, July 1993.

[4] C. J. Biggs, W. Brookes, and J. Indulska. "Enhancing Interoperability of DCE Applications: a Type Management Approach". In *Proceedings of the First International Workshop on Services in Distributed and Network Environments, SDNE'94*, 1994.

[5] A. Black, N. Hutchinson, E. Jul, H. Levy, and L. Carter. "Distribution and Abstract Types in Emerald". *IEEE Transactions on Software Engineering*, SE-13(1):65–76, Jan. 1987.

[6] W. Brookes, A. Berry, A. Bond, J. Indulska, and K. Raymond. "A type model supporting interoperability in open distributed systems". In *Proceedings of the First International Conference on Telecommunications Information Networking Architecture, TINA '95*, Feb. 1995.

[7] V. Cahill, R. Balter, N. Harris, and X. Rousset de Pina. *The COMANDOS Distributed Application Platform.* Volume 1. ESPRIT Research Reports (Project 2071). Springer-Verlag, 1993.

[8] L. Cardelli and P. Wegner. "On Understanding Types, Data Abstraction, and Polymorphism". *Computing Surveys*, 17(4):471–522, Dec. 1985.

[9] R. A. de By and H. J. Steenhagen. "Interfacing Heterogeneous Systems through Functionally specified Transactions". In D. K. Hsiao, E. J. Neuhold, and R. Sacks-Davis, editors, *Proceedings of IFIP DS-5 Semantics of Interoperable Database Systems*, volume 2, pages 38–45, Lorne, Victoria, Australia, Nov. 1992.

[10] S. J. Garland, J. V. Guttag, and J. J. Horning. "An overview of Larch". In *Functional Programming, Concurrency, Simulation and Automated Reasoning*, number 693 in Lecture Notes in Computer Science, pages 329–348. Springer-Verlag, July 1993.

[11] J. Indulska, M. Bearman, and K. Raymond. "A Type Management System for an ODP Trader". In *Proceedings of the IFIP TC6/WG6.1 International Conference on Open Distributed Processing, ICODP'93*, (Berlin, Germany, 13–16 September, 1993). IFIP Transactions, Elsevier, North Holland, 1994.

[12] ISO/IEC DIS 10746-3. Draft Recommendation X.903: Basic Reference Model of Open Distributed Processing — Part 3: Prescriptive Model, Apr. 1994. Output of Geneva editing meeting 14–25 February 1994.

[13] B. Liskov and J. M. Wing. "A New Definition of the Subtype Relation". In O. Nierstrasz, editor, *Proceedings of the European Conference on Object Oriented Programming, ECOOP '93*, number 707 in Lecture Notes in Computer Science, pages 118–141, Kaiserslautern, Germany, July 1993. Springer-Verlag.

[14] B. H. Liskov and J. M. Wing. "A Behavioural Notion of Subtyping". *ACM Transactions on Programming Languages and Systems*, 1994. (to appear).

[15] F. Manola and S. Heiler. "A "RISC" Object Model for Object System Interoperation: Concepts and Applications". Technical Report TR-0231-08-93-165, GTE Laboratories Incorporated, Aug. 1993.

[16] E. Najm and J.-B. Stefani. "A formal semantics for the ODP computational model". *Computer Networks and ISDN Systems*, to appear.

[17] Object Management Group. *Object Management Architecture Guide*, second edition, Sept. 1992. Revision 2.0, OMG TC Document 92.11.1.

[18] Object Management Group and X/Open. *The Common Object Request Broker: Architecture and Specification*, 1992.

[19] J. Palsberg and M. I. Schwartzbach. "Three Discussions on Object-Oriented Typing". *ACM SIGPLAN OOPS Messenger*, 3(2):31–38, 1992.

[20] J. Palsberg and M. I. Schwartzbach. *Object-Oriented Type Systems*. John Wiley & Sons, 1994.

[21] W. Rosenberg and D. Kenney. *Understanding DCE*. Open System Foundation, 1992.

[22] J. Shirley. *Guide to Writing DCE Applications*. Open System Foundation, 1992.

6

A General Resource Discovery System for Open Distributed Processing

Qinzheng Kong and Andrew Berry

CRC for Distributed Systems Technology, †
The University of Queensland,
Australia, 4072
<qin@dstc.edu.au>, <berry@dstc.edu.au>

The networks of today support a wealth of resources which can aid the daily tasks of a diverse user base. Resource discovery is a relatively new field that deals with the problems of finding, organising and accessing these resources in a global network. The architecture model proposed by the DSTC's Architecture Unit provides a way to specify open distributed application systems. A resource discovery system is an example of such an application. This paper uses the DSTC architecture model to specify a general resource discovery system.

Keyword Codes: C.2.4; H; H.2.c
Keywords: Distributed System; Information Systems; General Model

1 INTRODUCTION

The distributed systems of today support a wealth of resources which can aid the daily tasks of a diverse user base, ranging from school children to business professionals. The types of resources available include network services, documents, software, video, sounds and images. Resource discovery is a relatively new field that deals with the problems of finding, organising and accessing these resources in an open distributed system.

The Architecture Model [1] proposed by the DSTC's Architecture Unit provides a way to specify open, distributed application systems. A resource discovery system in a distributed environment is an example of such an application. This paper uses the DSTC Architecture Model to specify the architecture of a general Resource Discovery System.

In the remainder of the document, section 2 gives a brief description of the concepts defined in the DSTC Architecture Model. Section 3 gives a general architecture for resource discovery systems. Sections 4 and 5 describe how the concepts defined in the architecture model can be used to specify a general resource discovery system. Section 6 describes some fundamental requirements of a distributed environment for general distributed applications. Finally, section 7 discusses future research directions in the distributed resource discovery system area.

---

† The work reported in this paper has been funded in part by the Cooperative Research Centres Program through the Department of the Prime Minister and Cabinet of the Commonwealth Government of Australia

2   BASIC CONCEPT OF THE ARCHITECTURE MODEL

Distributed applications in the DSTC architecture model are specified in terms of objects, their interface types and bindings between the interfaces of objects.

2.1   Objects

In the DSTC architecture model, an object is an entity that encapsulates state, behaviour and activity (i.e. the ability to take independent action). Objects interact by exchanging strongly typed messages over a binding. Objects connect to a binding through their interfaces.

2.2   Bindings

A binding is an association between a set of objects that allows the objects to interact. Bindings are strongly typed—the binding type defines the roles of objects in a binding and the interaction that can occur between objects fulfilling those roles.

Roles of a binding are filled by the objects participating in the binding. For example, a binding to support RPCs must have a "client" role and a "server" role. Each role of a binding specifies an interface type that must be satisfied by objects fulfilling that role— objects participating in the binding must therefore instantiate an interface that is a compatible with their role in the binding.

The fundamental unit of interaction in a binding is a single, strongly typed message, although more complex, high-level interactions can be defined, for example, RPC or multicast.

2.3   Interfaces

An interface is instantiated to fulfil the role of an object in a particular binding. Interfaces are strongly typed—an interface type describes the structure and semantics for interactions of an object during a binding. In other words, an interface type describes the structure of messages and the object behaviour associated with messages sent and received by that object during a binding.

An object can offer many interface types and the set of interface types offered can vary over time.

3   AN ARCHITECTURE FOR GENERAL RESOURCE DISCOVERY SYSTEMS

A resource discovery system in today's distributed environment is a distributed application in general. On one hand, it shares many common features with all the other distributed applications. For example, its infrastructure supports naming and directory services, security facilities, underlying protocol transparency, alternative messaging routing, etc. On the other hand, it has its own characteristics in terms of the services provided to its user. Resource discovery services mainly include:

- identifying the possible location(s) of user interested information;

- selecting the information required by providing the searching and browsing facilities;
- presenting the information
- delivering the required information in user specified form to user specified destination.

In general a distributed resource discovery system consists of many sites. Each site can be described as having a *Server* and one or more *Clients*.

The *Server* is responsible for accessing local resources, communicating with other remote servers to access their resources, and providing common services required by its clients. The *Client* provides a mechanism for users to discover and access resources. The *Client* can be further decomposed into two major components: a *User* part which provides an interface specifically designed for the end-user environment; and an *Agent* part, which provides user specific services and communicates with the local server.

Each server can support more than one agent and each agent can support more than one user.

The architecture of such a system is illustrated in Figure 1:

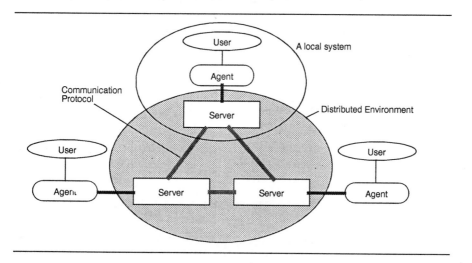

Figure 1: General Architecture of Resource Discovery System

Each local system can be refined as in Figure 2:

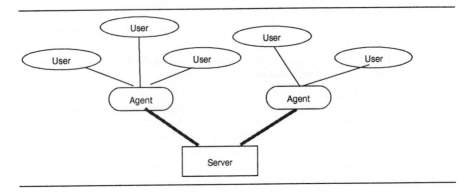

Figure 2: An example of a local system

4   THE SPECIFICATION OF A GENERAL RESOURCE DISCOVERY SYSTEM

As mentioned before, a resource discovery system in a distributed environment is only a particular case of a general distributed application. The DSTC architecture model can be used to specify a general resource discovery system.

The first step of specifying a distributed system using the model is to identify a set of basic *Objects*. Based on the architecture described in section3, there are three top level objects: a *User* an *Agent* and a *Server*.

The second step is to describe the interactions between these objects with a set of *Binding Types*. In our example of resource discovery system, three binding types are required, *User-Agent Binding Type*, *Agent-Server Binding Type* and *Server-Server Binding Type*. These binding types define the communication and interaction protocols used to support resource discovery.

From the binding types, we can identify and derive a set of *Interface Types* to be supported by these objects.

The interface types supported by the *User* object are *End-user Interface Type* and *Agent-request Interface Type*. The *End-user Interface Type* is used to specify the end-user interface, which might be a GUI interface. The *Agent-request Interface Type* defines the interactions a *User* can expect to have with an *Agent*.

The interface types supported by the *Agent* object are the *Agent-service Interface Type* and *Resource-user Interface Type*. The *Agent-service Interface Type* is used to specify the local services provided to clients of the agent and the parameters required for those services. The *Resource-user Interface Type* defines the interactions an *Agent* can expect to have with a *Server*.

The *Server* object supports two interface types, *Resource-service Interface Type* and *Remote-service Interface Type*. The *Resource-service Interface Type* is used to specify the services provided to an *Agent*, and the *Remote-service Interface Type* is used to specify the interactions that a *Server* can expect to have with other servers.

The relationship between the objects, interface type and binding types of a Resource Discovery System is illustrated in Figure 3:

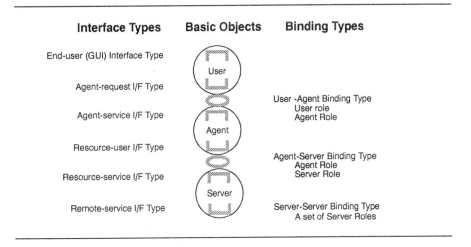

Figure 3: Object, Interface Types and Binding Types

Notice that separate interfaces are defined for the *User* and the *Agent* in the *User-Agent* binding type and similar separations in the other binding types. This is to allow for differences in what the user expects, and what the agent provides. An agent might be accessed by more than one user, and each user can have different expectations. Provided the expectations of the user are a subtype of the service provided by the agent, a binding between the user and agent is meaningful.

The following sub-sections describe the binding types in more detail. Detailed interface types are not described, to allow maximum flexibility in the definition of systems based on this architecture.

4.1   User-Agent Binding Type

The *User-Agent Binding Type* specifies the possible interactions between the user and the agent objects. It has two roles, *user* and *agent*. A number of interactions can take place between a user and an agent. For example, the user can issue a request that is delivered to the agent, and it should always be followed by a response from the agent delivered to the user. (The handling of the request is a local issue of the agent and is not discussed here.) Another example of the interaction might be that the agent can issue a message to a number of users when a certain event occurred.

This binding type can be expressed using an arbitrary syntax as follows:

```
User-Agent Binding type
      Roles are: user, agent
      Interactions are:
            query:      user.request -> agent.request;
                        agent.response -> user.response
            message:    agent.sending -> user.receiving
```

As mentioned before, a binding type can be used to derive minimal interface types required for different roles. Each action taken by different roles should be specified as an action in the corresponding interface type. For example, the *Agent-request Interface Type* must support the action of sending requests and the *Agent-service Interface Type* must support the action of sending responses.

The binding type for a general resource discovery system specified above can be used as the basis for defining binding types in a specific system. A binding type for a system (RDS) based on Z39.50[2], could be based on the general binding type as demonstrated below:

```
RDS-User-Agent Binding Type
Derived from User-Agent Binding Type
      Roles are: user, agent
            user has RDS User-Agent Interface using Z39.50
            agent has RDS Agent Interface using Z39.50
      Interactions are:
            query:init: ... ...
            query:search:... ...
            query:present:... ...
            query:delete:... ...
            query:scan: ... ...
            query:sort: ... ...
            ... ...
            segment:     ... ...
            ... ...
```

In which, the *init, search,* etc. are specific queries derived from the *query* of the general *User-Agent Binding Type* and *segment* is a new interaction. Detailed descriptions of each interaction can be specified but are not given in this paper.

4.2   Agent-Server Binding Type

The *Agent-Server Binding Type* may be expressed as:

```
Agent-Server Binding Type
      Roles are: agent, server
      Interactions are:
            query: ... ...
            service: ... ...
            message: ... ...
```

It has the similar form as the *User-Agent Binding Type*. When the server object is refined to capture more detailed services supported by the server (See "REFINEMENT OF

TOP LEVEL OBJECTS" on page 9.), the binding type will have more details.

4.3   Server-Server Binding Type

The binding type for interaction between servers is more complicated. The binding type can be specified either as a multi-party binding with many servers, or a set of 2-party bindings between servers.

In the case of multi-party binding, an instance of a binding type is a single binding which contains two or more servers. Each server in the binding has an equivalent role (i.e they are peers). Servers can choose to join and leave existing bindings. The following figure illustrates a multi-party binding:

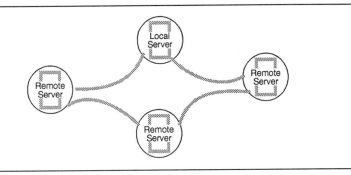

Figure 4: Multi-Party Binding Type

The *Multi-Party Binding Type* is often required in a real distributed system. A key problem of such a binding type is to define the policy of finding, then joining and leaving a binding. In a distributed environment, it might not be possible to have a centralised system to keep the information of all the existing bindings, so the policy should include how to find existing bindings in a distributed environment.

It is assumed that when a server wishes to join a binding, it should have the knowledge of at least one server which is already in the binding. With a join request, a set of conditions can be specified. When the server in the binding receives the join request, it is its responsibility to notify all the servers already in the binding that a new server has joined. If the server which receives the join request is a stand-alone server (not in any existing binding) then a new instance of the binding type is created which contains only two servers initially.

The *Multi-Party Binding Type* can be expressed as:

```
Multi-Party Server Binding Type
      Roles are: servset - a set of servers
      Interactions are:
            query:      ∃s:servset,
                        s.sending -> ∀r:servset-{s}: r.receive;
                        ∀r:servset-{s},
                        r.response -> s.response
```

```
join:          ∃s1:servset, s2: server,
               s1.join(s2) -> ∀r:servset-{s1}:r.newserv(s2)
leave:         ∃s: servset,
               s.leave(s2) -> ∀r:servset-{s}:r.leaving(s2)
    ... ...
```

A number of interactions can be added to the above specification, such as, combine two existing bindings, create a new binding, etc.

In the case of a *2-Party Binding Type* (it is sometimes also referred to as peer-to-peer binding), each binding contains only two servers, one is referred as *Local Server* and the other as *Remote Server*. The following figure illustrates *2-Party* bindings.

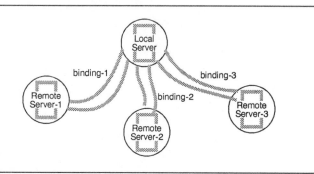

Figure 5: 2-Party (Peer-To-Peer) Bindings

Each binding instance in the above figure is a simple binding between a pair of servers. Although several servers are connected through indirectly by the *local server*, there are no direct bindings between these servers. If direct interactions between all servers are required, specific bindings have to be set up.

The *2-Party Binding Type* can be expressed as:

```
2-Party Server Binding Type
     Roles are: server1, server2
     Interactions are:
          query:      s:server1 or server2, r: not s,
                            s.sending -> r.receive
                      r.response -> s.response
          create:     s:server1 or server2, r: not s,
                            s.create -> r.create
          delete:     s:server1 or server 2, r: not s,
                            s.delete -> r.delete
     ... ...
```

It is possible that within one system, both the multi-party binding and the 2-party binding are used at the same time. From the descriptions above, it is apparent that multi-party bindings provide significantly more flexibility in describing the interactions at the expense of complexity.

5 REFINEMENT OF TOP LEVEL OBJECTS

The top level objects of the general resource discovery system only give an abstract view of the system. These objects can be refined to capture additional detail. For example, apart from normal query access, an agent object can provide some intelligent services. An example of intelligent agent service might be to register the interests of a user whenever that user issues a request. This data can be used by the agent to intelligently gather information for delivery to the user.

Another example of an additional service might for the server and the agents to provide an administrative interface for configuration and administration.

Hence, based on these object refinements, more binding types are required which can either be derived from existing binding types or be newly defined types. Some examples of these binding types are given in the following sections.

5.1 Intelligent Agent Binding Type

The intelligent agent binding type can be derived from the general *User-Agent Binding Type* with an extra operation performed by the agent to register user's interests:

```
Intelligent-Agent Binding type
      Derived from User-Agent Binding Type
      Roles are: user, agent
      Interactions are:
            query:        user.request -> agent.request and
                                agent.register;
                          agent.response -> user.response
            notify:       agent.notify -> user.notify
```

5.2 Admin-Server Binding Type

An administrator is a special user to the system. Apart from normal access through local agent interface, the administrator might have direct access to the Server for administration. A new binding type is required to capture these abilities.

A template of *Admin-Server Binding Type* can be given as:

```
Admin-Server Binding type
      roles are: admin, sever
      interactions are:
            insert:
            delete:
            update:
            backup:
            verify:
            security:
            management:
            ... ...
```

5.3 Refinement of the Server

The refinement of the server object can be achieved by specifying a set of *Service* objects embedded in the *Server* object. That is, each service provided by the server can be defined as an object. For a general resource discovery system, the following service objects might be defined:

- Security Service Object — This performs the service level security checking, including the checking of access made by local users or remote servers.
- Directory Service Object — This provides the X.500 Directory Services. The information in the directory may include the information about other servers in the system and information about existing Bindings.
- Trader Object — This provides a "yellow pages" service, for example, an ODP Trader[3][4][5].
- Register Object — This is an alternative to the Trader Object. It provides a simple mechanism for information providers to register themselves with the distributed system.
- Type Management Object — This object provides a service to determine the compatibility of types[6].
- Search and Retrieve Object — This is the search engine of the system. The search and retrieve are based on the system knowledge maintained and managed by other objects, such as Directory Object, Trader Object, Register Object and Type Management Object. User required information / resource can be discovered by the co-operation among these objects.
- Local Database Management Service — This might be any kind of persistent data storage management system. This service may be used by other Service objects such as the Directory Service objects, the Type Management objects and the Trader objects.
- Accounting Service Object — This provides resource control service for the user.
- Logging Service Object — This object records events of interest. A general logging service can be used for many different purposes, e.g. security logging, search event logging, etc.
- Federation Object — This object is responsible for the federation among different Servers. Its activities include joining and leaving the federation, communication agreement to other servers, security agreements and many other issues.

Some of the above services may be defined as common services (e.g. the *Accounting Service*, the *Logging Service* and the *Security Service)* shared with other applications running on the same node. Other services might be specific to the resource discovery system. Each of these services can be specified as an object. Figure 6 illustrates the relationship between these service objects. Different interface types and binding types among these objects are also illustrated:

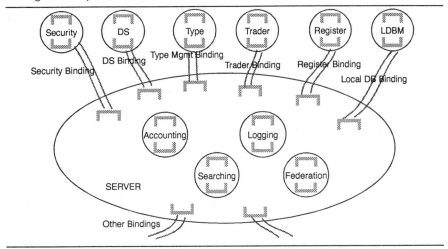

Figure 6: Service Objects and the Binding Types

6 DISTRIBUTED ENVIRONMENT

Ideally, a resource discovery system would be built on top of a standard distributed environment. Such an environment must provide location transparency and protocol independence. Each component of the resource discovery service can be defined as services supported by the distributed environment and can be dynamically added in the environment when needed. The abstract view of such an environment is illustrated in the following figure:

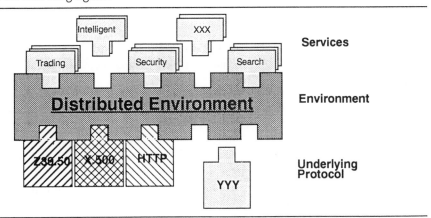

Figure 7: Resource Discovery Services in a Distributed Environment

It can be seen from Figure 6 that the resource discovery services can be built together with many other distributed applications in a sufficiently rich distributed environment.

7   DISCUSSION

A brief specification of a general resource discovery system using the DSTC Architecture Model has been given. It demonstrated that a resource discovery system is only a special example of a distributed application, and it can be built on top of an open distributed environment with other distributed applications.

Further refinement of this specification of a general resource discovery system will include the following tasks:

- Each binding type will be refined in order to capture all the possible interactions between different roles.
- Interactions in each binding type will be refined to specify the details of the actions and to give the allowed parameters in each action
- More detailed interface types will be derived from the binding types.
- Formal specification languages, such as LOTOS and / or Object-Z should be used to give more accurate definition of the objects, interface types and binding types.

ACKNOWLEDGEMENTS

Thanks to numerous research staff of the DSTC Resource Discovery Unit, the DSTC Architecture Unit and the University of Queensland Computer Science Department for useful discussions and suggestions. The work reported in this paper is funded in part by the Cooperative Research Centres Program through the department of the Prime Minister and the Cabinet of the Commonwealth Government of Australia.

REFERENCES

1.   A. Berry and K. Raymond, *The A1√ Architecture Model*, submitted to International Conference on Open Distributed Processing (ICODP) 1995.

2.   ANSI / NISO Z39.50-1994 Information Retrieval: Application Service Definition and Protocol Specification, Draft 1994.

3.   ISO / IEC JTC1 / SC21, *Draft Recommendation X.903: Basic Reference Model of Open Distributed Processing - Part 3: Prescriptive Model* (ISO / IEC DIS 10746-3.1), February, 1994.

4.   ISO / IEC JTC1 / SC21: *ODP Trader*, 1994.

5.   M. Bearman, *ODP-Trader*, Proceedings of the International Conference on Open Distributed Processing 93 (ICODP'93), Berlin, September 1993.

6.   J. Indulska, M. Bearman and K. Raymond, *A Type Management System for an ODP Trader*, Proceedings of the International Conference on Open Distributed Processing 93 (ICODP'93), Berlin, September 1993.

SESSION ON

Management – a Telecommunications Persepective

7

Management Service Design: From TMN Interface Specifications to ODP Computational Objects

A.Wittmann*, T. Magedanz°, T. Eckardt°

*GMD FOKUS, Hardenbergplatz 2, D-10623 Berlin, Germany
°Technical University of Berlin, Dept. OKS, Hardenbergplatz 2, D-10623 Berlin, Germany

This paper addresses the relationships between Telecommunications Management Network (TMN) and Open Distributed Processing (ODP) standards in respect to the methodologies for the development of Management Services (MSs). In the light of emerging object-orientation in the telecommunications environment the development of object-oriented MSs becomes a major issue. ODP standards represent the foundation of object-oriented service design, for both telecommunication and management services. TMN represents the world-wide accepted framework for the management in telecommunications. Although TMN concepts promote the object-oriented modeling of management information, the modeling of MSs is still function-oriented and in infancy. Within this paper we will illustrate, that TMN concepts for MS design can be combined adequately with ODP computational modeling methodologies. The basic idea of this "integrated" MS design methodology is to map TMN Management Functions onto ODP Activities for computational modeling. An example for the presented approach will be given for the modeling of a User Registration MS in the context of an advanced Personal Communication Support System, currently under development at the Technical University of Berlin for the Deutsche Telekom Berkom.

Keywords: *Activities, Computational Objects, Management Functions, Management Services, ODP, TMN*

1.   INTRODUCTION

The telecommunications world is changing its face. Progressing libralization and increasing competition on the one hand, and the development of international standards for the uniform provision and management of telecommunication services, such as *Intelligent Networks (IN)* [Q.1200] and *Telecommunication Management Network (TMN)* [M.3010] on the other hand, pave the way towards an open telecommunications environment. In such an open environment telecommunication services and their corresponding management services should be introduced efficiently and rapidly, where the reusage of existing services and service components represents an important prerequisite. This is the reason for the world-wide acceptance and global deployment of IN and TMN standards, since both allow for the uniform and rapid design of telecommunication and management services. However, the basic service design methodologies of both concepts are primarily function-oriented, which limits the power of these concepts. This was the starting point for investigating the application of object-oriented methodologies dreived from the computing world for telecommunication and management service design.

The application of object-oriented principles into the telecommunications environment, as promoted by emerging *Open Distributed Processing (ODP)* [X.90x] standards, is therefore currently in the focus of international research. In particular the harmonization of IN and TMN concepts in the light of the ODP Reference Model (RM-ODP) represents a challenging aspect of this research, where in particular the work of the international Telecommunication Information Networking Architecture Consortium (TINA-C) [TINA-C] could be seen as the major representative. In addition, several RACE and EURESCOM projects are currently investigating the evolution of TMN concepts, taking into account upcoming ODP standards. The problem is evident: management service development originating in the world of telecommunications with a functional modeling approach can not immediately be specified as an ODP conformant distributed application and thus cannot benefit from the research on ODP.

In this context this paper focuses on the integration of TMN and ODP concepts in the field of *Management Service (MS)* design, trying to bridge the gap between the function-oriented TMN MS design methodology, as defined in the TMN recommendation M.3020 [M.3020] and the object-oriented modeling approach of ODP. Based on the assumption, that the TMN Interface Specification Methodology corresponds to the ODP enterprise viewpoint, we propose a mapping of the identified *TMN Management Functions*, representing the smallest parts of a MS visible to the user, to *ODP Activities*, which consist of one or more Computational Objects within the ODP Computational Viewpoint.

The proposed approach is decribed in more detail in the following chapter, whereas chapter 3 provides a detailed example, which illustrates the design of an advanced "User Registration" MS in the context of a TMN-based *Personal Communication Support System (PCSS)* [Maged-95]. The PCSS represents a generic support platform which provides for personal mobility and service personalization to an open set of communication services based on the currnt advances in mobile computing and (universal) personal telecommunications. A short summary in chapter 4 concludes this paper.

2.   A GENERAL METHODOLOGY FOR THE DEVELOPMENT OF MANAGEMENT SERVICES

Management Service (MS) design always starts with an analysis of the requirements the service has to satisfy. These requirements arise at the business level of an organization and are investigated within the ODP Enterprise Viewpoint. The methodology to be followed in the development of an MS, is to relate these requirements to the systems processing capabilities. To do so, a set of terms and concepts is needed, appropriate to express the captured requirements from the user perspective on the one hand and from the system's viewpoint on the other. The RACE project PRISM, which is investigating Service Management based on TMN standards, has established a generic Management Service Model [PRISM-93]. This model is based upon the TMN Interface Specification Methodology [M.3020], and is briefly introduced below because we apply this model to our MS design methodology.

The PRISM MS model is visible at the ODP Enterprise Viewpoint through the type of functions the MS offers. A distinct boundary exists between the user who selects the functions to use and the system which implements them. The set of functions available may be considered to occur at an interface to the service. The user interface is made up of three types of components. These, in descending level of decomposition, are:

- *Management Service (MS)*, which is an offering by a MS provider to satisfy a specific communication management need of a management user.

- *Management Service Components (MSCs)*, which represent reusable components that can be assembled to form MSs.

- *Management Functions (MFs)*, which are the smallest parts of an MSC with the ability to cause effects in the management system.

All the three types of components are visible to the user, i.e. there is a representation of them at the user's interface. The relationships between these interface components, the user and the Management System are shown in Figure 1. It has to be stressed that the implementation of the MS is independent of the functionality visible at the user interface.

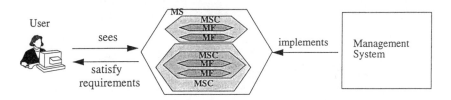

FIGURE 1.   Enterprise View of a Management Service

After introducing the PRISM service model we are now ready to describe our MS design methodology which starts at the Enterprise Viewpoint. At the business level of an organization requirements for managing telecommunication applications and systems arise and are discussed by the different actors of the enterprise. A subsequent requirements analysis identifies the management requirements in more structured manner. The concept used here to group the management requirements, are the Telecommunications Management Functional Areas (TMFAs) [H400]. An analysis within every single TMFA ensures a comprehensive requirements capture. The identified requirements have to be mapped on the functions and processing of appropriate MSs. A prose description of MSs satisfying the management requirements is the initial step towards an MS Enterprise Specification. A more detailed image of the MSs is gained by applying the PRISM MS model as a specification tool. The MS Specification advances to a composition of reusable components. MSs are composed of MSCs which themselves are built of lower level MSCs or MFs.

On the other side Computational Objects (COs) interacting in ODP Activities have to be identified within the ODP Computational Viewpoint, using the ODP Computational Viewpoint Language. COs are ODP objects contained in computational specifications and are defined in Part 3 of the RM-ODP [X.903]. ODP Activities are single-headed graphs of actions defined in Part 2 of the RM-ODP [X.902].

The control flow between the computational objects can be illustrated by *event trace diagrams* providing a high level view on the management activities. Another specification tool in the computational viewpoint is an *activity flow diagram* specifying the sequence of operation invocations of one management activity. Finally, a detailed computational specification is defined, using *object class* and *interface templates* for specifying computational objects, interfaces, and operations. An information model specifying the information objects used in the management system extends the MS specification. The information model may conform to the OSI Management standards [X.720] or may be specified with the terms and concepts of the ODP Information Viewpoint Language contained in RM-ODP Part 3 [X.903].

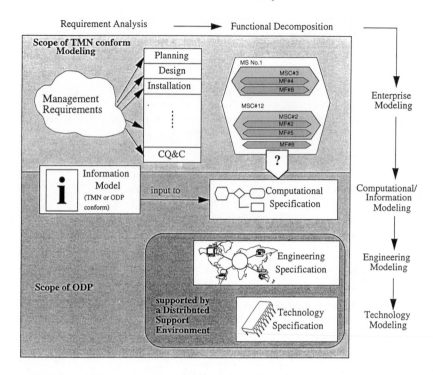

FIGURE 2.   Development Path of the proposed MS Design Methodology

Before starting with the modeling of the Engineering Viewpoint, ODP transparencies have to be defined. The computational objects are now mapped to *Basic Engineering Objects* and are distributed in the system. A Distributed Support Environment (DSE) which realizes the specified transparencies has to be used. Finally, technical issues are elaborated on in the Technology Viewpoint. However, the two latter viewpoints are not treated in this paper. Figure 2 shows the proposed design methodology, pointing out the path which starts from an management requirement analysis and leads to specifications at different ODP viewpoints.

2.1   Problem: The Gap in the MS development path

The MS design methodology starts with the Enterprise Viewpoint modeling using the TMN concepts for MS decomposition. This step ends with an identification of MFs representing the lowest level components of the user interface of an MS. This function-oriented approach of the TMN Interface Specification Methodology complicates a subsequent ODP conformant modeling since ODP is based on an object-oriented approach. The MF's capabilities cannot be directly modelled in the ODP Computational Viewpoint Language because there does not exist any concept associated to a TMN MF. This problem represents a conceptual gap in the design path of a MS illustrated by the question mark in the white arrow in Figure 2. This situation calls for a solution that integrates the functional approach and the object oriented approach and thus harmonizing the two worlds of TMN and ODP.

2.2 Solution: Mapping between TMN Management Functions and ODP Activities

The gap between the functional approach and the object-oriented approach, as identified above, is now closed by the thesis that every identified TMN MF can be modelled as an ODP Activity. There is a natural correspondence between a MF that changes something in the system and an ODP Activity which is composed of actions, i.e. something that happens in the system. Figure 3 focuses on this overlap of the TMN functional approach and the ODP object-oriented way.

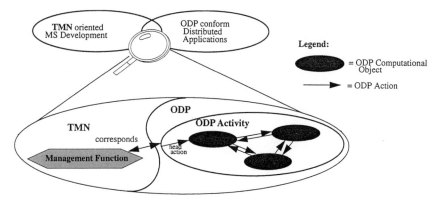

FIGURE 3.   Correspondence between TMN Management Functions and ODP Activities

A TMN MF which is the lowest user perceivable structure in the user interface of an MS is associated with an ODP Activity. The triggering of an MF corresponds with an operation invocation of an activity's head action. This occurs at an operational interface of an appropriate ODP computational object. The Activity may be composed of several chained actions which may involve the interaction of several computational objects. The tail action of that Activity, however, has to occur at the same object where the Activity has been started. We define an activity originating from an MF as a *Management Activity* in contrast to other internal activities in an ODP system. The MF, which could be specified in terms of pre- and post conditions, benefits from the modeling power of ODP which enables the integrated specification of functional aspects (e.g. operations), dynamic aspects (actions and activities) and structural aspects (relationships between computational object).

Unfortunately there is no general algorithm that takes us automatically from a TMN MS specification to an ODP Computational specification. In particular, the MS designer will need a lot of experience to transform a TMN MF into an ODP Management Activity and has to consider every individual case separately. Nevertheless, the following steps present a general design guideline for the transformation of a TMN MS functional decomposition to an ODP conformant specification:

1. Starting point is the functional decomposition of the MS under consideration, i.e. appro-priate MSCs and MFs have been identified.
2. Identifiy possible entities involved in the realization of a MSC/MF, referred to as "Mana-gement Process" and identify possible information flows by means of an Event Trace Diagram.

3. Identify candidate COs engaged in each MF/management activity being part of the iden-
    tified Management Process (from step 2) and draw an Activity Flow Diagram.
4. Give a detailed Computational Specification using an appropriate specification template.

3. MODELING OF A USER PROFILE MANAGEMENT SERVICE FOR A
    PERSONAL COMMUNICATION SUPPORT SYSTEM

This chapter provides an illustration of the proposed approach. The field of application of
the presented MS design methodology is a TMN-based *Personal Communication Support
System (PCSS)*, currently under development at the BERKOM II project "IN/TMN
Integration" performed by the Department for Open Communications Systems at the Technical
University of Berlin for Deutsche Telekom Berkom (•De•Te•Berkom•). The PCSS provides
IN-like personal mobility and service personalization capabilities by means of TMN concepts
in a generic way for a broad range of telecommunication services. In the following we will
provide a brief overview of the PCSS and address in particular the development of User
Registration MF being part of a User Profile Management Service, based on the presented MS
design methodology.

3.1   PCSS Overview

The PCSS has been specially designed for providing *personal mobility, personalization of
services* and advanced service interoperability. It enables users to configure their
communications environment according to their specific needs, with respect to parameters
such as *time, location, quality, medium, cost, accessibility and privacy* [Eckardt-94]. The PCSS
incorporates the most recent developments in the area of mobile computing and (universal)
personal telecommunications. However, the basic idea of the PCSS is to use advanced TMN
concepts for the realization of IN service feature related to universal personal communications
[Maged-95].

One of the most prominent principles of the PCSS is the largely person-oriented and
location-oriented operation of the PCSS in order to support real personal communications. This
attribute distinguishes the PCSS approach from other advanced telecommunication services
and IN services which are primarily based on (network) numbers when specifying features
such as "call forwarding on busy (CFBY)".

In a less strict usage of the term, the PCSS denotes the *PCSS infrastructure* comprising the
Core PCSS and elementary management services or enabling technologies provided as
specialized services. In this category of services, we find the different registration services
gathering the elementary location information of the various users to be written into the
respective "user profiles". An important registration service in this context is the *Automatic
Registration Service* which currently uses the Active Badge System of Olivetti. The *User
Profile Management Service* represents another important example of elementary management
services comprising the PCSS infrastructure. This management service provides a user
interface to the Service User Profile and is used to display any information contained in the user
profile as well as to modify certain data entries within the profile. In particular, different types
of registration will be supported by this services as will be described in the next section.

The individual components of the Core PCSS and the main applications interacting with
those components are displayed in Figure 4. The central data structure within the PCSS profiles
is the *Generic Service User Profile* containing all the information required to support personal
mobility, personalization of communication services, and advanced user information services.

An application programming interface (PCSS API) specifically defined for the context of the PCSS provides various types of teleservices and location-aware information services, such as an *User Location Information Service*, with personal communications-related information. More details on the PCSS can be found in [Maged-95].

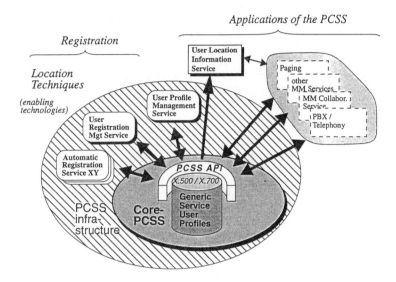

FIGURE 4.   Applications of the Core PCSS

3.2   Modeling of the PCSS User Profile Management Service

The requirements identified for the PCSS *User Profile Management Service* can be captured according to the PRISM Service Model, leading to the decomposition into MSCs and MFs as illustrated in Table 1. This corresponds to step one of the design guidelines given in section 2.2.

MS: User Profile Management
MSC No. 1: User Registration
MF No. 10:   Manual Registration
MF No. 11:   Automatic Registration (Enable/Disable)
MF No. 12:   Scheduled Registration (Enable/Disable)
MSC No. 2: Authentication
MF No. 20   User Authentication
MF No. 21   Provider Authentication

TABLE 1. Functional Composition of the User Profile Management Service

The User Profile MS is composed of several MSCs, where some MSCs have been defined within other MSs and will be reused here (e.g. MSC Authentication). Within the scope of this paper we want to concentrate on the MSCs *User Registration* and *Authentication*. The *User*

Registration MSC constitutes three MFs: the first, "Manual Registration", enables the user to explicitly specify a communication end point for all incoming calls, the two latter, "Automatic Registration" and "Scheduled Registration" provide the ability to select the registration mode, respectively. The MF "Automatic Registration" corresponds to the use of electronic location techniques. The MF "Scheduled Registration" allows the user to specify a time dependent routing of incoming calls to different communication end points e.g. to the office in the morning and to the conference room in the afternoon. The *Authentication* MSC provides for user authentication and provider authentication which is prescribed for accessing the User Profile according to the PCSS's security policy.

Before turning to the Computational Viewpoint Specification we want to provide an overview of the User Registration Process (which could be regarded as the computational modeling of the corresponding MSC) with the event trace diagram shown in Figure 5, thus following step two of the design guidelines.

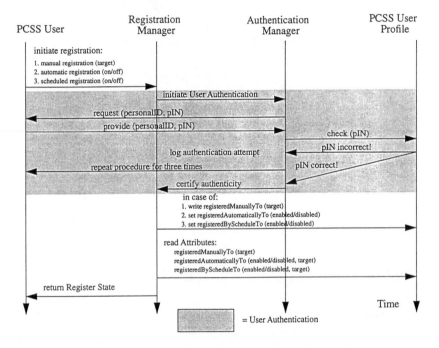

FIGURE 5.   Event Trace Diagram for the PCSS User Registration Process

There are four parties involved in the User Registration Process: The "User" who wants to register himself to the PCSS, a "Registration Manager" which is in charge of managing the User Registration Process, an "Authentication Manager" responsible for the user's authentication and the "User Profile" which holds registration and authentication relevant data. (Note that these "parties" are not Computational Objects. They are only used to obtain a general overview of the User Registration Process. However, the identified parties could serve as a basis for the later identification of COs.)

The User Registration process starts with the User initiating the registration via the

Registration Manager. He selects one registration type (i.e. manual, automatic, and scheduled) and in case of manual registration provides a target communication end point (i.e. a room number or a terminal address). The Registration Manager starts the authentication process at the Authentication Manager's interface. The Authentication Manager selects a PIN-based (Personal Identity Number) authentication mechanism according to the PCSS security policy. In case of success he certifies the user's authenticity to the Registration Manager who subsequently executes the registration at the User Profile Manager. After that the Registration Manager reads the actual values of the three registration attributes (i.e. manual, automatic, and scheduled) and reports the new registration state to the user.

Being now familiar with the User Registration Process as a whole, we can turn to the Computational Specification according to step three of the design guidelines. This process is composed of four Management Activities, corresponding to the MFs *Manual Registration, Automatic Registration, Scheduled Registration* and *User Authentication* respectively. In the following we focus only on the Management Activities for *Manual Registration* and *User Authentication* MFs. Figure 6 depicts a combined activity flow diagram for both Management Activities.

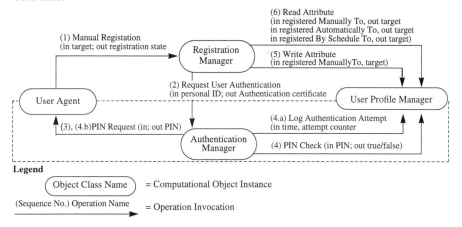

FIGURE 6.   Activity Flow diagram for the Manual Registration and User Authentication Management Activities

The Manual Registration Activity comprises the following Computational Objects (COs): *User Agent, Registration Manager,* and *User Profile Manager.* The User Authentication Activity, marked with the shaded area, additionally includes the *Authentication Manager* CO. The diagram illustrates the CO's interactions. Operation invocations at CO interfaces together with a sequence number constitute the Management Activities. The operation names are further specified by arguments but not in an exhaustive manner. The operations are now described in the order of their sequence number:

(1) Manual Registration: This operation invoked at the Registration Manager's interface is the head action of the Registration Management Activity. The User CO is the client object providing a target communication endpoint as input parameter. The operation is an interrogation, meaning that the User CO expects a reply message from the Registration Manager CO. The reply message is stored in the output parameter "register state" and informs the user about success of the Activity and the actual registration state.

(2) Request User Authentication: This is the head action of the User Authentication Activity initiated by the Registration Manager CO at the Authentication Manager CO's interface. Within this interaction the Registration Manager takes the client role and the Authentication Manager plays the server role. Again this operation is an interrogation. The Registration Manager CO provides the "personalID" of the user to be authenticated and expects an authentication certificate as a result parameter.

(3), (4b) PIN Request: This is the second action within the Authentication Activity. The Authentication Manager, acting as the client object, invokes this interrogation at the User CO's operational interface. The User CO starts an interactive dialog with the real user via some GUI to request the user's PIN. This interaction is modelled as internal action. The User CO, acting as the server object, finally delivers the requested PIN as an output parameter of this operation.

(4) PIN Check: This operation also belongs to the User Authentication Management Activity. The Authentication Manager CO invokes this interrogation at the User Profile Manager CO's interface, provides the user's PIN and expects a boolean output parameter stating whether the PIN is correct or not.

(4a) Log Authentication Attempt: Depending on the result of the PIN Check operation the Authentication Manager selects this operation or terminates the User Authentication Management Activity with an authentication certificate delivery to the Registration Manager CO. In the case of a negative result of the PIN Check, i.e. the user PIN was incorrect, the Authentication Manager invokes this operation at the User Profile Manager's interface, providing a time stamp and an integer value to increase the counter for the failed attempts. If the attempt counter reaches a threshold value, the Authentication Manager CO terminates the activity with a negative authentication statement.

(4b) PIN Request: This operation is the same as in sequence number (3) and is invoked after a negative result of the PIN Check. The user has the possibility to provide a new PIN, however the maximum number of failed attempts is limited.

(5) Write Attribute: This operation is an announcement, i.e. the Registration Manager CO invokes this operation at the User Profile Manager's interface without expecting a terminating operation, which delivers any results. This operation is only selected in case of a positive authentication result, otherwise the Activity is terminated. The parameters contain the attribute to be set ("registeredManuallyTo"), and the new value of that attribute (e.g. a "roomID").

(6) Read Attribute: The Registration Manager invokes this interrogation at the User Profile Manager's interface in order to retrieve the actual values of the provided Attributes (registeredManuallyTo, registeredAutomaticallyTo, registeredbyScheduleTo). The values of these attributes form the actual state of the Registration and are delivered to the User Agent CO. This action terminates the Registration Management Activity successfully.

The Activity Flow Diagram already provides a level of granularity identifying COs and interactions between them in a detailed manner. However, to overcome the descriptive fashion and to gain a formal and comprehensive computational specification of the identified COs, interfaces and operations, appropriate CO specification templates have to be used, such as those defined in the TINA-C Computational Modeling Concepts [TINA-93]. An example for the Registration Manager CO specification is given below. This represents step four of the proposed design guidelines:

object template Registration Manager;
operations
      void   ManualRegistration (**in roomID** target, **out string** registration state);
          /* This operation invoked by an User Agent starts the Registration Management
          Activity and delivers as a result the actual registration state */
      void   AutomaticRegistration (**in boolean** enabled, **out string** registration state);
      void   ScheduleRegistration (**in boolean** enabled, **out string** registration state);
initialization
      void   init (**out interfaceref** RegMInterface);
required interface templates
      UProfileInterface, AuthentMInterface;
behaviour
      "An instance of this object template executes the Registration Management Activity. It provides operations for three different registration modes. Within the course of the registration process it initiates an instance of the Authentication Management Activity and invokes depending on the authentication process's result, Write Attribute and Read Attribute operations at the User Profile Manager's interface. If the Registration Management Activity selects a success termination, the Registration Manager delivers the actual registration state to the invoking User Agent."
supported interface templates
      interface template RegMInterface;
          operations
             ManualRegistration, AutomaticRegistration, ScheduleRegistration;
          behaviour
             "This is the template of an interface provided by an object that provides operations to User Agents for PCSS Registration in three different modes. "

This specification represents the input for the Engineering Viewpoint modeling and directly prepares a subsequent MS implementation.

4. CONCLUSION

This paper has focused on the aspect of management service design taking into account the function-oriented interface specification methodology of current TMN standards and the object-oriented modeling concepts of emerging ODP standards. Since ODP represents the fundamental framework for the long term evolution of the telecommunications environment towards object-orientation, TMN concepts have to be aligned with ODP standards, in particular in the still evolving area of MS design and specification. Within this paper we have illustrated that both TMN and ODP concepts could be combined for the sake of object-oriented MS design, where TMN interface specification concepts will be used within ODP enterprise modeling, and ODP computational modeling concepts will be used for the object-oriented MS specification. In particular, the proposed mapping of TMN Management Functions onto ODP Activities bridges the gap between TMN and ODP modeling concepts. We thus demonstrated the common use of the TMN functional approach and the ODP object-oriented approach combined in one design methodology. An example modeling of a "User Profile Management Service" has been provided in order to illustrate our approach.

REFERENCES

[Eckardt-94]   T. Eckardt, T. Magedanz: "On the Personal Communications Impacts on Multimedia Teleservices", International Workshop on Advanced Teleservices and High-Speed Communications Architectures (IWACA), Heidelberg, Germany, September 26-28, 1994.

[H400]   RACE Common Functional Specification (CFS) H400: "Telecommunications Management Functional Specification Conceptual Models: Scopes and Templates", 1992.

[Maged-95]   T. Magedanz, R. Popescu-Zeletin, T. Eckardt: "A (R)evolutionary Approach for Modeling Service Control in Future Telecommunications - Using TMN for the Realization of IN Capabilities", International TINA Conference, Melbourne, Australia, February 13-16, 1995.

[M.3010]   ITU-T Recommendation M.3010: "Principles of a Telecommunication Management Network", Geneva, 1992.

[M.3020]   ITU-T Recommendation M.3020: "TNN Interface Specification Methodology", Geneva, 1992.

[PRISM-93]   RACE Project 2041 PRISM Deliverable 4: Service Management Reference Configuration, September 1993.

[Q.1200]   ITU-T Recommendations Q.12xx Series on Intelligent Networks, Geneva, March 1992

[Strick-94]   L.Strick et.al.: "ODP-A Framework for Defining Service Management Reference Configurations" from ICODP '93 Proceeding Papers, 1993.

[TINA-C]   Telecommunication Information Networking Architecture - TINA Consortium, Work Program Proposal, Draft Issue 4, January 1993.

[TINA-93]   TINA-C Doc. No. TB_A2.NAT.002_1.0_93: "Computational Modeling Concepts", TINA-C, December 1993.

[X.500]   ITU-T Recommendation X.500 I ISO/IEC 9594-1: Information Technology - Open Systems Interconnection - The Directory: Overview of Concepts, Models and Services, 1992.

[X.720]   ITU-T Recommendation X.720 I ISO/IEC 10165-1: Information Technology - Open Systems Interconnection - Structure of management information: Management information model., 1992.

[X.901]   ITU-T Draft Recommendation X.901 I ISO/IEC 10746-1: Basic Reference Model of Open Distributed Processing - Part 1 "Overview and guide to use", 1993.

[X.902]   ITU-T Draft Recommendation X.902 I ISO/IEC 10746-2: Basic Reference Model of Open Distributed Processing - Part 2 "Descriptive Model", 1993.

[X.903]   ITU-TS Draft Recommendation X.903 I ISO/IEC 10746-3: Basic Reference Model of Open Distributed Processing - Part 3 "Prescriptive model", 1993.

8

Analysis and design of a management application using RM-ODP and OMT

Erik Colban[a] and Fabrice Dupuy[b]

[a]Norwegian Telecom c/o Bellcore, NVC-1C115, 331 Newman Springs Road, Red Bank, NJ 07701, USA. Phone: + 1 908 758 2875, e-mail: erik@tinac.com

[b]France Telecom / CNET, LAA/EIA/BSA, Technopole Anticipa, 2 avenue Pierre Marzin, 22307 Lannion Cedex, France. Phone: + 33 96 05 36 65, e-mail: dupuy@lannion.cnet.fr

Abstract

This paper studies the assimilation of OMT, an object-oriented design method not particularly well suited for distributed applications development, into RM-ODP, a framework for distributed applications lacking a method. It reports on current work in this venture in the TINA Consortium.

Keyword Codes: D.2.2; D.2.7; D.2.10
Keywords: Software Engineering, Tools and Techniques; Distribution and Maintenance; Design

1. Introduction

Adhering to the RM-ODP viewpoints along with object-orientation is not as easy as it may seem from the reading of the standard and it can lead to different interpretations as to what the objects described mainly from the information viewpoint, the computational viewpoint and the engineering viewpoint really represent.

Since the standard to be RM-ODP [1] [2] [3] does not prescribe any method along with the architectural concepts, a system designer can for example choose to follow an object-oriented analy-sis and design method, without any additional consideration of the RM-ODP viewpoints, and be convinced (somehow rightly) that his/her analysis and design models correspond respectively to the RM-ODP information and computational models. Another designer can adhere to the functional model and describe the functional modules by means of the computational viewpoint and the information exchanged between them by means of the information viewpoint. RM-ODP would perfectly suit his/her needs.

The authors of this paper, and the TINA-C core team [4] working on the software architecture of the future telecommunications information networks, adopted another interpretation or method which consists of taking over an object-oriented analysis and design method, namely OMT [5] and adding a substantial set of RM-ODP concepts to the analysis and design phases in order to take more thoughtfully account of the distribution issues.

This paper describes first the application example that provides the ground for comparison. Then the main sections of this paper aim at showing how the example is examined using the OMT method on one hand and using a mixture of RM-ODP and OMT, as proposed by the TINA-C core team, on the other hand. The conclusion highlights the benefits of the latter method.

2. Example

All along this contribution, the same example, network resource management, will be taken. The problem statement that serves as an input to the analysis and design phases of the methods is proposed to be the following: A computer-based distributed system is required to manage a telecommunication network, regardless of the size and the type of the network (ATM, SDH, POTS,...), in order to react to its faults as seamless and transparently to its users as possible, and to control the correct establishment of end-to-end connections at its various end-points. A topological map of the telecommunications resources that make the network is stored in the system in order for the network manager to be kept updated about the network configuration, to easily locate the established connections or the faults, and to identify the actions to undertake.

It should be noted that the purpose of this contribution is of course not to come up with a complete problem statement, nor to completely show how the sketchy statement given above will be analyzed and turned into a complete object-oriented design. Only parts of this problem statement will be used to exemplify the discourse and therefore will undergo the process of analysis and design outlined below. The focus is mainly on the OMT method, its benefits and its shortcomings related to distribution concerns.

3. The OMT approach to the analysis and design

OMT is only used in this contribution to exemplify the discourse and because the core team gained experience in using it during two years (In no way, this choice should be considered an assessment of the various OOA&D methods).

The OMT object-oriented analysis and design method consists of three non compulsory sequential steps: analysis, system design and object design.

- It is proposed in OMT to analyze a system according to three related but different "viewpoints": an *object model*, a *dynamic model* and a *functional model*. The *object model* represents the static, structural "data" aspects of a system. The *dynamic model* represents the temporal "control" aspect of it. The *functional model* represents the transformational aspects.

- The OMT system design consists in grouping the objects identified in the analysis phase into subsystems, each subsystem being associated later with specific computing resources (DBMS, graphical user interfaces,...)

- The OMT object design consists in refining the objects identified during the analysis phase.

3.1 OMT analysis

During analysis, the OMT object model permits to describe the patterns of the objects, attributes, operations and links. In the case of our example, the object model enables to give a view of the telecommunications network structure composed of layer networks, subnetworks, termination points, to state that the responsibility of managing the network can be given to different kinds of network managers, to outline the difference between link connections established between two subnetworks and subnetwork connections established within a subnetwork. The Figure 3-1. below contains neither attributes nor operations for the sake of clarity.

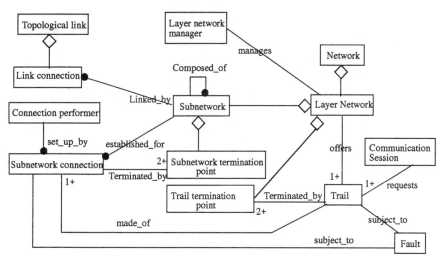

Figure 3-1. An OMT Object Model of the
network resource management system

A complete OMT object model would include, as stated earlier, a description of the attributes and operations of each object. For example, the subnetwork connection is composed of attributes named *direction* (a connection can be simplex or duplex), *cast* (a connection can be unicast or multi-cast), *Quality of Service* (throughput, delay, jitter), and it offers operations to delete the connection, to reserve a connection, to start transferring the media flow through the connection, to associate a connection with subnetwork termination points (SNWTPs).

The OMT Dynamic model permits to describe the states of the objects and the events sent by objects to stimulate the others. In our case, the dynamic model enables to examine the states of the objects identified in the object model and to relate any state change to events (mainly corresponding to operation invocations). The Figure 3-3. shows a state diagram of the subnetwork connection object (inspired from [6]).

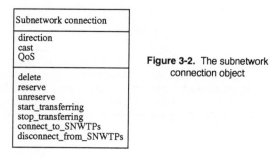

Figure 3-2. The subnetwork connection object

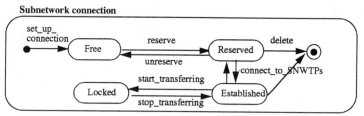

Figure 3-3. A bit of an OMT Dynamic Model of the network resource management application

The OMT Functional model specifies *what* happens, whereas the Dynamic model describes *when* it happens. The Functional model shows how output values in a computation are derived from input values. Consequently, this model describes data flows where the Dynamic model describes control flows. The Figure 3-4. shows an example with some functions associated to the object 'subnetwork connection'.

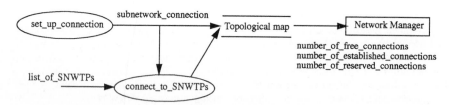

Figure 3-4. A bit of an OMT Functional Model of the network resource management application

3.2 OMT system design

During analysis, the focus is on what needs to be done. During design, the concern is on how the problem will be solved. The OMT **system design** consists in determining the overall structure of the system, a sort of high level design, in which the system is partitioned into subsystems and objects are packaged into these subsystems. The rationale behind partitioning into subsystems is to have the objects share some common properties: similar functionality, same physical location, or execution using the same computing resources.

For the network resource management example, it can be proposed to package the topological links and the link connections into a subsystem called transmission subsystem, the subnetwork, its subnetwork connections and its termination points into a switching subsystem, the trails, trail termination points and layer networks into an operations system, and the communication sessions into a telecommunication service package.

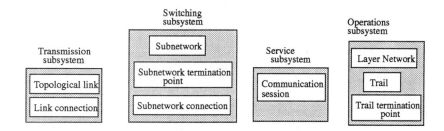

Figure 3-5. A system design of the network resource management system

One main remark is to be made concerning the system design step of the OMT method (this step consisting of packaging objects identified during the analysis phase into subsystems). Usually, when it comes to partitioning the overall system into packages, the packaged objects are executable units (or executable objects), which, by essence, constitute the basic units for structuring the computer-based system. Here, with the OMT method, the assembled objects are the objects directly derived from the analysis phase. As no intermediate step, in which the analysis objects would be turned beforehand into 'executable' objects, is explicitly proposed in OMT, it means these problem domain objects have also to be considered as executable units in OMT. Such a homomorphism between analysis and design objects cannot always hold: for example, the network manager 'human being' cannot be 'packaged' into a computer-based subsystem.

3.3 OMT object design

The OMT **object design** phase determines the full definition of classes, the implementation of the associations (with pointers to the associated objects for example), and the algorithms of the methods used to implement the operations. The objects discovered during analysis serve as a skeleton of the design, the object designer having to choose the ways to implement them, to add new objects for storing intermediate results or translating data representations,... These objects present in the analysis phase are usually carried directly to the design phase. Object design in OMT is then nothing more than a process of adding details and deciding how to implement.

For example, the object model discovered during analysis for the network resource management example is detailed during object design so that all operations and associations become implementable. The subnetwork connection object, for instance, is further designed (Figure 3-

6.) to offer access to its attributes (e.g., get_direction, get_cast, get_QoS) and to implement the relationships in which it takes place (e.g., give_termination_points). The type of the attributes is determined (e.g. *direction* of type enumeration).

Figure 3-6. The subnetwork connection object further refined during object design

3.4 The shortcomings of the OMT approach

In this approach, the OMT analysis stage seems to correspond to the RM-ODP information viewpoint, the OMT object design stage to the RM-ODP computational viewpoint: (the designer is more concerned, at this stage, by how the system works and what to do to optimize it), and the OMT system design phase to the RM-ODP engineering viewpoint (the system is broken down into subsystems or system components that share common properties).

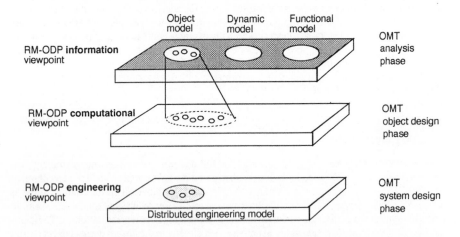

The shortcomings of this method can be essentially summarized as a lack of a framework for distribution:

- There is a suggested one-to-one mapping in OMT between the objects discovered during analysis and the objects eventually designed. It is deemed by the authors that this is not realistic, and too constraining. As RM-ODP does not prescribe such a mapping between information objects and computational objects, it should provide a more flexible framework.

- During the OMT system design, no distinction is suggested to be made between objects with interfaces for local use and objects with interfaces for possibly remote use. Therefore, the distributed processing infrastructure supporting the OMT object-oriented computational model (the OMT object/system design) has to handle all references to objects in the same manner, i.e. as if they were all distributable or none was at all. With the concept of "clusters", within which the communication mechanisms are left to the cluster manager, and between which communication relies on the processing environment kernel facilities (a sort of object request broker), the reference model of ODP provides a noticeable distinction that helps to reduce the distributed environment workload.

- During the OMT object design, all the burden implied by the distribution issues is left to the designer, who supposedly has to know how to deal with the different transaction semantics (which transaction model to take), with security (how to trade off between security and performance), with remote access (how to encode/decode the invocation parameters), and with quality of service procurement. It is clear that, on the subject, the RM-ODP computational model constitutes a substantial improvement of OMT.

- Also, this approach does not take parallelism or concurrent access into account. Besides an information model explaining by means of objects what the system does, further design decisions need to be taken in order to avoid too many accesses to the same object interface, or to allow concurrent processing.

4. The TINA-C approach of using OMT and RM-ODP

As the previous chapter suggests, OMT lacks concepts and principles, such as the ones included in the reference model of ODP, that can help a distributed system designer in his/her task, whereas it constitutes an object-oriented analysis and design method somehow efficient and promising. TINA-C aimed at using both RM-ODP and OMT to provide a complete and tool-supported framework. There are of course several ways in bringing these two together, not all as good. During 1993, TINA-C attempted to clarify these matters for three of the RM-ODP viewpoints: the information, computational, and engineering viewpoints.

4.1 Information Specifications with OMT

In 1994, TINA-C started to use OMT for the information specifications. Certain care must be taken in order to make an information specification design independent. Even if one restricts the exercise to what is traditionally seen as object-oriented analysis, computational aspects are often introduced mistakenly into the specification. Consider for instance, the *Subnetwork Connection* object type in Figure 4-5. A *connect_to_SNWTPs* operation is specified which associates *Subnetwork Termination Points* to a *Subnetwork Connection*. The operation does not pertain only to a *Subnetwork Connection,* but also to the *Subnetwork Termination Points.* One may therefore question why this operation has been assigned to the *Subnetwork Connection* object type and not to the *Subnetwork Termination Point,* or to the *Subnetwork.* As a matter of fact, assigning the operation to *Subnetwork Connection* is a design choice that belongs to the computational viewpoint. The *connect_to_SNWTPs* operation provides the means to add a *Subnetwork Termination Point* to the *Subnetwork Connect.* In the information viewpoint, the analyst should only be interested in stating that *Subnetwork Termination Points* can be associated with *Subnetwork Connections.* In Figure 4-1, this is accounted for by the multiplicity 2+ on the *Terminated_by* association.

For the information specifications, TINA-C suggests to only use the Object Model part of the OMT analysis phase. The reason for this is that it is difficult to reason about any event flows when no computational model is in mind. However, in specifying the information viewpoint, TINA-C goes far beyond simply providing a set of object diagrams.

Objects, in the information viewpoint as well, have a state. The state is defined by the values of the attributes of the object. Therefore, every object type is assigned a set of attributes. The state may change, but only according to a set of specified actions. Note that these actions only pertain to the object to which they belong, unlike the *connect_to_SNWTPs* operation mentioned earlier. If an operation can take place only under certain conditions, these are specified as pre-conditions to the operation. Any property that has to hold after the state change is specified as a post-condition. If there are any conditions that imply a change of state, they are specified as triggering conditions. Note that by specifying pre-, post, and triggering conditions, some of the expressive power of dynamic modelling is covered. Properties that hold regardless of the state of the objects are specified as invariants. In specifying all these conditions the analyst may refer to associated objects; the next paragraph gives more insight.

In order to account for dependencies between objects, we use the associations of OMT (referred to as relationships in TINA-C). These associations may have a state, in which case it is specified in the same manner as for object types. Also important, when specifying an association, are the constraints on the instances of the association or the associated objects. Multiplicities are one kind of such constraints, but there may be others. In specifying these constraints, we may refer to the class (in the sense of ODP) of the association and the associated objects.

In TINA-C an OMT tool has been customized in order to facilitate writing information specifications. With the customizing the tool allows the user to enter in a textual form any conditions that belong to the specifications. It also generates a report in a notation that has been chosen by the Core Team (the notation used is an adaptation of GDMO).

Figure 4-1 shows a small fragment of the TINA-C Network Resource Information Model. Note that the NRIM is under development and that it may change by the time the reader discovers it. Note also that currently the NRIM only includes information objects representing managed resources: it lacks information objects playing manager roles (as tentatively proposed in Figure 3-1.).

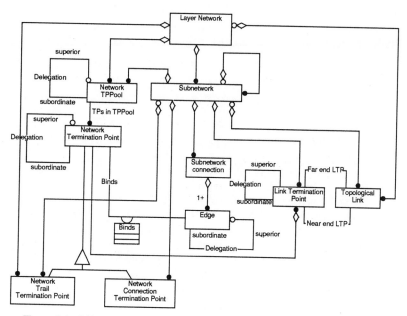

Figure 4-1. A Fragment of the TINA-C Network Resource Information Model

4.2 Computational Specifications with OMT.

Actually, OMT already includes a computational model: as mentioned in Section 3, all object-oriented methods assume a computational model; furthermore, as stated in Section 4, the OMT object design phase partly corresponds to the RM-ODP computational viewpoint. However, enhancements of this computational model are necessary relatively to the following points.

4.2.1 Correspondence between analysis and design objects

In object orientation, "real world" concepts are modelled as object types and phenomena as objects. The purpose is to get a "homomorphism" between the real world and the model or, in other words, to reflect the structure of the real world in the model. This homomorphism should be carried through to implementation. The value of doing this is to produce systems that are easier to maintain and to extend.

These principles of object orientation guide the information modelling and should, as far as possible, also guide the computational modelling. Therefore, a good starting point for the computational specification is to naturally map each information object type onto one computational object type (this is precisely the correspondence embedded in OMT) and, whenever the one-to-one mapping between information objects and computational objects can not be maintained, the designer has to remember that the homomorphism, and therefore one advantage of object orientation, are lost. In this homomorphous scenario, relationship types (corresponding to associations in OMT) may either be mapped onto separate computational object types or one may let the related objects maintain on their own the given relationship.

However, for several reasons given below, the one-to-one mapping is difficult to maintain and may lead to a non realistic distribution-driven computational object design:

- An information object may have to be split and distributed on several nodes thus providing interaction points at several locations. For example, the information object 'subnetwork' is split up into as many computational objects (also called hereafter subnetworks) as there are levels in the network structure recursiveness.

- Information objects that appear in great number can be stored in a data object container and retrieved through a single interaction point. For example, an image of all created subnetwork connections may be stored in that manner. The database manager, called hereafter topological map repository, is seen as one computational object whose interfaces are determined by the possible requests to the database.

- In order to perform certain operations that involve collection of objects, computational objects that have interfaces that offer such operations may be specified. These objects do not necessarily correspond to any information object.

- In an open application it is important to be able to set the limits on how open the application should be. One may need to prevent users from getting direct access to a piece of information to prevent users to tamper on it. The solution is to offer to the users computational interfaces that clearly defines and delimits the possible interactions that may occur.

- Trading is a costly operation that should be minimized. For this reason one might group operations, that can be related, into one interface instead of spreading them on several computational interfaces.

- A transaction is defined as an operation that spans more than one computational object. A transaction is also a costly operation and one should try to minimize the number of them in the same way as trading is minimized. This can be achieved by mapping several information objects into one computational object. For example, grouping all termination points related to a subnetwork within the same computational object 'subnetwork' allow to have the transactional operation *Connect_to_SNWTPs* not span any other computational object.

Due to these differences, separate object models need to be developed for each of the information and the computational viewpoints. Not only are the object diagrams different, but object types will also be specified differently in the computational object model than in the information object model. Object types in the computational object model will for instance be specified with other operations and attributes than the corresponding object types in the information object model. For instance, attributes that serve a object reference holders in the computational object model could correspond to associations in the information object model. See also previous comments about the operation *connect_to_SNWTPs* which is a perfectly valid operation in the computational object model, as shown in the figure below (inspired from [6] and [7]).

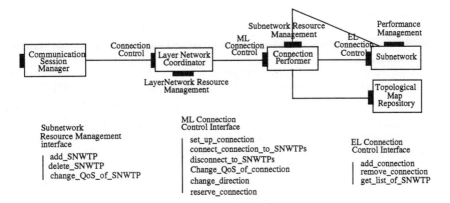

Subnetwork
Resource Management
interface

| add_SNWTP
| delete_SNWTP
| change_QoS_of_SNWTP

ML Connection
Control Interface

| set_up_connection
| connect_connection_to_SNWTPs
| disconnect_to_SNWTPs
| Change_QoS_of_connection
| change_direction
| reserve_connection

EL Connection
Control Interface

| add_connection
| remove_connection
| get_list_of_SNWTP

Figure 5-3. A possible Network Resource Management Computational Model

4.2.2 Extensions of the OMT computational model

In addition, the computational concepts in OMT are not fully complete to allow a good design of a distributed application. OMT lacks concepts to express, for instance:

- objects with multiple interfaces and dynamic instantiation of interfaces,
- stream interfaces through which continuous media flows are permitted (for a lower-level subnetwork called a matrix and conveying user data for instance),
- quality of service like transactional QoS, real-time QoS, availability,
- concurrency control.

In TINA-C, an Object Description Language (ODL) has been defined to enable the designer to specify these aspects of a computational object (ODL is an extension og OMG IDL). We believe that OMT should be enhanced with these concepts although the TINA-C core team has not yet totally carried out any experiments that validate the appropriateness of the language.

4.3 Engineering Specifications with OMT

TINA-C is currently working on the definition of a so-called Life-Cycle Service (as close to the OMG object life cycle as it can be) and on the deployment of computational objects into RM-ODP clusters. LCS and cluster management are used to master the computational object life cycle (creation, activation, deactivation, deletion) respectively at a fine grain (application level) and at a coarse grain (system level). Cluster management constitutes a major part of an engineering specification.

The TINA-C core team believes that the OMT system design is insufficient with respect to engineering modelling. In a one or two year time frame, it should be able to propose a language, aligned with the RM-ODP engineering concepts, that enhances OMT on this matter.

5. Conclusion

OMT is not a method that is targeted towards distributed software development. In order to use OMT in such a context, it is necessary to enhance it with concepts that belong to the area of distributed computing. TINA-C is currently pursuing this route by working on a software architecture intended to include the most interesting concepts and principles of both RM-ODP and OMT.

Further more, the core team is currently customizing an OMT tool so that the user can express various aspects or viewpoints of distributed computing as defined in RM-ODP. So far, the work has been done for the information viewpoint, but similar customizing is planed for the computational and engineering viewpoints. This customized OMT tool may constitute in the future a skeleton of a distributed software engineering tool that would enable to user-friendly and more easily follow the RM-ODP recommendations.

Acknowledgments

The authors would like to thank the core team of the TINA Consortium who contributed to the specifications of the TINA Architecture, Magnus Lengdell (Telia, Sweden)who tested the usefulness of the OMT tool customization and Valère Robin (France Telecom / CNET) who provided valuable comments on this contribution.

References

[1] ISO/IEC 10746-2.2 / ITU Recommendation X.901, *Basic Reference Model of Open Distributed Processing - Part 1: Overview and Guide to Use,* International Organization for Standardization and International Electrotechnical Committee, June 1993.

[2] ISO/IEC 10746-2.2 / ITU Recommendation X.902, *Basic Reference Model of Open Distributed Processing - Part 2,* International Organization for Standardization and International Electrotechnical Committee, June 1993.

[3] ISO/IEC 10746-3 / ITU Recommendation X.903, *Basic Reference Model of Open Distributed Processing - Part 3 - Prescriptive Model,* International Organization for Standardization and International Electrotechnical Committee, November 1992.

[4] Fabrice Dupuy, Gunnar Nilsson, *The TINA Consortium: Towards Telecommunications Information Networking Services,* to be published in the ISS'95 proceedings, Berlin, April 1995.

[5] James Rumbaugh, Michael Blaha, William Premerlani, Frederick Eddy, and William Lorensen, *Object-Oriented Modeling and Design,* Prentice Hall, Englewood Cliffs, N.J.:, 1991.

[6] ITU SG15 Q.30/15 D.272, *Information Specification of the subnetwork connection services,* May 1994.

[7] Bloem, J., Pavon, J., Oshigiri, H., Schenk, M., «*TINA-C Connection Management Architecture,* TINA'95 Proceedings, Melbourne, February 1995.

9

Distributing Public Network Management Systems Using CORBA

Brian Kinane
Ericsson Applied Research Laboratory - Network Management
Broadcom Eireann Research Ltd.,
Kestrel House, Clanwilliam Pl., Dublin 2., Ireland.
Email: bkinane@broadcom.ie

Abstract

In the search for competitive advantage, network management systems are becoming increasingly important to public network operators. In an attempt to reduce the cost of management systems, the telecommunications community have, for many years, been working on the standardisation of interfaces between systems to encourage a multi-vendor environment. As the computing and telecommunications domains converge, there is now a possibility to use distributed object technology to increase the cost-effectiveness of management systems. This paper discusses experiences of using the Common Object Request Broker Architecture (CORBA) as a basis for a public telecommunications network management system platform.

Keyword Codes: C.2.4; D.1.5; D.1.1
Keywords: Distributed Systems; Object-oriented Programming; Applicative (Functional) Programming

1. INTRODUCTION

The International Telecommunications Union (ITU) addresses the management of telecommunications networks through their Telecommunications Management Network (TMN) M.3000 series of recommendations [1]. Although these address the interoperability of management systems, they do not directly provide methodologies or guidelines for actual implementation of systems. Issues such as the infrastructural requirements of management systems including management service implementation and deployment are not addressed by TMN standards.

As the telecommunication and computing domains converge, technologies based on concepts from Open Distributed Processing (ODP) [2] are being applied to the design and implementation of telecommunication management systems. These technologies support the implementation of management applications as distributed heterogeneous systems.

At present, an on-going study, RACE-II project PRISM R2041 [3], is using TMN methodologies in conjunction with the ODP enterprise, information and computational viewpoints for the specification of management systems. The next logical stage is implementation of these specifications using distributed object technologies.

This paper discusses experiences associated with constructing TMN management systems using the Object Management Group's (OMG) Common Object Request Broker Architecture (CORBA) [4]. CORBA is a core component of the Object Management Architecture (OMA) - an architecture which reflects the computational and engineering viewpoints of ODP. The suitability of CORBA as the basis for a telecommunications management platform is evaluated through the implementation of a TMN-based security management application using Orbix [5] - a full implementation of CORBA.

CORBA is targeted at the implementation of systems using heterogeneous components but at present there are only C and C++ standardised language bindings available. As a result, use of CORBA implies use of C/C++. To explore use of other language paradigms with CORBA, Erlang [6], a declarative concurrent functional programming language (FPL) was integrated with CORBA through a partial language binding. It was used, in addition to C++, because of its high-level nature (i.e. it is more formal and closer to specification). The construction of management systems using C++ and Erlang components, integrated via an ORB, is discussed in this paper.

The paper consists of three sections. Firstly, a framework for integration of CORBA within TMN is identified. Based on the resulting framework, a case-study which uses the Orbix CORBA-implementation is described. Finally, the implementation is analysed and conclusions presented.

2. DEVELOPING MANAGEMENT SYSTEMS

While TMN guidelines are effective for specification of management systems and identification of points of distribution and interoperability between management applications, it does not provide a path which leads to implementation of the specification. It is suggested here that the CORBA may be used, in conjunction with TMN, as an implementation platform for TMN specifications.

2.1. TMN approach

A Telecommunications Management Network (TMN) provides monitoring and control of another network. The TMN may be separate or share facilities of the network it manages. According to the ITU-T recommendation X.700 [7] a management network should perform five functions - configuration, fault, security, accounting and performance management.

The TMN recommendations standardise some of the functionality and many of the interfaces of management software. This is intended to enable software from different vendors to interoperate within a TMN and to enable the exchange of management information between TMNs of different organisations. The reference architecture of a TMN is defined in the ITU base recommendation M.3010 [8] "Principles for a Telecommunications Management Network".

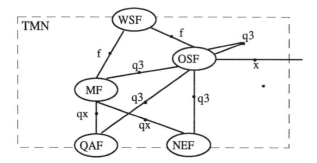

Figure 1 - TMN Functional Architecture showing Operations System, Mediation , Q-Adapter, Network Element, Work Station Function blocks and reference points.

The TMN functional architecture consists of a number of **function blocks** - the Operations System (OSF), Network Element (NEF), Q-Adaptor (QAF), Mediation (MF) and Work Station (WSF) function blocks (TMN functional architecture in figure 1) . Each of these blocks performs a particular function and cooperate to provide management services. Between each pair of communicating function blocks there exists a TMN **reference point.** The reference point (q3, qx, x, f in Figure 1) defines the 'service boundary between two management function blocks' and is a potential point of physical separation between systems (provided by different vendors or existing in different organisations).

Building blocks are actual systems which implement the above function blocks. Interconnection between these TMN building blocks is facilitated by a set of standard inter-operable **interfaces** . These interfaces are defined with respect to reference points and specify what information a building block should present and what means of communication it should use.

2.2. Distributing TMN systems

TMNs are physically distributed information processing software systems. Such systems are realised using heterogeneous technologies from different vendors and hence present substantial inter-operability problems. Specifying a management system using the TMN guidelines identifies function blocks which cooperate to provide particular services. According to TMN these blocks can physically exist separately. Support is provided to define standardised interfaces to these functions blocks in order to enable different function blocks to interoperate.

Building blocks can be further decomposed into functional components which are a set of simpler and more generic Management Software Components (MSCs). Little attention is given to defining interfaces, implementation and distribution of these components. It is contended, in this paper, that building blocks will be implemented as heterogeneous MSCs distributed over heterogeneous architectures. ODP addresses such distribution issues. By mapping these components to a model such as the ODP computational viewpoint, these logically distinct functional components can be defined and implemented, using ODP 'support environments' (e.g. CORBA, ANSAware), as physically distributed management software objects with well defined IDL interfaces. This enables a commodity environment where MSCs are re-used and combined to rapidly construct and customise management systems.

The Object Management Architecture (OMA) and specifically the Common Object Request Broker Architecture (CORBA) has the potential to be an implementation medium for TMN applications since it offers a location/technology transparent object distribution environment which overlaps with the computational, engineering and technology viewpoints.

2.3. CORBA

The goal of the Object Management Group is 'to develop a set of standard interfaces for inter-operable software components'. This is being accomplished through the OMG's Object Management Architecture (OMA) Reference Model which is a model for object management. In this model, application objects communicate with objects that provide common facilities and with low level object services through a communications infrastructure called the Object Request Broker (ORB). The ORB is a location transparent, distributed object platform and the Common Object Request Broker Architecture (CORBA) standard is a specification defining standard interfaces to the ORB.

2.3.1. Orbix

Orbix is a full implementation of the Object Management Group's Common Object Request Broker Architecture standard, developed by Iona Technologies Ltd. Orbix provides a C++ language binding for CORBA and is supported on SunSoft SunOS, Silicon Graphics IRIX, HP/UX and Windows NT. Orbix is being ported to Windows 3.1 and other Unix platforms. Orbix interoperates across all supported platforms.

2.4. Role of CORBA within TMN

In the TMN model, the distribution of management systems is at building block granularity. Each block provides particular services such as configuration or accounting. These services are accessed through an information model over the Common Management Information Service/Protocol (CMIS/P) [9]. The building block is considered a monolithic system which presents a TMN management interface.

CORBA extends this above concept of a building block. Instead of being monolithic, it is a set of cooperating objects distributed via CORBA. The TMN management interface is provided by the 'gatekeeper' object through which services offered by the building block are accessed (see Figure 2). Consequently, two layers of distribution exist;

- TMN distribution between systems belonging to different organisational domains or built by different vendors;
- CORBA within the building block inter-connecting both generic and specific MSCs which cooperate to provide the service offered by that building block.

The building block is a unit of vendor and domain inter-operability whereas CORBA provides component distribution and integration within the building block.

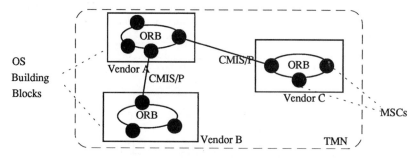

Figure 2 - TMN and CORBA Distribution

This approach raises a number of issues:

- Why not replace TMN distribution with CORBA?
- How does the TMN and the CORBA model interwork?
- How is the building block structured in terms of the OMA model?

TMN distribution is required as it is the most accepted mechanism for achieving shared management knowledge (SMK) between different systems. CMIP is standardised for the Q3 interface and is being used today. In addition, a large amount of work has been done in areas such as OSI security. CORBA, can however, support TMN distribution by providing a distribution and component integration platform for building blocks. It is likely that much of the new computing technology will be accessible over ORBs [10]. Using the ORB as a building block platform will enable the use of this technology in management applications.

In a CORBA-based system, OMA application objects interact in a client server model invoking methods from the operational interface of other application objects. Common Facilities (CF) objects and object services provide platform support to these management oriented application objects. For CORBA to interact with TMN, a mapping from CMIS requests on the management information model (MIM) to requests on server objects is required. This request initiates an activity (chain of operations on the MSC application objects) which upon completion will manipulate the TMN interface to cause a response to be sent to the manager application.

TMN describes the functional decomposition of building blocks in the form of functional components which include:

- **Message Communication Function** (MCF) which provides communication over a CMIS/P interface in either a manager or an agent role;
- **Management Information Base** (MIB) which contains the available management information;
- **Management Application Function** (MAF) which performs processing capability and implements the management services;

- **Presentation Function** (PF) and **Human Machine Adaptor** (HMA) which combine to transform and present management information in a human comprehensible format.

These functional components are implemented as application objects on the ORB platform. Building blocks should have the structure shown in Figure 3.

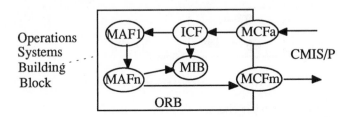

Figure 3 - Building Block as a set of objects

In this structure, an ORB object acts as an *MCF* (or gatekeeper) and in conjunction with the *MIB* object provides a TMN agent interface to the building block. An *MCFm* object is required for *MAF* objects to use services from other building blocks (manager role). When a request is received via the *MCFa* (agent role), it is transformed into an operation on a *MAF* object by the Information Conversion Function (ICF). This causes a chain of further requests (activity) on objects within the building block which upon completion will cause a response to propagate back to the *ICF* and be transformed to a TMN response. Both *MAF* objects and the *ICF* object can query and update the *MIB*. A specification process such as the PRISM methodology could be used to define the object interfaces and MIB specification which could also be used as the basis for the TMN Management Information Model.

In this model, the building block becomes a unit for deployment of objects which interact to provide a service. Certain attributes such as location, security and throughput are associated with the building block. Attributes such as throughput are requirements that must be met by the deployment environment (e.g. hardware, software). Others, such as security, relate to the interface of a building block. Since the building block is a closed set (objects are not directly accessible outside the building block), these attributes are determined by the TMN interface of the *MCFa* (gatekeeper) object.

2.5. Implementation with CORBA

As CORBA matures, standardised facilities such as Object Oriented (OO) database object adaptors will become available. These can be used to support MIB implementation. More language bindings such as Smalltalk will be standardised enabling a multi-language in addition to the present multi-hardware environment. At present, however, CORBA implementations offer only C and/or C++ standard bindings. C++ is a low-level language. As a result, programmers have to cope with low-level issues such as memory management instead of the logic of their application. While C++ can be used to develop robust applications, it is not, in the opinion of this author, an ideal language for implementation of management application functionality. This is partly due to its lack of built-in primitives for concurrency and fault-tolerance.

It was decided to address these issues by using C++ in conjunction with another language technology for implementation. Erlang, a functional programming language (FPL), was chosen because it provides comprehensive support for real-time concurrency and fault tolerance and because the declarative nature of FPLs reduces the gap between specification and implementation. CORBA purports to offer the capability of task driven language selection. This was investigated through the ORB-supported integration of Erlang and C++ components.

2.5.1. Erlang

Erlang is a concurrent, real-time, declarative, and functional language intended for the implementation of real-time industrial control systems. It was initially developed by Ericsson for the implementation of telecommunication systems.

Erlang provides a number of primitives which support real-time, fault tolerance, code modularisation and hot replacement of code. In addition, since Erlang is a functional programming language it has the property of variable single assignment. This has important ramifications for the application of formal methods to Erlang specifications and also code reliability. Function selection is made by pattern matching which leads to highly succinct code.

Erlang has a process based model of concurrency. Concurrency is an explicit and natural part of the language and encourages design of systems as numerous lightweight processes. Message passing between processes is asynchronous and is based on Communicating Sequential Processes. The use of processes overcomes a number of the problems associated with using functional languages.

Erlang provides a distribution mechanism based on TCP/IP which supports the location/ technology transparent distribution of processes. Application processes can be distributed over a heterogeneous network without affecting the semantics of the process interaction.

3. APPLICATION CASE STUDY

3.1. Scope

In order to evaluate the suitability of the above framework (i.e. ORB distribution within TMN and the resulting building block structure) a detailed case-study was conducted. This case-study consists of a TMN conformant security management application which provides management services that configure and monitor the security controls of a service layer management system. The demonstrator consists of the following:

- A Value Added Service Provider (VASP) Operations Systems (OS) building block providing end-to-end virtual leased lines to Customer Network Management (CNM) building blocks over an X interface;

- A Security Manager OS which provides security management services for the VASP OS's security mechanisms using Management Application Functions (MAFs). The Security Manager is implemented in Erlang and C++;

- A Security Manager WorkStation (WS) building block (implemented in Erlang) which provides a Graphical User Interface (GUI) to a human operator.

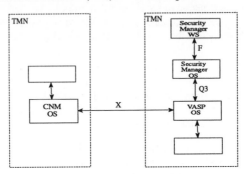

Figure 4 - Case-study scenario

Communication between the Security Manager OS and VASP OS is via the Q3 interface. The F interface enables communication between the Security Manager OS and WS.

3.2. Specification of the management application

Before any implementation could begin it was necessary to specify the requirements on the security manager. As TMN offers a framework for defining and specifying management functionality and their relationships, its guidelines were followed:

The services provided by the application are defined as follows:

- Visualisation and configuration of the access controls of the OS;
- Monitoring and reporting of access violation notifications emitted from the OS.

Once the services have been defined, the building blocks and the constituent functional components (which are needed to provide these services) are identified. The building block and functional component inter-relationships must also be specified. After this analysis phase, a reference configuration for this scenario was formed.

The chosen reference configuration for the Security Manager consists of two building blocks - an Operations Systems building block which provides the actual management services and a Work Station building block which enables the operator to interact with the provided services. These building blocks are composed of a number of functional components such as Message Communication Function (MCFs), Management Application Function (MAF) and Management Information Base (MIB) components.

Mapping functional components into objects

In this demonstrator the functional components were mapped directly into OMA application objects. These objects are grouped into packages based on the function block to which they belong. The relationship between the objects is client-server and the specific relationships are derived from the manager-agent roles used by TMN.

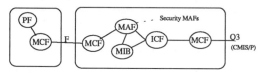

Figure 5 - Object specification for case-study management application

Applying this approach to the reference configuration the MIB, MCF and MAF functional components map to objects (see Figure 5). Where a function block is realised as a building block, the MCF application object provides the TMN management interface.

Defining the object interfaces

Once the application objects have been defined, interfaces for them can be specified. These interfaces are used by other objects to access the services offered by an object acting in a server role. The Object Management Group's Interface Definition Language (IDL) was used because it offers an object oriented abstract specification language and because all CORBA implementations support IDL. An IDL interface class was specified for each object - its methods defining services offered by that object.

3.3. Implementation of the object model

The transformation of the TMN management application specification into an object model (where each component has a clearly defined interface) enables the use of distributed object techniques to be applied to the implementation.

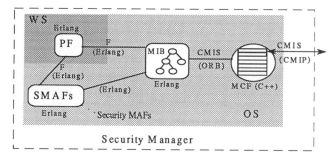

Figure 6 - Implementation Specification

In the implementation, which is shown in figure 6, each Management Software Component was implemented as either an Orbix object or an Erlang process interacting via a combination of Orbix and the Erlang distributed environments

The Presentation Function, Management Application Function and Management Information Base Functional Components were implemented as Erlang processes and the CMIS Message Communication Functional Component was implemented as a C++ object encapsulating a sourced C++ based ISODE CMIS/P stack [11] (This is further discussed in section 4).

3.4. Integration of the Erlang distributed process environment & CORBA

In order for Erlang to be used as part of the CORBA environment, a partial binding from Orbix/IDL to the Erlang environment had to be developed. This was possible due to the similarity of the models:

- Both models consist of active objects (processes in Erlang can be viewed as active objects);
- Communication between these active objects is via message passing (method invocation is a form of message passing) which can be either asynchronous or synchronous;
- Active objects present clearly defined interfaces.

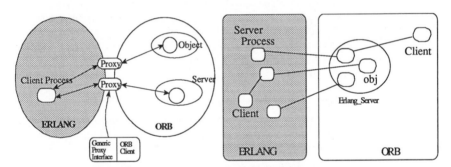

Figure 7- Representation (a) ORB objects to Erlang (b) Erlang processes to ORB

Representing resources in the ORB as native Erlang resources requires the mapping of objects which exist within the ORB environment into processes which exist within the Erlang environment. These processes act as Erlang proxies (i.e. a local entity which acts on behalf of a remote entity) for the remote objects (as shown in Figure 7(a)). Within the Erlang domain a proxy appears as a normal Erlang process. Simultaneously it appears as an ORB client in the ORB domain. When a process, acting as a proxy for a remote object within the ORB's domain, is created in Erlang, any messages sent to the process will be transformed into methods invoked upon the remote object. If the method returns a result then this result is mapped into an Erlang *term* (i.e. data structure) and is sent to the client process. The only ORB specific information an Erlang client needs is the name of the remote object which is required when spawning a proxy.

Access by non-Erlang clients to Erlang processes via the ORB is of equal importance. To achieve this, it is necessary to represent Erlang processes as objects in the ORB environment. Erlang processes make themselves available to ORB clients by registering as objects within the ORB domain as depicted in the figure 7(b). Any methods invoked on such an object will cause an equivalent message to be sent to the associated Erlang process. The return message, if any, will be sent to the ORB client application via the result of the method invocation.

4. ANALYSIS

Implementation of the demonstrator allowed a number of issues relating to the use of distributed environments in supporting TMN applications to be considered. These issues cover a range from management services development to implementation of management systems.

4.1. Developing application on a CORBA based platform

Any telecommunications platform should support development of management applications. By using CORBA, it was possible to use its support for object oriented software construction in management application development. IDL proved to be a very useful abstract language for defining interfaces to the Management Software Components (MSCs). It is independent of implementation. Once MSCs were associated with the interface, the ORB provided trouble free integration of these components. The ORB interface repository enabled the development of components to proceed without having to see the class hierarchy of the particular interface class being used. The main difficulty in using the ORB was providing a TMN (CMIS/P) interface to the application objects. This required mediation to convert requests on the management information models and requests on the application objects. Mediation was hardwired in the demonstrator (but more automatic mediation is being investigated by a joint Network Management (NM) Forum / OMG committee at present [12]).

Re-use of components is essential to rapid construction of systems and was well supported by the CORBA implementation used here. It is possible to bind to a required type of MSC using its interface name. The ORB locator will find an appropriate active object and will launch one if none are available. One feature lacking in Orbix was an interface repository browser. More tools such as browsers are required.

Flexibility is well supported through mechanisms like the Dynamic Invocation Interface. Objects can learn about new services and make use of them without recompilation. In addition, once an object is registered with the interface repository, it is available to all ORB clients in a black box (binary) format.

Support for new services can be achieved through the facility provided by CORBA of hot addition and replacement of MSCs. In this way a MSC that adds new functionality may be added at run-time. In addition, a new service may be developed by using different combinations of existing MSCs. Orbix allows numerous instances of an interface to exist. Because of its unique naming structure for objects, they can be easily distinguished.

Finally, CORBA enables the re-use of existing management tools and components such as the ISODE CMIS/P stack and manager/agent code through encapsulation with C++ wrapper classes. Re-use is not limited to management software. As new object services and third party Common Facility (CF) objects become available in the computing community, those telecommunications vendors using CORBA will have a more structured method for integration of new computing technologies and products into their platforms.

4.2. Orbix support for management platform requirements

Management applications impose a number of requirements such as distribution and fault tolerance, scalability, technology independence and security. Orbix, as a full implementation of CORBA, was evaluated against these requirements. The beta-version of Orbix, used in the demonstrator, provides the basis for the experiences outlined in the paper. Continued development of Orbix has since addressed a number of issues raised in this section.

Orbix provides a fully *distributed* location/technology independent platform. As it uses a XDR/TCP/IP communication protocol, full distribution via the internet is supported. Orbix also provides binding, brokering and limited trading services. This enables the platform to provide transparent distribution to management applications.

CORBA is clearly targeted at achieving *independence from the computing environment.* Orbix supports transparent access to objects over a variety of operating systems e.g. SunOS, Solaris, HP-UX, Windows NT. This is a very useful feature to management system vendors in reducing the costs of their management systems.

Controlled scalability is supported by the Orbix in the sense that services can be instantiated an arbitrary number of times. Extra support such as load-balancing objects is required to control when and where new instantiations should occur. In the demonstrator this is done manually by registering new object-servers on extra hosts. Vendors can, using a CORBA-based platform, scale the processing power (e.g. number of workstations) for their applications to suit the requirements of the customer's network.

Fault tolerance was not well supported by the ORB used in the demonstrator. Persistence for objects was not easily accomplished. The entrance of companies specialising in fault tolerance into the CORBA market (e.g. ISIS has since been integrated with Orbix) in conjunction with the OMG's efforts with object services should rectify this problem.

An important requirement of management platforms that Orbix lacked was *security* (e.g. authentication at binding). There are, however, other ORBs which provide object security as a value added feature. The OMG is looking at this and it will become a standardised object service. In the demonstrator, OSI-based security control was performed by the Message Communication Function agent object (which acts as a gatekeeper to the building block).

4.3. CORBA & Erlang

Erlang offers a fully distributed location/technology transparent process (active object) environment. It is not, however, open. There are no standardised interfaces for Erlang components, no standardised object model or standardised interface definition language, although the Erlang environment could potentially be used as a technology basis for a 'ODP support environment'.

From the perspective of this paper, Erlang as an adjunct (via an IDL to Erlang binding) offers a more effective method than C++ for implementing the control functionality of systems. This is, primarily, due to the declarative and functional paradigm to which Erlang conforms.

Using a functional approach, programs are implemented by specifying the relationships between the input set and output set of a function rather than describing imperatively how the function accomplishes its state transformation. With Function Programming Languages, data is not stored explicitly by the programmer. Instead, garbage collection is done by the environment. This form of environment has advantages for implementing MSCs.

- Due to the declarative nature of Erlang, the code required and development time required in this demonstrator to implement MSCs is less than C++ (e.g. Erlang provides automatic memory management);

- As a MSC implemented in Erlang is defined in terms of function clauses, the state of the MSC and the possible transitions are specified explicity and transparently. Consequently, the behaviour of the MSC is defined in a more formalised manner and is more tractable. In the demonstrator, Vienna Design Method (VDM) specifications for Directed Acyclic Graphs (DAG) were used for the Management Information Tree (MIT) operations.

It was found that the transparent integration of C++ and Erlang via CORBA was useful. Existing OSI/TMN software utilities (e.g. ISODE) could be encapsulated within C++ classes and made accessible over the ORB, in a manner analogous to a toolbox. Erlang was used for the complex logic of the Management Application Functions. Erlang was also used for the Management Information Base. In retrospect, C++ or an Object Oriented database would have been a superior solution due to Erlang's lack of support for inheritance (although there is work on-going at present at developing an OO extension to Erlang).

5. CONCLUDING REMARKS

The Object Management Architecture and specifically CORBA provides a viable platform for implementation of TMN conformant management systems. The benefits of this approach are two-fold. By using CORBA, management specification methodologies based on the synergy of ODP and TMN (e.g. PRISM) can be mapped more easily to physical implementation. This requires the convergence of the OMA with the computation and engineering viewpoints of ODP. Secondly, because CORBA is an accepted middleware standard, the use of CORBA as a telecommunications platform will support the incorporation of advanced (and out-sourced) computing technology (e.g. object oriented databases) into management systems. More work is required, however, on the integration of TMN specifications and the OMA object model.

The relationship between GDMO and the OMA object model (IDL), including mediation, is being addressed by a joint Network Management Forum / OMG task force at present.

New language bindings need to be developed for CORBA IDL to take advantage of languages that provide more formalised methods for specifying object behaviour. The use of Erlang in this project has shown the benefit of using declarative FPLs for implementing MSCs. The combination of Erlang, C++ and CORBA is a useful advance in the development of telecommunication management applications. We found that the use of Erlang significantly improved development productivity and code reliability of the Management Application Functions.

REFERENCES

[1]   K. Shrewsbury, "TMN in a Nutshell", Network Management Forum, 1994

[2]   ISO/IEC JTC 1/SC 21/N 7053 (CCITT X.901) Draft Recommendation, Basic Reference Model for ODP - Part 1: Overview and Guide to Use, December 1993

[3]   RACE R2041 Prism Deliverable 8: Reports on Selected Areas of the Service Management Reference Configuration, September 1994

[4]   Common Object Request Broker : Architecture and Specification OMG Document Number 91.8.1

[5]   "Orbix - A Technical Overview", Iona Technologies Ltd., July 1993

[6]   J. Armstrong et al., "Concurrent Programming in Erlang", Prentice Hall 1993

[7]   X.700 Recommendation I ISO/IEC 7498-4 "OSI - Basic Reference Model - Part 4: Management Framework"

[8]   ITU-T Recommendation M.3010, "Principles For A Telecommunications Management Network", November 1992

[9]   ITU-T Recommendation X.710, "Common Management Information Service For CCITT Applications"

[10]   J. Stikeleather, "Why Distributed Object Computing is Inevitable", Object Magazine, March-April, 1994

[11]   G. Pavlou, "Implementing OSI Management - A Tutorial", Dept. of Computer Science, University city London

[12]   C. Ashford, Comparison of the OMG and ISO/CCITT Object Models, The Report of the Joint NM Forum/OMG Taskforce on Object Modelling, April 1993

SESSION ON

Trading

10

Designing an ODP Trader Implementation using X.500

Andrew Waugh[a] and Mirion Bearman[b]

[a]CSIRO Division of Information Technology,
Research Data Network Collaborative Research Centre[*]
[b]Faculty of Information Science and Engineering, University of Canberra,
Collaborative Research Centre for Distributed Systems Technology Centre[*]

This paper describes how an ODP Trader can be implemented using the X.500 Directory Standard. It differs from previous work in this area as we concentrate on the details of the X.500 interface for an ODP Trader, rather than the implementation of the Trader agent itself. We take the ODP specification and describe how the functionality can be supported using X.500. Because of the different data and operations models it is not possible to directly replace an ODP Trader with X.500. Instead, X.500 is used to store the Trader information, including the Service Offers, Links to interworking Traders, and information about the Trader itself. The Trader functionality is implemented using a specialised X.500 user agent, the T-DUA. The T-DUA uses the X.500 operations, particularly search, to interrogate and manipulate the Trader database. Use of X.500 in this way simplifies the implementation of the Trader.

Keyword Codes: C.2.4
Keywords: Distributed Systems

1. INTRODUCTION

The Reference Model for Open Distributed Processing (RM-ODP) defines a "coordinating framework for the standardisation of ODP by creating an architecture which supports distribution, interworking, interoperability and portability." [5]. The RM-ODP identifies a set of functions that are required to support Open Distributed Processing. One of these is the Trading Function. It is realised by an ODP Trader, which matches objects providing services to objects wishing to use those services. Thus, a trader provides the means for advertising service offers and the means to discover service offers by service requests. Part of the requirements for the progression of the ODP Trading Function standard to Committee Draft status was the inclusion of an Annex which describes how an ODP Trader could be implemented using the X.500 Directory standard. This paper describes the work which led to that Annex.

The underlying data models of both X.500 and the ODP Trader are very similar. Both represent information as type/value pairs. In both, these type/value pairs are collected into objects which can be manipulated as units. This similarity means that much of the complexity of an ODP Trader implementation can be avoided if X.500 is used to store the Trader information.

Our design principle was to attempt to make maximum use of the features of X.500 while implementing exactly the functionality required by an ODP Trader. However, we quickly recognised that the functionality required by the Trader was more sophisticated in many areas than could be directly supported by X.500. For example, Trader service properties may be 'indirect', that is, the Trader does not store the value of the service property, but retrieves the

---

[*]   The work reported in this paper has been funded in part by the Cooperative Research Centres Program through the Department of the Prime Minister and Cabinet of the Commonwealth Government of Australia

value from the Exporter when required. Such differences in functionality led to an implementation design that uses X.500 as a database to store the Trader Information Object. A special Directory User Agent, the Trader DUA (T-DUA), implements the functionality required by an ODP Trader on top of X.500. This permits maximum use of the features of X.500 while still implementing the functionality required by an ODP Trader.

The X.500 Trader Information Object stores the information in one of four types of entry:

† the Trader Entry, which stores information about the characteristics of the Trader itself,

† the Trader Policy Entry, which stores information about the policies which govern the behaviour of the Trader,

† the Trader Link Entries, which store information about other Traders with which this Trader interworks,

† the Service Offer Entries, which store information about the Service Offers known to this Trader.

The X.500 protocol is used to search and manipulate this information. The use of X.500 in this fashion means that the T-DUA does not have to implement an object based database, nor to implement functions to perform complicated manipulations on objects in this database. This simplifies the implementation of an ODP Trader.

The functionality which could not be implemented directly using X.500 includes:

† The Trader Operations. The T-DUA is responsible for mapping each Trader Operation into one or more X.500 operations.

† Indirect Properties. These are retrieved by the T-DUA which is consequently responsible for the final evaluation of candidate Service Offers retrieved from the X.500 Trader Information Object.

† The Selection Preference. As X.500 provides no support for ranking search results, the evaluation of the selection preference is the responsibility of the T-DUA.

We plan to implement the Trader design presented in this paper. Apart from validating our design, this will assist in defining and clarifying concepts in an ODP Trader specification. These include, for example, the functionality required of: access control, the Type Manager, and interworking (federation) of Traders.

2.   ODP TRADER

The concepts and terms used to describe the Trader and its interactions in this paper are taken from the ODP Trading Function Standard [1]. The following description introduces this Trader model. Only those concepts needed in this paper are described.

An object which advertises a service for use by other objects *exports* a description of that service to a *Trader*. This effectively forms an advertisement for that service. The Trader stores these description in its *Trader Information Object*. *Importing* involves an object needing a service sending a description of the desired service to the Trader. The Trader searches its Information Object for matching Service Offers. These are returned to the importer, which selects one and connects directly with the server. The Importer may request the Trader to sort the returned service offers, or even select the 'best'.

A Service Offer describes the type of service on offer and the *Properties* associated with that service. In addition, a service offer can include *Service Offer Properties* that are associated with the offer itself. For example, a service offer of a printing service can have a Service Property describing the quality of the output and an Offer Property describing the expiration date of the offer. The properties are described by name/value pairs where several values may associate with a name. The properties which can be included in a particular Service type are controlled by the *Type Manager*, which is logically separate from the Trader.

Some properties included in a particular Service Type may be defined in the Type Manager to be *indirect*. The Exporter does not include values for these properties when a Service Offer is exported. Instead, it specifies an interface at which the values can be dynamically obtained. This interface is known as the *Service Offer Evaluation Interface* (SOE Interface). When the

Trader needs to evaluate whether or not a particular Service Offer satisfies a particular Import request, it contacts the SOE Interface and retrieves the values of the indirect properties.

Traders may be linked together so that an Import request to one Trader may result in the return of a Service Offer exported to a different Trader. This is referred to as *interworking*. To enable interworking, a Trader also stores a database of *Trader Links* in addition to the database of Service Offers. Links contain properties which describe the linked Traders. These properties are evaluated when a Trader decides to follow a link to another Trader when attempting to fulfil an Import request. Figure 1 shows the interactions between the Importer, Traders, and Exporters.

The Importer can specify the following in an import request:

†   *Service Description* - specifies the type of Service required.

†   *Service Property Names* - specifies a list of Service Properties to be returned with the selected Service Offers. The default is no Service Properties to be returned.

†   *Service Offer Property Names* - specifies a list of Service Offer Properties to be returned

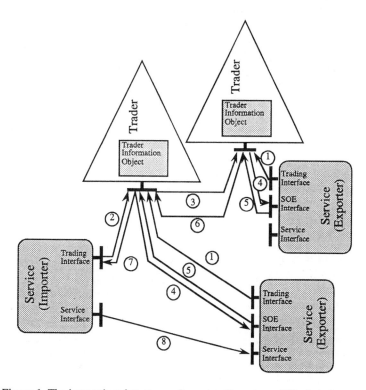

Figure 1. The interactions between an Importer, Exporter and Traders. 1: Exporter exports Service Offer to Trader. 2: Importer requests Service Offer. 3: Trader forwards Import request to Interworking Trader. 4: Trader retrieves values of indirect properties. 5: Exporter responds with indirect properties. 6: Interworking Trader responds with Service Offers. 7: Trader responds with ordered list of Service Offers. 8: Importer requests service from Server.

with the selected Service Offers. The default is no Service Offer Properties to be returned.

† *Import Policy Constraint* - describes the expectations of the scope of the services imported. Little detail is specified for this constraint in the Trading Function Standard. We found it useful to separate Import Policy Constraints into three categories, which we have named:

- Importer Trader Constraints. These are constraints on the Trader using Trader properties, for example the ownership of the Trader. If the Trader cannot satisfy these constraints it does not undertake the Import request
- Importer Link Constraints. These are constraints on any Trader which is passed the Import request by the original Trader when interworking. Again, an example is the ownership of the linked Trader. These constraints are expressed as link properties and are applied by the first Trader, not by the interworking Trader.
- Importer Import Constraints. These are constraints on the search of the local Service Offers, for example an upper bound on the number of offers to be returned.

† *Matching Constraint Criteria* - specifies the filter to be applied to the Service Offers of the desired type to determine whether the Offer is of interest. This is a boolean expression on the Service and Service Offer properties.

† *Selection Preference* - specifies an ordering function to be applied to the matched Service Offers. This function specifies which matched Service Offers are to be preferred. This is an expression on the Service and Service Offer properties.

3.   X.500

X.500 [2, 6] is a standard that allows the construction of very large, distributed, directories. An X.500 directory is implemented using two entities: DUAs (Directory User Agents) and DSAs (Directory System Agents). The DSAs are the servers and store the directory information. The DUAs are the user agents.

X.500 is a directory and stores information about objects: in a Trader, objects include Service Offers, Trader Links, and the Trader itself. These objects are represented as entries. Each entry contains the information about one object. Entries are equivalent to records in a conventional database. The information in an entry is represented by attributes. Each attribute contains a fact about the object. Typical facts are the object's name, it's location, and description. Attributes are similar to fields in a conventional database; however they have some differences. An attribute may have more than one value. The telephoneNumber attribute, for example, can store several telephone numbers. The attributes that may be contained in an entry of a particular type are controlled by the X.500 schema. Changing the schema does not require shutting down the entire directory, nor need the entire X.500 schema be changed at once. X.500 allows new attributes and types of entries to be defined at any time.

Entries are arranged hierarchically and form nodes in a Directory Information Tree (DIT). Typically, this tree can reflect the structure of the directory information. Within an organisation, for example, the tree can represent the organisational hierarchy with the following levels: country, organisation, organisational units, staff. Underneath the root of the tree, for example, there can be entries for each country. Then, under a country entry there can be entries for each organisation in the country. This hierarchical organisation allows the directory information to be distributed amongst the many DSAs (servers), and yet still allows quick identification of the server which holds a particular entry.

X.500 provides operations that allow (among other things) a user to read the contents of an entry, to search a portion of the tree for entries which satisfy certain criteria, to create entries, and to modify the contents of an entry.

4.   RATIONALE

Using X.500 to support an ODP Trader has some significant attractions. These include:

†   The re-use of existing infrastructure. X.500 implementations exist and are being deployed in organisations. A significant amount of investment has already been made in X.500 technology by both software developers and organisations using X.500. Using X.500 as a base to support an ODP Trader allows us to exploit this investment, potentially significantly reducing the cost of deploying an ODP Trader system.

†   The underlying information model supported by X.500 is similar to that required by an ODP Trader. The concept of X.500 attributes, for example, is very similar to the Trader concept of properties. In particular, an X.500 attribute may contain multiple values, which is different from most conventional databases.

†   X.500 was designed to allow administrators considerable flexibility in changing the schema without either needing to bring down the directory, or requiring that the schema be changed throughout the distributed directory at one time. This would allow similar flexibility in the administration of an ODP Trader Service Type definitions.

†   An ODP Trader could use the general X.500 infrastructure to look up network addresses of linked Traders and Clients. It could also use the security features of X.500 to authenticate users and control access to Trader information.

If an existing X.500 system is used as the base for the implementation of an ODP Trader, it is unnecessary to re-implement these functions in the Trader. This significantly simplifies the work of implementing the ODP Trader.

The goal of this project was to design an ODP Trader implementation using X.500. Thus, the functionality to be supported was fixed by the Trader group and we could not change it. Similarly, the functionality supplied by X.500 was also fixed. The design has to map between what is supplied by X.500 and what is required by an ODP Trader specification. In our design, we aimed to maximise the use of X.500 functionality.

5.   USING X.500 AS THE TRADER

Our first design option was to use X.500 as the Trader. Unfortunately, this option was not feasible as the model used by X.500 and the model used by the ODP Trader differ significantly in a number of ways. These include:

†   The functional model. The fundamental unit in X.500 is the entry. The X.500 operations consequently operate on entries, for example, reading information from the entry, adding new entries, or searching for entries which match a specified filter. The fundamental units in an ODP Trader model are Service Offers, Trader Links, Trader Policies and Trader Properties. The Trader operations consequently manipulate these units. There are several distinct operations which add information to the Trader Information Object. Examples of these operations are: Export, Add Link, New Trader Property and New Trader Policy. X.500, on the other hand, has just one such operation: AddEntry. It is consequently necessary to map between the Trader protocol and the X.500 protocol.

†   The method for supporting distributed operations. In X.500 the distribution of information amongst the servers is fixed by the standard and is based on the hierarchical structure of the tree. The user has no control on which server a particular entry is placed: the user specifies the entry's DN and the entry is placed on the server which stores that particular portion of the tree. All DSAs in an X.500 directory co-operate to implement this single hierarchical distribution mechanism. Distribution in an ODP Trader is much less fixed. The user exports a Service Offer to a specific Trader. The administrator of that Trader may arrange for it to interwork with other Traders, in which case the exported service is available to a greater range of users. The standard, however, does not specify any structure for these interworking Traders.

†   The source of information. All information in an X.500 directory is assumed to reside on X.500 servers. The server could 'hold' this information on another database and retrieve it when required, but this is transparent to the X.500 protocol. Properties in a service offer held by an ODP Trader, however, may be specified to be 'indirect' and the values of these properties are not held by the Trader. When it is necessary to evaluate the service

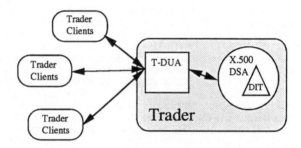

Figure 2: The Trader with its components and clients

offer (during an Import, for example), the Service Offer Evaluation Interface is called to obtain the current values of those properties.

†   Ordering of Results. An importer may specify how the returned set of Service Offers is to be sorted. X.500 does not support the ordering of a set of results.

6.   USING X.500 AS THE TRADER DATABASE

6.1. Trader structure

The approach adopted was to implement an ODP Trader as two components, a combination of an X.500 DUA and an X.500 DSA. The DSA stores the Trader Information Object (i.e. the information that the Trader knows). This information is accessed by the T-DUA (Trader DUA). The T-DUA implements the aspects of an ODP Trader specification which cannot be directly supported using X.500. Requests from trader clients (importers, exporters, and administrators) are mapped by the T-DUA into operations on the X.500 database. Figure 2 shows the components of an ODP Trader and its interactions with the clients.

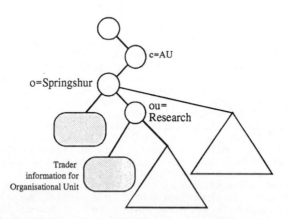

Figure 3: Placement of the Trader Information Object within the X.500 global directory.

6.2. Proposed X.500 directory information tree

The information stored by the X.500 directory includes:
†   The Trader properties (i.e. information about the Trader itself),
†   The Trader policies (i.e. rules to determine and guide Trader behaviour)
†   The set of Service Offers (i.e. information used by the Trader when acting as a server to undertake import or export operations)
†   The set of Trader Links (i.e. the information used by the Trader when acting as a client accessing other Traders when undertaking an import operation)
†   Type Manager Information (i.e. the information used by the Trader to control the form of Service Offer exports)

The Type Manager, and its information, are conceptually separate from the Trader and will be discussed in a later section of this paper.

The remaining four categories of information (trader properties, trader policies, service offers, and trader links) collectively form the Trader Information Object. This information is represented as a subtree of the X.500 directory tree. The subtree can be attached anywhere in the X.500 tree, but we expect that the Trader subtree would be most commonly attached beneath entries representing Organisations and Organisational Units (representing, respectively, the information known to organisational and organisational unit traders). This is shown in Figure 3.

Note that no attempt is made to link these packages of information together using the inherent X.500 distribution mechanism. Interworking between Traders is solely performed by the T-DUA using the Trader Link information.

Each Trader Information Object is composed of four types of X.500 entries. For simplicity, these entries are arranged in a flat tree as shown in Figure 4. The four types of entries are:
†   The Trader Entry contains information about the Trader itself. The information consists of:
-    Trader Properties, such as the owner of the Trader
-    Configuration Information, such as the name of the Trader and its network address
-    Access control information
-    Authentication information for the Trader Administrator
-    Limit information, such as the amount of resources to be consumed in an Import search.

†   The Trader Policy Entry contains the details about the Trader policies and is located immediately underneath the Trader Entry. The entry contains the trader policies,

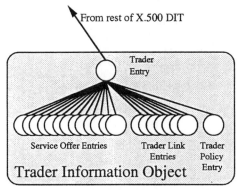

Figure 4: How the Trader Information Object is represented using X.500 entries.

expressed as a collection of X.500 attributes. Logically, this information is actually part of the information about the Trader and so should be stored in the Trader entry. Storing the Trader policy information in a separate entry, however, simplifies implementation of the separate Trader operations which manipulate Trader Policies and Trader Properties. If both the policies and properties were stored in one entry it would be difficult to ensure that a Trader Policy operation could only manipulate policies and not properties (and vice versa).

† A Service Offer Entry contains the information about one Service Offer. Service Offer entries are also located immediately beneath the Trader Entry. The basic attributes which are contained in each Service Offer entry are:
- Name: The name of the Service Offer, assigned by the Trader
- Exporter: The X.500 name of the Exporter of the service
- ServiceInterface: The X.500 name of the interface of the service being offered
- ServiceDescription: An X.500 auxiliary object class identifier
- SOEInterfaceId: The X.500 name of the interface from which the current value of the indirect properties are to be retrieved.

In addition to these basic attributes, a Service Offer entry contains attributes which represent the Service Properties. These properties depend on the type of service the Service Offer describes. The Service description is represented by a Service Type Identifier in this implementation. This is represented by an X.500 auxiliary object class that controls (via the X.500 schema) which attributes (i.e. properties) must be present and which attributes may be present in the Service Offer.

The Service Offer may contain Service Offer Properties. These are also represented as X.500 attributes in the Service Offer entry. Depending on the Trader Importer Policy these Service Offer Properties may contain ValidFrom and ValidTo properties which define the period over which the Service Offer is valid.

The Service Offer entry only contains direct properties. The handling of indirect properties is implemented by the T-DUA. This will be described in the section on Import and Export.

Finally, the Service Offer entry may contain other attributes which contain access control information, human readable information about the service offer (e.g. a description), and rules that describe an exporter's expectations of a service offer (i.e. exporter policy).

We considered structuring the Service Offer Entries into a tree based on a service hierarchy. However, this did not produce any functionality required by the Trader, nor did it provide any implementation advantages.

† A Trader Link Entry contains the information about one Trader Link. A Trader Link entry contains:
- Name: The name of the Trader Link. This is assigned by the Trader
- TraderInterface: The X.500 name of the linked Trader.

It also contains a set of Trader Link properties, which are expressed as attributes. The Trader Link entry may contain other attributes which contain access control information, human readable information about the trading service offer of the linked Trader, attributes from the linked Trader Entry, and the period over which this link is active. How these properties will be used to find Service Offers on remote Traders is currently not defined in the RM-ODP standard. This is one area we wish to investigate in the future.

7.   TYPE MANAGER

It was decided not to design a sophisticated Type Manager, primarily because it was not clear from the ODP Trader document exactly what functions a Trader would require of a Type Manager. Also, the Type Manager is conceptually separate from the Trader and hence outside the strict scope of our work. It was necessary, however, to design a minimal Type Manager which would be able to perform the following functions:

† Control the Service Properties and Service Offer Properties present in a Service Offer. This involves ensuring both that mandatory properties are present and that the exporter has not included any extraneous properties. This part of the Type Manager is handled by the X.500 schema which can control the mandatory and optional attributes (properties) present in an entry (Service Offer) of a particular object class (Service Description).

† Store the set of indirect properties to be included in a Service Offer of a particular type. This is handled by defining a new type of X.500 entry which represents Type Manager information. One such entry exists for each different type of Service defined. Each entry includes two attributes. One contains the names of any mandatory indirect properties (i.e. those that the Service Offer must supply) and the other contains the names of any optional indirect properties (i.e. those that an Service Offer may supply).

These entries could form the basis of a more sophisticated Type Manager. (These entries are not shown in figure 4 as they are not part of the Trader Information Object.)

We intend to extend this minimal Type Manager as we determine what functions are required by an ODP Trader.

8. TRADER OPERATIONS

To give the reader an idea of the interactions between the T-DUA and the DSA, we describe the Import and Export Operations. The description is a fairly high level overview, intended to highlight the issues involved in implementing the ODP Trader.

8.1. Export

The Export Operation adds a Service Offer to the Trader Information Base. Specifically, the Export Operation is mapped onto an X.500 Add Entry operation to add a new Service Offer entry to the X.500 database.

The mapping is straightforward since the Service Description is an X.500 object class and the Service Properties and Service Offer Properties are X.500 attributes.

It is necessary to check that the exported Service Offer conforms to the template in the Type Manager. This check is performed in two parts. The first part is undertaken directly by the T-DUA, which reads the template entry for that Service Type from the Type Manager (using X.500). A successful read operation confirms that the Trader does handle this type of Service Offer and indicates whether this Service Type has indirect properties. If indirect properties exist then the T-DUA checks that the Service Offer includes the mandatory indirect properties and is not proposing to include any extraneous optional indirect properties (note that the Trader does not store which optional indirect properties the Exporter is providing at the SOE Interface). All the indirect properties are then stripped from the Service Offer and the result is then stored in the X.500 database. If the Service Type includes indirect properties, the T-DUA checks that the Service Offer includes an SOE Interface.

The second part of the check is an X.500 schema check. This checks that, given the Service Description, the Service Offer contains all the mandatory properties and does not contain any extraneous properties. This check occurs when T-DUA attempts to add the Service Offer entry to the X.500 database.

A Service Offer is added to the database only if it passes both checks.

8.2. Import

Import is a complicated Trader operation because of:

† The necessity to check the Importer Policy
† The potential for searching interworking Traders
† The presence of indirect attributes in the Service Offers.

The first step in processing an Import request is for the Trader to decide if the Importer Trader Constraints allow it to answer the request at all. This is achieved by the T-DUA performing an X.500 Search Operation on the Trader entry using the Importer Trader

Constraint as the filter. If the search operation is successful, then the Trader satisfies the constraint.

The second step is to decide if any of the Interworking Traders should be searched. This is achieved by the T-DUA performing a second X.500 search. Here, the target entries are the Trader link entries and the filter is the Importer Link Constraints. A set of Trader Links (possibly empty), which satisfy the Importer's constraint, are returned. The Import operation should be forwarded to these linked Traders, which may be done in parallel to the local search, before the local search, or after it.

The third step is to perform a search of the local database of Service Offers to see if any of them will satisfy the Import request. The search can be complicated by the presence of indirect properties. As these are not stored with the Service Offers, evaluating the Matching Criteria must be a three phase process:

1)   The T-DUA retrieves from the Type Manager (using an X.500 read) the list of indirect properties in a Service Offer for this particular Service Type.

2)   The Matching Criteria are rewritten to produce a new filter that does not include any tests on indirect properties. This filter finds all the Service Offers which *might* satisfy the Import request. An X.500 search is then performed for Service Offers that satisfy the requested Service Description and the rewritten filter. A set of potential matches is returned.

3)   The Matching Criteria are then rewritten to produce a new filter which only includes the indirect properties. For each Service Offer returned in the previous phase, the specified SOE Interface is contacted and the values for the necessary indirect properties are obtained. The indirect property values are applied to the rewritten filter and that Service Offer is either accepted or eliminated from the set of matches.

The fourth step is to combine the set of Service Offers found locally with the set of Service Offers found though searching the linked Traders, and to rank these Offers using the Selection Preference. The fifth and final step is to select the 'best' matches using the Selection Preference. This may involve evaluating indirect properties. The selected matches are then returned to the Importer.

9.   A SIMPLE EXAMPLE

To define a new type of service, the Trader administrator has to construct the X.500 schema definitions for the new service properties associated with the new service type. For example, the X.500 schema definition for the properties of a printer service type might be:

```
printerServiceProperties OBJECT-CLASS ::= {
        SUBCLASS OF        top
        KIND               auxiliary
        MUST CONTAIN       {printerType}
        MAY CONTAIN        {locationRoom | locationBuilding | costPerPage |
                           languagesSupported | pagesPerMinute | pageSize | dotsPerInch |
                           colourCapable | driverName}
        ID                 {2.7.1000.5.34678.3.5}}
```

This definition states that an exported offer of a printer service must contain one property ('PrinterType'), and may contain ten other properties, including 'costPerPage'. Properties are defined as X.500 attributes, the 'costPerPage' property, for example, might be defined as:

```
costPerPage ATTRIBUTE ::= {
        WITH SYNTAX                Integer
        EQUALITY MATCHING RULE     IntegerMatch
        ORDERING MATCHING RULE     IntegerOrderMatch
        SINGLE VALUE
        ID                         {2.7.1000.5.34678.4.1}}
```

which states that a 'costPerPage' property contains a single integer. With most X.500 DSA implementations, adding such simple definitions only involves editing configuration files and restarting the DSA. The attribute definitions also need to be added to the T-DUA configuration files and the T-DUA restarted.

When an entry is created in the X.500 database representing an actual Service Offer for a printer service, the entry will contain a number of additional attributes. These represent the Service Type and Service Offer Properties and include, for example, the Service Interface Identifier, the identity of the Exporter, and the period of validity of the Service Offer.

Finally the Type Manager entry needs to be added to the X.500 Database. This entry only contains the name and identifier of this new service.

Importers can search for these offers by specifying a Service Description of the OID '2.7.1000.5.34678.3.5' (which is usually aliased to a more user-friendly string such as 'printerServiceOffer' in the T-DUA) and a Matching Criteria, such as

printerType = "Laserwriter" and dotsPerInch > 600 and languagesSupported = "Postscript"

10.   INDIRECT PROPERTY EXAMPLE

Assume that the importers wished to check whether a print service was actually available (i.e. not out of paper, out of toner, or jammed). This could be represented by an indirect property 'printerAvailable'. It is not necessary to modify the 'printerServiceProperties' definition given in the previous example as the information to be stored in the X.500 database for each Printer Service offer will not be changed. However, the Service Type entry for the Printer Service must be altered to include the new indirect attribute. The contents of this entry might be:

```
{commonName          = printerService
 serviceType         = {2.7.1000.5.34678.3.5}
 indirectProperties  = printerAvailable }
```

To import a print service, the user might specify the following Matching Criteria:

printerType = "Laserwriter" and languagesSupported = "Postscript" and printerAvailable

The T-DUA first reads the printerServiceOffer Service Type entry and discovers that 'printerAvailable' is an indirect property. It rewrites the Matching Criteria to remove that property to give

printerType = "Laserwriter" and languagesSupported = "Postscript"

Evaluating this X.500 query will return information about all the Print Services matching the filter known to the Trader. The T-DUA must then contact the SOE Interface of each of these offers to determine whether the 'printerAvailable' property is true, and hence whether the Service Offer is actually selected by the Matching Criteria.

11.   RELATIONSHIP TO OTHER WORK

A number of papers have proposed, or developed, implementations of Traders using X.500. These include Popien & Meyer [3], and Pratten et al [4]. Unfortunately, other papers are confidential and their contents cannot be publicly disclosed. The overall design of the Traders described in [3, 4] is similar to that described in this paper. X.500 is used for the storage of information, while the Trader functionality is implemented using the equivalent of the T-DUA. Popien & Meyer specifically identify the difference in distributed operations between X.500 and an ODP Trader as a reason for this design.

Both Popien & Meyer and Pratten describe the implementation of an ODP Trader using X.500. However, there is little overlap between this paper and the earlier papers as both earlier papers concentrate on the internal design of the equivalent of the T-DUA and give little detail on the X.500 interface. Furthermore the operations supported by the Traders are different from those in this paper, although this is probably explained by the development of the ODP Trader model subsequent to the writing of the earlier papers.

12. CONCLUSIONS

It appears to be possible to implement an ODP Trader using X.500 as a base. X.500, however, is not powerful enough to directly implement the functionality required by an ODP Trader, and so cannot replace an ODP Trader. Instead, it is necessary to provide a special X.500 DUA, which we call the T-DUA, to implement these additional functions. Among other things, the T-DUA implements:

†   Support for the different operational models used by the Trader and X.500, including mapping between the X.500 and ODP protocols.
†   Support for the distributed Trader operations (interworking).
†   Support for indirect attributes.
†   Support for ordering the Service Offers returned from an Import operation.

The underlying data models of both X.500 and an ODP Trader are very similar. Information in the Trader is represented as Properties which have a type and one or more values. This directly maps into the equivalent X.500 attributes. The collections of properties and policies in the Trader (Service Offers, Trader Links) map easily into X.500 objects. The X.500 operations for accessing, searching, and modifying information provide similar functionality to those functions required by the Trader.

The use of X.500 to support an ODP Trader consequently should mean a significant simplification in the implementation of the Trader. This is particularly so as X.500 products are now available off the shelf. In addition, using X.500 provides access to the features of X.500 which are currently poorly defined in the Trader. This includes authentication of users, access control of information, methods for modifying and maintaining the underlying data schema, and methods for mapping names of objects in the network to addresses.

We have presented a description of how the Trader Information Object can be represented in the X.500 Directory Information Tree, and have sketched how the Import and Export operations can be implemented.

The exercise of designing an implementation of the ODP Trader has also highlighted areas of the Trader document which need more definition, for example, the specification of the Importer Policy. Ultimately, however, it is an open question as to how detailed the Trader specification should become. Where does defining a general model end, and defining a specific implementation start?

The work described in this paper is continuing. We plan to implement the T-DUA. Apart from validating the design, a prototype will assist in validating the specification of the ODP Trader. We also hope to extend areas of the design, particularly those concerning the Type Manager.

REFERENCES

[1]   ODP Trading Function, ITU/ISO Committee Draft Standard ISO 13235/ITU.TS Rec.9tr (1994) (To appear)
[2]   Information Technology - Open Systems Interconnection - The Directory, ITU/ISO Standard X.500/ ISO 9594 (1994) (To appear)
[3]   Federating ODP Traders: An X.500 Approach, C. Popien, B. Meyer, Proc. of the ICC'93, Geneva, Switzerland, May 1993.
[4]   Design and Implementation of a Trader-Based Resource Management System, A. Warren Pratten, James W. Hong, J. Michael Bennett, Michael A. Bauer, Hanan Lutfiyya, (Private Communication).
[5]   Reference Model of Open Distributed Processing: a Tutorial, Kerry Raymond, Open Distributed Processing II, ed J. de Meer, B. Mahr, and S. Storp, North-Holland (1994)
[6]   X.500, the directory standard and its application, Douglas Steedman, Technology Appraisals Ltd (1993)

11

An evaluation scheme for trader user interfaces

Andrew Goodchild*

Department of Computer Science, The University of Queensland, QLD 4072, Australia

This paper provides a set of criteria to evaluate the usability of trader user interfaces with respect to resource discovery. The criteria were identified because trader user interface developers sometimes make unrealistic assumptions about the information seeking behavior of their users, leading to interfaces that are difficult to use for resource discovery. This paper uses some insights from the cognitive aspects of the information science field to justify the usability criteria. The results of this paper are general enough to aid the developer of a user interface to any resource discovery tool.

Key Codes: C.2.4; H.3.3; H.5.2
Keywords: Distributed Systems; Information Search and Retrieval; User Interfaces

1 INTRODUCTION

Resource discovery tools are systems which enable users to search for and employ resources in very large networks [1]. Traders provide an advertising/matchmaking service for objects in a distributed environment [2]. Naturally there is a desire to use the functionality of a trader for resource discovery by attaching a user interface (UI) to it.

Current research in the trader has largely come out of the OSI arena, where the main problems considered have been technical ones. The paradigm behind such research may not be totally suitable for resource discovery. Application of rational methods, as seen in current trader research, leads to systems that suffer from a strong technological bias; and as Ellis [3] likes to describe it: "the user is equated to an peripheral input/output device." It is natural for developers when they attach UIs to traders to want to avoid this problem. However, when developers implement UIs for traders, they make assumptions about the way users seek information. Some of these assumptions maybe unrealistic.

The purpose of this paper is to question the developers assumptions about usability by providing a treatise from the information science literature. This paper does not seek to solve the traders usability problems, nor does it suggest specific technical solutions to existing problems. What this paper does is to introduce an evaluation scheme from the users point of view for trader UIs.

Ideally this usability evaluation scheme should be useful for aiding developers in isolating possible failure points in existing trader UIs and help in identifying new areas for trader UI development. A full set of usability criteria would cover an immensely broad

---

*email: andrewg@cs.uq.oz.au

set of issues, including general user interface design principles [4, pages 65-88]. The focal point of this work is to identify a set of usability criteria derivable from a representative model of the general cognitive behavior of information seekers. In other words, rather than base this work on the human computer interaction (HCI) literature, this work is based upon librarians' observations of how people find information. Information science provides a useful starting point as librarians have been dealing with resource discovery problems for centuries now, and to ignore it would be inviting rediscovery of much existing knowledge.

This paper is related to and supports the quality of service issues raised by Milosevic et al [5]; hence references to it will be made from time to time.

2　THE TRADER UI USABILITY CRITERIA

2.1　Interconnected searches

The typical model, as Bearman [2, page 38] illustrates, for the client (importer) interaction with the trader involves first a service request (import request) followed by a reply from the trader with a match to the most appropriate service offer (import reply). Taken at face value, which I think many trader UI developers are, this model is much like the classical representative model [6] in information retrieval.

The classical model, as illustrated in Figure 1, is concerned with the user's actions within a single episode. The episode starts by the user first identifying an information need, formulating a query (i.e. like an import request) which is matched against the database contents, and finally produces a single output set (i.e. like an import reply).

Unfortunately such a model does not fit well to real-life searches involving people. Salton [7] observed that before people arrive at a final result set, they gradually refine their query. As a result, he enhanced the classical model by introducing interactive feedback, or query reformulation, to improve the output. Salton's model is illustrated in the center of Figure 1.

The first interpretation is also at fault as it presumes that the information need leading to the query is the same, no matter what the user might learn from the documents in the preliminary retrieved set. Hancock [8] disproved this presumption when observing students search in a library; she found evidence that searching is an adaptive process. In real-life searches in manual resources, end users may begin with just one feature of a broader topic, and move though a variety of sources. Each new piece of information they encounter gives them new ideas and directions to follow and, consequently, a new conception of the query. Thus the point of feedback is not only to improve the representation of a static need; but, to provide information that enables a change in the information need itself.

Furthermore, with each new conception of the query, the user may identify useful information and references. In other words, the query is satisfied not by a single retrieved set, but by a series of selection of individual references and bits of information at each stage of the search. A better interpretation of query reformulation is illustrated on the

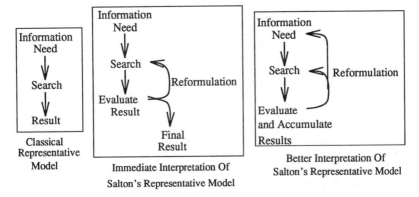

Figure 1: Representative models of information seeking

right hand side of Figure 1. Bates [6] and O'Day and Jefferies [9] also confirmed the same behavior. Hence, the users interaction with the trader is more than Bearman's [2, page 38] model suggests.

When the trader UI has poor provision for interconnected searches, users are often left dissatisfied with the tool. Immediately, it can be seen that responsiveness and reliability are an important criteria. For example, when online catalogs were batch oriented frequent users of such catalogs were often frustrated by the one-shot query orientation of the system and preferred continuous feedback [10]. In general, as Rushinek [11] observed, a lack of responsiveness usually leads to user dissatisfaction. This suggests:

Usability Criteria 1 *the trader UI must be reliable with quick response times.*

Milosevic et al [5] are indeed right when they suggest that characteristics of quality services are responsiveness and reliability. Users would find a trader UI with long delays for import requests and replies or traders which are down quite often frustrating to use simply because they cannot "interact" with the trader to perform interconnected searches.

Another case of poor provision for interconnected searches is not acknowledging that its the accumulation of results that matters and not the final retrieved set. For example the Z39.50 standard [12] until recently didn't make this acknowledgment. Z39.50 version 1 was stateless, meaning that the server could discard the results as soon it had sent them to the client. Version 2 supports state, allowing the user to perform queries upon queries in an interconnected fashion. This leads to:

Usability Criteria 2 *the trader UI must support the accumulation of search results.*

The result of this criterion is that it maybe inappropriate to assume that the binding between the UI and the trader is stateless. Users need to be able to refine their searches by performing additional searches on their existing search. In addition, a positive step would be to provide the ability to store search results for later use or for combining with several other searches. Management and refinement of search results is a feature of a trader UI that is underdeveloped and can be improved.

2.1.1   Changing information needs

Supporting changing information needs in a trader UI is a little more difficult. O'Day and Jefferies [9] made a key observation in trying to understand changing information needs. Their observation was that there was some decision that led to the next step in the interconnected search: a "trigger". These triggering actions where the result of encountering something interesting, or explaining a change, or finding missing pieces or pursuing a plan. The motivation for triggers was identified by Kuhlthau [13]. Kuhlthau described information seeking as a process which moves the user from a state of uncertainty to understanding. She asserted that it was the anxiety associated with uncertainty that motivated users to search for information.

O'Day and Jefferies [9] observed that interconnected searches stopped when there were no more compelling triggers for further searching or in a few cases, there were specific inhibiting factors like lack of time, money, or expertise in a field to allow effective searching. O'Day and Jefferies labeled this behavior as encountering "stop conditions". Again, Kuhlthau provides some insight into the motivational aspects. She asserted that a sense of confidence is associated with the understanding that search results bring. If users start experiencing redundancy in their search results, their confidence grows until they are satisfied with their search. If users experience many different unique search results their confidence shrinks until they are disappointed with their search.

Hence stop conditions are strongly related to anxiety in the searcher, and as a result inducing artificial ones will prematurely terminate searches. For example Bysouth [14, page 60] noted that complicated connect time pricing schemes for online databases induce artificial stop conditions, inhibiting interaction. Thus we must take great care in designing tools and ensure that:

Usability Criteria 3 *the trader UI must not induce artificial stop conditions.*

All features of a trader UI that users are likely to encounter must be readily understandable; otherwise, the anxiety of use is likely to cause the user to stop using the trader prematurely. This is in alignment with Milosevic et al [5] concept of "simplicity" as one of the factors determining the quality of service. An example of artificial stop conditions would be complicated charging schemes which result in the users perceiving a high cost, thus deterring them from extended searches in attempting to locate their resource.

Lack of sufficient triggering information can also lead to users prematurely stopping their search, in many cases falsely satisfied, even though there is much more relevant material to be found. For example Bates [15] observed this behavior in subject catalog users. This problem is directly linked to the fact that the library she studied could not afford the resources to put "see also" entries in their subject card catalog, thus inhibiting the production of triggers. If an initial search is unsuccessful, then the users biased feeling that no relevant material exists is confirmed, artificially inducing a stop condition.

This problem is related to Bates' PRINCIPLE OF VARIETY [16]: "The variety of query formulation must be as great as the variety of document (resource) descriptions for search success."

Authors and indexers produce a great variety of terms in their indexing, so to cope successfully searchers must also produce an equal variety in formulating searches on any given topic. For example some interdisciplinary topics like artificial intelligence can be classified into many different areas in the library. As a result you may find books in areas where you do not normally expect them; so the searcher is now responsible for generating many ways to effectively search. The variety that searchers can generate in their queries will inevitably be much less than the variety of resource descriptions, so to help the users cope the trader UI developer could:

1. *Reduce the variety of resource descriptions.* Traditionally this has been achieved by vocabulary control, i.e. reduce the variation by eliminating variety in words by morphological analysis to reduce different word forms to common "stems", for example: removing plurals, verb conjugations, synonyms, etc. More sophisticated techniques based upon natural language understanding have yet to be shown to be cost effective [17]. The simpler techniques have had considerable success in improving search performance – but it has its limits. We can decrease the variety in language used to describe information, but we cannot reduce the variety of information itself without defeating the purpose of the information retrieval system.

2. *Increase the variety of users queries.* The principle mechanism for achieving this is to use cross referencing. Users can increase the variety in their search by following links between terms. Another technique is to have users expand queries by selecting additional terms from lists suggested by the system. Belkin and Croft [17] noted that this technique was not effective. The reasons for these differences are not obvious, although it appears that using only system suggestions is too restrictive and does not make full use of the users domain knowledge.

Traders with poorly structured offer spaces will provide inadequate triggering information, and may result in the user prematurely stopping their search even though there are many relevant resources available. Thus using type management of service type facilities like relationships and subtyping to provide syndetic structure (*see also* entries) should be strongly encouraged. In addition, tools like morphological analysis should be a capability of the query interface. Milosevic et al [5] characteristic of accuracy in exporting is symptomatic of failing to meet this criterion. Hence following these design principles leads us to:

Usability Criteria 4 *the trader search interface/offer space must respect the principle of variety.*

2.2   Querying and browsing

There is a tendency among resource discovery tool developers to see browsing a casual, don't-know-what-I-want behavior that one engages in separately from "regular" searching

[6]. This emphasis that browsing is not "regular" leads to a bias held by many information retrieval tool developers including trader UI developers and that is: a bias towards querying as a main search interface.

This bias is unrealistic as browsing is a rich and fundamental part of human information seeking behavior – one may even argue that it is queries that are irregular. The notion that a well defined query is possible was challenged by Nicholas Belkin [18]. Belkin claimed that for a user to state their information need, they have to describe what they do not know. In effect users do not naturally have "queries"; but, rather they have what is Belkin calls an "anomalous state of knowledge"[18, page 62].

Browsing is a fundamental behavior as it is the users attempt to deal with what Bates calls the PRINCIPLE OF UNCERTAINTY: "Document description and query development are indeterminate and probabilistic beyond a certain point." This principle reflects the difficulty users have in formulating queries. Essentially, if a user is not looking for something they already know exists; then, the more specific the query is the more likely that it will fail. A related example would be to ask two people to describe a very specific resource; the principle of uncertainty manifesting itself as them using very different terms. Whereas, if you were to ask them to describe a fairly general resource then they are more likely to use more similar terms.[†]

Focusing on tools which just support browsing behavior only is not the answer either. Conklin [20] described that early hypertext tools lacked support for indices, which became a problem for users of large hypertexts. Users often complained that they became "lost in space" – not knowing how to get to a node that they know, or think, exists. It wasn't until the ability to query an index of the hypertext was possible that being lost became less of a problem. To help users perform both general and specific searches the tool needs both querying and browsing facilities. Queries can help users get closer to their desired information faster, but only by browsing will they be able to locate the most specific items, unless of course they already know where it is. Acknowledgment that querying and browsing are both necessary leads to:

Usability Criteria 5 *the trader UI must seamlessly provide for both querying and browsing.*

The designers of trader UIs cannot expect users to identify their specific resources by performing just queries, to do so would be violating the principle of uncertainty. In addition, it is too much to expect the users to find their resources by browsing all the time. Thus there should be a certain degree of search flexibility provided by the trader UI. Rather than supporting only queries, as has been popular in the past, both querying and browsing facilities should be supported in an integrated fashion.

[†]As an aside, one interesting effect of the uncertainty principle is it allows authors like Salton to make rather counter-intuitive statements like [19]: "Evidence from available studies comparing manual and automatic text-retrieval systems does not support the conclusion that intellectual content analysis produces better results than comparable automatic systems."

2.3   The role of strategy in searching

Developers maybe tempted to interpret the interconnected search undertaken in subsection 2.1 as simply a series of queries of the classical sort. This interpretation is common in traditional information retrieval systems, as they tend to exploit their file structure in only one way. In contrast, users conduct searches using many different techniques in endless variation. From the standpoint of effectiveness in searching, as Bates [6] pointed out, the searcher with the widest range of search strategies available is the searcher with the greatest retrieval potential. Bates [21] provides a system of classification of different search strategies:

Moves are identifiable thoughts or actions that are part of information searching and are the atoms considered in this model, in much the same way as the entity is an atom in conceptual modeling. Moves can be part of a plan with a specific goal in mind, or part of a formless effort by a user who doesn't know what they are doing. An example move may remove unwanted elements from a retrieved set by introducing the **and not** operator. The moves identified by Fidel [22] provide many more examples.

Tactics are one or more moves made to further a search. A tactic represents the first level at which strategic considerations are primary. Tactics are utilized with the intention of improving search performance in some way, either in anticipation of problems, or in response to them. An example tactic for dealing with a retrieved set that is too large maybe to reject unwanted elements when reformulating the query. Bates' [23, 24] work on tactics provides many more examples.

Stratagems are larger, more complex set of thoughts and/or actions than tactics. A stratagem consists of multiple tactics and/or moves, all designed to exploit the file structure of a particular search domain thought to contain the desired information. Tools which allow footnote chasing and citation searching are examples of systems which exploit stratagems. The search methods identified by Ellis [25, 26] and Bates [6] fall into this category.

Strategies are plans which may contain many moves, tactics and/or stratagems, for an entire information search. Another analogy for a strategy is an "overall plan of attack" [27, page 408]. A strategy for an entire search is difficult to state in any but the simplest of searches, because most real-life searches are influenced by the information gathered along the way of the search. The solution to a given search problem may be characterized by a single encompassing search strategy, but many require application of interim tactics and stratagems to accomplish the overall goals of the searcher. The research done by O'Day and Jefferies [9] was completed at the strategic level.

Currently the majority of information retrieval systems communicate with the user at the move level. Strategic behavior must almost always be exercised by the human searcher. Little or no operational capability of a strategic nature is provided by current systems to the user. Lack of strategic support can make systems very difficult to use. For example, Belkin and Croft [17] noted that users find boolean queries extremely difficult to generate. This is because the user is thinking at a strategic level and the system is demanding that the user operate at a move level. These points lead to the next criterion:

Usability Criteria 6 *The trader UI should allow the user to easily exercise many different strategic search choices.*

This criterion highlights the need for trader UI developers to go beyond simple attribute searching of the offer space and provide a variety of search strategies. The most novel approaches to this problem of providing alternative search strategies have come from the information exploration tools research area [4, pages 395-438]. Famous examples include hypertext and multimedia. Other less well known examples are things like graphical boolean queries [4, pages 423-428], graphical fisheye views [28], information crystals [29], and document landscapes [30]. The limitation of the new search strategies is of course how easy the new techniques are to understand. Features which are difficult to understand will quite often be under-utilized by users, thus reducing the possible benefits they may offer.

2.4   Support for searching

Support for searching can come in many forms: no support at all, decent online help and meaningful system messages, intelligent decision support systems for information retrieval, and intelligent agents. Each of these options describes the amount of user involvement in the search and is complemented by the involvement of the system – that is to say, the more system involvement, the less the user has to do in the actual search process. Waterworth and Chignell [31] described this spectrum of involvement as the "degree of mediation".

Riddle [32] identified an important assumption about mediation that many developers like to make. Software developers often believe human time and resources are extremely scarce and costly resources, much too expensive to be wasted upon resource discovery problems. Consequently, developers try to solve every problem by automating it. Bates likes to describe the same belief as [21, page 575]: "if part of the information search process is not automated, it is only because we have not yet figured out how to automate it."

While effective systems will doubtlessly be produced, not all users will want this kind of response from an information system. For example Shniederman [4, page 66] points out that differing levels of user expertise require different UIs. Novice users prefer hand-holding menu driven interfaces and expert users prefer command line systems. In this case the degree of mediation is decreasing as the user expertise increases.

Many developers are tempted to increase the degree of mediation because they can see the apparent order in user's searching. While on the surface searches are ordered; deep down they are chaotic systems. Researchers who try to mimic human searching on computers discover that the implementation rapidly becomes too complex [3]. Bates reflects this problem in her PRINCIPLE OF COMPLEXITY [16]: "Entry to and use of an information system is a complex and subtle process." Thus we cannot make any assumptions about a user's current searching preferences and this leads to:

Usability Criteria 7 *the trader UI must respect the principle of complexity by being flexible*

Flexibility for customization is a key issue for trader UIs. The trader UI needs to provide options to let users customize the service to their own needs. Not providing potential for customization assumes that user will want to use the trader in a particular fashion; and thus would violate the principle of complexity.

3   CONCLUSION

The purpose of this paper was to identify some of the key issues in making trader user interfaces easier to use for resource discovery. Existing research has a problem in that it has been conducted largely in isolation from usability issues, particularly with regard to resource discovery. By using established knowledge from the information science field various assumptions that trader UI developers make about usability can be explored and in some cases made more realistic. To aid trader UI developers in examining their assumptions a usability evaluation scheme was developed establishing seven usability criteria.

Problems with current trader UIs stem from the assumption that searches by users upon a trader are independent of each other. But, as noted in section 2.1, searches are interconnected, as each search provides the user with new ideas and directions to follow in their next search. Thus, user satisfaction is linked to the response time of the search interface; faster response times allow users to perform interconnected searches more fluidly (see usability criteria 1). Also, it is the accumulation of search results that matters to the user and not the final retrieved set. As a result the UI should support results management and query refinement (see usability criteria 2). In addition the trader UI should not stifle the fluidness of an interconnected search, as a result the UI should be simple to understand (see usability criteria 3).

More problems stem from the way resources are cataloged in trader offer space. The trader search interface should aid the user in navigating thru the offer space by providing morphological analysis of queries and judicious use of typing capabilities to provide cross-referencing (see usability criteria 4). A popular bias among trader UI developer is to only provide a query interface; such a bias limits the users ability to effectively explore the offer space. Provision for both querying and browsing should be a feature of the trader UI (see usability criteria 5). Querying and browsing should not be the only search strategies available to the user - infact the more alternative search interfaces available to the user the more effective they can be at exploring the offer space (see usability criteria 6). The last assumption overturned by this paper is that UI developers will try to automate as many parts of the search process as possible; such an assumption is open to creating many problems as the searching process is known to be far more complex than we currently understand. As a result the trader UI should provide a great deal of flexibility and customization to allow users to search in their own peculiar ways (see usability criteria 7).

As an aside, due to the general nature of the work presented in this paper, it is not surprising to observe that the criteria are not limited in application to trader UIs, but they are versatile enough to be applied to other resource discovery tools. For example, they have been applied to Virtual Libraries on the World Wide Web [33].

ACKNOWLEDGMENTS

The author would like to thank Bob Colomb, Ashley Beitz, Renato Iannella and the anonymous referees for their helpful comments and suggestions. In addition, the author would like to acknowledge the assistance provided by the staff of the Distributed Systems Technology Centre (DSTC) and the Research Data Network Collaborative Research Centre (RDN-CRC). The author is supported by an Australian Postgraduate Award (APA).

REFERENCES

[1] C. Bowman, P. Danzig, and M. Schwartz. Research problems for scalable internet resource discovery. Technical Report CU-CS-643-93, University of Colorado at Boulder, March 1993.

[2] M. Bearman. Odp-trader. In J. De Meer, B. Mahr, and S. Storp, editors, *Open Distributed Processing, II*, IFIP Transactions C-20, pages 37–51. North-Holland, 1993.

[3] D. Ellis. The physical and cognitive paradigms in information retrieval research. *Journal of Documentation*, 48(1):45–64, March 1992.

[4] B. Shniederman. *Designing the User Interface*. Addison Wesley, second edition, 1992.

[5] Z. Milosevic, A. Lister, and M. Bearman. New economic-driven aspects of the odp enterprise specification and related quality of service issues. In *DSTC Trader Workshop*, 1993.

[6] M. Bates. The design of browsing and berrypicking techniques for the online search interface. *Online Review*, 13(5):407–424, 1989.

[7] G. Salton. *Automatic Information Organization and Retrieval*. McGraw-Hill, 1968.

[8] M. Hancock. Subject searching behaviour at the library catalogue and at the shelves: Implications for online interactive catalogues. *Journal of Documentation*, 43(4):303–321, December 1987.

[9] V. O' Day and R. Jefferies. Orienteering in an information landscape: How information seekers get from here to there. In *INTERCHI' 93 "Human Factors in Computing Systems"*, pages 438–445, 1993.

[10] S. Allen and J. Matherson. Development of a semantic differential to assess users' attitudes towards a batch mode information retrieval system (eric). *JASIS*, 27(5):268–272, September 1977.

[11] A. Rushinek and S. Rushinek. What makes users happy? *Communications to the ACM*, 29(7):594–598, 1986.

[12] National Information Standards Organization. *Information retrieval application service definition and protocol specification for open systems interconnection : an American National Standard.* NISO Press, 1993.

[13] C. Kuhlthau. A principle of uncertainty for information seeking. *Journal of Documentation*, 49(4):339–355, December 1993.

[14] P. Bysouth. *End User Searching: The effective gateway to published information.* Aslib, The association for information management, 1990.

[15] M. Bates. Factors affecting subject catalog search success. *JASIS*, 28(3):161–169, May 1977.

[16] M. Bates. Subject Access in Online Catalogs: A Design Model. *JASIS*, 37(6):357–376, 1986.

[17] N. Belkin and W. Croft. Information Filtering and Information Retrieval: Two sides of the Same Coin? *Communications to the ACM*, 35(12):29–38, 1992.

[18] N. Belkin, R. Oddy, and H. Brooks. ASK for information retrieval: Part I. Background and theory. *Journal of Documentation*, 38(2):61–71, 1982.

[19] G. Salton. Another look at automatic text-retrieval systems. *Communications to the ACM*, 29(7):648–656, July 1986.

[20] J. Conklin. Hypertext: An introduction and survey. *IEEE Computer*, 20(9):17–41, September 1987.

[21] M. Bates. Where should the person stop and the information search interface start? *Information Processing and Management*, 26(5):575–591, 1990.

[22] R. Fidel. Moves in online searching. *Online Review*, 9(1):61–74, 1985.

[23] M. Bates. Information search tactics. *Journal of American Society for Information Science*, 30(4):205–214, July 1979.

[24] M. Bates. Idea tactics. *Journal of American Society for Information Science*, 30(5):280–289, September 1979.

[25] D. Ellis. A behavioural approach to information retrieval system design. *The Journal of Documentation*, 45(3):171–212, September 1989.

[26] D. Ellis, D. Cox, and K. Hall. A comparison of the information seeking patterns of researchers in the physical and social sciences. *Journal of Documentation*, 49(4):356–369, December 1993.

[27] S. Harter and A. Peter. Heuristics for online information retrieval: a typology and preliminary listing. *Online Review*, 9(5):407–424, 1985.

[28] M. Sarker and M. Brown. Graphical fisheye views of graphs. In *CHI'92 "Human Factors in Computing Systems"*, pages 83–91, May 1992.

[29] A. Spoerri. Infocrystal: A visual tool for information retrieval and management. In *CIKM'93 "Information and Knowledge Management"*, pages 11–20, 1993.

[30] M. Chalmers. Visualisation of complex information. In *EWHCI'93*, pages 152–162, 1993.

[31] J. Waterworth and M. Chingell. A model for information exploration. *Hypermedia*, 3:35–58, 1991.

[32] P. Riddle. Library culture, computer culture, and the internet haystack. http://is.rice.edu/~riddle/dl94.html, 1994.

[33] A. Goodchild. Issues in building virtual libraries. Technical Report 310, Department of Computer Science, The University of Queensland, Australia., September 1994.

12

AI–based Trading in Open Distributed Environments

A. Puder, S. Markwitz, F. Gudermann and K. Geihs
{*puder,markwitz,florian,geihs*}@*informatik.uni-frankfurt.de*

Department of Computer Science
University of Frankfurt
D–60054 Frankfurt, Germany

Abstract

An open distributed environment can be perceived as a service market where services are freely offered and requested. Any infrastructure which seeks to provide appropriate mechanisms for such an environment has to include mediator functionality (i.e. a trader) that matches service requests and service offers. Commonly, the matching process is based upon some IDL–based service type definition, and the types of the various services have to be "standardized" and distributed *a priori* to all potential participants. We argue that such well defined "standards" are too inflexible and even contradict the idea of an open service market. Therefore we propose a new type notation based on *conceptual graphs*. The trader maintains a knowledge base about service types in form of conceptual graphs. During the trader operations the service type knowledge evolves as it is continuously refined and extended. Users of the trading service interact with the trader and formulate queries in a corresponding notation that allows for a conceptual specification of the desired service type. Adequate matching algorithms and protocols have been implemented.

Keywords: Trading, type specification, conceptual graphs.
Classification: C.2.4; D.1.5; I.2.4

1   Introduction

The emerging Reference Model of Open Distributed Processing (ODP) provides an architectural framework for the standardization of distributed system technology. It defines abstract concepts that are appropriate to reason about and specify general distributed systems.

The basic goal is to enable the interworking of heterogeneous systems. Furthermore, it is addressing the question of application portability and distribution transparencies.

One of the functions that will be standardized as part of the ODP activities is the trading function (see [ISO94]). It is concerned with the matching of service requesters and service providers. The matching is done based on the notion of a *service type*, which (informally) is something that expresses properties of an object. The trading function is provided by a component called *trader*. A service provider exports its service offer to the trader (called *service export*). The trader maintains a database of service exports. A service requester makes an inquiry to the trader for a particular service offer and — if available — receives a reference to a suitable service exporter. This is called *service import*.

The concept of a service type plays an important role in such environments. The notion of a type is well known from conventional procedural programming languages, where types are used in order to aid error checking and software maintenance. Static typing is the dominant approach in these languages. In object–oriented languages the notion of a type is somewhat more flexible because of subtype relationships (see [Ame90] and [BJ93]). Nevertheless, the programmer of an application has a rather precise knowledge about what kind of types to be used.

In a general, large, open distributed system with a variety of different service providers, service requesters and service types, there is much less knowledge about the set of (service) types that will be available during the lifetime of an application program. Clearly, an application needs to understand the basic semantics and the access rules of the services it is going to work with. However, in an open service environment many different "flavors" of a particular service type may be offered over time by different service providers using the same or very similar service interfaces.

Consequently, in such dynamic environments service providers and requesters need means to specify service types and to *learn* about new service types at runtime. We have developed a notation for expressing the knowledge about service types and thus to support the trading function in open distributed systems. Our approach is based on a knowledge representation technique called *conceptual graphs*. A conceptual graph captures the knowledge about a service type and allows the specification of a type using a powerful, extensible notation. The trader matches service imports and exports using the information contained in the conceptual graphs.

This paper motivates our approach and demonstrates its strengths. In Section 2 we describe our assumptions that result from the envisaged trading environment. Section 3 introduces the conceptual graph technique. We present an example and a formal theoretical framework that is adapted to the trading requirements. The specification of a type may evolve over time. Therefore, an algorithm is presented describing how the interacting entities can incrementally acquire more knowledge about a service type. In Section 4 we give an overview of the trading protocol which allows for an interactive service type negotiation. A matching algorithm and protocol have been implemented and are available. Section 5 contains further details as well as our conclusions.

2   Environment

We use a basic model called the *object graph* to motivate our definitions of type notations within open distributed environments. The discussion of object graphs in this section serves as a starting point for the conceptual graphs, which will be presented in the following sections.

2.1   Object graph model

Our model is based upon the classical definition of an object, as it can also be found in the ODP RM [ISO93a], i.e. *an object is characterized by its behaviour and, dually, by its state. An object is distinct from any other object.* Using this definition, a problem domain may be decomposed as a set of interacting and co–operating objects. A snapshot of such an object-based computation may be visualized as a directed graph, where nodes represent objects and arcs represent references. A reference (or arc) is therefore a referral of an object's identity. The direction of the arc determines whose identity is known to whom. For an object to hold a reference to another object means to know about the existence of this particular instance, allowing operation invocations (also commonly called *method invocations*). Thus a directed arc between two nodes (objects) represents the ability to invoke operations along the direction of this arc (i.e. the service provider is at the arc head, and the requester is at the tail). Service providers are also called *server objects* and service requesters are called *client objects*. The directed graph will be called an *object graph*. In terms of level of abstraction a *client* may be an object in the common sense or a human user interacting with a client object. The terms *client* and *user* will be used synonymously throughout this paper.

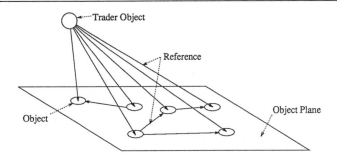

Figure 1: Trader object bridges the knowledge/visibility gap.

An important consequence of the object model is that an object encapsulates data and code. The role of a type specification is therefore crucial in the sense that it should provide enough information to describe an object's behavior, yet conceal any implementation specific details. We assume that both references and objects are typed. Implementation details of a server object are irrelevant to a client. From a client object's perspective (the one holding the reference) a reference guarantees a treaty that the server object must fulfill. Polymorphism here occurs when the type of the reference is a super–type of the object to which it points. The server object therefore is a specialization to what the client expects, if it can fulfill the treaty[1].

One can distinguish two different cases with respect to *when* a type specification is required: at *compile time* or at *runtime*. For compile time type notations there exists a wide range of notations based on interface signatures defined in some *interface definition language* (IDL for short). A type specification written in some IDL commonly lists a set of methods which are implemented by the server object. Special tools generate so–called *stubs* or *proxies* which eventually get linked to the client object. In terms of level of abstraction, an IDL is intended for programmers and is transparent for users of the client object at runtime.

In contrast a type specification notation used during runtime must build upon different mechanisms. The level of abstraction of the underlying objects is higher in the sense that the *user* determines at runtime the kind of service he or she wishes to use. A common technique for building such systems are *generic client objects* which are able to communicate with an a priori unknown server. Examples for such systems are the World Wide Web (WWW), OLE2, OpenDoc or COSM (see [BL+94], [Mic93], [Lab94] and [MML94] respectively).

All these systems have in common that different services can be provided at runtime without the need for a specific service interface definition at compile time. Instead, a service provider has means to dynamically convey its particular user interface to the client via some sort of *graphical user interface (GUI)* descriptions. The generic client is able to interpret these descriptions and to build and present an appropiate GUI for the end user. The user may interact with the generic

---

[1]Polymorphism is often referred to as the "principle of substitutability" where an object of type A may be substituted by an object of type B without anything "bad" happening.

client to invoke operations and to provide parameters embedded as widgets such as edit fields or checkboxes appearing in the GUI. These parameters are transfered to the service provider who takes appropriate actions to perform the request. For example, within the WWW system a GUI description is based on the *hypertext markup language* (HTML) which the generic client is able to translate into a visual presentation.

Via such mechanisms there may be a rich variety of different services accessible. However, there arise questions such as:

- How does a user specify its service requirement?

- How can a user find a suitable service provider for a desired service type?

It is not immediately clear what a notation for a type specification should look like. A type specification in this environment is more abstract and vague than an IDL based specification. In particular it should support the cognitive domain of the users and not of the programmers. In the following sections we propose a technique, called *conceptual graphs*, which is appropriate for runtime type specification.

2.2   Trading and the dualism of type definitions

In this section some consequences with regard to the role of a trader will be discussed. We assume the need for runtime type specifications as discussed above. Furthermore the object graph will be seen in the context of an open distributed environment. By *open* we mean an environment where all participating service providers are not known *a priori*. Thus, the object graph and its modifications are to be seen as an abstraction of a service market, where services are freely provided and requested by independent parties.

In a distributed environment the object graph will generally be partitioned[2]. A client object has only a limited view on the object graph, as global knowledge of it's structure is generally impossible to acquire. As references between objects induce a "knows–about–relation", it is also clear that without appropriate support from an underlying infrastructure a client can't see beyond a transitive closure of the references it holds (i.e. the partition of the object graph in which the client is embedded).

These considerations have led to proposals like the ODP Trader or CORBA's Request Broker (see [ISO93b] and [Gro91] respectively), which serve as a mediator between service requesters and service providers and therefore bridge the knowlegde/visibility gap. The trader matches service requests with previously stored service offers and thereby helps to establish references in the object graph. The match process heavily depends on the precise definition of the type specification notation. With respect to the categorization made in the previous section, current traders primarily match compile time type information.

It is important to think about the role of a type in such an environment. In order for the match algorithm within the trader to succeed, a type description must conform to a kind of "standard", which all participating parties have to agree upon *a priori*. This standard has to be defined well enough to be matched unambiguously against other types. Current traders, like the aforementioned ODP Trader, base their match algorithm mainly upon syntactic features of the interface. The implication is that the exact syntactic structure of a particular service signature must be communicated to all parties.

---

[2]Note that in *one address space* object systems the usual approach is that all except one designated root partition are subject to a garbage collector, as in Smalltalk.

The requirements of the definition of type specification notations can therefore be characterized as follows:

1. A notation should be based upon a *precisely defined syntax* to avoid ambiguities.

2. A notation should be *open* enough to avoid the need for an a priori standardization of service descriptions.

The first requirement originates from the fact that the trader must have solid grounds for a matching algorithm. This leads to an *explicit definition* of an object's type which necessarily must be known by all potential clients in advance. The second requirement on the other hand stems from a pragmatic point of view, whereas a client object should not need the a priori knowledge on *how* a server object has chosen to describe its type. This leads — contrary to the first requirement — to an *implicit definition* of a server object's type. The latter requirement clearly would be desirable as it would avoid the need to standardize every object type in advance.

We call the obvious contradiction the *duality* of the requirements of the notation for a type specification in open distributed environments. We have previously proposed a formal framework to solve this duality for compile time type notations (see [Pud94]). In the following section, we present a notation suitable for runtime typing, which in particular addresses the dualism mentioned above.

3 Towards AI–based trading

In contrast to compile time types, which are handled by a programmer, a type specification suitable for runtime represents an information artifact which is dealt with by a user. A notation therefore must adhere to the world of discourse of the user community with much less precisely defined syntax. On the other hand the notation should be flexible enough to allow for a broad expressiveness for a large variety of services as the experience with the WWW has shown.

Our approach — which copes with the aforementioned dualism — is based upon techniques which originated in the field of machine learning. There exists a wide range of literature on machine learning and various proposals have been made (see [Bol87] for an overview). Concerning the problem of AI–based trading, we have decided to build our framework upon a knowledge representation method called *conceptual graphs* (see [Sow84]). We have devised our own theoretical framework for conceptual graphs to suit the particular needs of a trader. In the following subsection the notion of a conceptual graph and a machine learning algorithm will be presented from a pragmatic point of view. Then a formal specification will be given.

3.1 AI–based trading: a pragmatic example

Conceptual graphs have been developed to model the semantics of natural language. Service descriptions based on conceptual graphs are therefore intuitive in the sense that there is a close relationship to the way human beings represent and organize their knowledge. From an abstract point of view a conceptual graph is a finite, connected, directed, bipartite graph. The nodes of the graph are either *concept* or *relation nodes*. Due to the bipartite nature of the graphs, two concept nodes may only be connected via a relation node.

A concept node represents either a concrete or an abstract object in the world of discourse. As for the context of service types a concept may be a concrete object such as PRINTER, COMPILER or DATABASE including specific instances (e.g. HP-Laserjet, GCC, Ingres, etc), as well as an

abstract object such as PRINTING-SPEED or PROGRAMMING-LANGUAGE with no physical representation. Whereas concepts model objects of our perception, a relation node expresses a specific
relationship between concept nodes. In the following examples a concept node is surrounded by
square brackets and relations by round brackets, respectively.

The following conceptual graph labeled as CG1 describes an object oriented language called
C++[3]. The informal semantic of the concept is: "Something which is a superset of a programming
language called C, supports classes which themself consist of methods and a state. Furthermore
the methods describe the behaviour of classes." The syntax of the following examples is according to a grammar which we have defined for AI–based trading and can be processed by our
implementation.

```
CG1:  [OO-LANGUAGE:{"C++"}] -
            -> (SUPERSET-OF) -> [PROGRAMMING-LANGUAGE:{"C"}],
            -> (SUPPORTS) -> [CLASSES] -
                              -> (HAVE) -> [METHODS] -> (DESCRIBE) -> [BEHAVIOR],
                              -> (HAVE) -> [STATE].
```

A concept can be recursively defined via subconcepts. The concept CG1 is therefore *explained*
by two subconcepts which are connected to the *root concept* C++ with the relations SUPERSET-OF
and SUPPORTS. A concept node itself is divided into a *type* and a possibly empty list of *instances*
for that type. The root concept of CG1 therefore defines C++ as an instance of type OO-LANGUAGE.
If the concept CG1 is regarded as a service which is offered by some provider, then the following
conceptual graph would represent a query which matches with the previous service description:

```
CG2:  [SOMETHING:*] -> (SUPPORTS) -> [CLASSES]
```

The informal semantics of CG2 is: "I need something which supports classes." As CG1 has
previously been defined as something which actually does support classes, the trader would match
these two descriptions. It should be noted that queries and service descriptions are formulated
using the same notation. The root concept of CG2 [SOMETHING:*] introduces two new notions.
The asterisk "*" denotes a *generic object* which will be matched with any other object. On the
other hand it is not clear how SOMETHING is to be matched with OO-LANGUAGE. A concept node is a
typed entity which may have arbitrary number of *instances*. In our notation a type is written left
of a colon whereas the (possibly empty) instance list is written inside curly brackets to the right
side. The set of all types T form a lattice with a partial ordering $\leq_T$ which denotes *specialization*.
The type lattice used for this example is shown in figure 2.

The type SOMETHING as the top element of the lattice is *generic* in the sense that all other
types are specializations of it. The matching process that the trader must perform can specialize
types in a query. In order to match CG1 and CG2, the type SOMETHING is specialized or *reduced* to
OO-LANGUAGE. Next consider a different query called CG3:

```
CG3:  [SOMETHING:*] -> (ENCAPSULATE) -
                          -> [STATE],
                          -> [BEHAVIOR].
```

Some client wishes "something which encapsulates state and behaviour." As can been seen
easily, even after a proper reduction of the root concept [SOMETHING:*], the requirement formulated in CG3 does not match CG1 although from a intuitive point of view they should. There is no

---

[3]For the purpose of this example we assume no further refinement of this service description (i.e. whether the
service CG1 represents a language reference, product information or other).

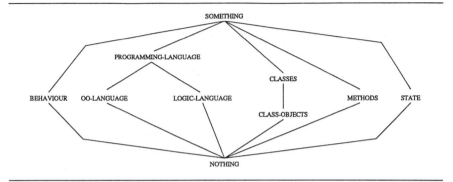

Figure 2: A possible explication of a type lattice.

way the trader can possibly match those two concept graphs because it doesn't have any notion of the underlying semantics. But if the trader were told that the two descriptions denote the same concept, then it could enhance CG1 by *learning* the new features of the concept called C++:

```
CG4: [OO-LANGUAGE:{"C++"}] -
            -> (SUPERSET-OF) -> [PROGRAMMING-LANGUAGE:{"C"}],
            -> (SUPPORTS) -> [CLASSES] -
                                    -> (HAVE) -> [METHODS] -
                                                -> (DESCRIBE) -> [BEHAVIOR].,
                                    -> (HAVE) -> [STATE].,
            -> (ENCAPSULATE) -
                    -> [STATE],
                    -> [BEHAVIOR].
```

Obviously the query CG3 will match the description in CG4. Next consider a different service provider registering a new service called Objective–C. The initial concept graph describing the service might look like:

```
CG5: [OO-LANGUAGE:{"Objective-C"}] -
            -> (SUPERSET-OF) -> [PROGRAMMING-LANGUAGE:{"C"}].
```

The previous concept graph is a subgraph of CG4 and therefore the trader will ask the new service provider whether Objective–C is merley another instance of the type OO-LANGUAGE along with C++. If this should be the case, the trader will simply add the new instance to the root concept node. For the purpose of this example the service provider considers C++ different from Objective–C. In doing so he must augment his original conceptual graph by appropriate subconcepts which distinguish it from CG4. This augmentation results in the following graph CG6 which states that "Objective–C is a superset of C and supports class objects":

```
CG6: [OO-LANGUAGE:{"Objective-C"}] -
            -> (SUPERSET-OF) -> [PROGRAMMING-LANGUAGE:{"C"}],
            -> (SUPPORTS) -> [CLASS-OBJECTS].
```

As the subconcept [CLASS-OBJECTS] distinguishes the two concepts, it is added as a counter example to CG4 which yields the following new conceptual graph for C++:

```
CG7: [OO-LANGUAGE:{"C++"}] -
          -> (SUPERSET-OF) -> [PROGRAMMING-LANGUAGE:{"C"}],
          -> (SUPPORTS) -> [CLASSES] -
                          -> (HAVE) -> [METHODS] -
                                      -> (DESCRIBE) -> [BEHAVIOR].,
                          -> (HAVE) -> [STATE].,
      -> (ENCAPSULATE) -
              -> [STATE],
              -> [BEHAVIOR].,
      -> (NOT SUPPORTS) -> [CLASS-OBJECTS].
```

As the previous discussion suggests, a conceptual graph *explains* through an amalgamation of examples and counter examples. The trader can increase the quality of a concept over time as it incorporates new subconcepts. The quality of the matching process performed by the trader will therefore increase in the same way.

3.2   Formal Specification

The previous subsection has presented an extended example to demonstrate the power of a trader employing AI–techniques. In this section a formal framework for conceptual graphs, the join of two graphs as a learning mechanism and finally a match of two graphs will be given. We will start by defining the basic sets of the formal model:

Types T: Let T be a set of all types. The types in T and the partial ordering $\leq_T$ form a lattice $(T, \leq_T)$ with SOMETHING $\in T$ the top element and NOTHING $\in T$ the bottom element.

Relations R: Let R be the set of all relations. The relations R and the partial ordering $\leq_R$ form a lattice $(R, \leq_R)$ with LINK $\in R$ the top element and NO-LINK $\in R$ the bottom element.

Objects O: Let O be the set of objects of our perception. The objects are to be seen as instances of one or more types from T.

Concepts C_n: Let $C_n = T \times 2^O$ the set of all concepts. A concept is a tuple of a *type* and a subset of the set of all *objects*. The *generic object* (denoted by $*$) is a representative for any object and defined as O for formal reasons.

The set of *relations* is also organized in terms of a lattice with the partial ordering $\leq_R$. This will allow greater flexibility for the match operation. The set of all concepts may not be true with respect to the world of discourse. Therefore we introduce a *conformity relation* which provides a link to a higher order knowledge base. The relation *Conf* is not meant to be implemented, rather as a formal framework to argue about the *truth* of concepts. But there are nevertheless some properties which must hold. The decision of the meaningfulness of a concept eventually can only be decided outside the scope of the trader.

Conformity Relation: Conf $: C_n \rightarrow \{true, false\}$

Let $(t, o) \in C_n$ with $t \in T$ and $o \in 2^O$.

1. $\Big(\text{Conf}((t, o)) = true\Big) \Rightarrow \Big(\forall t' \in T : (t \leq t') \Rightarrow (\text{Conf}((t', o)) = true)\Big)$.

2. $\text{Conf}((\text{SOMETHING}, o)) = true$.

3. $\text{Conf}((\texttt{NOTHING}, o)) = false$.

4. $\text{Conf}((t, *)) = true$.

5. $\text{Conf}((t, \emptyset)) = false$.

If an object is an instance of a type, then it must also be an instance of all it's super types (i.e. more general types). All objects are instances of the top type SOMETHING and no object is an instance of the bottom type NOTHING. Finally every type has at least the generic object as an instance.

One important transformation of concept nodes is that of a *restrict operation*. A restrict specializes two concepts to their least common ancestor in terms of their types and instance lists. It is important to note that the restrict operation *does not* necessarily preserve truth (i.e. the result of a restrict operation on two true concept nodes, with respect to the conformity relation, must not necessarily be true). The join and match operation will use the restrict to transform a query for building maximal common subgraphs of two concepts.

The result of Restrict_C is the *minimal common subtype* (i.e. the least subtype which can be obtained by specializing two types). Restrict_R denoting the *minimal common relation* is defined analogously.

Restrict$_C$: Let $\text{Restrict}_C : C_n \times C_n \to C_n$ where

$$\text{Restrict}_C((t_1, o_1), (t_2, o_2)) =_{df} \begin{cases} (t, o) & : \begin{aligned} &(t \leq_T t_1 \wedge t \leq_T t_2) \wedge ((\forall d \in T) \\ &((d \leq_T t_1 \wedge d \leq_T t_2) \Rightarrow d \leq_T t)) \text{ and} \\ &\text{with } t \neq \texttt{NOTHING} \text{ and } o = o_1 \cap o_2 \end{aligned} \\ (\texttt{NOTHING}, \emptyset) & : \text{otherwise} \end{cases}$$

Restrict$_R$: Let $\text{Restrict}_R : R \times R \to R$ where $\text{Restrict}_R(s, t) = u$ iff $u \leq_R s$ and $u \leq_R t$ and $(\forall w \in R)((w \leq_R s \wedge w \leq_R t) \Rightarrow w \leq_R u)$.

The first major definition is that of a *conceptual graph* as a graph containing concept and relation nodes.

Conceptual graph G: Let $N \subseteq \mathbb{N}$ be a finite, not empty set of node numbers and K be a set consisting of concepts and relations with $K \subseteq C_n \cup R$. There must be at least one concept in K and K is finite ($C_n \cap K \neq \emptyset$ and $|K| < \infty$). Let $m : N \to K$ be a total, not necessarily surjective numbering function. Let $V \subseteq N \times N$ be a set of vertices.

Let $G = (N, K, V, m)$ be a rooted, connected, acyclic and bipartite digraph with

(i) $\left(\forall (n_1, n_2) \in V \right) \left(\left| \{(n, n_1) | (n, n_1) \in V \wedge n \in N \} \right| \leq 1 \right)$

(ii) $\left(\exists n \in N \right) \left(\left| \{(n_1, n) | (n_1, n) \in V \wedge n_1 \in N \} \right| = 0 \right)$ (n is called the root node number of the conceptual graph ($root(G) = n$))

(iii) $\left(\forall (n_1, n_2) \in V \right) \left((m(n_1) \in C_n \wedge m(n_2) \in R) \vee (m(n_1) \in R \wedge m(n_2) \in C_n) \right)$

(iv) $\left(\forall (n_1, n_2) \in V \right) \left(\left| \{(n_2, n) | (n_2, n) \in V \wedge n \in N \} \right| = 0 \Rightarrow m(n_2) \in C_n \right)$.

The set of all conceptual graphs is denoted by CG.

The join operation which is defined next merges two conceptual graphs into one. The join is not possible if the root concept nodes of the two graphs can't be restricted. Otherwise the resulting graph is obtained by recursively trying to overlay subconcepts as much as possible. The merging of two graphs is minimal in the sense that the joined graph is the smallest possible. The join operation is the basis for a machine learning algorithm. It should be noted that the result of a join necessarily has to be checked against the conformity relation. A join of two graphs can therefore only be a tool provided by the trader to aid a service provider augmenting and refining one of his or her service descriptions.

Join operation $Join : CG \times CG \rightarrow CG \cup \{nil\}$. The result of $Join(G_1, G_2)$ with $G_i = (N_i, K_i, V_i, m_i)$, w.l.o.g. $N_1 \cap N_2 = \emptyset$, $n_i = root(G_i)$, $k_i = m_i(n_i)$ the root concept node of G_i, $i = 1, 2$ is:

(i) *nil* if $\text{Restrict}_C(k_1, k_2) = (\text{NOTHING}, \emptyset)$ or

(ii) $G_J = (N_J, K_J, V_J, m_J)$. $\tilde{k} = \text{Restrict}_C(k_1, k_2)$ the new root node of G_J and $\tilde{n} = \max\{N_1 \cup N_2\} + 1$ the new root node number.

   (a) $N_J =_{df} N_1 \cup N_2 \cup \{\tilde{n}\} \setminus \{n_1, n_2\}$.

   (b) $K_J =_{df} K_1 \cup K_2 \cup \{\tilde{k}\}$.

   (c) $V_J =_{df} V_1 \cup V_2 \cup \{(\tilde{n}, n) | \exists (n_1, n) \in V_1 \vee \exists (n_2, n) \in V_2\} \setminus \{(\hat{n}, n) | (\hat{n} = n_1 \vee \hat{n} = n_2) \wedge n \in N_1 \cup N_2\}$.

   (d) Define m_J as follows:

   $$m_J(n) =_{df} \begin{cases} m_1(n) & : & n \in N_1 \setminus \{n_1\} \\ m_2(n) & : & n \in N_2 \setminus \{n_2\} \\ \tilde{k} & : & n = \tilde{n} \end{cases}$$

(iii) Set $\hat{n} = \tilde{n}$.

(iv) If there exists direct successor nodes of $\hat{n}$: for all direct successors $\hat{n}_1, \hat{n}_2$ of $\hat{n}$:

   (a) if $m_J(\hat{n}) \in C_n$ and $\text{Restrict}_R(m_J(\hat{n}_1), m_J(\hat{n}_2)) = k' \neq \text{NO-LINK}$ then define $m_J(\hat{n}_1) =_{df} k'$ and connect all direct successors of $\hat{n}_2$ with $\hat{n}_1$. Define $m_J(\hat{n}_2) =_{df}$ undef. $\hat{n} = \hat{n}_1$. Go to (iv).

   (b) if $m_J(\hat{n}) \in R$ and $\text{Restrict}_C(m_J(\hat{n}_1), m_J(\hat{n}_2)) = k' \neq (\text{NOTHING}, \emptyset)$ then define $m_J(\hat{n}_1) =_{df} k'$ and connect all direct successors of $\hat{n}_2$ with $\hat{n}_1$. Define $m_J(\hat{n}_2) =_{df}$ undef. $\hat{n} = \hat{n}_1$. Go to (iv).

The match operation is one of the key mechanisms of the AI–based trading concept. A match takes two conceptual graphs as input and produces their intersection.

Match operation $Match : CG \times CG \rightarrow CG$: Let G_S be a conceptual graph for a service description, G_Q a conceptual graph for a query. $Match(G_S, G_Q) = G_M$ the match graph is constructed as follows:

(i) $G_M =_{df} G_Q$, $n_i = root(G_i)$, $i \in \{S, M\}$

(ii) If $m_S(n_S) \in C_n$ then

   (a) if there exists direct successor nodes of n_S:

if the number of direct successor nodes of n_M is greater 0 then for all direct successor nodes n'_M of n_M: if there exists no node n'_S (direct successor of n_S) with $\text{Restrict}_R(m_S(n'_S), m_M(n'_M)) \neq \text{NO-LINK}$ then delete n'_M and all existing successor nodes of n'_M else set $n_S = n'_S$ and $n_M = n'_M$, go to (ii)

(b) else delete all existing successor nodes of n_M; if $\text{Restrict}_C(m_S(n_S), m_M(n_M)) = (\text{NOTHING}, \emptyset)$ then delete n_M.

(iii) If $m_S(n_S) \in R$ then

(a) if there exists direct successor nodes of n_S:

if the number of direct successor nodes of n_M is greater 0 then for all direct successor nodes n'_M of n_M: if there exists no node n'_S (direct successor of n_S) with $\text{Restrict}_C(m_S(n'_S), m_M(n'_M)) \neq (\text{NOTHING}, \emptyset)$ then delete n'_M and all existing successor nodes of n'_M else set $n_S = n'_S$ and $n_M = n'_M$, go to (ii)

(b) else delete all existing successor nodes of n_M; if $\text{Restrict}_R(m_S(n_S), m_M(n_M)) = \text{NO-LINK}$ then delete n_M.

The trader decides the quality of a match by evaluating the result of a match operation according to some metric (for an in–depth discussion on this topic see [PMG95]). As we have just finished a prototype of the AI–based trader, this metric will be subject to modifications as we gain more experience. It should be clear that *wrong* answers to a query are possible if the quality of the service description or the query itself isn't sufficient. This will be discussed in greater detail in the following section.

4   AI–trading protocol

In this section we focus upon the *trading protocol* which embeds the trader as well as client and server objects into one framework. The justification for a designated protocol becomes clear when compared with the traditional task of service trading based on compile time type notations. The proposed type notation, introduced as *conceptual graphs*, does not rule out that the trader may make mistakes due to unprecise service descriptions. The interaction with a trader therefore goes beyond the one time matching of service requests. An AI–based trader may have to backtrack and refine previously stored descriptions through learning of new concepts and offering a client different services. It should be noted that a human user (i.e. *not* some software component) eventually recognizes a wrong service which may lead to further interaction with the trader providing a refined description.

We will discuss the protocol only on an informal level. Three distinguished roles may be identified which participate in the trading process: a *client* (respectively a *user*), a *server* and the *trader* itself. Each of these roles will be discussed separately.

Trader: Conceptually the trader maintains a database of all service providers who have registered themselves previously. The database holds tuples each containing a *conceptual graph* as one argument and *addresses* of one or more server objects providing the service described by the conceptual graph as another. If a match of a request and a service offer succeeds, the trader uses the address to construct a reference which will be given to the client as the result. The precise structure of an address lies outside the framework.

Service provider: A service provider implements some service and wishes to export it through the trader. A suitable service description is formulated as a conceptual graph. The service provider may have to adjust his or her concept upon request from the trader. Eventually the quality of the service description will increase as the conceptual graph is refined over time.

Service requester: A service requester seeks a particular service and consults the trader for an appropriate reference. The desired service is specified again via a conceptual graph. As the trader's knowledgebase is incomplete initially, a service requester may not get what he or she intended. In this case the user has to further interact with the trader.

With respect to the AI–based trading protocol there are two distinct interactions with the trader: the *export* of a new service and the *import* of a particular service. A service provider initially exports a conceptual graph describing the service along with an address. The trader tries to match this graph with those already stored in its database. The service provider is presented a list of possible matches, i.e. services which are similar. The service provider either has to refine his initial description to increase the semantic distance to those matches or can decide that his service is just another instance of a service already registered.

A service requester formulates a query in terms of a conceptual graph which the trader tries to match with those services previously stored in its database. If there is no match the requester has to browse all services manually. If this search leads to the desired service, the trader then forwards the original query to the provider. It is the provider's task to augment his own conceptual graph accordingly, such that the formerly unsatisfied query will produce a match. In this case the learning process occurs on the side of the service provider.

If the service requester notices that an inadequate service was given to him or her then the query has to be re–formulated and posted again to the trader. Eventually the service requester will get the desired service. The learning process here occurs on the requester's side who has learned to precisely define what he or she wants. This scenario suggests that in terms of machine learning terminology the trader assumes the role of a teacher as well as a student (when new service descriptions are being taught), the client assumes the role of a student and the server may both act as a teacher and as a student.

5   Conclusion and outlook

Open distributed environments may be seen as a service market where services are freely offered and requested. The mediation of these services is done by a designated system component known as a *trader*. Current traders primarily base their matching algorithm of services upon IDL–based type notations. In this paper we have proposed a new type notation which allows for abstract descriptions of arbitrary services. This notation — building upon techniques from the domain of machine learning — supports the cognitive domain of the users. The trader maintains a knowledgebase which is refined over time as the trader *learns* various ways of describing a service. The quality of a match therefore increases in the same sense, thus solving what we call the *dualism* of type notations.

We have implemented the algorithms and the protocol described in this paper. The complete source, using various C++ PD–class libraries, are placed in the public domain and may be obtained from the first author. Our implementation of an AI–based trader maintains a database of *uniform resource locators* (URL) of the World Wide Web. Future work will include a more comfortable GUI–based front end for conceptual graphs as well as experiments with various metrics for the match algorithm.

Acknowledgements

We thank Scott M. King (Thinkage LTD, Canada) and an anonymous referee for their comments and discussions on this article.

References

[Ame90] P. America. Designing an object oriented programming language with behavioural subtyping. In *REX School/Workshop, LNCS 489*. Springer, May/June 1990.

[BJ93] B. Liskov and J. Wing. A new definition of the subtype relation. In O. M. Nierstrasz, editor, *ECOOP'93: Object–Oriented Programming*. Springer, 1993.

[BL+94] Tim Berners-Lee et al. The World–Wide Web. *Communication of the Association for Computing Machinery*, 37(8):76–82, August 1994.

[Bol87] Leonard Bolc, editor. *Computational Models of Learning*. Springer, 1987.

[Gro91] Object Management Group. *The Common Object Request Broker: Architecture and Specification Revision 1.1*. 1991.

[ISO93a] ISO/IEC. Information Technology – Basic Reference Model of Open Distributed Processing – Part I. ISO/IEC COMMITTEE DRAFT ITU-T RECOMMENDATION X.902, 1993. ISO/IEC CD 10746–2.3.

[ISO93b] ISO/IEC. *ODP–Trader, Document Title ISO/IEC JTC 1/SC 21 N 8192*. 1993.

[ISO94] ISO/IEC. Working Document - ODP Trading Function, January 1994. ISO/IEC JTC1/SC21 N8409.

[Lab94] Component Integration Laboratories. Shaping tomorrow's software (white paper). Technical report, cil.org:/pub/cilabs/tech/opendoc/OD–overview.ps, 1994.

[Mic93] Microsoft. OLE 2.0 Design Specification. Technical report, ftp.microsoft.com /developr/drg/ole–info/OLE–2.01–docs/OLE2SPEC.ZIP, 1993.

[MML94] M. Merz, K. Müller, and W. Lamersdorf. Service trading and mediation in distributed computing environments. In *Proceedings of the International Conference on Distributed Computing Systems (ICDCS '94)*. IEEE Computer Society Press, 1994.

[PMG95] A. Puder, S. Markwitz, and F. Gudermann. Service trading using conceptual structures. In *3rd International Conference on Conceptual Structures (ICCS'95)*, Santa Cruz, University of California, 14–18 August 1995. Springer.

[Pud94] Arno Puder. A Declarative Extension of IDL–based Type Definitions within Open Distributed Environments. In *OOIS'94: Object–Oriented Information Systems*, South Bank University, London, 1994. Springer.

[Sow84] John F. Sowa. *Conceptual Structures, information processing mind and machine*. Addison–Wesley Publishing Company, 1984.

SESSION ON

Interworking Traders

13

A Model for a Federative Trader

Luiz Augusto de Paula Lima Jr.[1]
Edmundo Roberto Mauro Madeira[2]

This work proposes a model of a TRADER considering specially the aspects related to the establishment and management of FEDERATIONS OF TRADERS inside the framework of the Reference Model for Open Distributed Processing (RM-ODP). The modules of the proposed model are commented and the operations at the interfaces are listed. Finally, a protocol for the communication between the trader and its administrator is defined and the federation establishment process is discussed. A prototype of this model was implemented inside the Multiware Platform which is being developed at the University of Campinas.

Keyword Codes : C.2.3; C.2.4; D.4.4.
Keywords : Network Operations; Distributed Systems; Communications Management.

1. INTRODUCTION

The advances in the communication technology with high transmission rates, improved security and low error rates, the future (and present) needs of the users for integration of automation islands, cooperative work (CSCW) and architectures for open distributed services and the progress in the standardization (RM-ODP) made an environment for open distributed processing not only possible, but also indispensable to attend the new demands of the users (e.g., decision-making systems [1], distributed artificial intelligence (DAI), etc.).

The open environment has the advantage of being unlimited and free to admit any kind of user, component or application. Therefore, this environment is characterized by heterogeneity and decentralization, but, at the same time, it must assure the autonomy of each component or local environment.

In an Open Distributed System, it is highly desirable the existence of a means to make dynamic selection of computational services that satisfy certain properties. For this reason, the

1.Institut National des Télécommunications - LOR - 9, rue Charles Fourier - 91011 - Evry Cedex - France
E-mail : lima@hugo.int-evry.fr or lima@dcc.unicamp.br
2.Universidade Estadual de Campinas - DCC - Cx. Postal 6065 - 13081-970 - Campinas - SP - Brazil
E-mail : edmundo@dcc.unicamp.br

trader is a key component in such an environment. Its role is to manage the knowledge of the currently available services and to find service offers that match the clients' requirements [2].

The grouping of traders in *Federations* allows that distinct distributed systems work together but, at the same time, keep control of their own domains [3].

In this work, we present a model which can be used as a basis for an implementation of a trader, and we give special attention to the aspects related to Trader Federations. This is done according to the standard of the ISO for ODP [4][5][6][7][8][9][10]. This means that autonomy, decentralization and encapsulation are the fundamental principles that will lead all the following proposals.

A prototype of the model was implemented inside the framework of the Multiware Platform [11], which aims at providing support for the creation of distributed applications.

2. A MODEL FOR A TRADER

There are basically two groups of activities performed by a trader :

* the management of the data base related to the (static and dynamic) information held by the trader; and
* the execution of the operations using the information.

We propose a model that consists of the refinement of these two "modules", considering separately local functions and those related to the establishment of federations and federated operations, and also putting in different modules the management of the static and dynamic information.

Figure 1 presents a model for the trader where we can identify three basic components (agents, from the enterprise viewpoint). They are :

* the client which is the importer or exporter of service offers;
* the administrator - the *local* one that deals with local management and the *federation* one which deals with the establishment of federations (section 3);
* the trader.

2.1. The description of the modules

Two modules are responsible for storage and information retrieval in the trader. They are :

* *Static Information Module*

This module concentrates the operations that handle the static information which corresponds to service types (in the Type Repository[1]) and service offers (in the Directory).

1.The *Type Repository* can be located outside the trader and this module can be used to deal with the efficient communication with the "real" type repository.

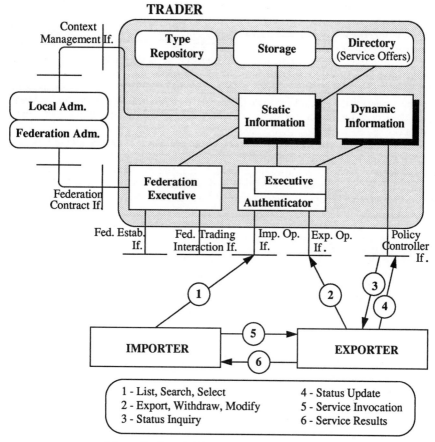

Figure 1 : A model for a trader. (if. = interface)

• *Dynamic Information Module*

It is responsible for updating the dynamic properties of a service offer. This module contacts the *Policy Controller* of the exporting object to obtain this information. If the identifier of the policy controller in the service offer stored as a static information is null, no operation is done.

The *Type Repository* supports the storage (using the *Storage Module*) and the dynamic matching of service types to indicate, for example, when two types are equivalent or when a type is a sub-type of another, according to the ISO standard [6]. As we said before, the type repository can be an autonomous object with its own interfaces and operations. For simplicity, we placed it here.

The *Directory* stores all the service offers exported to the trader. It can be implemented using the X.500 of the ITU-T, for example (see [12]).

The *Executive Module* is responsible for executing the local operations of the trader using the information available. The *Federation Executive* analyses the requests that arrive through the interface of federated interaction considering the respective exporting contract and it may send the request to the executive module (in case the constraints in the contract are satisfied).

The federation executive is also responsible for the process of establishing the federation contracts.

The *Authenticator* is the module that checks the permission of the importer or exporter to execute a given operation.

The *Local Administrator* uses the Static Information Module to create and remove contexts, and to deal with the security aspects related to the static information. The *Federation Administrator* uses the Federation Executive to establish contracts that are stored as special static information.

2.2. An example - federative search

When the trader receives an import request, it checks through the *authenticator* the access permissions. Then, the *executive* tries to find service offers from the modules of *static* and *dynamic information* that match the importer requirements. To do so, first of all, the types are checked (*Type Repository*) and the information is then retrieved from the directory. If there is a policy controller interface defined, the *dynamic information* is searched. If there is no service offers locally, then a request for a federated import is passed to the *federation executive* module that checks the existence of suitable importing contracts in the search scope and then sends the request for a federated search to the remote trader.

In the remote trader, the search request arrives in the *Federation Trading Interaction* interface. Then, the *federation executive* applies the restrictions, rules and mapping functions which are specified in the respective export contract to the parameters received with the remote *search*. Once all the constraints are satisfied, a new operation with possibly new (transformed) parameters is issued to the *executive* module that performs the normal operation and returns the found values (if any) to the *federation executive* module.

For example, when an operation

    search (..., service_type, matching_criteria, scope, ...);

is performed in some *Federation Trading Interaction* interface, the *federation executive* applies the rules in the associated export contract transforming (if necessary) some parameters to meet the requirements specified in the contract and to make them understandable for the local trader. Then an operation

    search (..., service_type', matching_criteria', scope', ...);

is issued to the *executive* module which goes on with the normal procedure.

2.3. Interfaces

From this model, we identify a set of interfaces that are needed to allow trader management and to attend the functionality that is required. The interfaces with the respective operations are presented in figure 2.

Interfaces	Operations
Export Operations Interface	EXPORT, WITHDRAW, REPLACE
Importer Operations Interface	LIST_OFFER_DETAILS, SEARCH, SELECT
Policy Controller Interface	STATUS_INQUIRY, EXPORTER_POLICY
Context Management Interface	CREATE_CONTEXT, DELETE_CONTEXT, LIST_CONTEXT, LIST_CONTEXT_CONTENT, AUTHORIZE
Federation Contract Interface	ESTABLISH_FEDERATION, DISTRIBUTE_CATALOGUE, REQUEST_CATALOGUE
Federation Establishment Interface	EXCHANGE_CONTRACT
Federated Interaction Interface	EXPORT, WITHDRAW, REPLACE, SEARCH, LIST_OFFER_DETAILS, SELECT

Figure 2 : Table of interfaces and their operations.

The interfaces for *importing* and *exporting operations* correspond to the *trading interface*. This is the most used interface in a *Trading Community* because it contains the basic operations.

The interfaces of *Context Management* and *Federation Contract* are used only by the local and federation administrators, respectively.

The interface of the *Policy Controller* allows the trader to obtain the dynamic information of a service offer.

The interface of *Federation Establishment* offers to other traders the possibility of establishing federation contracts between them.

Finally, the interface of *Federated Interaction* receives requests of federated imports and exports from other traders that are acting on behalf of their clients and with whom there is a federation contract established. There is a distinct interface of Federated Interaction for each exporting contract of the trader. All the operations in this interface have the same semantics and the same parameters that are found in the operations of the *Export* and *Import Operations* interfaces. But here all of these operations are submitted to the restrictions specified in the associated export contract.

3. THE ADMINISTRATOR

The local and federation administrators are responsible for the definition and the enforcing of the trading policies in the local and federation levels, respectively. The *local administrator* adds new service types to a given context, creates, destroys and renames service offer contexts and authorizes clients to use (for a search or an export) the several existing contexts. The role of the *Federation Administrator* is to prepare the catalogue, to request a catalogue from other trader, to decide when and with whom it should establish a federation, to accept, refuse or make proposals for federation contracts defining the policies to establish such federation. A *catalogue* contains information on the service types and on the extension of the service offer data base that the trader will make available to other traders. It also specifies the permitted operations through federation (exporting trader policy) and the linking extension to other traders, i.e. if the exporting trader will pass on requests to other traders or not.

The administrators are located outside the trader to add the possibility of several traders being managed by only one administrator, which is desirable if we consider several traders in a same enterprise, for example. In other words, a trader must have only one administrator, but an administrator can manage several traders (as suggested also in [14]).

Our administrator is divided into two parts as can be seen in figure 3.

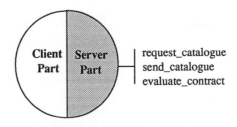

request_catalogue
send_catalogue
evaluate_contract

Figure 3 : The administrator.

• The *Server Part*, that offers the following services to the trader :

  - *send_catalogue* : using this operation the trader can inform the administrator about a received catalogue;
  - *request_catalogue* : asks the administrator to create a catalogue;
  - *evaluate_contract* : asks the administrator to evaluate a proposal for a federation contract.

• The *Client Part* that uses some operations in the trader's interfaces. These operations are :

- in the *Type Repository Interface* :

• *add_service_type* : adds a service type to the type repository;
• *display_types* : shows the service types in the type repository;
• (other operations needed to manage the type repository).

- in the *Context Management Interface* :

• *create/delete_context* : creates/destroys a given context in the directory of service offers;
• *authorize* : defines the set of available interfaces, operations and contexts for a client.
- in the *Federation Contract Interface* :

• *distribute_catalogue* : asks the trader to advertize its catalogue to some other trader.
• *request_catalogue* : asks the trader to obtain the catalogue from some other trader.
• *establish_federation* : asks the trader to try the establishment of a federation contract with other trader providing a contract proposal.

3.1. Establishing federation contracts

To establish a federation contract it is necessary the intervention of the Federation Administrator, because the decisions to create a federation and to evaluate and propose contracts are taken by this object.

In figure 4 it is represented the scenery for negotiation and establishment of a federation contract between two traders, where one performs the role of *exporter* and the other the role of *importer* of services.

Beforehand, in order to establish a federation contract, it is necessary that the importing trader receives the catalogue of the exporting trader. And there are two ways to do this. In the first one, the *administrator 1* decides to send its catalogue to a potential importing trader. When the exporting trader receives the request to distribute the catalogue (*distribute_catalogue* operation) from its administrator, it acts as a normal client of the importing trader using the "*export*" operation to send the catalogue as service properties and to announce the identifier of its *Federation Establishment* interface. As soon as the importing trader identifies an export request with the standard service type of *federation establishment*, it warns its administrator (*send_catalogue* operation) about the arrival of a catalogue. What is returned to the importing trader, to the exporting trader and to the *administrator 1* is only an indication that the *administrator 2* is informed about the exporting trader's catalogue. The other way to notify the *administrator 2* about the catalogue of the exporting trader consists of simply requesting the catalogue and the process is similar.

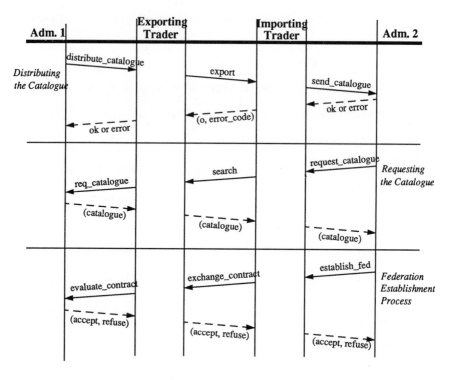

Figure 4 : Scenery to establish a federation.

Once the catalogue of the exporting trader is known to the *administrator 2*, it can now begin the negotiation of the federation contract. If it is interested in the entries in the catalogue, it can start the contract establishment process. The *administrator 2* creates a federation contract proposal (based on the information held in the catalogue) and sends it to the *administrator 1* (*establish_fed*, *exchange_contract* and *evaluate_contract* operations). The *administrator 1* may accept, refuse, or refuse the proposal and suggest a reduced contract. When a federation contract is agreed, a new *Federated Interaction* interface associated with the exporting contract is created in the exporting trader and the federated operations can be now executed.

The operations that allow a trader to communicate with its administrator are presented in the table of figure 5.

3.2. Interworking considerations

In our work, we have identified basically two phases involved in the interworking of traders :

Operation	Parameters	Return Values
distribute_catalogue	location of the imp. trader, catalogue	ok or error
send_catalogue	catalogue, fed. establishment interface id.	ok or error
request_catalogue	location of the exp. trader	catalogue (possibly empty)
req_catalogue		catalogue
establish_fed	fed. establishment interface id., location of the exp. trader, proposal of a federation contract, local context to store the imp. contract	accepted or refused
evaluate_contract	proposal of a federation contract	accepted or refused

Figure 5 : Operations between the trader and its administrator.

• the phase of "federation establishment" (or "link establishment", or "contract establishment");
• and the phase when the federated operations are performed.

There are also two approaches in attempting to solve the problem of *service types* crossing type domain boundaries : either we leave the "heavy work" to be made in the importing phase (when a federated search is performed), or we leave it to the federation establishment phase.

From our point of view, it is better to do all the "heavy work" in the federation establishment phase for the following reasons :

• the trading operations through federation become lighter, that is, faster, and performance aspects are more important here than in the federation establishment phase;
• we will not need a universal standard language for expressing service types, but just some (type description) language agreed between each pair of traders in the federation establishment phase;
• we will not need to know, for example, the interface identifier of the remote type repository.

3.3. Implementation aspects

In our prototype, all of the operations of figure 5 were implemented together with those necessary to the federation establishment process (e.g., *export, search, exchange_contract,* etc.) and other security functions. The prototype was implemented using Remote Procedure Calls over the SunOS and it will be soon migrated to the Multiware Platform which is based on the ORB.

Our approach in attempting to solve the "interworking problem" is to rely on the federation establishment phase to define the equivalence between service types that belong to different "service type domains" (this is done using *mapping functions*) and to define the searching scope by means of the agreement of a number of available contexts via federation. This information is stored in the *exporting contract* in the exporting trader side and in the *importing contract* in the importing trader side.

The contexts of the remote (exporting) trader are associated to the local context space through the importing contract which is stored in some local context. An importing contract contains a list of the accessible contexts (for reading and/or writing) of the remote trader and it always belongs to at least one context of the importing trader. The appropriate context where it is stored is specified by the administrator in the federation contract establishment process. Therefore, the relation between the local and remote context spaces is only "logical", it is neither reciprocal nor hierarchical and, in fact, this relation is very "free", in a sense that it "associates" a context in one trader to a *list* of contexts in the remote trader, which may be or not part of a directed acyclic graph structure.

4. RELATED WORK

Many traders are currently being implemented around the world (DSTC - Australia, University of Hamburg, the trader of the BERKOM project, etc.) and some approaches have been proposed (integration with X.500, implementation over DCE/OSF, etc.). Here, we will comment briefly some aspects related to using the X.500 directory with the trader and some features of the trader of the BERKOM project.

The Directory presented in figure 1 can be implemented using the X.500 directory, as suggested in [12], but in our implementation we did not use X.500. Instead, we built a simple directory using C++ subroutines. As pointed out in [12] there are certain problems which cannot be solved inside the X.500 service (e.g., type checking) and it is impossible to model some relations between traders forming a federation. For this reason, in our model, the X.500 is suitable only with a limited "power", that is, only to store the (possibly distributed) static information which concerns only one trader. The fact that the structure of the X.500 directory is hierarchical can impose some undesirable constraints, since the directed graph required by the ODP documents [13] is a more flexible structure. To sum up, our trader is not based in the X.500 ideas because we think they may be restrictive in some cases, but we agree that the X.500 could be used to implement the directory depicted in figure 1.

In the trader of the Y Platform [15] the imports always return only the best service offer. In other words, their *search* corresponds to our *select*. This means that the trader must verify the availability of the found service in every import in order to assure that the client will not need another option (one that it cannot give, since it returns just "the best" one). Performance aspects are seriously considered [2] to reduce the importing time (e.g., *cache* memory). They will be added into our prototype that was implemented using the model of figure 1. There is no mechanism for federation in the analysed version of the trader of the Y Platform.

5. CONCLUSION

The model described in this work is suitable for an Environment for Open Distributed Processing, where autonomy, decentralization and encapsulation are basic principles. To sum up, the main contributions of this work are:

- the clear specification of the modules inside the trader leads to an easier implementation;
- the definition of the basic protocol between the trader and its administrator is very important (it shall be improved in the future) to allow interworking;
- one administrator may manage several traders in different sites (but inside a certain domain);
- a scenery for the federation contract negotiation phase was proposed.

Indeed, by now we cannot evaluate how expensive our proposal for interworking can be and future work is needed to find out if this approach is reasonable. It will certainly require more "human interaction" (by means of the administrator), but even some operations of the administrator can be automatized.

The problem of making *new service types* available to the remote trader is the same problem of the contract federation updating and in this stage of the work we do not have a proposal for a solution.

Following this discussion, there are some aspects that need a more detailed definition and can be subjects for future research :

- formalizing the "mapping function" concept;
- deciding on how to update a federation contract;
- deciding on how to negotiate the agreement of types, services, scope, etc. in an efficient way (defining the scope of "human interaction").

Our approach to interworking requires a greater participation of the trader administrator because most of the details for interworking are agreed in the federation establishment phase.

The Multiware Platform will incorporate the prototype that was implemented following the concepts of the proposed model.

6. ACKNOWLEDGMENTS

The work reported in this paper was funded by CNPq and FAPESP.

7. REFERENCES

[1] *Mendes, M. J.; Loyolla, W. P. D. C.; Madeira, E. R. M.* - "Demos : A Distributed Decision-Making Open Support System" - 4th IEEE Workshop on Future Trends in Distributed Computing Systems - Sept. 1993 - Lisboa - Portugal - pp 208-214.

[2] *Tschammer, V.* - "Trading in Open Distributed Systems" - Proceedings of the Invited Papers - XI SBRC - Campinas - Brazil - May 1993.

[3] *Bearman, Mirion; Raymond, Kerry* - "Federating Traders: An ODP adventure", IFIP 1992, pp 125-141.

[4] "ISO/IEC JTC 1/SC 21/N 7053" - Basic Reference Model of ODP - Part 1: Overview and Guide to Use.

[5] "ISO/IEC DIS 10746-2 - ITU-T Draft Rec. X.902 - Feb. 1994" - Basic Reference Model of ODP - Part 2: Descriptive Model.

[6] "ISO/IEC JTC 1/SC 21/N 10746-3.2 - ITU-T Draft Rec. X.903 - Feb. 1994" - Basic Reference Model of ODP - Part 3: Prescriptive Model.

[7] "ISO/IEC 21/N 7056" (Recommendation X.905) - Basic Reference Model of ODP - Part 5: Architectural Semantics.

[8] "ISO/IEC JTC 1/SC 21/N 7047, 1992-06-30" - Working Document on Topic 9.1 - ODP Trader.

[9] "ISO/IEC JTC 1/SC 21/N 7057, 1991-06-20" - WG7 Project Management Document: ODP list of open and resolved issues - June 1992.

[10] "ISO/IEC JTC 1/SC 21 - Nov. 1993" Information Technology - ODP Trading Function.

[11] *Loyolla, W. P. D. C; Madeira, E. R. M.; Mendes, M. J.; Cardoso, E.; Magalhães, M. F.* - "Multiware Platform : an Open Distributed Environment for Multimedia Cooperative Applications" - COMPSAC'94 - IEEE Computer Software & Application Conference, Taipei, Taiwan, Nov. 1994.

[12] *Popien, C.; Meyer, B.* - "Federating ODP Traders : An X.500 Adventure" - Proceedings of the International Conference on Communications 1993 (ICC'93), Geneva, Switzerland, pp. 313-317.

[13] *Bearman, M.* - "Tutorial on Trading Function" - Part of ISO/IEC 13235 : 1994 or ITU-TS Rec X.9tr - WG7 Committee Draft ODP Trading Function Standard - Sept. 1994.

[14] *Popien, C.; Heineken, M.* - "Object Configuration by ODP Traders" - Proceedings of the IFIP TC6/WG6.1 International Conference on Open Distributed Processing, Berlin, Germany, 13-16 Sept. 1993, Publ. North-Holland, pp. 406-408.

[15] *Popescu-Zeletin, R.; Tschammer, V. and Tschichholz, M.* - "Y Distributed Application Platform" - Computer Communications, vol. 14, 6 - July/August 1992.

14

Enabling Interworking of Traders

Andreas Vogel[a]   andreas@dstc.edu.au
Mirion Bearman[b]   myb@ise.canberra.edu.au
Ashley Beitz[a]   ashley@dstc.edu.au

[a] CRC for Distributed Systems Technology
DSTC, Level 7, Gehrmann Laboratories
University of Queensland, 4072, Australia

[b] CRC for Distributed Systems Technology
Faculty of Information Sciences and Engineering, University of Canberra
Bruce 2616, Australia

Abstract

To enable the interworking of traders, two type based problems have to be solved: the transformation of type identifiers when an interworking operation crosses a type boundary and a typed definition of policy rules to enable the passing of rules between interworking traders. Our solution to the first problem is based on a refined type definition for type identifier, a protocol to transform type identifiers between different type description languages and an identification mechanism for type managers. To resolve the second problem we have formalised the definition of policy rules. We show its application to specify importer, trader and global search policies. The proposed solutions are being prototyped in the framework of a global project for interworking amongst heterogeneous traders.

Keyword Codes: C.2.4; H.4.3
Keywords: Distributed Systems; Information Systems Application

1. Introduction

1.1. Trader overview

Distributed systems span heterogeneous software platforms, hardware platforms, and network environments. In order to utilise services in such systems, service users have to be aware of potential services and service providers. Furthermore, the locations and versions of services change quite frequently in large distributed systems, which makes late binding between service users and service providers a useful feature. To support late binding, mechanisms to locate and access services dynamically have to be provided. ODP's trading function provides a mechanism for dynamically finding services, and is realised by a trader.

A trader is a third party object that enables clients to find information about suitable services and servers. Services provided by service providers can be advertised in, or exported to, a trader. Such advertisements, known as service offers, are stored in the trader's database. The advertising object (the service provider or object acting on behalf of the service provider) is called an exporter. Potential service users (importers) can import from the trader, which means obtaining information on available services and their accessibility. The trader tries to match the importer's request against the service offers in its database. After a successful match the importer can interact with the service provider.

More detailed information about the trader can be found in the ODP Trading Function Standard, [14] which has reached the '*committee draft*' status and in a tutorial on the trader [2]. A number of prototypes of the trader have been implemented for a number of middleware platforms, e.g. DCE [4,19,22], ONC [17], ANSAware [1], and CORBA [16]. The trader has also been specified with various formal description techniques [11, 9].

1.2. Interworking traders

A trader usually covers the service offers of a particular domain. Domain boundaries can be administrative (e.g. organisations or divisions), topological (e.g. a local area network), technological (e.g. a DCE cell), etc. These boundaries need to be crossed to locate and use services in a large, potentially world-wide, distributed environment. Interworking (formerly called federation) of traders enables the discovery of services outside of a trader's domain, with domain boundaries being transparent to a service user. Interworking trader protocols have been proposed since 1991, see e.g. [3, 18,21,16].

The current Trading Function Standard[14] documents the basic agreements on interworking. Operations for interworking are provided at the trading interface. The operations are: *import, export, withdraw,* and *modify.* Links contain interface identifiers to linked traders for interworking. Information on the capability of the linked traders and information on what is on offer at these traders are expressed as link properties.

Additionally, it has been agreed that there are trader policies and importer policies that control interworking. Trader policies that affect interworking include global search policy, resource consumption policy, and domain boundary crossing policy. Importer policies define an importer's intended scope on an import operation. These policies are expressed as an importer policy parameter of the import operation. The resultant policy used to effect an import operation across traders is the unification of an importer's import policy with the trader policies of individual traders linked for interworking.

1.3. Motivation, problem description and outline

To successfully interwork two interacting traders must have the same understanding of the parameters in the interworking operations. The type definitions of these parameters are defined in the trading interface specification. However, there are the following two typing problems:

Type identifier

In ODP systems, a type repository function manages a repository of type specifications and type relationships [12, 15]. A type specification defines the syntax and the semantics of the type and is defined using a type description language. In addition, a type identifier is associated with each type specification and acts as a pointer into the type repository for the retrieval of the type specification. The type repository function can be provided as part of a trader, e.g. in the ANSAware trader [1] and a X.500 based trader [23] or it can be provided as a separate object, e.g. in the DSTC's infrastructure [4, 7]. In this paper, we call the object that provides the ODP type repository function, a type manager.

Type identifier problems arise from interworking traders because identifiers can originate from different type managers. Each type manager may

- use internally different naming systems for type identifiers,
- support different type description languages.

In Section 2 we propose a type definition for type identifiers and a protocol between type managers and traders to solve the problems listed above.

Policy rule

In the computational viewpoint, polices are expressed as rules. Each rule represents a single policy and is expressed as a proposition over both link properties and trader properties. Specifically, policy rules are specified as *'a proposition that a named property either exists or has a specified relationship to a stated value'* [14]. However, in order to pass such policy rules as parameters and to unify rules from different traders, such rules and rule types have to be formally specified. In Section 3 we introduce a formalisation of rules by the definition of a minimal rule language and demonstrate how this rule language can be used to specify importer policy rules and trader policy rules.

In order to validate our interworking ideas, a project on interworking of heterogeneous trader implementations has been initiated. Section 4 gives an overview on design decisions and results reached so far for the project. Finally, Section 5 concludes the article and outlines some future work.

2. Interworking over type domain boundaries

In this section, we first introduce a refined type definition for type identifiers to facilitate the crossing of type boundaries. We then suggest a protocol between traders and type managers to obtain a common understanding of types between interworking traders. We illustrate this protocol using service type identifiers. The term 'type domain' is used to describe the scope of a particular type manager.

2.1. Type identifier

To accommodate the identification of type definitions for a specific type repository we suggest the following definition for a type identifier. A type identifier consists of two components:

- an interface identifier which points to a type manager
- an opaque identifier which points to an entry within the type manager.

The opaque identifier determines the type definition. It is raw data to be interpreted by the type manager addressed by the interface identifier. Interface identifiers are discussed separately in the following section.

2.2. Interface Identifier

The Trading Function Standard does not further specify what a (computational) interface identifier is. We propose the following working definition:

Interface identifiers are divided into direct identifiers and indirect identifiers. A direct interface identifier determines an interface in a particular middleware environment. Currently we support (in the interworking trader project) the DCE and the CORBA environment. A DCE identifier can be represented either by a string binding or a Cell Directory Service CDS entry name. A CORBA identifier is represented by an object reference.

Indirect identifiers have two components. One component contains the actual identifier as raw data, the other component determines the type of this raw data. The latter component is a type identifier as described in the previous section.

To avoid endless levels of indirection during interworking, the interface identifier of the type manager is always a direct interface identifier.

More detailed research on service interface identifiers is reported by the authors in [5].

2.3. Example

The following example shows a type identifier. It uses DCE unique universal identifiers as internal identifiers. The interface identifier of the type manager is given as a direct identifier within a DCE environment using a CDS entry name.

* interface identifier of the type manager:
```
kindInterface = direct
middleware = DCE
dceKindBinding =  bindByCds
dceCdsBinding =  "/.:/a2/andreas/tm_server"
```
* opaque identifier within the type manager:
```
internalIdentifier = "2e49c9ea-3105-11ce-b3ca-08002bbceeee"
```

2.4. Type identifier transformation protocol

In this section we introduce a protocol to transform type identifiers between different type domains. We illustrate the protocol by the import operation and the service type identifier.

Figure 1 illustrates a scenario with two type domains represented by two type managers, *Type Manager 1* and *Type Manager 2*. The two traders, *Trader 1* and *Trader 2*, are in the respective type domains and *Trader 1* has a link to *Trader 2*. An exporter has exported a service with a certain type to *Trader 2*. The service type is expressed by an identifier Service-TypeIdX, which points to a type description stored in *Type Manager 2*. An importer requests a service offer of the type identified by ServiceTypeIdY, a pointer into the type repository of *Type Manager 1*. The service types to which ServiceTypeIdX and ServiceTypeIdY point are assumed to be equivalent.

Given the above scenario, interworking (illustrated for the import operation) can be achieved for the following cases:

(i)   *ignorant*
```
import( ..., ServiceTypeIdY, ...)
```
Trader 1 is not concerned about crossing type domains. It passes the requested type identifier over to *Trader 2*, and *Trader 2* has to deal with this type identifier.

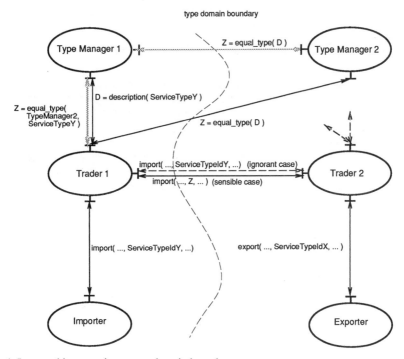

Figure 1. Interworking crossing a type domain boundary

(ii) *sensible*
```
import( ..., Z, ...)
```
Trader 1 produces a type identifier valid in the type domain 2 before interworking with *Trader 2*.

In both cases, the given type identifier ServiceTypeIdY has to be transformed into one which is understood by the *Type Manager 2*. The transformation process has two steps. The first step, performed by *Type Manager 1*, is the de-referencing of the type identifier ServiceTypeIdY into type description D. In the second step, *Type Manager 2* finds a type iden- tifier in *Type Repository 2* which matches the type description D.

Computationally, there are two ways to do the transformation:

explicit
the trader obtains the type description from *Type Manager 1* and asks *Type Manager 2* of a matching type identifier. In Figure 1 this is indicated by the black lines between *Trader 1*, *Type Manager 1*, and *Type Manager 2*.

type manager interworking
the trader asks its type manager to obtain an identifier for an equivalent type. The type manager employs a type manager interworking protocol to get the requested type identi- fier, as illustrated by the gray lines in Figure 1.

To apply any of the above variants of type domain crossing mechanisms, the issue of a type description language shared by the two type managers still needs to be addressed.

2.5. Identification of a type manager

A trader has to know the type manger of the other domain to understand service type identifiers as motivated in the above scenario. The issue here is how a trader acquires knowledge about the interface identifier of another domain's type manager. In the given scenario, either *Trader 1* (*ignorant* case) or *Trader 2* (*sensible* case) has to know the interface identifier of *Type Manager 2* or *Type Manager 1*, respectively. In the first case, this knowledge could be expressed as a link property.

We define properties as a list of triplets containing the property name, the property type and the property value. The interface identifier of the linked trader's type manager can be expressed as a link property. An example of a value specification has been given in Section 2.3.

```
property name = "type manager interface identifier"
property type = InterfaceIdentifierType
property value = ...
```

2.6. Type description languages

To enable type identifier transformations, two interworking traders/type managers must agree on a common type description format. The most obvious method is to use a standardised type description language which is supported by all type managers. However, a number of such type languages are already established in their respective domains, for examples, OMG's CORBA-IDL [20], OSF's DCE-IDL [24], OSI's ASN.1 [13], and Microsoft's MIDL [10]. It therefore seems infeasible, and contrary to the ODP approach, to aim for a single standardised type description language. Our approach accommodates the co-existence of a variety of type description languages. That is, we acknowledge the existence of the above type languages and use them to form a set of exchangeable type languages.

There remains the issue of how to agree on a common type description format between two traders (and their type managers). In the case of an assumed type manager interworking protocol, the issue is delegated to this protocol. For the *sensible* case the set of type description languages supported by a trader (and its type manager) could be expressed as another link property, e.g.

```
property name = "supported type description languages"
property type =  TypeDescriptionLanguageSetIdentifier
property value = { DCE-IDL, CORBA-IDL } .
```

Another solution is to directly enquire the type manager for the set of supported type description languages.

The bootstrap problem, i.e. the definition of unique names for type description languages, can be solved by registration of the names with a standardisation organisation.

2.7. Discussion

The proposed type identifier transformation protocol allows for a number of options. We favour the *sensible* case over the *ignorant* one. The decision is due to the asymmetry of links, i.e. a trader stores properties of the trader that it is linked to but not vice versa. The use of link properties simplifies the type identifier transformation protocol.

From the architectural point of view of a distributed system, the type manager interworking has the advantage of hiding the type identifier transformation from the interworking traders. However, there is neither a standard for type management nor a standard for type manager interworking protocols. For prototyping in the trader interworking project, the explicit transformation option is used. However, within the DSTC's architectural model approach [6], the option of type manager interworking will be supported.

3. Policy Rules

In this section, we first introduce our minimal rule language for the specification of policy rules. This is followed by examples of applications of the language to important policies required for interworking.

3.1. Minimal rule language

Policy rules are specified in the computational language as *'a proposition that a named property either exists or has a specified relationship to a stated value'* [14]. We have formalised the proposition by the definition of a minimal rule language. The language contains two fundamental boolean expressions:

```
exist( property )
```

which is true if the property `property` exists; and

```
relationship( relationship, property, value )
```

which is true if the value of property `property` and the value `value` satisfy the relationship `relationship`. We have already introduced the type of a property in Section 2.5. The type of `value` has to be of the same type as that specified for the property. Additionally, the logical operators NOT, AND, and OR can be used to construct more complex expressions. This minimal rule language has been included in the type definition of the trading specification used in interworking trader project[1].

Associated with each relationship is a relationship type which is managed by the type manager. The typing of relationships removes the need for the standardisation of any relationship and provides freedom in the definition of rules. Within our type definitions a relationship is expressed by a type identifier as defined and illustrated in Sections 2.1-2.3. An example of a relationship specification (as stored in the type manager) is shown below:

name: ==
types: integer, integer
semantics: is true if x and y have the same value, otherwise false

---

[1] The DCE IDL version can be accessed at the following URL:
http://www.dstc.edu.au/public/iwt/proxy.dce.idl .

3.2. Importer policy rules

Importer policy rules specify constraints on the scope of the import operation such as

- on the set of traders to be searched,
- on the use of resources, and
- on the domains to be crossed.

The following examples illustrate the use of our minimal rule language to express an importer policy.

An example of constraints on the set of traders to be visited is where the importer only wishes to visit traders which trade a certain category of service types (which is also known as abstract service types). The corresponding link property is here referred to as abstract-ServiceType and has the service identifier type. If the importer is only interested in meteorology services identified by, say, weatherServiceId and has the service identifier type, then this can be expressed as:

```
relationship( subtype, abstractServiceType, weatherServiceId )
```

An example of use of resources is where an importer wants to make sure that no fees are required for searching a trader. This can be expressed as:

```
(exist( fee ) AND relationship (==, fee, 0)) OR NOT exist(fee)
```

An example of domains to be crossed is where an importer wants to ensure that only traders located in Australia are used. If a trader property traderLocation is assumed, then the constraint can be expressed as:

```
exist( traderLocation ) AND relationship( EQ, traderLocation,
"Australia" )
```

3.3. Trader policy rules

Trader policy rules that impact on interworking include:

- resource consumption policy rules and
- domain crossing policy rules.

An example of a resource consumption policy is where the total time for an import operation must be less than 10 seconds which can be expressed as:

```
relationship( <, searchTime, 10)
```

An example of a domain crossing policy is to limit the crossing of organisational domain to a single domain.

The corresponding trader property is assumed to be organisation. The constraint to interwork only with traders within the DSTC can be expressed as:

```
relationship( EQ, organisation, "DSTC")
```

3.4. Global search policy rules

Global search policies specify how to traverse the graph of linked traders. Global search policies can be included in importer policies and in trader policies. To illustrate the specification of global search policy rules, we assume that the predefined properties searchDirectionPriority, searchLocationPriority and searchMethod have the following enumeration types, respectively:

search direction priority = { depthFirst, widthFirst }
search location priority = { localFirst, remoteFirst }
search method = { sequential, broadcast }

An example of a valid global search policy specification would be:

```
relationship(IN, searchDirectionPriority, depthFirst) AND
relationship(IN, searchLocationPriority, localFirst) AND
relationship(IN, searchMethod, sequential)
```

When traversing the graph of linked traders, loops can occur. There are two methods to avoid loops. One method is for a trader to keep track of the operations that have visited it. This requires that operations must be uniquely identifiable. The other method is for an operation to keep track of the traders that it has visited. This information can be stored in the importer policy parameter of the import operation as it initiates a search on another trader. This parameter can be expressed using the minimal rule language as follows:

```
NOT relationship( IN, link.traderId, visitedTraders )
```

where link.traderId is the trader identifier to which a certain link is pointing and visitedTraders is the set of identifiers of traders which have already been visited by the operation.

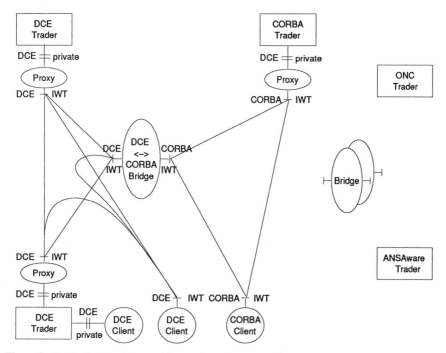

Figure 2. Test-bed of the interworking trader project

4. Prototyping

To validate the proposals made in the previous sections the interworking trader project[2] has been initiated to achieve and demonstrate interworking of traders existing in heterogeneous environments.

The trader prototypes provided by the project partners differ in both the specification of the interface (due to the frequency of change in versions in trader working documents) and the heterogeneity of middleware domains that the traders are implemented in. Interface definitions present a dilemma since private interfaces serve well in particular systems but interworking requires common interfaces. To overcome this, an ad hoc proxy approach is used. A proxy supports two interfaces:

- a refined 'standard' one (which includes all the proposals made in this paper) and

- a 'private' one (which supports the local trader implementation).

For interworking over different middleware domains, we suggest a similar ad hoc solution - bridges which provide interfaces to two middleware domains. Such bridges seem to be easily implementable, but the problem of a general ODP infrastructure [8] remains unsolved. Proxies as well as bridges transform parameters between their respective interfaces.

Figure 2 illustrates the architecture of the project's test-bed based on proxies and bridges where the project defined interfaces are labelled IWT (InterWorking Traders). The IWT interface contains all the proposals on type refinements made in the previous sections. There are equivalent specifications of the IWT interface in both, DCE IDL and CORBA IDL. The transformation between them has been facilitated by a tool set currently under development at the DSTC. Within the DSTC, the DCE proxy has been implemented The implementation of a DCE-CORBA bridge is under development at DSTC.

5. Conclusions and Future Work

We have identified two type based problems which hinder interworking of traders:

- the transformation of type identifiers when an interworking operation crosses a type boundary and

- a formal definition of policy rules to enable the passing of policy rules between interworking traders.

Our proposed solution to the first problem is based on a refined type definition for type identifiers, a protocol to transform type identifiers between different type managers, and an identification mechanism for type managers. To resolve the second problem we have formalised the definition of policy rules and have shown its application to specify importer, trader and global search policies.

To validate the proposals we have initiated the interworking trader project. Within this project, a refined trading interface specification, which includes all the type refinements proposed in this article, has been defined. Furthermore, interworking through this refined interface has also been achieved.

The implementation of further infrastructure objects, i.e. proxies and bridges, will continue and will enable more traders to participate in the interworking project.

---

[2] Find details about the project at: http://www.dstc.edu.au/public/iwt/welcome.html .

Apart from the interworking trader project, we are working on the enhancements of our type manager, in particular in the transformation of type descriptions from one language into another. Tools for DCE IDL↔CORBA IDL transformations are nearly completed.

Finally, we propose to the ISO standardisation body for the Trading Function that the requirements we have found for interworking be considered for inclusion in the Trading Function.

6. Acknowledgements

The work reported in this paper has been funded in part by the Cooperative Research Centres Program through the department of the Prime Minister and Cabinet of Australia.

The authors wish to thank our colleagues at DSTC, in particular, Jaga Indulska, Kerry Raymond, and Andy Bond, for the many fruitful discussions. Also thanks to all who helped to get the interworking trader project off the ground, in particular to Kay Müller (Univ. of Hamburg), Mike Beasley (APM) and Richard Soley (OMG).

References

1.   "The ANSA Reference Manual.," Architecture Projects Management Limited, Cambridge (1989).

2.   M. Bearman, "ODP-Trader" in *Open Distributed Processing, II,* ed. J. de Meer, B. Mahr, and S. Storp, pp. 19-33, North-Holland (1994).

3.   M. Bearman and K. Raymond, "Federating Traders: an ODP Adventure." in *Open Distributed Processing,,* ed. J. de Meer, V. Heymer, and R. Roth, pp. 125-143, North-Holland (1992).

4.   A. Beitz and M. Bearman, "An ODP Trading Service for DCE" in *Proccedings of the First International Workshop on services in Distributed and Networked Environments,* pp. 34-41, IEEE Computer Society Press, Los Alamitos, CA (1994).

5.   A. Beitz, M. Bearman, and A. Vogel, "Service Location in an Open Distributed Environment," Technical Report #21, CRC for Distributed Systems Technology, Brisbane (1995).

6.   A. Berry and K. Raymond, "The A1! Architectural Model" in *Proceedings of 2nd International Conference on Open Distributed Processing,* ed. K. Raymond and J. de Meer (1995, in press).

7.   C.J. Biggs, W. Brookes, and J. Indulska, "Enhancing Interoperability of DCE Applications: A Type Manager Approach" in *Proccedings of the First International Workshop on services in Distributed and Networked Environments,* pp. 42-49, IEEE Computer Society Press, Los Alamitos, CA (1994).

8.   A. Bond and D. Arnold, "An Approach to Implementing an ODP Infrastructure" in *Proceedings of 1994 ACS Asia-Pacific Conference,* pp. 9-16, Gold Coast, Queensland, Australia (September 15-18, 1994).

9.   J.S. Dong and R. Duke, "An Object-Oriented Approach to the Formal Specification of ODP Trader" in *Open Distributed Processing, II,* ed. J. de Meer, B. Mahr, and S. Storp, pp. 312-323, North-Holland (1994).

10. G. Eddon, *Network Remote Procedure Calls: Windows NT, Windows, DOS,* P-H (1993).

11. J. Fischer, A. Prinz, and A. Vogel, "Different FDTs Confronted with Different ODP-Viewpoints of the Trader" in *FME'93: Industrial Strength, Formal Methods,* ed. J.C.P. Woodcock and P.G. Larsen, pp. 332-350, LNCS 670 Springer Verlag (1993).

12. J. Indulska, K. Raymond, and M. Bearman, "A Type Management System for an ODP Trader" in *Open Distributed Processing, II,* ed. J. de Meer, B. Mahr, and S. Storp, pp. 169-180, North-Holland (1994).

13. ISO, *Information Processing - Open systems Interconnections - Specification of Abstract Syntax Notations One (ASN.1).,* IS 8824, International Standard.

14. ITU/ISO, "Reference Model of Open Distributed Processing - Trading Function," Committee Draft ISO/IEC/JTC1/SC21 N807 (October 1994).

15. ITU/ISO, "Reference Model of Open Distributed Processing - Part 3: Prescriptive Model," Draft International Standard 10746-3, Draft ITU-T Recommendation X.903 (1994).

16. L. A. de Paula Lima Jr. and E. R. Mauro Madeira, "A Model for a Federated Trader" in *Proceedings of 2nd International Conference on Open Distributed Processing,* ed. K. Raymond and J. de Meer (1995, in press).

17. L. Kutvonen and P. Kutvonen, "Broadening the User Environment with Implicit Trading," in *Open Distributed Processing, II,* ed. J. de Meer, B. Mahr, and S.Storp, pp. 157-168, North-Holland (1994).

18. L. Kutvonen, "Federation transparency in ODP Trading function," Technical Report C-1994-32, Department of Computer Science, University of Helsinki (May 1994).

19. M. Merz, K. Müller, and W. Lamersdorf, "Service Trading and Mediation in Distributed Computing Systems" in *Proceeding of the 14th International Conference on Distributed Computing Systems,* pp. 450-457, IEEE Computer Society Press, Los Alamitos, CA (1994).

20. Object Management Group and X/Open, *The Common Object request Broker: Architecture and Specification* (1991).

21. C. Popien and B. Meyer, "Federating ODP Traders: An X.500 Approach" in *Proceedings of the International Conference on Communications,* pp. 313-317, Geneva.

22. A. W. Pratten, J. W. Hong, J. M. Bennett, M. A. Bauer, and H. Lutfiyya, "Design and Implementation of a Trader-Based Resource Management System" in *Proceedings of CASCON'94,* Toronto (1994).

23. A. Waugh and M. Bearman, "Designing an ODP Trader Implementation using X.500" in *Proceedings of 2nd International Conference on Open Distributed Processing,* ed. K. Raymond and J. de Meer (1995, in press).

24. X/Open Company Ltd., *X/Open Preliminary Specification X/Open DCE: Remote Procedure Call* (1993).

15

An Explorative Model for Federated Trading in Distributed Computing Environments

Ok-Ki Lee and Steve Benford

Communications Research Group, Department of Computer Science, University of Nottingham, Nottingham, NG7 2RD, United Kingdom

We propose a model for trading in very large-scale distributed computing environments which is based on the gradual evolution of a federated trading space through a process of continual exploration and evaluation, rather than on the imposition of a strictly managed hierarchy. In our model, each trader autonomously acquires local knowledge of the trading space, called trading knowledge, through a process of distributed resource discovery. Trading knowledge typically consists of trader links which reference other traders. Trader links also contain a measure of affinity: a strength of attraction based on a comparison of so called service and interest profiles, perhaps combined with a history of how useful and reliable other traders have proved to be. This notion of affinity helps a trader to decide how to resolve import requests which cannot be satisfied locally. It also helps a trader to decide which trader links to retain as it periodically and autonomously explores the trading environment. Instead of being concerned with the detailed management of individual trader links, human managers can then control the evolution of the trading space through a number of high level management policies. These include service and interest profiles, definitions of affinity and instructions on when and how exploration should occur. Our paper also describes a reference implementation of the model within the ANSAware distributed processing environment called the Explorative Trading Service (ETS).

Keyword Codes: C.2.4; H.3.3
Keywords: Distributed Systems, Federated Trading, Resource Discovery

1. INTRODUCTION

As the number of computer systems involved in distributed computing increases, one centralised trader will simply not be enough to accommodate a large number of exporters and importers. It therefore becomes necessary that trading be performed in a distributed manner. However, in the research community it has been an issue how to organise and manage such a distributed trading service [ANSA 89, ISO Trader 92, Maffeis 93, Linden 92].

One proposed approach has been to use global naming services such as the X.500 Directory [ISO DS 88] to support distributed trading [ISO Trader 92]. This approach aims to establish a global naming/trading hierarchy with traders being linked through the naming service. Global naming hierarchies are intended to simplify query resolution, ease administration and are also claimed to be representative of the underlying structure of human organisations. However, along with others [Linden 92], we have grave doubts about the feasibility of the hierarchical approach:

- Although hierarchy is technically appealing due to its relative simplicity to implement in software, it is politically difficult to implement in practice as it requires the establishment of powerful, maybe global administrative authorities. Such authorities may be particularly hard to establish in a climate which favours deregulation of telecommunication services.
- It is not at all clear that organisations are hierarchies. Various, more recent, models have been proposed for human organisational structures including organisations as networks, matrices and organisms.
- Hierarchical information structures tend to be unwieldy and resistant to change. Indeed, one can argue that this is why the hierarchical database model has generally been superseded by the relational and object oriented models. It was an early assumption of X.500 that the top levels of the global hierarchy would be stable. However, a brief look at international events since 1988 (the date of the first X.500 standard) show this to be generally untrue.

For these reasons, we propose to adopt a more ad-hoc and explorative approach to organising a federated trading service. In particular, we propose that a trading space should grow and evolve through a process of continual exploration and that each trader should autonomously maintain a personalised view of the trading space. This approach has been motivated by the observation that the real world is not necessarily well-organised and often supports autonomy and ad-hoc organisation [Saltzer 1978]. We have also been greatly influenced by the recent success of a number of NIR (Network Information Retrieval) projects, particularly the work of Netfind [Schwartz & Tsirigotis 91], Internet Gopher [McCahill 92] and Prospero [Neuman 92]. We suspect that their rapid spread may be at least partly due to the autonomous and de-centralised nature of their organisation.

Section two of our paper clarifies the design goals behind our model. Section three then describes the model in detail. Finally section four describes a reference implementation of the model in the ANSAware [ANSA 92] environment.

2. DESIGN GOALS

The design goals of any distributed trading service might include the following:
- **Partitioning responsibility** - responsibility for managing many different services may be divided between different traders.
- **Fault tolerance through redundancy** - replicating multiple instances of a service across multiple traders may enable fault tolerance in the light of both service and trader failures.
- **Decentralised ownership** - services will be provided, owned and managed by different organisations. These organisations may wish to retain autonomous management of their own services, even when they cooperate with each other, let alone when they compete.

In addition to these general goals, the development of our trading model has been driven by a number of more specific requirements.
- **Realistic assessment of others** - when a service is requested but not available locally, a trader might rely on other traders to find it on its behalf (the basic idea behind distributed trading). However, the issue of trustworthiness then arises. In the real world, it is not always the case that organisations or people actually provide the service or quality of service that they claim to. Why should it be any more true of their traders? Realistic

appraisal of the claims of other traders should be a part of distributed trading. In other words, we should support traders in distinguishing "good" friends from "bad" ones. Furthermore, this process might involve some notion of learning from past experience (for example, lookup success rate, actual quality of service delivered and availability).

- **Explore, adapt and evolve** - our model should not assume a consistent or stable environment. Instead, in a large system, services and traders will constantly appear, change and disappear. Such changes cannot be instantly conveyed to all interested parties. Traders should be therefore be supported in exploring the trading space, interrogating each other about services offered and other traders known about. Furthermore, they should be prepared for inconsistencies in trading knowledge and must be able to adapt. The overall picture of the trading environment will therefore be one of a constantly evolving system with no one trader ever understanding the complete picture. Furthermore, the ability for traders to cope and adapt quickly will be critical to their success.

- **Ease of human administration** - As seen in NIR, human administrators are required to spend a significant amount of time maintaining their servers. In our model, human administrators should only be required to specify high level management policies. The traders themselves should then take care of administrative details as much as possible.

Having clarified the design goals behind our model, the following section now introduces the model itself in terms of its key concepts.

3. THE EXPLORATIVE MODEL FOR FEDERATED TRADING

Our model assumes as its starting point the existing concepts of service providers, service consumers and traders. On top of these we define the concepts of trader knowledge, trader links, distributed request resolution, trader knowledge management, affinity, service, interest and reputation profiles, exploration, announcement and management policies.

3.1. Trader links and trader knowledge

We conceive of a distributed trading system as involving two logical planes, a service plane in which service providers and consumers exist and a trading plane in which multiple traders exist. As a result of trading, the service consumers and providers become connected by service links, thus enabling the distribution of applications. In turn, the traders are connected by **trader links**, thus enabling the distribution of the trading process itself. This structure is shown in figure 1.

A trader link tells a local trader how to connect to a given remote trader. It therefore defines a unidirectional ordered pair of traders. In order to achieve its purpose a trader link must contain at least minimal addressing information (it may contain other information as we shall see below). For example, it might take the form of a Uniform Resource Locator (URL) [Berners-Lee]. In a very large and dynamic distributed system, it is unlikely that a trader will be able maintain links to all other traders. Instead, each trader will utilise a local pool of trading links called its **trader knowledge.** Consequently, the general structure of trading space is that of a directed graph.

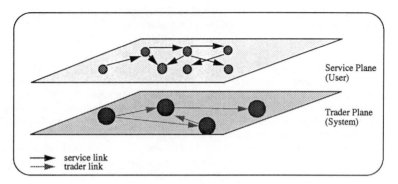

Figure 1: The trading universe

3.2. Distributed request resolution

Distributed request resolution refers to the way in which a group of traders resolves a service import request. In our view, this process should be as transparent as possible to the service providers and consumers. In other words, a service provider should only be required to contact one trader in order to resolve an import request.

When an importer makes a request to a trader for a service, the initial trader contacted will first search its local trading space. If the service is found, the trader will of course return the reference (address) of the service to the importer. However, if no local service is found, then the trader will contact its **adjacent traders** by following its local trader links. These adjacent traders will then search their local trading spaces. If the service is still not found, the initial trader may interrogate its adjacent traders to find out about their links to others and may follow these in search of the service. This process continues until either the service is found or there is good reason to cease (e.g. exceeding a cost/time limit or running out of unexplored trader links).

Notice that we make the local trader responsible for following all trader links and do not adopt a policy of "chaining" whereby requests are recursively passed along a chain of traders via their trader links. We believe that this allows the local trader to retain tighter control over the resolution process, thereby avoiding problems with loops caused by cycles in the trading space and also makes it easier to realise constraints such as time-outs and cost limits.

3.3. Trader knowledge management

Trader knowledge management describes the process by which trader links are created, maintained and destroyed over time. A basic approach to this problem might be to provide human administrators with a protocol and management tool for maintaining individual links. However, such fine grained management would impose a potentially large overhead on human managers and leaves unanswered the question of how these managers find out about other traders in the first place. Instead, our model makes the traders themselves responsible for locating other traders of interest and for evaluating how useful they might be. This is achieved through a process of exploration. The traders themselves might then be responsible for maintaining individual links and human managers could be provided with a set of higher level management policies for controlling the exploration process. Thus, the trader knowledge management problem involves a number of sub-issues:

- Basic mechanisms for traders to manage trader links (e.g. establish, release, propagate);
- Techniques for assessing the usefulness of other traders;
- Mechanisms to support automated exploring;
- Definition of suitably high level management policies for humans.

3.4. Affinity

Associated with each trader link is a notion of **affinity**. The affinity of a link is intended as a measure of how useful the local trader believes the remote trader to be. In the simplest case, affinity might be expressed as a numeric value normalised to be in some range (e.g. between 0 and 1) which expresses the degree of overlap between the interests of the local trader and the services supported by the remote trader (more of this later). Thus, the trading space now becomes a weighted directed graph. Figure 2 shows a snapshot of a simple trading space. In this example we see that the trader Tf currently has trader links with four other traders, Th, Ti, Tj and Td with affinities ranging from 0.9 to 0.7.

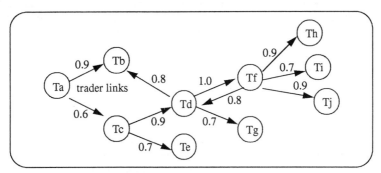

Figure 2: An example trading space with affinity

Affinity plays two roles in our model. First, it provides a way for a trader to decide whether or not to retain an on-going link to another trader. For example, a trader might choose to remember all links whose affinity exceeds some threshold. Alternatively, if a trader can only store a limited number of links (perhaps for reasons of space or speed of searching) it might compare the affinity of a newly discovered link with its least useful current link in order to decide which of the two to keep. Second, affinity may enhance the process of distributed request resolution by allowing a trader to prioritise the order in which it contacts other traders. The assumption here is that a high affinity link offers the best chance of a quick resolution.

3.5. Service and interest profiles

The concept of affinity requires that a trader can assess how the services offered by another trader match its own interests. This idea is more formally captured in the concepts of service profile and interest profile. A **service profile** is a description of the services which one trader is prepared to offer to others (note that a trader may not be prepared to offer all of its services). Service profiles might range from simple lists of service types to more complex descriptions (e.g. involving offers of quality of service). An **interest profile** describes those services which a trader is interested in finding in other traders and may be described in the same way as the service profile (in this case quality of service would refer to desired quality). Interest profiles might be automatically generated from logs of clients' import requests over a period of time. Similarly, service profiles might automatically be generated from interface specification code (e.g. from IDL in ANSA).

The degree of difference between a single trader's service and interest profiles will reflect is motivation for entering into distributed trading. A strong overlap suggests that fault tolerance through redundancy is of prime importance whereas a weak overlap suggests a partitioning responsibility for managing different services.

The affinity that one trader has for another is then a function of the interest profile of the former and the service profile of the latter. The exact nature of such an affinity function will of course depend on the nature of the profiles. For example, were the profiles to be simple lists of services then a basic function expressing the affinity of trader A for trader B might be the size of the intersection of the interest profile of A and service profile of B as a fraction of the interest profile of A.This delivers the value 0 when B provides none of A's desired services and 1 when it provides them all. However, many other functions might be possible. In fact, we might usefully draw on techniques from the domain of information retrieval systems when constructing such functions. (e.g. we might apply weighting factors to the different entries in service and interest profiles).

3.6. Reputation profile

Affinity as described above compares the stated interests of one trader against the promises of another. We argue that it is important that this be tempered by a trader's actual experience. For example, if a remote trader continually fails to resolve requests, the local trader might reduce its affinity for it. Conversely, a low affinity link that continually delivers useful results might increase in strength. The quality and reliability of the actual services found should also modify affinity. For example, if the actual quality of service encountered is frequently worse than that promised, affinity might drop. In other words, the affinity associated with a given link should be dynamically modified by a trader's actual experiences both in terms of successful resolution of requests and quality of end services located. Thus, a trader can learn by experience. These ideas are captured in the concept of a **reputation profile**, a collection of statistical information gleaned through the process of interacting with other traders and which serves as a measure of the reputations of these other traders in the eyes of the local trader.

3.7. Exploration and announcement

So far we have introduced concepts to allow a trader to adopt an informed view of other traders it encounters. However, how does it find out which other traders are out there in the first place?

Our model supports this discovery through the processes of exploration and announcement. Many of our ideas here have been strongly influenced by work on resource discovery within the Network Information Retrieval (NIR) community.

Exploration involves a trader in periodically navigating the trading space by following trader links. This process is similar to distributed request resolution as described earlier, in that a trader follows its links to other traders and then follows their links to yet others. For each trader encountered, it invokes its affinity function. At the end of the exploration the trader then invokes some local **decision procedure** in order to update its trader knowledge (e.g. deciding which links to keep and which to drop).

The obvious similarity between distributed request resolution and exploration relate to a distinction between **on-demand** and **off-demand** exploration.

- On-demand exploration means that a trader starts exploring when a service is requested and the service is not found locally. In other words, it refers to a process of distributed request resolution which might also include evaluation of other traders en route.

- Off-demand exploring means that a trader explores automatically as a separate process, probably during periods when other activity is low (e.g. nightly).

In contrast to exploration, **announcement** involves traders who are involved in some change actively spreading this news. Typical changes might involve the creation of new traders, the removal of existing ones or significant changes in service profile. Announcements, which may include revised service profile and addressing information, are then passed from trader to trader over a special protocol using trader links. Thus, announcement is the converse of exploration.

It might also be useful to combine exploration and announcement into specialised exploration agents whose job it is to autonomously explore the trading space, collate results and announce them in some digested form (similar to Robots and Spiders in NIR systems [Koster]).

3.8. Management policies

Many different aspects of our model might be parameterised in order to control the manner in which the distributed trading process works. These include: controlling distributed request resolution (e.g. by specifying time or cost limits); defining service and interest profiles; defining affinity functions; constraining the amount of local trader knowledge to be held at any one time; defining decision procedures; defining reputation profiles; and configuring the exploration and announcement processes (e.g. saying when and how often they should occur and how far they should spread).

Between them, these parameters represent a set of management policies that allow human managers to influence the trading process and configure local traders without having to manipulate individual trader links. These 'high-level' policies might then be stored in the local trader and be created and updated through a trader management client using a specialised protocol.

This concludes our presentation of the basic model. The following section describes how we have implemented it in the ANSAware environment.

4. REFERENCE IMPLEMENTATION

This section describes a reference implementation of our model called the **Explorative Trading Service (ETS)**. Our aim here has been to demonstrate how the model can be realised in a typical ODP environment and to explore the components and interfaces required to support it. Future work will involve using our implementation to test the model through large scale simulation experiments. Our reference implementation has been developed on top of the ANSAware 4.0 platform. The implementation environment consists of a collection of SUN 3s and

SUN 4s which run ANSAware 4.0 under SunOS 4.1.3. The languages to be used are C, C++ and Tcl/Tk [Ousterhout 93].

ETS consists of several components: the **Explorative Trader (ET), Trader Browser, Trader Manager** and **Interface Repository.** These are shown in figure 3 along with a summary of the interfaces they offer to other components. Full specification of these components is beyond the scope of this paper, but can be obtained from the authors on request. Instead, the following paragraphs briefly summarise each in turn.

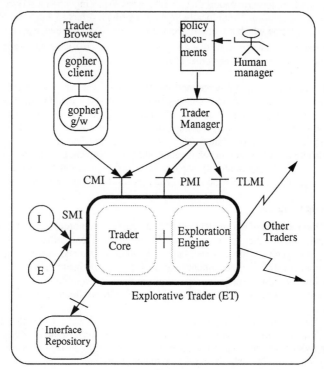

Figure 3: ETS Architecture

4.1. Explorative trader

The Explorative Trader is the main component of ETS and consists of two further sub-components: the **trader core** which provides a standard localised trading service and the **exploration engine** which is responsible for distributed request resolution and trader knowledge management. ET provides four interfaces to other components:

- Service Management Interface (SMI) - defines operations which can be directly used by an application (i.e. a collection of importers and exporters). An importer and an exporter make use of this interface to import and export a service.

- Context Management Interface (CMI) - is used to manage internal organisation of a trader. Specifically it provides operations for managing trader contexts.
- Trader Link Management Interface (TLMI) - consists of operations which are related to trader link management. It supports trading in exploring the trading space and in establishing links to other traders. The intention here is that, via a process of exploration, traders can acquire knowledge of other traders and so engage in 'federated trading'. The interface plays a key role in our exploration model for trading.
- Policy Management Interface (PMI) - provides operations for imposing policies on a trader. This interface will affect the static and dynamic behaviours of a trader. Categorically, the interface supports profile management policy and trader link management policy. For example, the trader manager makes use of this interface to change the exploration time-out or depth of search etc.

4.2. Trader browser

The Trader Browser allows a human manager to browse and explore the local trading space within a specified trader via the Context Management Interface. It has been our intention to re-use existing tools for ETS wherever possible. In particular, the Internet Gopher [McCahill 92] has attracted our attention, because its protocol is relatively simple and there already exist a number of popular user interfaces such as xgopher or Mosaic available across a variety of platforms. Consequently, we have developed a Gopher to Trader gateway, which allows any gopher client to browse the trading space. It will be extended to support service and link level operations in the near future.

4.3. Trader manager

The trader manager is an ETS component which allows a human administrator or an application program to manage the Explorative Trader via a set of given interfaces. Mainly, the trader manager is involved in the following:

- Manual trader link management when it is required
- Providing trader link management policy for ET - for example, maximum number of trader links to explore, when to start/stop exploring
- Managing trader contexts inside ET

4.4. Interface repository

ETS includes an interface repository which is mainly intended to store IDL definitions. As an example service, we have developed a dictionary interface[1] which supports the following operations: lookup, synonym and antonym.

5. SUMMARY

Our paper has been concerned with an explorative approach to distributed trading. Our main design goals have been to support autonomy, decentralisation, learning by experience, gradual evolution of a trading space, realistic assessment of other trader's capabilities and high-level

---

1. In fact, an ANSAware "wrapper" is used to encapsulate an existing dictionary service on the network.

human administration. These goals are reflected in our explorative trading model which is summarised by figure 4.

The figure shows that trader space takes the form of a weighted directed graph consisting of traders connected by trader links. A trader link contains some addressing information as well as a measure of affinity derived by comparing the interest profile of a local trader with the service profile of a remote one. Trader links are used as part of distributed request resolution where a local trader coordinates interaction with traders to which it is linked (and perhaps other ones to which they are linked) until a request is resolved or terminated. Trading knowledge is acquired and updated through the processes of exploration and announcement, influenced by the concepts of affinity and also reputation, the latter allowing a trader to learn from experience. Finally, humans manage the system through a set of high level management policies which include the profiles mentioned above as well as policies for controlling how and when exploration and announcement take place.

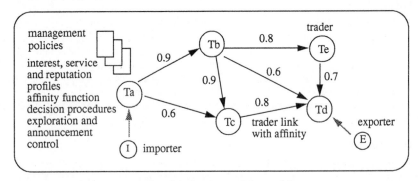

Figure 4: Summary of the model

Finally we have described a reference implementation of the model called the Explorative Trading Service (ETS), based on ANSAware. ETS shows some of the components and interfaces which may be required to support the model in future large scale distributed systems.

REFERENCES

[ANSA 89] *The ANSA Reference Manual*, Architecture Projects Management Ltd., Cambridge, UK, 1989.

[ANSA 92] *ANSAware 4.0: Application Programming in ANSAware*, Architecture Projects Management Ltd., March 1992.

[Berners-Lee] Tim Berners-Lee, *Uniform Resource Locators*, URL: ftp://info.cern.ch/pub/ietf/url4.ps,(.txt).

[ISO DS 88] ISO and CCITT, *Information Processing Systems - Open Systems Interconnection - The Directory*, ISO 9594-108, CCITT X.500-X.521, 1988.

[ISO Trader 92] *Working Document on Topic 9.1- ODP Trader*, ISO/IEC JTC1/SC21/WG7/N743, November, 1992.

[Koster] Martijn Koster, *List of Robots*, URL: http://web.nexor.co.uk/mak/doc/robots/active.html.

[Maffeis 93] S. Maffeis, *Electra - Making Distributed Programs Object-Oriented*, Proceedings of the Symposium on

Experiences with Distributed and Multiprocessor Systems IV, USENIX, San Diego, CA, USA, September 1993.

[McCahill 92] M. McCahil, *Internet Gopher: A Distributed Server Information System*, ConneXions - The Interoperability Report, 6(7), pp. 10-14, Interop Inc., July 1992.

[Neuman 92] B. C. Neuman, *Prospero: A Tool for Organizing Internet Resources*, Electronic Networking: Research, Applications, and Policy, 2(1), pp. 30-37, Meckler Publications, Wesport, CT, USA, Spring 1992.

[Linden 92] R. J. van der Linden, *Naming and Federation*, in Naming Facilities in Internet Environments and Distributed Systems (Special Issue of the Distributed Processing Technical Committee Newsletter), Bharat Bhargava (ed.), Volume 14, Number 1, pp 21-27, IEEE Technical Committee on Distributed Processing, June 1992.

[Ousterhout 93] John Ousterhout, *An Introduction to Tcl and Tk*, Addison-Wesley, 1993.

[Saltzer 78] J. H. Saltzer, *Research Problems of Decentralized Systems with Largely Autonomous Nodes*, in R. Bayer, R.M. Graham and G. Seemüller (eds.) Operating Systems: An Advanced Course, Lecture Notes in Computer Science 60, pp. 583-593, Springer-Verlag, 1978.

[Schwartz & Tsirigotis 91] M. F. Schwartz and P. G. Tsirigotis, *Experience with a Semantically Cognizant Internet White Pages Directory Tool*, J. Intenetworking: Research and Experience, 2(1), pp. 23-50, March 1991.

16

Cooperation policies for traders

C. Burger

University of Stuttgart, Institute of Parallel and Distributed High-Performance Systems (IPVR),
Breitwiesenstr. 20-22, D-70565 Stuttgart, Germany
E-Mail: cora.burger@informatik.uni-stuttgart.de

In ODP a general mechanism has been constructed, how cooperation between trader entities can be performed. But little has been said about when and in which cases such a cooperation should be initiated. Furthermore, the above mentioned cooperation mechanism requires an enormous amount of computer and communication resources, counteracting the profit of cooperation, e.g., by degrading the mean response time of traders. As a consequence, policies are needed to resolve such conflicts.

In the following, a general cooperation model is introduced and applied to traders. From this general model cooperation policies for traders are derived. One of these policies can be adapted to optimize the quality of trader service by observing the dynamic behaviour of traders, their clients and servers.

Keyword Codes: C.2.4, H.4.3
Keywords: Computer Systems Organization, General; Distributed Systems; Information Systems, General; Communications Applications

1 INTRODUCTION

Because of increasing size and complexity of distributed systems, traders are needed to bring those objects together that want to build a client-server-pair. This topic has been widely studied before (see, e.g., [ANSA91], [Burg90], [KoWi94], [ISO/93]).

Even in small distributed systems, more than one trader can exist to avoid performance bottlenecks and increase fault tolerance. By connecting a number of such smaller systems to enable cooperation among organizations, a set of traders arises automatically. I. e., the existence of more than one trader in a distributed system is quite reasonable. By cooperating and combining their facilities in a suitable way, traders can ameliorate their functionality as well as their quality of service and stay autonomous nevertheless.

Former studies on cooperation between traders concentrated on the mechanisms needed to achieve cooperations. We classify three approaches, that have been used to cooperate in the case, that a certain client request cannot be treated locally:

- light weight cooperations by acting as a client of another trader ([ISO/93]),
- negotiations for importing exactly one service for a certain period of time ([NiGo94]),

- federations to enable cooperation for a certain set of services during a certain period of time including the handling of heterogeneity between traders ([BeRa91], [ISO/93], [SPM94]).

The only cooperation policy proposed so far, is the one to ask for cooperation help each time, a trader fails to answer a certain client request ([NiGo94], [DIN92]). To motivate further treatment of this topic, let us consider two examples of daily life. While searching for a very important reference in a local library, a scientist would be happy to get the desired article or book from another library, even if it is located at the other end of the world. On the contrary, if a single client asks about an unknown product in a supermarket, the manager would certainly not run to his or her telephone and try everything to purchase the desired product. From these examples it can be concluded, that the above mentioned cooperation policy leaves open a number of important questions:

1. In which cases is it reasonable to cooperate for each failing answer to a client request? Are there further cases in which a trader should ask other traders for cooperation help? Both questions concern the so-called IMPORT case.
2. In which cases should a trader offer cooperation help to other traders (the so-called EXPORT case)?
3. In which cases should a trader withdraw from a cooperation?
4. Which services should be included in a federation between two traders?
5. Which cooperation form (light weight, simple negotiation or federation) should be applied in which cases?
6. Which are the criteria, the answers to the above questions should be based on?

With these questions a new and complex area with a large variety of aspects is entered that must be studied by an incremental approach. In the following, we base the answers to the first five questions on the detection of cooperation need in a general cooperation model, on service usage by clients and on quality of service parameters of traders and servers. The consideration of cost and security is beyond the scope of this paper and remains a topic for future research.

The general cooperation model and its application to traders are introduced in section 2. These studies result in an answer to the first and second question by stating how cooperation need is detected. From this cooperation policies for traders are derived in section 3. The third, fourth and fifth of the above posed questions are answered by realizing policies and possibly adapting them according to the dynamic behaviour of clients, traders and servers. Section 4 describes the current state of the project and outlines future research topics.

2   TRADERS AS COOPERATING SUBJECTS

To study trader cooperation on a broader base, we introduce a general cooperation model (cf. [BuSe94]). A cooperation is built by a number of subjects. Depending on their facilities and goals, subjects decide how to join a cooperation. Thus, the behaviour of subjects with respect to a cooperation can be defined by the following terms that depend on time:

subject = ((facilities, cost), (goals, weights))

In this model, facilities comprise

- those actions that the subject can perform,
- the resources being needed to perform actions,
- the subject's knowledge about itself, about other subjects and its environment.

Facilities can be offered to other subjects by charging cost, a term, that is presented for the reason of completeness (cf. section 4). Goals can be divided in those concerning the execution of tasks and in qualitative ones. To order them for an execution schedule or to resolve conflicts, goals can be weighted.

In the following, this general model is applied to traders by describing their facilities and goals.

2.1 Trader facilities

The kinds of trader facilities as described above can be derived easily from the well known functionality of traders.

Trader actions concern the handling of clients requests and servers offers. In both cases a number of options like e.g., search scope and selection criteria in requests or like criteria to restrict offers to a set of importers can be involved. Because each option requires a different action, we have a wide spectrum of possible trader actions. Normally, not each trader is able to perform all those actions. For example, not each trader can take into account all existing selection criteria.

Trader resources comprise computing and communication resources, especially disk space, as well as the algorithms being used to perform actions. Some resources have limited capacities and therefore restrict the amount of facilities. This plays an important role especially with respect to the trader's response time and the number of service offers that can be stored. As a consequence, traders with different resource capacities exhibit different parameters of quality of service.

Trader knowledge can concern histories of service usage by clients, properties of registered servers and properties of other traders. A trader either knows all traders in a system or a subset, e.g., all neighbours in a spatial relationship (the possibility of trading for traders, i.e. a recursive trading, is not considered here). To avoid isolated traders in the second case, it has to be guaranteed, that the union of all subsets is equal to the set of all traders willing to cooperate. Furthermore, knowledge of traders does not only differ with respect to registered servers but also to other traders.

From the descriptions above, we conclude that traders with very different facilities can exist. If they combine them by means of a cooperation, each of them can profit. It should be noted that actions and knowledge are published explicitly whereas resources are shared only in an indirect way. The amount of offering and exporting its own facilities to other traders, depends on goals that are described in the next subsection.

2.2 Trader goals

In the following, only those trader goals are considered, that are relevant to cooperations. Furthermore, cost and security aspects are not taken into account (cf. section 4). Cooperation relevant goals can be divided into three different types and partially subdivide in functional and qualitative goals. Some of these goals are in conflict. The following informal description of these goals is quantified in subsection 3.1.

- Client biased goals of traders:
  - Answer each client request.
  - Optimize allocation between client and server according to client biased criteria, e.g., by minimization of server response times.
- Server biased goals of traders:
  - Offer to a large set of potential clients to increase probability of client requests.
  - Optimize allocation between client and server according to server biased criteria, e.g., throughput of one server or of a group of servers.
- Trader specific goals:
  - Exhibit a high quality of trader service.
  - Cooperate with other traders if necessary.

How these goals are weighted, has to be defined by the administrator of the system. In general, traders can differ in the set of supported goals and in weights of these goals.

2.3 Cooperation between traders

Partners want to cooperate if they detect a discrepancy in a positive or negative sense between their facilities and some or all of their goals. For traders whose cooperation relevant goals are included in the set as described in subsection 2.2, the following discrepancies can appear:

- Lack of knowledge about servers or traders.
- Set of servers too small to optimize according to client biased criteria.
- Set of clients too small to have a reasonable number of allocations to servers or to optimize according to server biased criteria.

As can be seen, there are four reasons to favour a cooperation, three for importing and one for exporting facilities. Thus, answers to the first and second of the questions about cooperation need are found (cf. introduction). Furthermore, it is obvious how to answer the third question: if the discrepancy between facilities and goals has been removed, the cooperation is ended.

If another goal contradicts the desire for cooperation, this conflict has to be resolved by weighting and modifying goals. Depending on their weights, either the cooperation willingness or the contradicting goal has to be modified.

The effort to initiate and maintain a cooperation can degrade the quality of trader service substantially. Especially response times are influenced negatively, even for those requests, that can be treated locally. Thus, the goal to exhibit a high quality of service by optimizing the response time of a trader, is an example for a goal that can contradict cooperation desires.

3   COOPERATION POLICIES

A cooperation policy determines weights of goals and the cooperation mechanism to be used. For the trader specific goals as defined in subsection 2.2, the policy function takes the following form:

$$policy = w_{coop} \bullet CooperationWillingness + w_{QoS} \bullet QualityofService$$

where the weights $w_{coop}, w_{QoS} \in [0,1]$ and $w_{coop} + w_{QoS} = 1$. Depending on these weights, two cases can appear:

- $w_{coop} = 1, w_{QoS} = 0$: The resulting behaviour exhibits as much cooperation as necessary to achieve all client and server biased goals according to their weights, whereas the quality of the trader service is not taken into account. This case coincides with the example of a scientist searching for a very important reference in a local library (cf. introduction).

- $w_{coop} < 1, w_{QoS} > 0$: As long as the quality of trader service is not affected, this case coincides with the last case. If the service quality decreases because of too much resource consumption by cooperations, the cooperation willingness is adapted according to weights. For $w_{coop} < \frac{1}{2}$, this case can be compared to the example of an unknown product in a supermarket being asked about by a single client (cf. introduction), where quality of service is weighted higher than cooperation willingness.

In both cases, a suitable cooperation mechanism has to be chosen depending on the actual situation. In the second case, adaptations have to be performed to modify the cooperation willingness. For the realization and especially the adaptation of cooperation policies, suitable parameters and mechanisms are needed. They are described in the following.

3.1 Quantification of goals

To be able to realize the cooperation policies depicted above, the goals of subsection 2.2 must be quantified. To this end, each of these goals has to be examined to find suitable parameters. Furthermore, for the second case of cooperation policies the adaptability of parameters is required.

The client biased goal to answer each request of clients is a fixed one, i.e. it cannot be adapted. Therefore it is replaced by the following weaker but variable goal:

- If a certain number N_{fail}^{max} of client requests concerning the same unknown service type S is reached during a certain time interval, a cooperation must be initiated to find a trader who is responsible for this service type.

In this goal, the parameter N_{fail}^{max} can be varied. For the case $N_{fail}^{max} = 1$, the weaker form coincides with the original one.

Both optimization goals, the client biased as well as the server biased optimization, could result in load balancing across organisation boundaries. Because of organizational problems and perhaps of problems with heterogeneity, we propose to use load sharing by applying a simple mechanism with thresholds (cf. the classification of allocation mechanisms in [Burg90]). To this end, the following parameters are needed:

- Upper threshold $UT_{server}(S)$ for load of servers of a certain service type S.

- Lower threshold $LT_{trader}(S)$ for client requests concerning a certain service type S and lower threshold $LT_{server}(S)$ for load of servers of this type.

- The period of time T, expressions with the above defined parameters must have been valid before being taken into account.

The degree of cooperation willingness is defined by taking all parameters of this subsection together. The usage of these parameters is described in subsection 3.2. Furthermore, they are subject to modifications that are needed to adapt the cooperation willingness (cf. subsection 3.4).

3.2 Realization of cooperation policies

The algorithm to realize the cooperation policies introduced above proceeds straight forward. The following tasks are accomplished:

- Count those client requests concerning a certain unknown service type S that appear with temporal distance less than T. Compare this counter with N_{fail}^{max}. If both values are equal, initiate a cooperation to try to import S.

- Compare the load of servers of a certain service type S with $UT_{server}(S)$. If the load exceeds this value during the whole period T, initiate a cooperation to import S.

- Compare the usage counter for a certain service type S with $LT_{trader}(S)$. If the counter is below this value during the whole period T, compare the load of servers of this service type with $LT_{server}(S)$. If the load is below this value during the whole period T, initiate a cooperation to export S.

At start time, N_{fail}^{max} is set to one and thresholds are set to predefined values. By modifying these values, the cooperation willingness of a trader can be influenced (cf. subsection 3.4).

3.3 Choice of cooperation form

In the last subsection, the conditions for a cooperation initiation have been described. Figure 1 shows three different forms that can be used to perform such initiations (cf. introduction): one trader acting as a client of another trader (light weight cooperation), traders negotiating with respect to one service (simple negotiation) and traders negotiating the amount of mutual use of facilities and solving heterogeneity problems (federation). These forms differ as well in the amount of resource consumption as in their abilities.

Figure 1 : Cooperation forms

To be able to make a reasonable choice that depends on the current situation, we classify according to the following criteria:

- Homogeneity respectively heterogeneity of service types of those traders that participate in a cooperation.
- Number of services that are going to be exported or imported.
- Frequency of service usage.

The algorithm to choose a cooperation form takes the one with lowest resource consumption, that is powerful enough to cope with all characteristics of the current situation. In the following we shortly describe the conditions for each cooperation form.

For light weight cooperation, service types of participating traders must be homogeneous. Furthermore, no more than one service type may be concerned and it may not be requested very frequently. Otherwise the cooperation form with simple negotiation should be used. It applies to the homogeneous case too and concerns one service as well. But the service is requested more frequently. Because federations are the most heavy mechanism to cooperate, they should only be used if necessary. Thus either the heterogeneous case applies or more than a certain number of servers is involved or both.

There are two parameters which care for differentiating between usage of different cooperation forms: the limit concerning frequency of requests divides light weight cooperation and simple negotiation, the maximum number of servers that are handled by simpler mechanisms separates them from the federation mechanism. Both parameters can be subject to an adaptation to influence the quality of trader service.

3.4 Adaptation mechanism

According to weights of goals in the second class of coordination policies, we base the modification of cooperation willingness on the observation of the quality of trader service: if the quality of trader service increases, its cooperation willingness increases too, whereas it decreases for the case of decreasing quality of trader service. The amount of increase or decrease can be defined as a constant or it can depend on the gradient of the quality of trader service. In figure 2 this relationship is shown by taking the mean response time as an example for the quality of trader service. The modification of the cooperation willingness is realized by modifying its parameters (cf. subsection 3.1).

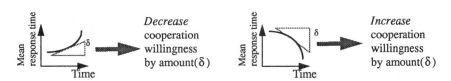

Figure 2 : Adapting cooperation willingness

For the separation between usage of different cooperation forms we proceed in an analogous way: if the quality of trader service decreases, usage boundaries are shifted to be able to choose a cooperation form consuming less resources. To this end, the frequency limit or the number of services required for federation are adapted.

To demonstrate the mechanism more clearly, we study a number of examples. At start time, there is a maximum cooperation willingness, i.e. each client request that cannot be answered locally leads to a cooperation. If response time degrades in the following time interval T, cooperation willingness is decreased by increasing N_{fail}^{max}, the number of failing client requests that are necessary before cooperation is initiated.

The second example copes with lightly loaded servers. If a cooperation has been initiated to offer such a server and the response time of the trader degrades in the following time interval T, cooperation willingness is decreased by decreasing $LT_{trader}(S)$ and $LT_{server}(S)$. If one or both of these thresholds reach zero, this means a very low cooperation willingness.

If the quality of trader service behaves poorly after usage of the light weight cooperation form, the frequency limit should be decreased to allow the use of simple negotiation earlier. In a similar way the number of services needed for federation is increased, if the quality of trader service degrades after a federation. A bad reaction to the usage of simple negotiation can be answered by increasing the frequency limit or by decreasing the number of services needed for federation. These possibilities should relieve one another to avoid loops.

After having described the mechanism of adapting, it must be considered when such an adaptation has to be performed. To achieve this, traders have to observe and analyse their behaviour. Cases in which the quality of trader service decreases independently of any cooperation, have to be filtered out. For example, the response time of a trader without cooperation has to be estimated to find out the reason for an increase.

4   PROJECT STATE AND FUTURE RESEARCH

The functionality of an isolated trader has been implemented in the MELODY project (Management Environment for Large Open Distributed sYstems) and is documented in [KoWi94]. The architecture consists of two main parts, the trader user agent (TUA) that is bound to clients and servers and the trader server agent (TSA). The agents are written in C++ and communicate via DCE-RPC, therefore object adapters (OA) are needed to map from the object oriented to the procedural model (cf. figure 3). MELODY traders take into account dynamic properties of servers. This is achieved by management mechanisms that observe and collect dynamic information. These mechanisms can be applied to the realization of cooperation policies also.

The MELODY traders are now extended to cooperate with one another. The corresponding architecture is shown in figure 3 (cf. [KoBu95]). The first step consists in implementing a

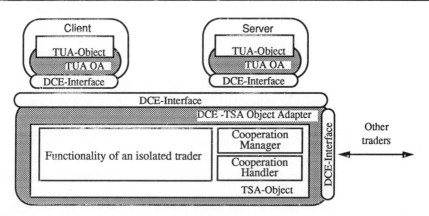

Figure 3 : Architecture of cooperating MELODY traders

cooperation handler which makes available the three cooperation forms as described in figure 1. By intercepting the case that no suited server has been found by the functionality of an isolated trader, the cooperation manager is invoked to choose one of the operations of the cooperation handler. To start with, this choice will be configured statically. Measurements of the performance of the different cooperation mechanisms will deliver criterions for the adaptive policy, i.e. for thresholds, period of validity and the limits between usage of different cooperation forms.

Two approaches are pursued with respect to cooperation policies: in one case the trader cooperation is controlled by the system administrator, in the other case the policies that have been described in section 3 are applied. This means answering the questions of the introduction either by the system administrator or automatically.

In a next step, cost and security aspects will be included. To prepare this, cost have already been included in the general model. The need for security could either be appended to each of the trader's facilities or introduced as a global goal.

5 SUMMARY

By considering a general cooperation model, we found a large variety of different types of traders with respect to their facilities, goals, and weights of goals and identified two main cases of cooperation policies. These cooperation policies can be realized by means of quantifying the goals of traders and observing and analysing the behaviour of clients, servers and traders. Furthermore, an adaptation mechanism for the second class of cooperation policies has been defined.

Collecting those aspects, answers to all questions posed in the introduction have been found:

- Determination of conditions leading to a cooperation respectively to its end (subsection 2.3 and 3.2).
- Finding out the service types to be imported or exported by the cooperation (subsection 3.2).
- Determination of the cooperation form to be applied (subsection 3.3).

Thus, traders are equipped with policies to decide on cooperation automatically. This contributes directly to ODP trading.

ACKNOWLEDGEMENT

This work is granted to the German Hochschulsonderprogramm II. Furthermore, the author wishes to thank E. Kovacs for his explanations of the MELODY trading system and for helpful discussions.

REFERENCES

[ANSA91] ANSA. *ANSAware 3.0 Implementation Manual*. Architecture Projects Management Limited, February 1991.

[BeRa91] M. Bearman and K. Raymond. *Federating Traders: An ODP Adventure*. International IFIP Workshop on Open Distributed Processing, October 1991.

[Burg90] C. Burger. *Performance optimization in heterogeneous distributed systems by dynamic allocation of user tasks (in german)*. Doctor thesis, Universität Karlsruhe, Fakultät für Informatik, February 1990.

[BuSe94] C. Burger and F. Sembach. *A survey of mechanisms for computer supported cooperation (in german)*. Technical Report 7, Universität Stuttgart, Fakultät für Informatik, July 1994.

[DIN92]   German DIN. *A structural specification of the ODP trader with federating included.* Technical Report N1-21.1.1/55-92, ISO/IEC JTC1/SC21/WG7, 1992.

[ISO/93]   ISO/IEC JTC1/SC21/WG7 and ITU-TS SG7.Q16. *Working Document on Topic 9.1 - ODP Trader*, July 1993.

[KoWi94]   E. Kovacs and S. Wirag. *Trading and Distributed Application Management: An Integrated Approach.* In Proceedings of the Fifth IFIP/IEEE International Workshop on Distributed Systems: Operations & Management. IFIP/IEEE, October 1994.

[KoBu95]   E. Kovacs and C. Burger. *Support for usage of distributed systems (in german).* Technical Report 8, Universität Stuttgart, Fakultät für Informatik, February 1995.

[NiGo94]   Y. Ni and A. Goscinski. *Trader cooperation to enable object sharing among users of homogeneous distributed systems.* computer communications, 17(3):219 – 229, March 1994.

[SPM94]   O. Spaniol, C. Popien, and B. Meyer. *Services and Service Trading in Client/Server Systems (in German),* volume 1 of Thomson's aktuelle Tutorien. International Thomson Publishing, Bonn, tat edition, 1994.

SESSION ON

Realizing ODP Systems

17

Charging for information services in ODP systems

M.R. Warner

Telecom Australia Research Laboratories, 770 Blackburn Rd, Clayton, Vic, 3168, Australia.

This paper describes a banking service which enables charging for information services in open distributed systems. A novel two phase payment protocol is proposed which overcomes many of the shortcomings of other distributed accounting systems. The banking service is specified using the ODP viewpoint languages. The behaviour of the banking service is formally specified in the information viewpoint using the Z notation, while the interface signatures are defined using ANSA's IDL.

Keyword Codes: H.1.m, H.3.5; H.4.0; K.6.m
Keywords: Charging; Accounting; Online Information Services; Viewpoint Specifications

1. INTRODUCTION

The development of ODP standards should allow a wide variety of information services to be provided over public networks, without the need for specialised hardware or software for the clients of such services. Individuals and organisations will make use of information services provided by others, thereby eliminating the need to duplicate these services. Charging for service usage in such a system will be of great importance, particularly in stimulating the market for such information services.

In theory, any user in an ODP system may provide information services to any other user by developing applications and publishing the interfaces to them. Therefore the number of potential clients and servers in the system is extremely large. Servers must be able to charge for service usage by any client in the system, most of whom will not be known to the server in advance.

2. DISTRIBUTED ACCOUNTING SYSTEMS

Distributed accounting systems have received comparatively little attention, perhaps reflecting the lack of emphasis given to management and operational aspects in experimental distributed systems. Two distributed systems that have included accounting functions are Amoeba [1,2] and the Distributed Academic Computing Network Operating System (DACNOS) [3]. These systems introduce the notion of a trusted banking service which maintains bank accounts for clients and servers in the system. Both meet most of the requirements for resource control in a distributed system; however, neither offers sufficient flexibility nor security to allow automatic charging in a commercial distributed system, particularly one which spans multiple administrative domains.

Both systems require the client to make funds available to the server before a service request is carried out, either by depositing funds into the server's account (Amoeba) or creating an account from which the server can withdraw (DACNOS). Either way, there is nothing to prevent the server taking the money and running. The client can limit the potential loss by only depositing sufficient funds for a single transaction, although this increases the overheads since funds must be transferred before each transaction. Furthermore, if the price of an individual transaction is high, the client's potential loss will also be large.

Another problem occurs where the price of the service is not known in advance. If the client deposits too little the server must either trust the client to deposit additional funds, or withhold service until the funds have been transferred. If too much is deposited, there is no way for the client to retrieve the excess, since both systems prevent the client from withdrawing funds deposited into the server's account. Were this not the case, the client could defraud the server by depositing funds before the request and withdrawing them again before the transaction is completed.

Since the concept of service covers a prolonged binding with potentially many operation invocations, the combination of high price, which is not fixed *a priori* is the norm rather than the exception. Were separate payments made for individual invocations, the overhead associated with the charging system would become excessive.

Many of the problems associated with charging, particularly where the price is high and/or unknown before service is completed, can be overcome by introducing an intermediate stage to the payment process, analogous to the authorisation process used in credit card transactions. Essentially it provides a half-way point from which neither party can withdraw without the other's consent, although the transaction can be stalled, thereby ensuring that neither party obtains an advantage by dishonouring the initial terms. Stalled transactions can then be settled manually.

3.   ENTERPRISE SPECIFICATION

The enterprise specification of an ODP function identifies the objects which participate in the provision and use of that function, defining the roles that the various objects can take with respect to the function, and the activities which they perform. Deontic relationships (ie. permission, obligation and forbiddance) should be specified between the different roles and activities. Some of these relationships will be generic, while others may be specific to a particular management policy.

3.1   Community rules

The purpose of the banking community is to allow servers to charge clients for the use of their services. The basic roles within a banking community are:

Purchaser: an object which is the client of the service being produced, and which instructs a bank to transfer funds from its account into the vendor's account to pay for the service.

Vendor: an object which is the owner of the resource providing the service, and has funds transferred into its account as payment for the service. The vendor may not necessarily provide the actual service.

Administrator: an object which manages (part of) the account space of the banking service.

Bank: an object which provides the banking service.

The purchaser and vendor roles are relative to a particular transaction. An individual object may act as both a purchaser and vendor, perhaps even simultaneously if the object is engaged in multiple transactions. The additional role of Account Holder may be identified as being a super-type of the Purchaser and Vendor roles. As the name suggests, this role includes any object which holds an account with the banking service, regardless of the role it takes with respect to particular transactions.

It is assumed that the objects which perform the administrator role for a bank are identified when the bank is created. These administrators then assign roles to different objects with respect to particular accounts.

3.2 Activity rules

The basic activity within the banking community is that of purchasing a service. In addition there will be a number of activities for managing the account space, including activities to create and delete accounts, deposit and withdraw funds, and to obtain information about the state of accounts. These will be not be addressed in this paper; however, they are described in [4].

Purchase Service may be initiated at any time by a purchaser, but relies on the co-operation of the vendor and the bank.

The purchase service activity is a composite activity which is performed by a number of different objects in different roles co-operating with each other. The composition of the purchase activity is defined by a contract between the purchaser, vendor and banking service. In general, the sub-activities which are required to implement the contract are:

Request Contract: The purchaser requests the formation of a contract by sending a signed authorisation request for an amount A to the vendor. Guidelines for choosing the value of A will be included in the service offer exported by the vendor to the trading service.

Check Contract: The vendor examines the contract to determine if A is sufficient. This will depend on the charging policy of the vendor, the level of trust between the vendor and purchaser and the expected cost of the vendor's service. If A is sufficient, the vendor may forward the authorisation request to the banking service in order to obtain an authorisation number to guarantee the creditworthiness of the purchaser. This indicates acceptance of the contract.

Authorise: The banking service checks the balance of the purchaser's account. If the purchaser has sufficient funds the banking service deducts the amount A, and issues an authorisation number to the vendor. Details of the outstanding authorisation are kept until payment is made. The receipt of an authorisation number obliges the vendor to provide service up to the value of the authorisation.

Request Service: The purchaser, after requesting the formation of the contract, may begin issuing service requests to the vendor.

Serve: After receiving an authorisation number from the banking service, the vendor must service the requests it receives from the purchaser, maintaining an account of the value of service provided so far. Service is provided up to (or exceeding) the value of the authorisation amount A, or until the purchaser requests that service be terminated.

Check Service: The purchaser checks the service results as they are returned to ensure that the service has been performed satisfactorily.

Bill: The vendor computes the final price B, and issues a bill for this amount to the purchaser.

Request Payment: If satisfied with the service, the purchaser must send a signed payment request for the amount B to the vendor.

Check Payment: The vendor checks that the details of the payment request, including the amount, are correct. It then forwards the payment request to the banking service, including the previously obtained authorisation number.

Payment: The banking service adds the initial amount A back to the purchaser's account, and transfers B to the vendor's account before confirming the successful payment to the vendor.

Any of the above sub-activities may fail, either as a result of an active decision by the entity performing that activity, or because of some equipment or other failure. Examples of the former are the vendor's decision that she can no longer provide service, or the banking service's refusal to provide an authorisation because the purchaser has insufficient funds in her account. Whenever a sub-activity does not terminate successfully, the entire sequence should be backtracked, reverting the system to its state prior to the initiation of the activity. This backtracking may not always be possible as the service may already have been performed, and requires the co-operation of all the entities involved in the transaction.

Where the failure was caused by a dispute an entity may refuse to backtrack, thereby causing the entire transaction to stall. In general, a stalled transaction will provide neither party with a gain, since the vendor will not receive payment, while the amount previously authorised will not be available to the purchaser. Stalled transactions may be identified by the banking service when outstanding authorisations reach a certain age. The procedure for recovering from stalled transactions cannot be completely automated as the parties involved may well dispute the terms of the original contract, as well as the actual fact of what service (if any) was provided. Where any automatic procedures fail to resolve such disputes they must be resolved by a human arbitrator (perhaps a magistrate) according to applicable policies and, perhaps, the common law of restitution.

3.3   Policies

The terms of the contract for purchasing a service will be determined by the following policies:

Charging Policy: The charging policy of the vendor will determine what value of authorisation A is acceptable. This policy must take into account the degree to which the purchaser is trusted. Where this level of trust is low, the vendor will demand that the value of the authorisation is high, perhaps exceeding the expected service price B. A high authorisation ensures that the purchaser will gain little from withholding payment, since the authorised amount will be unavailable until the dispute has been resolved. If the vendor trusts the purchaser then the charging policy may allow the authorisation phase to be omitted. In such cases the vendor accepts the risk of the purchaser defaulting on payment.

Payment Policy: The payment policy of the purchaser will determine whether a particular service offer is accepted. Both the amount demanded as an authorisation, and the expected service price, must be considered when determining whether an offer will be accepted. If the purchaser trusts the vendor, the policy may allow for pre-payment for service. In this case the purchaser pays for the service before obtaining it, thereby accepting the risk of the vendor defaulting on the service.

Credit Policy: The credit policy of the bank will dictate whether or not the bank allows authorisation and/or payment operations whose value exceeds the purchaser's current balance.

Security Policy: The bank's security policy will dictate a method for generating signatures for authorisations and payments.

3.4 Security

The proposed system relies on a number of generic security services to prevent fraudulant transactions. These services include data origin authentication and integrity services which prevent an attacker (including the vendor) from modifying or replaying messages signed by the purchaser. Note that since payment requests from the purchaser are passed to the banking service via the vendor, a generic digital signature facility must be available in addition to link by link data origin authentication.

4. INFORMATION SPECIFICATION

The information specification defines the functionality of the banking service by specifying the information elements of which it consists, and the transformations that occur to these elements as a result of the activities described in the enterprise specification. The Z language [5] is used to provide a formal description of the information specification. Activities performed by the purchaser or vendor are not specified.

4.1 Information elements

We begin by introducing the primitive information elements which are needed to describe the banking service. Unique account numbers are used to distinguish between the many different accounts, and authorisation numbers are needed to identify particular authorisations. Time-stamps may be used to detect replays of transaction requests and to identify stalled transactions. The way in which these elements are represented is not considered in the information specification, but will be covered in the computational viewpoint. They are therefore introduced as primitive types:

[AcctNum, AuthNum, TimeStamp]

Money is represented by a natural number, which precludes negative balances. Fixed amounts of credit may still be incorporated into the system by depositing funds into an account before they have been paid for.

Money ::= N

As described earlier the transaction protocol is based on the concept of authorisations. The details recorded about the outstanding authorisations must include the value of the authorisation, the time it was issued, and the account into which the payment will eventually be credited. An outstanding authorisation may be represented by a simple schema with the following three components:

```
 _Authorisation_____
| value : Money
| payee : AcctNum
| date  : TimeStamp
|_____
```

Using these basic types, we may now define an account. An account will consist of the balance which is an amount of money. The set of outstanding authorisation numbers is maintained, and a function is defined which maps them onto the details of the authorisations.

```
 _Account_____
| balance : Money
| outstanding : $\mathbb{P}$ AuthNum
| auth : AuthNum $\rightarrowtail$ Authorisation
|_____
| outstanding = dom  auth
|_____
```

4.2   Dynamic schemata

Activities in the enterprise specification are associated with information transformations in the information specification. The following schemata define the transformations of a particular account.   Note that in addition a number of initialisation and error checking schemata are required. These are omitted here for brevity and clarity. They may be found in [4].

We begin with the operations deposit and withdraw. These operations are used to define the payment operation. They are not normal banking operations (and therefore do not appear in the Teller interface in the computational specification) since money cannot be created or destroyed, but only transferred between accounts. Apart from the balance, the remaining components of the account should remain unchanged as a result of these schemata.   A precondition of the withdrawal schema demands that the existing balance in each currency be no less than the amount to be withdrawn.   In addition the enquiry operation which yields the state of an account may be defined.

```
 _Enquiry_____
| $\Xi$Account
| balance! : Money
|_____
| balance! = balance
|_____
```

```
 _Deposit_____
| $\Delta$Account
| amount? : Money
|_____
| balance' = balance + amount?
| outstanding' = outstanding
| auth' = auth
|_____
```

```
 _Withdraw_____
| $\Delta$Account
| amount? : Money
|_____
| balance $\geq$ amount?
| balance' = balance - amount?
| outstanding' = outstanding
| auth' = auth
|_____
```

The authorisation schema must add the new authorisation to the set of outstanding authorisations. The only restriction placed on the authorisation number issued is that it is not already in use.

$$
\begin{array}{|l}
\text{\_\_Authorise} \\
\hline
\Delta Account \\
amount? : Money \\
dest? : AcctNum \\
time? : TimeStamp \\
number! : AuthNum \\
\hline
balance \geq amount? \\
number! \notin outstanding \\
balance' = balance - amount? \\
outstanding' = outstanding \cup \{number!\} \\
auth' = auth \cup \{number \mapsto \langle\!| \, amount?; dest? ; time? \, |\!\rangle\} \\
\hline
\end{array}
$$

Two separate cases of payment may be distinguished depending on whether an authorisation has already been obtained. The simpler case of an unauthorised payment may be specified simply as a withdrawal and a deposit operation. Component renaming [5, page 18] is used to allow the same variable (*amount?*) to be used in the withdrawal and deposit schemata which have been decorated with subscripts to distinguish between the two different accounts on which they operate.

$$
\begin{array}{|l}
\text{\_\_UnAuthPay} \\
\hline
Withdraw_1[amount?/amount?_1] \\
Deposit_2[amount?/amount?_2] \\
amount? : Money \\
number? : AuthNum \\
\hline
number? = null \\
\hline
\end{array}
$$

The authorised payment is more complex since the authorisation must be checked, the authorised amount (*value*) added back to the account, and finally the authorisation number must be removed from the list of outstanding authorisations. Due to these additional changes to the state of the source account, the authorised payment operation cannot conveniently be based on the withdrawal schema.

$$
\begin{array}{|l}
\text{\_\_AuthPay} \\
\hline
\Delta Account_1 \\
Deposit_2[amount?/amount?_2] \\
amount? : Money \\
dest? : AcctNum \\
number? : AuthNum \\
Authorisation \\
\hline
number? \in outstanding \\
number? \neq null \\
\theta Authorisation = auth_1(number?) \\
dest? = payee \\
balance_1 + value \geq amount? \\
balance_1' = balance_1 + value - amount? \\
outstanding_1' = outstanding_1 - \{number?\} \\
auth_1' = auth_1 \triangleleft \{number?\} \\
\hline
\end{array}
$$

The two pay operations may now be combined into a single operation:

Pay $\triangleq$ *AuthPay* $\wedge$ *UnAuthPay*

4.3   Banking service

So far we have considered operations on particular accounts. The actual banking service will consist of a set of accounts referenced by their account numbers.

```
┌─ BankingService ─────────────────────────────────────────
│ account_space : ℙ AcctNum
│ bank : AcctNum ⇸ Account
├───────────────────────────────────
│ account space = dom bank
└─────────────────────────────────────────────────────────
```

Again, initialisation and management schemata have been omitted here, but may be found in [4].

A promotion schema [7, pages 24-25] may be defined which allows us to refer to a particular account in the banking service, while leaving the remaining accounts unchanged.

```
┌─ ΦBankingService ─────────────────────────────────────────
│ ΔBankingService
│ src? : AcctNum
│ ΔAccount
├───────────────────────────────────
│ src? ∈  account_space
│ θAccount = bank(src?)
│ bank' = bank ⊕ {src? ↦ θAccount'}
│ account space' = account space
└─────────────────────────────────────────────────────────
```

Using this, the previously defined operations may be performed on the appropriate accounts within the banking service as a whole.

BankAuthorise $\triangleq$ Φ*BankingService* $\wedge$ *Authorise*
BankEnquiry $\triangleq$ Φ*BankingService* $\wedge$ *Enquiry*

The payment operation affects two accounts in the banking service. The promotion of these two accounts is incorporated with the actual operation to produce the following schema:

```
┌─ BankPay ─────────────────────────────────────────────────
│ ΔBankingService
│ src?,dest? : AcctNum
│ ΔAccount₁,ΔAccount₂
│ Pay
├───────────────────────────────────
│ {src?,dest?} ⊆ account_space
│ θAccount₁, = bank(src?)
│ θAccount₂ = bank(dest?)
│ bank' = bank ⊕ {src? ↦ θAccount₁', dest? ↦ θAccount₂}
│ account space' = account space
└─────────────────────────────────────────────────────────
```

4.4 Error checking and security

As mentioned earlier, the error and security checking schemata have been omitted here. In [4] error checking is incorporated by defining special error reporting schemata which are combined with the operational schemata via schema conjunction. Security checking is incorporated in a similar way, although additional components must be added to the Account schema to describe digital signatures and an authorisation function. Again, more detail can be found in [4].

5. COMPUTATIONAL SPECIFICATION

The computational specification of the banking service defines the interfaces to which clients of the banking service bind. The banking service is offered via the teller interface. This interface is defined below using ANSA's Interface Definition Language (IDL)[8]. In addition, banks will provide an administration interface through which the account space can be managed[4].

As defined in the enterprise specification, all authorisation and payment requests from the purchaser are passed to the banking service via the vendor. Therefore only the vendor need form a binding to the banking service's teller interface during the course of a transaction. In addition the vendor will provide a service interface to which the purchaser will bind. This interface, or a separate accounting interface, will require a number of operations to exchange accounting information such as the signed requests. Although it may be possible to standardise these accounting operations, this is considered outside the scope of this work. Furthermore, since the purchaser and vendor must define a common service interface, it is reasonable for them to also define an appropriate accounting interface, particularly since its form may depend on the nature of the service provided.

5.1 Type definitions

We now define how the information elements identified in the information specification are represented using the basic types defined in the IDL. Only those information elements which are passed across interfaces need to be defined. The way in which accounts and authorisations are represented is an implementation issue.

Account numbers will be composed of two parts: one indicating the "branch" of the bank, and the other a unique account number within the branch. Cardinals are used to represent amounts of money since the information specification defined only positive sums of money. Since authorisation numbers are always used in the context of a particular account, a simple cardinal representation may be used. The null value used for unauthorised payments may be represented by the value zero. Finally the success or failure of an operation may be defined as an enumerated data type. If the IDL supported exceptions these could be used to report the failure of an operation.

```
BankTypes : INTERFACE =
BEGIN

    BranchName    : TYPE = STRING;
    Name          : TYPE = STRING;
```

```
AccountNumber : TYPE = RECORD [branch : BranchName,
                                number : CARDINAL];
Money         : TYPE = CARDINAL;
AuthNumber    : TYPE = CARDINAL;
Report        : TYPE = {Ok,InvalidAuthorisation,PermissionDenied,
                        AccountOperational,InsufficientFunds,
                        UnknownAccount, AccountExists}
END.
```

In addition a number of types are required for security. These include a signature used to
validate authorisation and payment requests. This could be constructed using a symmetric
key algorithm such as DES to generate a message authentication code for the concatenation of
the parameters in the operation. Under such a scheme, signatures would be represented by a
64 bit block, conveniently represented by an array of octets. A time stamp may be
represented by a cardinal.

```
SecurityTypes : INTERFACE =
BEGIN
    Signature    : TYPE = ARRAY[8] OF OCTET;
    TimeStamp    : TYPE = CARDINAL;
END.
```

5.2   Teller interface

The teller interface is used by the vendor to initiate authorisations and payments after having
received the appropriate request from the purchaser. The operations supported by this
interface are enquiry, authorise and payment. Each of these operations requires a number of
security related parameters indicating the identity of the user requesting the operation, a time
stamp and a signature. Note that in the case of the authorise and payment operations, the
identity and signature refer to the purchaser rather than the vendor who actually invokes the
operation. Any security service between the vendor and bank may require additional
parameters. Other parameters are as indicated in the information specification. These
comprise a source account number, a destination (ie. the vendor's) account number and an
amount for the authorise and payment operations, as well as an authorisation number for the
payment operation. In addition to the results described in the information specification, a
status result is returned, indicating success or the reason for a failure.

```
Teller : INTERFACE =
NEEDS BankTypes;
NEEDS SecurityTypes;
BEGIN
    Enquiry : OPERATION [account : AccountNumber,
                         id : Name,
                         time : TimeStamp,
                         sig : Signature]
             RETURNS [result : Report,
                      balance : Money];
    Authorise : OPERATION [account : AccountNumber,
                           dest : AccountNumber,
                           amount : Money,
                           id : Name,
                           time : TimeStamp,
                           sig : Signature]
               RETURNS [result : Report,
                        auth : AuthNumber];
    Payment : OPERATION [account : AccountNumber,
```

```
                    dest : AccountNumber,
                    amount : Money,
                    auth : AuthNumber,
                    id : Name,
                    time : TimeStamp,
                    sig : Signature]
         RETURNS [result : Report];
END .
```

5.3   Distribution of banking service

A single, centralised bank can provide all the functionality specified above; however, apart from in relatively small, isolated systems, such an implementation is not practical. In a large scale ODP environment, it is inevitable that a decentralised banking service must be provided by a number of co-operating banks. There are several reasons for this. Since the provision of a banking service for charging is likely to be profitable, it must be assumed that a number of different organisations will compete to provide the banking service. For account holders with one bank to have access to servers whose accounts are held with another bank, the two banks must co-operate in some way, yet they must be able to maintain their autonomy. The way in which multiple banks interact to provide the complete banking service is described in [4].

7.   SUMMARY

This paper has provided the outline of a specification for a banking service which allows charging for information services. The banking service makes use of a novel two stage payment protocol, analogous to the authorisation and payment phases of a credit card transaction. The proposed system has a number of significant advantages over both Amoeba and DACNOS. These include:

- reduced risk of fraud for both client and server, particularly in cases where the price is high or is not known prior to service completion;
- no need to create a sub-account for every client-server pair;
- fewer steps to protocol when client uses a new service;
- fewer steps before service completion, thereby reducing delay;
- fewer communicating parties therefore reducing the binding and communications overheads.

Furthermore, by allowing the client to pay for larger amounts of service usage without the risk of fraud, the authorisation mechanism may reduce the overheads involved with the charging system in comparison with either Amoeba or DACNOS.

The banking service was specified using the ODP viewpoint languages. In particular, enterprise, information and computational specifications have been provided. As such this paper also provides an example of the structure and content of the viewpoint specifications. The Z notation is used to provide a formal specification of the behaviour of the banking service. The full specifications of the banking service (of which only a subset are included in this paper) have been formally verified for completeness and consistency[4]. Interface signatures are defined using ANSA's IDL.

An initial demonstration system has been implemented over ANSAware.   This
implementation allows authorisation and payment operations to be performed over an account
space partitioned into an arbitrary number of banks.   The banks interwork via the same teller
interface used by their account holders.   While the initial implementation only includes
rudimentary security features, a more elaborate implementation based on DCE is being
considered.   This enables the use of DCE's secure RPC to provide additional security.

ACKNOWLEDGEMENTS

The permission of the Director, Telecom Australia Research Laboratories, to publish this
paper is hereby acknowledged.

REFERENCES

1. Mullender, S.J., "Accounting and Resource Control", *Distributed Systems*, S.J. Mullender
    (ed), ACM Press, New York, 1989.

2. Mullender, S.J. and Tannenbaum, A.S., "Protection and Resource Control in Distributed
    Operating Systems", *Computer Networks*, Vol. 8, pp. 421-432, 1984.

3. Harter, G., and Geihs, K., "An Accounting Service for Heterogeneous Distributed
    Environments", *Proceedings of 8th International Conference on Distributed Computing
    Systems*, 1988.

4. Warner, M.R., "Charging and Resource Control in Open Distributed Systems", *PhD
    Thesis*, Cambridge University, 1993.

5. Spivey, J.M., *The Z Notation - A Reference Manual*, Prentice Hall International Series on
    Computer Science, 1989.

6. Hayes, Ian (ed.), *Specification Case Studies*, Prentice Hall International Series in
    Computer Science, 1987.

7. Macdonald, R., "Z Usage and Abusage", *Royal Signals and Radar Establishment Report
    91003*, February, 1991.

8. Architecture Project Management Ltd, *ANSAware 4.1: Application Programming in
    ANSAware,* Document RM.102.02, February 1993.

18

Intercessory Objects within Channels

Barry Kitson

Telecom Research Laboratories, 770 Blackburn Road, Clayton, Victoria, 3168, Australia

The management of interactions between components of a distributed system necessitates mechanisms for monitoring activity, and influencing behaviour, within *channels* binding those components. A low-level mechanism, based on *intercessory objects*, is considered here as a foundation for the construction of higher level management functions.

The role of intercessory objects in a distributed system is discussed, with emphasis on the underlying distributed processing environment and on the structure of the channel. The presence of intercessory objects introduces additional management requirements to the system, including a need for the distributed processing environment to provide support for the configuration of these objects within the channel. Restrictions placed on this configuration by distributed processing environment implementations are discussed, as is the value of generic interfaces for the implementation of reusable intercessory objects.

Keyword Codes: C.2.4; K.6.4
Keywords: Distributed Systems; System Management

1. INTRODUCTION

A *distributed processing system* (or *distributed system*) is a software application, typically consisting of components supported by a *Distributed Processing Environment* (DPE). A DPE provides basic infrastructure support for a distributed system, and is usually sufficiently generic to support a wide range of applications. Notable emerging DPEs include APM's ANSAware [1], OSF's DCE [2] and CORBA [3] implementations such as IBM's DSOM, Digital's ObjectBroker, Hewlett-Packard's ORB Plus and IONA's Orbix.

A DPE provides basic management facilities and resources to the system it supports and, more significantly, embodies aspects of an architectural model or framework which influences the system structure and system management. The architectural models by which distributed systems are constructed and managed are extremely important in realistic applications, as they provide a basis for integration and interworking between independently developed distributed systems. The standardisation of these models for open distributed processing systems is a significant activity in organisations such as the International Standards Organization (ISO) [4] and the Object Management Group (OMG).

One of many significant requirements of a management model for distributed systems is the need to manage run-time interactions between system components. To monitor and control these interactions, a management system requires notification of significant events during each interaction, and also requires access directly into the *channel* [5] binding system components. Each of these management capabilities can be provided by introducing the concept of an *Intercessory Object* (IO) into the management model.

Intercessory objects are inserted into channels to providing basic access to information collection and control functions. Higher-level management functions spanning OSI fault, configuration, accounting, performance and security management functional areas [6] may be constructed from these basic services. IOs may also be used to provide base functionality for debugging distributed applications, and may be seen as a generalisation of the *transparency object* concept from work on the Reference Model for Open Distributed Processing (RM-ODP) [5].

Section 2 gives a brief overview of important features of object-based DPEs, and describes a model of channels into which intercessory objects are introduced in Section 3. These models are based heavily on RM-ODP work. Some early thoughts on implementation and configuration management issues are outlined in Section 4, before the application of IOs, and their relationships to transparency objects and other concepts, are discussed in Section 5.

2.   DISTRIBUTED PROCESSING ENVIRONMENT FEATURES

A distributed processing environment provides the infrastructure on which components of a distributed system operate, and supports an architecture describing the nature of these components. Object-based architectures, such as that of RM-ODP, encapsulate components of distributed systems, allowing external access to internal state only via well-defined interfaces. Conforming DPEs support this object model and provide the communications infrastructure necessary for objects comprising a distributed application to interact. This includes support for addressing and binding functions.

2.1.   The object model

The object model defined in architectures such as ISO's Reference Model for Open Distributed Processing (RM-ODP) [5] and adopted into the Telecommunications Information Networking Architecture (TINA) [7] is depicted in Fig. 1. Objects encapsulate state, and provide services to other objects via the interfaces they support. Typically, each operational interface consists of at least one *operation*, equivalent to a member function or method in traditional object models. The *interface type* specifies the operations available on instances of that type, and an *operation type* specifies the types of argument and return parameters. Interactions in these models commonly adopt Remote Procedure Call (RPC) semantics, blocking the invoking thread during normal execution of remote operations. Non-blocking operations, equivalent to message passing, are also provided.

For each interaction, a *client* and a *server* object may be identified. The client object initiates an interaction by invoking an operation on an interface of the server. Communications between these objects is handled by the DPE, via a *channel* [5]. The channel may consist of a number of important component objects, as outlined in Section 2.3.

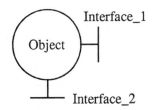

Figure 1: A multi-interface object. This shows the object model of TINA and RM-ODP.

2.2. Generic interfaces

Within OMG work, the Common Object Request Broker Architecture introduces the concept of a Dynamic Invocation Interface (DII) [3], and recent contributions have defined the closely related Dynamic Skeleton Interface (DSI) [8]. The DII allows objects, acting as clients, to access interfaces without requiring that associated client stubs be available. The DSI provides analogous capabilities on the server side of a binding. A server can use the DSI to service calls to arbitrary interface types.

In programming languages such as C and C++, the DII and DSI each appear as an Application Programming Interface (API) expressed in the native language. The DII includes functions to specify an interface to be used, operations to be called, arguments to be passed and values to be returned. The DSI is similar, but with slightly different addressing semantics. In languages where types are treated as first-class entities, such as Scheme, there is much greater scope for representing all interfaces consistently at the application level, but the underlying facilities are invariant across language mappings.

These APIs are implementations of a class of operational interface which will be termed a *generic interface* (GI) here. A generic interface may be specified in CORBA Interface Definition Language (IDL), for example, and has the important property that it is capable of representing operation invocations on any other interface type that can be represented in the model. The operations of a GI are equivalent to those at the lower levels of stub marshalling routines. Specifically, the marshalling routines for base and standard types, and for operation invocations, are represented in a GI.

For management systems, a standard GI provides a significant capability. At the simplest level, it allows information specific to the interactions between particular managing and managed systems to be represented independently of the interfaces by which they communicate. The GI is capable of supporting interactions with any interface type. It is the property of genericity, and the fact that mappings between any *application-specific* interface type and the GI type can be made, that is the value of a GI in the context of this discussion. This will be expanded in Section 4.1.

2.3.   Binding objects and the channel

The channel between the client object and the server object is created by the distributed processing environment during the binding of the client to the server's interface. However, it should be noted that many of the channel components are produced at compile time and may be shared for implementation efficiency. The channel will typically contain marshalling and de-marshalling functions and other protocol-specific communications mechanisms. Although these components are not normally implemented with all the support given to application level objects by the DPE, it is convenient to depict them as logical objects, as shown in Fig. 2.

The channel objects of Fig. 2 form stacks of functionality associated with the client and server sides of the SInt service interface. A particular DPE may implement only some of these objects and may not, for reasons of efficiency, encapsulate functionality into objects identifiable as those shown in the figure. However, the objects of Fig. 2 represent specific channel capabilities germane to discussions in following sections.

The Proxy and GI Proxy objects shown in Fig. 2 provide encapsulations of the addresses of, or pointers to, the next objects in the sequence supporting interactions in the channel. As such, the proxies identify points were control flow may be redirected. This concept helps clarify discussions below.

The Proxy objects are dependent on the type of the interface being bound. In the example, the service interface is identified as having type SInt. Each Proxy object supports an interface of type SInt and makes use of a similar interface. Each GI Proxy object is similarly associated with a generic interface type GI, which is standard throughout the DPE.

A Proxy object, or more correctly the SInt interface of a Proxy object, will typically appear to the application programmer as a programming language dependent representation of the server, instantiated within the client code. The Proxy object makes use of the marshalling stub object, Mshl Stub, to perform the high level marshalling of arguments and de-marshalling of return parameters associated with the operation invocation. The exact nature of the Mshl Stub object is implementation dependent. In most currently available DPEs, these stubs involve relatively complex operations determined at compile time, based on information contained in the SInt interface type definition. However, given appropriate programming language support there is no reason why this same functionality cannot be developed at run-time.

The Mshl Stub object is responsible for mapping the high level SInt representation of the interface into a low level representation such as that provided by a GI. Some DPEs, particularly those based on CORBA, already support an explicit DII at the application level. Use of such a DII is equivalent to direct use of the GI interface by the client, as shown in Fig. 2. Even if the DPE does not make a DII available to applications, functionality equivalent to the GI interface will exist in the marshalling structures of the channel.

In some DPE implementations, it may only be possible to implement the Mshl Stub object if the SInt type is known at compile time. This depends on programming language support, but it should be noted that some implementations will allow the construction of this object at run-time, and others will not.

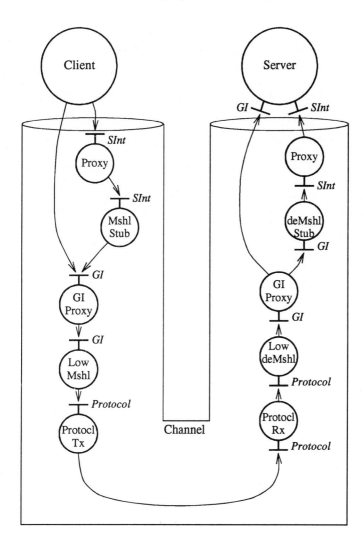

Figure 2: Channel binding a client to a server interface. The channel consists of a number of logical objects forming stacks below the client and server. A particular distributed processing environment may implement the functionality of only some of these objects when binding the client and server of service interface type SInt.

The GI Proxy object performs a function similar to the Proxy object described above, allowing another level of indirection. The addressing encapsulated by this object may be used to select protocols for communications, for example. The Low Mshl object provides a protocol-specific mapping from the GI interface type to the representation used by the underlying communications protocol object (Protocol). Such protocol objects are responsible for the actual transmission of information across the network.

On the server side of the channel, there is a similar stack of logical objects. It should be noted that Proxy and GI Proxy objects on the server side perform identical functions to their counterparts on the client side. In fact, the implementations of these objects on each side of the channel need not be different.

Another important point is that the server in the example may be offering the SInt interface in its own right, with a complete mapping of the interface type into the application code, or alternatively, as a specific set of supported GI functions. In the former case, a high level de-marshalling stub, deMshl Stub is required, while in the latter, this and the Proxy above are not required.

For example, a server might implement a large range of functionality internally, and may decide at run-time to make only some of this functionality available to clients. The exact functionality, or operations available on an interface for the use of clients, may be dependent on the nature of those clients and security or other constraints. The server can use the GI to give access to this functionality as determined at run-time. The client may be unaware of this, and may view the interface as an instance of the SInt type, for example, binding to it via the channel shown in Fig. 2.

3.   INTERCESSORY OBJECTS

The purpose of intercessory objects is to allow almost arbitrary functionality to be added to the channel binding clients to server interfaces. Ideally, an application programmer should be able to implement an intercessory object just as any application level object in the distributed system, to provide arbitrary complexity of function. It should also be possible to implement an IO independently of any particular distributed system, so that reusable libraries of IOs may be assembled and effectively utilised. These requirements place restrictions on the implementations of intercessory objects themselves, and also on the mechanisms by which they may be inserted into the channel on the client or server side.

As an example, consider an intercessory object which notifies a managing system whenever RPC invocations of operations are commenced or completed on an interface of type SInt. An IO which performs this function is depicted in Fig. 3.

Such an intercessory object may be inserted into the channel on either the client or server side, and may perform actions without affecting the operation of the client or server themselves.

When inserting an intercessory object into a channel stack, it is necessary to allow for the inclusion of additional intercessory objects in future. The mechanism for including objects in a sequence supporting a channel involves breaking the sequence, redirecting calls to the intercessory object, and then directing calls from the intercessory object back through the objects in the channel after the insertion point. This redirection is central to

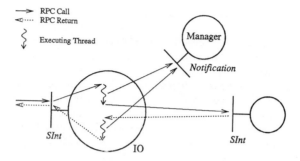

Figure 3: Intercessory object used to monitor interactions. The intercessory object IO executes a short sub-thread, or sequence of operations, before and after passing an invocation on to another object in the channel stack. In each of these sub-threads the management system is notified via a call on a Notification interface. (The returns from these RPCs are not shown for simplicity.)

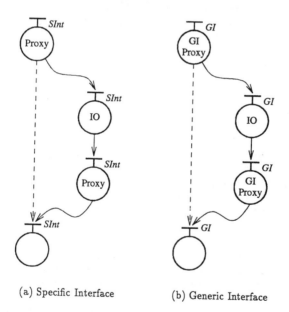

(a) Specific Interface   (b) Generic Interface

Figure 4: Insertion of an intercessory object. The insertion of an intercessory object IO in a client stack is shown (a) where the object IO is implemented in a fashion specific to the SInt interface type bound by the channel and (b) where the IO object is generic, and is implemented in terms of the generic interface type GI. The initial sequence of objects in the channel is indicated by the dashed arrow.

the functionality of the Proxy objects described in Section 2.3. Insertion can only take place after such a Proxy object, and in order to allow for future insertions further in the channel after a given intercessory object, another Proxy object should be added after each inserted intercessory object. The insertion process for this object is shown in Fig. 4(a).

An intercessory object to perform activities such as monitoring would typically be constructed with a view to reuse. The requirement for such an object is so common that a reusable version may even be supplied as part of the DPE management infrastructure. In any case, the object implementation is independent of the actual interface type being bound, and would therefore be most easily implemented as an object which both supports and uses interfaces of the generic GI type. Insertion of such an object into the stack supporting the client involves redirection of the thread of control to the intercessory object. The redirection is performed by a GI Proxy object in this case, since a GI Proxy is capable of calling the GI interface supported by the intercessory object. The insertion process is shown in Fig. 4(b).

Similar insertion of intercessory objects is obviously possible in the server stack, by redirecting Proxy or GI Proxy objects to call intercessory objects. In the case of the monitoring example considered above, it would obviously be usual to implement the relevant intercessory objects in terms of GI interfaces to assist with reuse.

The same approach may be taken in the case of interaction management activities which involve the intervention of the management system in interactions themselves. Such intervention may include adding and checking authentication information for security purposes, adding and using control information for management of transactions, re-binding in the event of server failure, or duplicating an interaction and directing it to multiple servers, possibly collating and combining results.

4. IMPLEMENTATION AND MANAGEMENT ISSUES

At the lowest level, the configuration of intercessory objects in the channel is controlled by the use of proxy objects. A proxy object should support an interface which allows the redirection of invocations to other objects in the client and server stacks and a reference to this interface should be made available to the system managing the channel. However, there are other implementation issues which affect configuration.

4.1. Generic interfaces and intercessory objects

As described above, there are two basic forms an intercessory object implementation can take. These forms are distinguished by the interfaces supported by the intercessory object. It should be noted that each particular IO will use the same interface type as a client that it provides as a server. An IO may be implemented in terms of a specific interface type or in terms of a generic type. The choice is a pragmatic one, but it is important to note that the limitations of particular DPEs and programming languages will limit the options available.

For programming languages which treat types as first-class entities, where it is possible to dynamically bind to representations of specific interface types, the definition of a GI type is not a significant issue, as equivalent functionality can be achieved within the language. But in many languages this is not the case, and the reuse of generic IOs is limited by the acceptance of the corresponding GI specification.

4.2. Sequencing

It is clear that the positioning of IOs in the stacks of both clients and servers is significant. Consider, for example, a replication IO which duplicates any invocation it receives to multiple servers, and a notification IO which notifies a management system of any interaction occurring within the channel. If the replication IO is placed after the notification IO, a single message will be received by the management system as the client calls the replicated interface. However, if the notification IOs are placed after the replication IO, multiple notifications will be received for each invocation.

Another sequencing issue is caused by the relationship between IOs based on the generic interface types and IOs based on application-specific interface types. In the case of channel objects depicted in Fig. 2, interface-specific IOs must be placed above generic IOs, on both the client and server sides of the channel, unless de-marshalling stubs are available on the client side and marshalling stubs on the server side. These stubs may be used to map specific interface types to generic interface types and vice versa.

As the semantics of IO functionality may depend on the sequence in which the IOs are invoked within a particular interface interaction, there is a requirement that this structure be managed in some way. Constraints, placed on these structures by IO implementations, complicate the management process.

4.3. Channel management

As the number of intercessory objects in a channel may change dynamically, and the sequence of IOs in the channel may change, it will be convenient to use a *binding object* to manage the internal configuration of the channel itself.

In many cases, intercessory objects will have additional interfaces specifically related to the operations they are performing or to the functionality they provide. An example is a replication IO which may be redirected to multiple servers, where these replicas are dynamically created and destroyed at run-time. The replication IO needs to be informed about the creation of new server replicas and deletion of old ones along with other reconfigurations which may take place in the replicated system. Information of this nature may be passed to the replication IO via such additional interfaces.

In any case it is possible, and indeed highly likely, that intercessory objects will require a status similar to that of application objects, requiring the ability to create and destroy interfaces dynamically. The types of these interfaces cannot be known to the client or server objects if the intercessory object functions are transparent to them. Therefore, management paradigms cannot make assumptions about these types.

5. APPLICATIONS

A number of management and related areas require the capabilities provided by intercessory objects. Transparency objects from RM-ODP, and *filters* and *smart proxies* from Orbix [9], the CORBA implementation from IONA Technologies, satisfy similar requirements.

5.1.   RM-ODP transparency objects

RM-ODP [5] includes definitions of eight *transparencies* which may be provided to distributed systems by conforming environments. Each of these transparencies represents a possible feature of distribution which may be masked from an application by the underlying infrastructure or DPE. Examples include *location transparency*, which masks from a client the location of its server, and *replication transparency*, which masks from a client the fact that it interacts with multiple servers on each invocation.

These transparencies, at least five of which are directly related to interactions between system components, were to be implemented by partially standardised *transparency objects*. A replication transparency object, for example, may intercept an invocation from the client, and send a copy to each of a number of servers. The object may then collate the server responses for return to the client.

Following some work in this area, it became clear that the variability inherent in aspects of this function, such as collation strategies and behaviour in the event of reconfiguration, make standardisation of the associated transparency objects inappropriate.

However, consider the case that the replication transparency object is implemented as an intercessory object, with associated mechanisms for insertion into the channel. Now, the variable functionality could be seen as useful differentiation, and specific functionality could be defined by the developer of the application or the management system, the DPE vendor, or by a third party. Although the intercessory and transparency objects may behave identically, the IO concept does have the advantage of a model where mechanisms are specified for its insertion into the channel.

5.2.   Orbix filters and smart proxies

Orbix, the CORBA implementation from IONA Technologies, also provides capabilities similar to those of intercessory objects. The channel model for this CORBA implementation is slightly complicated by the concept of an address space, corresponding to an operating system process, but access to the communication mechanisms is possible. In particular, Orbix filters provide direct access into the channel for server objects, and Orbix smart proxies provide similar access to the client stack. One difference though, is that a smart proxy will be bypassed if the CORBA DII is used for interactions. In that sense, a smart proxy is similar to the interface-specific IO defined in Section 3.

Of particular interest are the uses to which these capabilities are put in Orbix. Suggested applications [9] include caching of data within smart proxies to avoid unnecessary communications overheads, the creation of threads on operation invocations, and the passing of authentication information between clients and servers. These functions are all useful management applications, and the facilities provided by Orbix are essentially equivalent to those which could be provided by IOs of specific types.

5.3. Other management applications

The ability to monitor events taking place within the channel binding two objects has numerous management applications. The types of events which might be monitored include initiation and completion of an interaction on the client side, and initiation and completion of an operation call on the server side. The parameters passed during interactions are also of interest to the management system, as are outright failures and timeouts of interactions.

Monitoring events is useful for debugging and location of faults, and allows the measurement of interaction frequency and data flow, and the determination of usage for accounting and performance management. The management system may base the distributed system configuration on this information, in addition to an understanding of physical resources in the network supporting the distributed system.

Other examples of interaction management include the maintenance and transfer of authentication information for security purposes, and the use of alternative protocols to improve performance of data transmission.

Through the use of intercessory objects, management of this form may be undertaken by management components of the distributed system, without any significant impact on core application operation. The distributed application may be subdivided into managing and managed subsystems in such a way that management operations can be tuned, or even replaced, without making changes to any core application components.

More significantly, management requirements such as these are providing strong motivation for the development of facilities to introduce arbitrary functionality into the channels between objects comprising distributed systems. The value of IOs does not lie in their standardisation, but in the standardisation of the mechanisms which surround them.

6. CONCLUSIONS

Intercessory objects allow the insertion of control and monitoring capabilities into the channel binding objects in a distributed system, and this may be totally transparent to the objects involved. However, the concept of an intercessory object is not new. The transparency objects of ISO's RM-ODP, and the filters and smart proxies of IONA's Orbix, for example, offer similar functionality. The distinctions between IOs and these concepts, and the areas where standardisation work might be valuable, lie in the models describing the configuration of IOs.

With the current emergence of a number of commercial DPEs, and the desire in computing, telecommunications and other industries to build large-scale distributed systems, it is important that work on management of these systems develop further. Intercessory objects offer a mechanism on which many higher-level management activities can be constructed. Libraries of intercessory objects could provide a valuable tool set to perform a wide range of activities, and IOs could be selectively included into various points in the distributed system. Debuggers, and test harnesses and stubs, could be of particular value.

Structures and functions for management of distributed software systems are not yet well defined or understood, but it appears that intercessory objects may have an important role to play in the management of interactions, while introducing their own management requirements.

ACKNOWLEDGEMENTS

The permission of the Director, Telecom Research Laboratories, Telstra Corporation Limited, to publish this paper is hereby acknowledged. The author also wishes to thank Leith Campbell, Geoff Wheeler, Ajeet Parhar and Michael Warner of Telecom Research Laboratories, for their comments on drafts of this paper.

REFERENCES

[1] Architecture Projects Management Ltd., "Application Programming in ANSAware," RM.102.02, Cambridge, February 1993.

[2] Open Software Foundation, in *Introduction to OSF DCE*, Prentice Hall, New Jersey, 1992.

[3] Object Management Group, "The Common Object Request Broker: Architecture and Specification," 91.12.1, 10 December 1991.

[4] ISO/IEC and ITU-T, "Open Distributed Management Architecture — First Working Draft," ISO 8801, July 1994.

[5] ISO/IEC and ITU-T, "Draft Recommendation X.903: Basic Reference Model of Open Distributed Processing — Part 3: Prescriptive Model," 10746-3.1/X.903, February 1994.

[6] ISO/IEC ITU-T, "Information processing systems – Open Systems Interconnection – Basic Reference Model – Part 4: Management Framework," ITU-T X.700, 1989.

[7] William J. Barr, Trevor Boyd & Yuji Inoue, "The TINA Initiative," *IEEE Communciations Magazine* (March 1993).

[8] Object Management Group, "Universal Networked Objects," OMG TC Document 94.9.32, 28 September 1994.

[9] IONA Technologies, *Orbix Advanced Programmer's Guide, Version 1.2*, Dublin, Ireland, February 1994.

19

A Fault-Tolerant Remote Procedure Call System for Open Distributed Processing

Wanlei Zhou, School of Computing and Mathematics, Deakin University, Geelong, VIC 3217, Australia

This paper is concerned mainly with the software aspects of achieving reliable operations on an open distributed processing environment. A system for supporting fault-tolerant and cross-transport protocol distributed software development is described. The fault-tolerant technique used is a variation of the recovery blocks and the distributed computing model used is the remote procedure call (RPC) model. The system incorporates fault tolerance features and cross-transport protocol communication features into the RPC system and makes them transparent to users. Our system is small, simple, easy to use and also has the advantage of producing server and client driver programs and finally executable programs directly from the server definition files.

Keyword Codes: *C.2.4, D.4.4, D.4.5.*
Keywords: *Open distributed processing, Fault-tolerant computing, distributed systems, remote procedure calls, client/server model.*

1 INTRODUCTION

The advances in computer technology has made it cost-effective to build distributed systems in various applications. Many experts agree that the future of open distributed processing is the future of computing. *The network is the computer* has become a popular phrase [5].

Remote Procedure Call (RPC) is perhaps the most popular model usd in today's distributed software development and has become a de facto standard for distributed computing. To use it in an open distributed environment effectively, however, one has to consider the cross-protocol communications because user programs built on top of different RPC systems cannot be interconnected directly. Typical solutions to this problem are:

1. Black protocol boxes: protocols used by RPC programs are left as black boxes in compiling time, and are dynamically determined in binding time [1].

2. Special interfaces [15] or RPC agent synthesis systems [7] for cross-RPC communications.

However, one issue is still outstanding in building RPC systems for open distributed systems: the fault-tolerance features.

An open distributed system consists of many hardware/software components that are likely to fail eventually. In many cases, such failures may have disastrous results. With the ever increasing dependency being placed on open distributed systems, the number of users requiring fault tolerance is likely to increase.

This paper is concerned mainly with the software aspects of achieving reliable operations on an open distributed processing environment. A system for supporting fault-tolerant and cross-transport protocol distributed software development is described. The system design is aimed toward application areas that may involve heterogeneous environment and in which requirements for fault-tolerance are less severe than in, for example, the aerospace field, but in which continuous availability are required in the case of some components failures [4]. The application areas could be, for example, kernel/service pool-based distributed operating systems, supervisory and telecontrol systems, switching systems, process control and data processing. Such systems usually have redundant hardware resources and one of the main purpose of our system is to manage the software redundant resources in order to exploit the hardware redundancy.

The reminder of this paper is organised as following: In Section 2, we summary some notable related work provide the rationale of our work. In Section 3, we describe the architecture of the SRPC system. Then Section 4 describes the syntax and semantics of the server definition files and the stub and driver generator. In Section 5, we present an example to show how this system can be used in supporting fault-tolerant, open distributed software development. Section 6 is the remarks.

2   RELATED WORK AND THE RATIONALE

There have been many successful RPC systems since Nelson's work [11]. But few of them consider fault tolerance and cross-protocol communication in their design, or they relay on users to build up these features.

Notable works on incorporating fault tolerance features into RPC systems are the Argus [10] and the ISIS [2] [3]. The Argus allows computations (including remote procedure calls) to run as *atomic transactions* to solve the problems of concurrency and failures in a distributed computing environment. Atomic transactions are serialisable and indivisible. A user can also define some atomic objects, such as atomic arrays and atomic record, to provide the additional support needed for atomicity. All the user fault tolerance requirements must be specified in the Argus language.

The ISIS toolkit is a distributed programming environment, including a synchronous RPC system, based on virtually synchronous process groups and group communication. A special process group, called *fault-tolerant process group*, is established when a group of processes (servers and clients) are cooperating to perform a distributed computation. Processes in this group can monitor one another and can then take actions based on failures, recoveries, or changes in the status of group members. A collection of reliable multicast protocols are used in ISIS to provide failure atomicity and message ordering.

However, when a server (or a guardian in the Argus) fails to function well, an atomic transaction or an atomic RPC has to be aborted in these systems. This is a violation of our continuous computation requirement. The fault-tolerant process groups of the ISIS can cope with process failures and can maintain continuous computation, but the ISIS toolkit is big and relatively complex to use.

Typical solutions to the cross-protocol communication in RPC systems are the black protocol boxes of the HRPC [1], the special protocol conversion interface [15] and the RPC agent synthesis system [7] for cross-RPC communications.

The HRPC system defines five RPC components: the stub, the binding protocol, the data representation, the transport protocol, and the control protocol. An HRPC client or server and its associated stub can view each of the remaining components as a "black box." These black boxes can be "mixed and matched." The set of protocols to be used is determined at bind time   long after the client and server has been written, the stub has been generated, and the two have been linked.

The special protocol conversion interface proposed in [[15]] uses an "interface server" to receive a call from the source RPC component (client or server) and to convert it into the call format understood by the destination RPC component (server or client).

The cross-RPC communication agent synthesis system proposed in [[7]] associates a "client agent" with the client program and a "server agent" with the server program. A "link protocol" is then defined between the two agents and allow them to communicate. The server and the client programs can use different RPC protocols and the associated agents will be responsible of converting these dialect protocols into the link protocol.

But none of the above cross-protocol RPC systems consider fault-tolerance issues. If the server fails, the client simply fails as well.

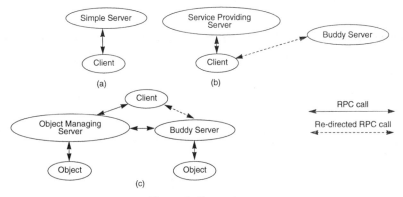

Figure 1: Server types

Incorporating both fault tolerance and cross-protocol communication into RPC systems is clearly an important issues for using RPCs efficiently and reliably in open distributed environments. In this paper we describe a system, called SRPC (Simple RPC) system, for supporting development of fault-tolerant, open distributed software. The SRPC incorporates fault tolerance features and protocol converters into the RPC system and makes them transparent to users. A *buddy* is set up for a fault-tolerant server to be its alternative. When an RPC to a server fails, the system will automatically switch to the buddy to seek for an alternate service. The RPC aborts only when both the server and its buddy fail. The clients and servers can use different communication protocols. To obtain these fault tolerance and automatic protocol converting services, users only need to specify their requirements in a descriptive interface definition language. All the maintenance of fault tolerance and protocol conversion are managed by the system in a user transparent manner. By using our system, users will have confidence on their open distributed computing without bothering with the fault tolerance details and protocol conversion. Our system is small, simple, easy to use and also has the advantage of producing server and client driver programs and finally executable programs directly from the server definition files.

3   SYSTEM ARCHITECTURE

The SRPC is a simple, fault-tolerant and cross-protocol remote procedure call system [16]. The system is small, simple, expandable and it has facilities supporting fault-tolerant computing and cross-protocol communication. It is easy to understand and easy to use. The SRPC only contains the essential features of an RPC system, such as a location server and a stub generator, among other things. The SRPC system has been used as a distributed programming tool in both teaching and research projects for three years.

The SRPC system has another interesting feature. That is, the stub compiler (we call it the *stub and driver generator*, or SDG in short) not only produces the server and client stubs, but also creates remote procedures' framework, makefile, and driver programs for both server and client. After using make utility, a user can test the program's executability by simply executing the two driver programs. This feature will be more attractive when a programmer is doing prototyping.

3.1   Server Types

The *client/server model* [13] is used in the SRPC system. An SRPC program has two parts: a server part and a client part. Usually the server provides a special service or manages an object. The client requests the service or accesses the object by using the remote procedures exported by the server.

There are three types of servers in the SRPC system: *simple servers, service providing servers* and *object managing servers*. Figure 1 depicts these three types of servers.

A simple server (Figure 1(a)) is an ordinary server possessing with no fault-tolerant features. When a simple server fails, all RPCs to it have to be aborted.

A service providing server (Figure 1(b)) has a buddy server running somewhere in the network (usually on a host different with the server's), but no communication between the server and its buddy. When a service providing server fails, an RPC to this server will be automatically re-directed to its buddy server by the system. As object changes in the server will not be available in its buddy, a service providing server usually is used in applications such as pure computation, information retrieval (no update), motor-driven (no action memory), and so on. It is not suitable to be used to manage any critical object that might be updated and then shared by clients.

An object managing server (Figure 1(c)) also has a buddy running in the network. It manages a critical object that might be updated and shared among clients. An RPC to such a server, if it will change the object state, is actually a nested RPC. That is, when the server receives such a call from a client, it first checks to see whether the call can be executed successfully (e.g. if the necessary write-locks have been obtained or not). If the answer is no, the call is aborted. If the answer is yes, then the server will call its buddy server to perform the operation as well. When the buddy returns successfully, the call commits (the server and its buddy actually perform the call) and the result returns to the client. To ensure the consistency of the objects managed by the server and its buddy, a two-phase commit protocol [6] is used when executing the nested RPC.

Like a service providing server, when an object managing server fails, an RPC to this server will be automatically re-directed to its buddy server by the system.

All buddy servers are simple servers. That means, when a server (service providing or object managing) fails, its buddy server provides alternative service in a simple server manner. Also, when a buddy server fails, a service providing server or an object managing server will be reduced into a simple server.

3.2   The Architecture

The SRPC has the following three components: A *Location Server* (LS) and its buddy (*LS buddy*), a *system library*, and a *Stub and Driver Generator* (SDG). This section describes the system architecture from a user's point of view. As server buddies are generally transparent to users, we will omit their descriptions here.

From a programmer's viewpoint, after the SDG compilation (see Section 5), the server part of an SRPC program is consisted of a server driver, a server stub, and a file which implements all the remote procedures (called *procedure file*). The server buddies are transparent to users. The server part (or a server program as it is sometimes called) is a "forever" running program which resides on a host and awaits calls from clients. The client part (or a client program) consists of a client driver and a client stub after the SDG compilation. It runs on a host (usually a different host from the server's host) and makes calls to the server by using the remote procedures exported by the server.

When the client driver makes a call, it goes to the client stub. The client stub then, through the system library, makes use of the client protocol for sending the calling message to the server host. Because the client and the server may use different communication protocols, a client-server protocol converter is used to convert the client's protocol into server's protocol. The calling message is then sent to the server. At the server host side, the server's protocol entity will pass the calling message to the server stub through the system library. The server stub then reports the call to the server driver and an appropriate procedure defined in the procedures file is executed. The result of the call follows the calling route reversely, through the server stub, the server protocol, the system library of the server host, the client-server protocol converter, the system library of the client host, the client stub, back to the client driver. This is called a *direct call* as the pre-condition of such a call is that the client knows the address of the server before the call.

With the help of the Location Server, the run-time address of a server can be easily

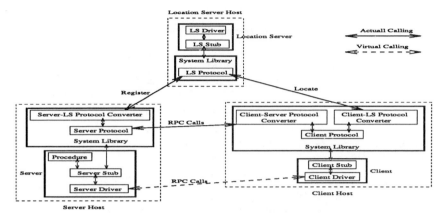

Figure 2: System architecture and a typical RPC

accessed. Figure 2 depicts the system architecture using a typical RPC. The dashed line represents the RPCs from the user's viewpoint.

In this project, cross-protocol communication requires an individual converter for each pair of different protocols. It has been noted that this solution is only reasonable for a few protocols. For a large number of protocols, an intermediate protocol description would be better.

3.3   The Location Server

One way of hiding out the implementation details is the use of the Location Server (LS). The LS is used to hide the server locations from users. It maintains a database of server locations and is executed before any other SRPC program is started. After that, it resides on the host and awaits calls from servers and clients.

The Location Server is an object managing server and has a buddy of its own. It has a well-known location, and this location can be easily changed when necessary. The LS itself is implemented by the SRPC system, using the direct calling method.

Usually there should be one LS (called local LS) running on each host for managing locations of that host, and these local LSs report to the "global LS" (like the NCA/RPC's local and global location brokers [14] [9]). In that case the locations of all LSs can also be hidden from users. We have planned to implement this facility.

The following call is used by a server to register itself to the LS:

```
int registerServer(sn, buddy, imp)
char *sn;          /* server name */
char *buddy;       /* buddy's name */
struct iinfo *imp; /* implementation info. */
```

where `imp` is a type `struct iinfo` structure and contains many implementation details, such as the server's host name, protocol, and so on. Because the call updates the LS database, it is also directed to the LS buddy. If the call returns OK, the location has been registered and a client can use the following call to find out the location of a server from the LS:

```
int locateServer(sn, buddy, imp)
char *sn;          /* server name */
```

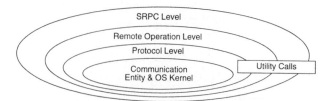

Figure 3: Relationships of system library levels

```
char *buddy;        /* server's buddy name */
struct iinfo *imp;  /* implementation info. */
```

If the call returns OK, the location of the server sn is stored in imp and the name of the server's buddy is stored in buddy for later use. This call does not affect the LS database state, so there is no hidden LS server and LS buddy communication here. Before a server is shut down, the following call must be used to un-register the server from the LS:

```
int unregisterServer(sn)
char *sn;  /* server name */
```

If the call returns OK, the server and its buddy (if any) are all deleted from the LS database. The system also provides other LS calls for maintaining the LS database.

All the usages of these functions in a server or a client program are automatically generated by the stub and server generator. A user does not need to look into the details of these calls if he or she is satisfied with the generated program sections.

3.4   The System Library

The system library is another way of achieving transparency. The library contains all the low-level and system- and protocol-oriented calls. Its main functions are to make the low-level facilities transparent to the upper-level programs and make the system as portable as possible.

The server and client programs must be linked with the system library separately. Reference [16] contains detailed descriptions of the library calls. All the library calls can be divided into the following call levels and Figure 3 depicts their relationships:

1. SRPC Level: This is the highest level. It contains calls that deal with RPC-related operations.

2. Remote Operation Level: It contains calls that deal with remote operations. These remote operations follow the definitions of the OSI Application level primitives [8].

3. Protocol Level: It contains calls that deal with protocol-specific operations.

4. Utility Calls: It contains all the utility calls used in different levels.

4   THE STUB AND DRIVER GENERATOR

4.1   Syntax

The purpose of the stub and driver program generator is to generate stubs and driver programs for server and client programs according to the *Server Definition Files (SDF)*. Listing 1 is the syntax of a server definition file.

We use a modified BNF to denote the syntax of definition files. The "variable", "integer", "string", "constant", and "declarator" have the same meanings as in the C programming language. Comments are allowed in the definition file. They are defined the same as in the C programming language (using /* and */).

Listing 1. Server definition file syntax

```
<SDF>      ::= BEGIN                       <BUDDY>    ::= Buddy <BDYTYPE>: variable;
                 <HEADER>                                 Using: <LANGUAGE>;
                 [ <CONST> ]               <BDYTYPE>    ::=      Auto | Forced
                 <FUNCS>                   <LANGUAGE>   ::=      C | Pascal
               END                         <CONST>    ::= constant
                                           <FUNCS>    ::= RPC Functions: <RPCS>
<HEADER> ::= Server Name: variable;        <RPCS>     ::= <RPC> { <RPC> }
             Comment: string;              <RPC>      ::= Name: string [Update];
                                                            <PARAMS>
             [Using: <LANGUAGE>;]          <PARAMS> ::= { <PARAM> }
             Server Protocol: variable;    <PARAM>    ::= Param: <CLASS>: declarator;
             Client Protocol: variable;    <CLASS>    ::= in | out
             [<BUDDY>]
```

4.2   Semantics

Most of the descriptions of Listing 1 are self-explanatory. We only highlight the following points:

1. The server's name is defined as a variable in the C language. This name will be used in many places. For example, it is the key in the LS database to store and access server entities. When the client asks the LS to locate a server, it provides the server's name defined here. The name is also used as a prefix in naming all the files generated by the SDG. The default language used in the server is the C language.

2. Different protocols can be defined for the server and the client respectively. The buddy, if it is defined, uses the same protocol as the server does. Currently, only three protocols are allowed: Internet_datagram (the UDP protocol), Internet_stream (the TCP protocol), and XNS_datagram (the XNS packet exchange protocol).

3. The <BUDDY> part is optional. If it is not specified, the generated server will be a simple server, otherwise it will be a service providing server or an object managing server, according to some definitions in the <RPCS> part (described below). The <BUDDY> part has a buddy definition and a language definition. The buddy definition defines that whether the buddy's name and execution is to be determined by the system (Auto) or to be determined by the programmer (Forced). If Auto is defined, the system will generate the buddy server's name (ServerNameBdy, used for registering and locating it), the buddy's driver and stub files as well as the *makefile*, and will treat the following variable as the name of the buddy's procedure file. Then, the buddy program will be compiled and executed together with the server program. The host of the buddy program will be determined by the system at run time.

   If Forced is defined, the generator will not generate any buddy's program file and will treat the following variable as the name of the buddy server used for registering and locating. The programming and execution of the buddy server will also be the programmer's responsibility.

   The language definition Using within the BUDDY part defines which language does the buddy program use. The key issue of software fault-tolerant is the *design diversity* or version independent, and one way of achieving design diversity is through the use of multiple programming languages [12]. Currently only the C programming language

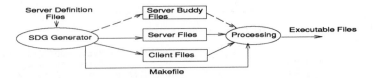

Figure 4: Processing structure of the stub and driver generator

is supported in the SRPC system. We have planned to support the Pascal language implementation soon.

4. The <FUNCS> part defines the remote procedures of the server. At least one remote procedure must be defined. Each remote procedure is defined as a name part and a parameter (<PARAMS>) part. The name of a remote procedure is simply a variable, with an optional Update definition. The latter definition distinguishes an object managing server with a service providing server. That is, if the <BUDDY> part is defined and the Update is defined in any one RPC definition, the server is an object managing server. If only the <BUDDY> part is defined but no Update part is defined in any RPC definition, the server is a service providing server. The meaning of the Update definition is: if an Update is defined following an RPC procedure name, that procedure must be maintained as a nested RPC affecting both the server and the buddy by the server program (See Section 3.1).

There can be zero or several parameters in a procedure, each consisting of a class and a declaration. The class can be in or out, which tells the SRPC system that the parameter is used for input or output, respectively. The declaration part is the same as in the C language. In this version, only simple character string is allowed in parameter definitions. Further extensions are under way.

4.3   Implementation Issues

After a programmer sends a server definition file to the generator, the generator first does syntax checking. If no errors are found, several program source files and a makefile are generated. The subsequent processing is specified by the makefile. That is, when using the make utility, the executable files of both the server and client will be generated. Figure 4 indicates the structure of the processing, The dashed lines represent optional actions.

At least one server definition file must be input to the SDG. If there are more than one server, their SDFs can be input to the SDG simultaneously. If there is only one SDF file, then the generated client driver can execute the server's procedures one by one. If the buddy part is also specified, the generated client can also call the buddy procedures directly (this is useful in testing the client-buddy communication).

If there are more than one SDF file, then for each server, the SDG will generate one set of server files, one set of client files, and one set of buddy files (if the buddy is defined), respectively. These files are the same as the servers being processed in single file input described above. One additional set of client files, the *multi-server client* program, will also be generated in this case. The client driver is called a *multi-server client driver*. It can call all the procedures of all the servers one by one. A further improvement is under way to let the client call these procedures in parallel.

The performance of an RPC in the SRPC system varies, according to which server type is used. Table 1 lists the null RPC performance on a network of HP and SUN workstations, where the server program runs on an HP 715/33 workstation and the server buddy and the client run on two separate SUN 4/75 ELC (33MHZ) workstations. The server (and the buddy, of course) uses the Internet_datagram protocol and the client uses the Internet_stream protocol. We are still investigating the system performance under various circumstances.

Server Type	Time
Simple	3.22±0.02ms
Service-providing	3.37±0.02ms
Object-managing	5.12±0.04ms

Table 1: Null RPC Performance

5 AN APPLICATION EXAMPLE

We use a simple example to show the application of the SRPC system. Suppose we have
a server definition file called sf.def. It defines a "send-and-forward" system in that the
server acts as a message storage and the client acts as both message sender and receiver.
Next is the server definition file:

Listing 2. Server definition file example

```
/* Store and forward: server definition file */

BEGIN                                      RPC Functions:
  Server Name:  sf;                          Name:  storeMsg Update;
  Comment: Store and forward system;          Param:  in receiver: char receiver[MXNAML];
  Server Protocol: Internet_datagram;         Param:  in msg: char msg[MXMSGL];
  Client Protocol: Internet_stream;           Param:  out stat: char stat[MXSTRL];
  Buddy Auto:    sfBdyOps.c;                 Name:  forwardMsg Update;
       Using:  C;                             Param:  in receiver: char receiver[MXNAML];
                                              Param:  out msg: char msg[MXMSGL];
#define MXNAML 64                            Name:  readMsg;
#define MXMSGL 500                            Param:  in receiver: char receiver[MXNAML];
#define MXSTRL 80                             Param:  out msg: char msg[MXMSGL];
                                            Name:  listMsg;
                                      END
```

When this file is input to the generator, the following files will be generated:

```
sf.h Header file, must be included by server,
     its buddy and client drivers and stubs.
sfSer.c     Server driver file.
sfStubSer.c Server stub file.
sfOps.c     Frameworks of server procedures.
sfCli.c     Client driver file.
sfStubCli.c Client stub file.
sfBdy.c     Server buddy driver file.
sfStubBdy.c Server buddy stub file.
makefile    Make file.
```

After using the make utility (simply use "make" command), three executable files are
created:

```
sfSer       Server program.
sfCli       Client program.
sfBdy       Server buddy program.
```

Note that the `sfOps.c` file only defines the frameworks of the remote procedures (dummy procedures). Their details are to be programmed by the programmer. The `sfBdyOps.c` file should be the same as the `sfOps.c` file (the only possible difference happens when the server buddy uses another programming language such as the Pascal, then the affix of the file would be `.pas`).

The server driver is simple. It does the initialisation first. Then it registers with the LS and invokes the buddy program on a neighbouring host because the buddy is defined as `Auto` in the SDF file. After that it loops "forever" to process incoming calls until the client issues a "shutdown" call. In that case the server un-registers from the LS and exits. The "un-register" call will automatically un-register the buddy from the LS as well. The incoming calls are handled by the server stub and underlying library functions. Following is the pseudocode listing of the server driver:

```
Listing 3. Server driver pseudocode
Initialisation (including invoke the buddy);
/* Register the server to the LS */
registerServer("sf", "sfBdy", imp);
while (1) {
  wait for client calls;
  /* comes here only if a client called */
  fork a child process to handle the RPC;
  if the call is "shutdown"
    break;
}
unregisterServer("sf");
```

The server buddy driver works in the same way as the server program, except that it does not invoke a buddy program. Also the buddy is a simple server and all calls to the buddy will not be nested.

The generated client driver can execute the server's remote procedures one by one. If the server driver is running and the client driver is invoked, the client driver first lists all the remote procedures provided by the server, and asks the user to chose from the list. The following is the menu displayed for this example:

```
Available calls:
  0      sf$Shutdown
  1      sf$storeMsg(receiver, msg, stat)
  2      sf$forwardMsg(receiver, msg)
  3      sf$readMsg(receiver, msg)
  4      sf$listMsg()
Your choice:
```

After the selection, the input parameters of the named remote procedure are then input from the keyboard. After that, the driver program does some initialisation and the remote procedure is executed and returned results displayed. The actual calling and displaying are handled by the client stub and underlying library functions. The format of all the four RPCs in the client program are the same as the the format listed in the above menu. That is, if the client wants to send a message to a receiver, it does the following call after the receiver's name and the message are input into `receiver` and `msg` variables, respectively:

```
sf$storeMsg(receiver, msg, stat);
```

Note that the remote procedure's name is named as a composition of the server's name `sf`, a $ sign, and the remote procedure's name `storeMsg` in the SDF file. Similarly, if the client wants to receive messages, it does the following call after the receiver's name `receiver` is obtained:

```
sf$forwardMsg(receiver, msg);
```

Before each RPC, a `locateServer("sn", buddy, imp)` call is issued to the LS to return the location of the server and the name of its buddy. The server location is stored in `imp` and the buddy name is stored in `buddy`.

The fault-tolerant feature of the system is completely hidden from the user. For this example, all the remote procedure calls from the client program will be first handled by the server. A nested RPC is issued if the incoming call is either `sf$storeMsg(receiver, msg, stat)` or `sf$forwardMsg(receiver, msg)`. This is because the two RPC functions are marked as `Update` in the SDF file. The nested RPC will ensure that actions of the incoming call will be made permanent on both the server and its buddy if the call is successful, and no actions of the incoming call will be performed if the call fails. Other two incoming calls, `sf$readMsg(receiver, msg)` and `sf$listMsg()`, will be handled by the server only.

If the server fails (that is, the RPC to the server returns an error), the client program will send the RPC to the server's buddy. The location of the buddy will be determined by another call to the LS:

```
locateServer(buddy, "", imp)
```

where `buddy` is the server buddy's name obtained during the first call to the LS, and `imp` stores the location of the buddy.

The cross-protocol communication is also hidden from the user. All the interfaces to the protocol converters (client-LS, client-server, and server-LS) are generated by the SDG (in the stub files) and used automatically by the stubs. If a user only deals with the RPC level, he or she will never notice the underlying protocols used by the server and client programs.

6 REMARKS

A system for supporting fault-tolerant, open distributed software development is described in this paper. The system is simple, easy to understand and use, and has the ability of accommodating multiple communication protocols and tolerating server failures. It also has the advantage of producing server and client driver programs and finally executable programs directly from the server definition files. The system has been used as a tool of distributed computing in both third year and graduate level teaching, and has been used by some students in their projects.

In tolerating server failures, similar efforts can be found in the RPC systems that provide replicated server facilities, such as NCA/RPC [14]. But in these systems, the user, instead of the system takes the responsibility of maintaining and programming the functions for object consistency. This is a difficult job for many programmers. Our approach in achieving fault tolerance is similar to the approach used in the ISIS toolkit (of course, ours is more simplified and less powerful). But our system is simple, easy to understand, and easy to use. In our system, we provide a server buddy to tolerant the server's failure. When the server fails, the client, instead of aborting, can access the server buddy to obtain the alternative service. Also in our system, it is the system, instead of the user, that is responsible of maintaining the consistency of the managed objects.

Providing server and driver programs directly from the server definition file (similar to the interface definition files of other RPC systems) is also an interesting characteristic of our system. It is related to the rapid prototyping of RPC programs [17]. The driver programs are simple, but yet have the advantages of testing the executability of the RPC program immediately after the designing of the SDF file. It is especially useful if the user makes some changes in the SDF file or the procedure file. In that case, these changes will be automatically incorporated into other related program files if the program is re-generated by the stub and driver generator. This will avoid a lot of troubles in the maintenance of consistency of program files.

References

[1] B. N. Bershad, D. T. Ching, E. D. Lazowska, J. Sanislo, and M. Schwartz. A remote procedure call facility for interconnecting heterogeneous computer systems. *IEEE Transactions on Software Engineering*, 13(2):880–894, August 1987.

[2] K. P. Birman and T. A. Joseph. Reliable communication in the presence of failures. *ACM Transactions on Computer Systems*, 5(1):47–76, February 1987.

[3] K. P. Birman, A. Schiper, and P. Stephenson. Lightweight causal and atomic group multicast. *ACM Transactions on Computer Systems*, 9(3):272–314, August 1991.

[4] M. Boari, M. Ciccotti, A. Corradi, and C. Salati. An integrated environment to support construction of reliable distributed applications (CONCORDIA). In *Parallel Processing and Applications*, pages 467–473. Elsevier Science Publishers (North Holland), 1988.

[5] D. Cerutti. The network is computer. In D. Cerutti and D. Pierson, editors, *Distributed Computing Environments*, pages 17–26. McGraw-Hill, New York, 1993.

[6] J. Gray and A. Reuter. *Transaction Processing*. Morgan Kaufmann Publishers, San Mateo, California, USA, 1993.

[7] Y.-M. Huang and C. V. Ravishankar. Designing an agent synthesis system for cross-rpc communication. *IEEE Transactions on Software Engineering*, 20(3):188–198, March 1994.

[8] B. N. Jain and A. K. Agrawala. *Open Systems Interconnection: Its Architecture and Protocols*. Elsevier Science Publishers B.V., The Netherlands, 1990.

[9] M. Kong, T. H. Dineen, P. J. Leach, E. A. Martin, N. W. Mishkin, J. N. Pato, and G. L. Wyant. *Network Computing System Reference Manual*. Prentice-Hall, Englewoods Cliffs, New Jersey, 1990.

[10] B. Liskov. Distributed programing in ARGUS. *Communications of the ACM*, 31(3):300–312, March 1988.

[11] B. J. Nelson. Remote procedure call. Technical Report CSL-81-9, Xerox Palo Alto Research Centre, May 1981.

[12] J. M. Purtilo and P. Jalote. A system for supporting multi-language versions for software fault tolerance. In *Proceedings of the 19th International Symposium on Fault Tolerant Computing*, pages 268–274, Chicago, USA, 1989.

[13] A. Sinha. Client-server computing. *Communications of the ACM*, 35(7):77–98, July 1992.

[14] L. Zahn, T. H. Dineen, P. J. Leach, E. A. Martin, N. W. Mishkin, J. N. Pato, and G. L. Wyant. *Network Computing Architecture*. Prentice-Hall, Englewoods Cliffs, New Jersey, 1990.

[15] W. Zhou. A remote procedure call interface for heterogeneous computer systems. In *Proceedings of the Open Distributed Processing Workshop*, Sydney, Australia, January 1990.

[16] W. Zhou. *The SRPC (Simple Remote Procedure Call System) Reference Manual*. Department of Information Systems and Computer Science, National University of Singapore, 1992.

[17] W. Zhou. A rapid prototyping system for distributed information system applications. *The Journal of Systems and Software*, 24(1):3–29, 1994.

Managing Distributed Applications

20

Towards a Comprehensive Distributed Systems Management[1]

Thomas Koch and Bernd Krämer
FernUniversität
58084 Hagen, Germany
Phone: (++49) 2331 987 371, Fax: (++49) 2331 987 375
E-Mail: {thomas.koch, bernd.kraemer}@fernuni-hagen.de

Abstract

This paper describes a hybrid approach towards distributed systems management. The rule-based software development environment Marvel is combined with conventional management tools and a new adaptive resource management approach. This approach improves distributed system management in at least two aspects: Formalized management policies are automatically enforced and standard management tasks can be delegated to the system. The encapsulation feature of Marvel allows an easy integration of separate management tools under a common policy. The description of a resource management application illustrates our approach.

Keyword Codes: C.2.4; K.6.4; D.2.9
Keywords: Distributed systems management, production rules, policy enforcement, ANSAware

1 INTRODUCTION

The availability of more and more distributed processing environments, like DCE[1], CORBA[2], CORDS[3] or ANSAware[4], has disclosed the limitations of conventional system management strategies. Especially the heterogeneity of nowadays distributed systems imposes a great challenge to system administrators. Two reasons for management heterogeneity can be identified: One is the heterogeneity of the system components and the other lies in the specialization of the available management tools. Most commercial management tools are limited in scope (e.g. networking or configuration only) and in the number of supported platforms. This situation will become even worse in the future, because distributed systems are expected to grow rapidly.

To keep large distributed systems manageable, the system should support the human administrator with the autonomous execution of simple management tasks according to a given policy. This "management by delegation" approach requires the formalization

---

[1]This research was sponsored by the European Union under contract no. ECAUS003 as an European-Australian collaborative project on Information Systems Interoperability.

of management policies with the additional advantage that policies can be checked for consistency.

In this paper we review an experimental study which we recently undertook with a flexible and partially automated approach to distributed systems management. In particular, we investigate the application of an existing process-centered environment, Marvel [14], to provide management support in the domain of distributed processing environments. Marvel was originally designed as an environment that assists software development and evolution by guiding and aiding individual programmers and helping coordinate software development teams. It uses a variant of production rules to model development tasks or design methodologies including knowledge about design data and the role of tools in individual development steps. Task specifications are executable due to a dynamic forward and backward chaining of rules based on an interpretation of their pre- and postconditions. Activities are automatically performed by the environment when it knows that their results will be required by the user.

The paper is organized as follows: Section 2 describes a management concept for distributed systems. In Section 3 we introduce Marvel and its specification language MSL. This language is used in Section 4 to define an executable model of the management process including the consideration of different management policies.

2   MANAGEMENT MODEL

Management of any component of a processing system helps to control the component in such a way, that it can meet its requirements in service provision. The management of stand-alone systems is usually performed in an informal style, where the administrator changes the system or any component whenever needs arise. With the interconnection of individual components to distributed systems the informal management style is no longer appropriate due to the openness of the whole system and the partial autonomy of individual components. A software upgrade on one system, for example, could create problems on a remote component. The scale of complexity soon reaches a level where the conventional ad hoc style produces disastrous results.

Therefore the activity of distributed systems management needs to be organized in a more systematic way to ensure a consistent management strategy. We chose an object based approach for our management model where all the components of a distributed system are defined as instantiations of different object classes. This object oriented approach is most widely used and serves also as a basis for evolving international standards [5, 6, 7].

2.1   Object concept

Any entity which is a target of management activity is termed **managed object** and the object that initiates the management activity is called **manager**. The basic difference between both classes of objects is that manager objects have a degree of authority to ask for information and to perform management tasks. One object can be in both groups at different times, e.g., a scheduler acts as manager when it asks for load information, but it is a managed object when the administrator asks for statistical data about the scheduling activity.

The types of interaction between managers and managed objects can be organized in three groups [8]:

- *control* actions,

- *requests* for information and

- *notifications* of events by the managed object.

2.2 Domains and policies

To deal with the size of large distributed systems, the management model groups objects into **domains**. Domains are used for modularization. They also provide different views on the same system. One object can be in different domains at the same time, depending on the viewpoint [9].

As already mentioned, we propose to use formal rules to organize the management process. A rule that describes a management activity is called a **policy**. A policy can give either an **authorization** or a **motivation** for an activity. Motivations can be both, positive or negative. Any management activity must be motivated and the manager must be authorized to initiate an action [10].

A typical example for a verbally expressed motivation policy could be:

"Every source file must be registered in the revision control system."

Policies are considered as instantiations of special object classes. Every member of a class has at least the following attributes:

Modality: describes authorization or motivation. In the quoted example a positive motivation is given.

Policy subjects: defines the objects which the policy applies to, here "source file".

Policy target object: states which object class the policy is directed to, here subject and target object are identical.

Policy goal: defines abstract goals or specific actions of the policy, in our above example "registration in the revision control system" is the objective.

Policy constraints: defines conditions that must all be satisfied before the policy can become active. No constraint is given in the example because the policy applies to "every" source file.

Note that this example just gives the motivation to register a source file in the revision control system. Authorization would be expressed by an additional statement like:

"Every user may use the revision control system."

The scope of a policy is given by the domain the policy is defined for. Here the object oriented approach gives another advantage: Policies can be specialized to minor domains by means of object inheritance. With this approach, it is automatically ensured that an inherited local policy is not in conflict with a global policy.

2.3   Management by delegation

The complexity of large distributed systems requires at least a partly automated management environment that provides the human administrator with more support from the system than just collection and presentation of management data. It is quite natural in a hierarchical structure that the assignment of tasks becomes more specialized and less comprehensive on the way from the top to the bottom of the structure. Especially highly specialized tasks with a strictly limited scope are promising candidates for automation. The delegation of tasks is well supported by our approach, because the Marvel environment interprets the policies and is able to activate management tools without intervention of human administrators.

3   SOFTWARE PROCESS MODELLING AND MARVEL

Software process modelling has assumed considerable importance in discussions of software engineering. In particular attention has been paid to the use of software process modelling in the construction of software development environments.

3.1   What is software process modelling?

Essentially, software process modelling is the construction of an abstract description of the activities by which software is developed. In the area of software development environments the focus is on models that are enactable, that is executable, interpretable or amenable to automated reasoning. A particular "instance" of the software development process – the development of a particular piece of software – can be seen as the "enaction" of a process model. That model can be used to control tool invocation and cooperation. A software development environment for a particular development is thus built up around (or generated from) an environment kernel which is essentially a vehicle for constructing and enacting such software process models.

3.2   Functionality of marvel

Marvel is a rule-based environment that was designed to assist users with the software development process automating well understood and formalized tasks. To construct a Marvel model, the developer must produce a data model and a process model. The data model describes the objects to be managed during the process of software development such as specifications, design documents, test data, code or project data. The process model describes the activities carried out on those objects by the developers and tools involved in the specified development.

Marvel uses the data model to generate an objectbase in which all artifacts created during a development are held. The objectbase also maintains history and status of the objects. The data model gives the types, or classes, of the objects involved, their attributes and the relationships between them. The objectbase is implemented straightforwardly as a Unix file structure. Each object instance has associated with it a unique directory, and directories are structured according to the relationships between the object instances.

```
ENVELOPE withdraw;
SHELL    sh;
INPUT    string : HOSTNAME;
OUTPUT   none;
BEGIN
  prog_answer='/home/TUNIX/gerald/ANSA/ChangeExport -H $HOSTNAME -e'
  prog_status=$?;
  if [ prog_staus -ne 0 ]
  then
    echo "Problems with ChangeExport!"
    RETURN "1";
  else
    echo "Exportpolicy on $HOSTNAME changed."
    RETURN "0";
  fi
  RETURN "1";
END
```

Figure 1: Tool Envelope

The process model is given in the form of production rules. Each rule defines a) object classes affected by the rule, b) the precondition which must be satisfied if the activity is to be carried out, c) the activity, and d) the effects of the activity execution on the objectbase in terms of an alternative list of assertions reflecting different outcomes of an activity. Exactly one effect becomes true when an activity is completed. Activities are carried out by tools made known to the objectbase via tool envelopes.

The Marvel kernel provides a means of enacting the process model. It does so in an "expert-system-like" manner by opportunistic processing. If the precondition of an activity is satisfied that activity will be invoked. This may in turn result in the satisfaction of the precondition of further activities, and by forward chaining they will be invoked too. If a particular activity is chosen by a user and is not eligible for invocation, the Marvel kernel will try to build a backward chain of activity whose activation provides the precondition necessary for the selected activity to be performed. Chaining does not introduce a computability problem as the predicates occurring in preconditions of rules are built over attributes of objects, the collection of objects to be checked is always finite, and Marvel verifies the consistency between process and data model prior to enaction.

Both data and process model are expressed as "strategies" in the Marvel Strategy Language. Strategies can be imported into a main strategy. It is standard to define the data model in a single strategy but have multiple process model strategies for related tool sets.

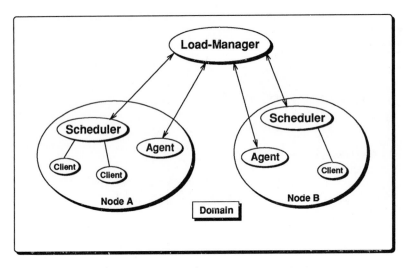

Figure 2: Resource Management Architecture

3.3   Tool integration

Typically, the activities of a Marvel model are performed by tools. Marvel is an open environment that allows the appropriate tools for the development to be added to the environment by the users. To integrate a tool into a Marvel environment it is necessary to create an "envelope". Tool envelopes are written in the Shell Envelope Language [15]. An envelope specifies types of inputs and outputs of tool invocations; in its kernel it uses Unix shell script (whose type is indicated in the head of the envelope) to activate one or more tools in their proper environment and matches the tools' relevant return codes with the effects of the invoking rules. That is, if the tool may return with two different codes, like the example in Figure 1, the corresponding rule must provide two postconditions listed in the order of the corresponding codes.

4   AN EXECUTABLE MODEL OF MANAGEMENT POLICIES

Our experiment started from the hypothesis that Marvel's capabilities are suited for maintaining management tools of a distributed environment based on policies that are formalized in terms of Marvel rules and are then enforced by the Marvel kernel. To test the viability of our approach we built and used a Marvel environment for some sample management tools.

We shall now introduce a resource management application to illustrate our approach. Figure 2 shows a generic architecture for an adaptive resource management tool. We use a multi level approach to handle the complexity of large distributed systems as proposed in [11], but it is sufficient for the purpose of this paper to restrict our description to the lowest level which corresponds to a single domain, as depicted in Figure 2.

Every node of a distributed system has a scheduler, giving recommendations to potential clients, and an agent to collect the required data. One centralized load-manager serves as management interface for all schedulers and agents of a domain, that is, the load-manager is not involved in the scheduling business but is responsible for all management tasks. A more detailed description of the architecture as well as the scheduling algorithm can be found in [12].

This management architecture is now embedded in a Marvel environment. As a consequence, all management commands will be issued to the Marvel interpreter and then Marvel decides which tool to use in what situation, based on the formalized policy and on the current state of the system.

4.1 Product model

As explained in Section 3, the managed objects of the system are described in terms of object classes, their attributes, and relations. Figure 3 shows the class definitions for our example. The most generic class is named *MO* (short for Managed Object). Every object in the system inherits the standard attributes from *MO*. Multiple inheritance is supported in Marvel, but not used in this example. An attribute with the keyword *link* in the type definition refers to a named and typed relation to an instance or a set of instances of the given class.

During a session and beyond, the Marvel object base maintains a set of uniquely identifiable objects whose attributes are determined by the corresponding class definition. Possibly initial attribute values can be declared in the class definition, e.g. `NotChecked` in Figure 3. For every process in our example, Marvel maintains an object instantiation of the corresponding class.

4.2 Process model

Every management policy has to be encoded by Marvel rules. All rules together build the model implementing the automated management assistant. As an example, we shall investigate the following informally stated policy for our resource management system:

"Every **authorized** user is allowed to withdraw his host from the pool."

According to the object model described in Section 2.1, we can identify the following attributes:

Modality: Permission

Subjects: Every user

Target object: Resource agents (the command is given to the load-manager, but the agents are affected)

Goal: Withdraw target host from the pool

```
MO :: superclass ENTITY;
  name   : string;
  status : (Up, Down, NotChecked) = NotChecked;
end

DOMAIN :: superclass MO;
  hosts        : set_of HOST;
  users        : set_of USER;
  loadmanager : LOADMANAGER;
end

HOST :: superclass MO;
  agent     : AGENT;
  scheduler : SCHEDULER;
  loc_user  : link USER;
end

USER :: superclass MO;
  auth : (True, False, NotChecked) = NotChecked;
end

AGENT :: superclass MO;
  export   : (Exported, Withdrawn, NotChecked) = NotChecked;
  chg_user : link USER;
  todo     : (Wait, Check, Nothing) = Nothing;
end

SCHEDULER :: superclass MO;
  algorithm : (SER, LMS) = LMS;
end

LOADMANAGER :: superclass MO;
  export : (Export, Withdrawn) = Export;
  host   : link HOST;
end
```

Figure 3: Description of managed objects

Constraints: User must be authorized and logged on the target host

The translation of a policy object into the Marvel environment can be done in three steps:

1. The constraints must be transformed into preconditions. The check if a constraint is satisfied or not, could require the use of a special tool. In this case a separate rule must be provided for the activation of the checking tool.

2. At least one rule is needed to reach the policy goal. The activation of a tool in order to reach the goal is not mandatory, although it is necessary in most cases. Some policies could reach their goal just by affecting the attributes of other objects, without tool invocation.

   If the target consists of instances of differently specialized managed objects, it could be necessary to create several similar rules taking different attribute sets into account to reach the goal.

3. The attributes must be changed after the rule was activated. Otherwise the Marvel interpreter would fire the same rule again and again.

One way to formalize our example policy is given by the following rule (line numbers are added for reference only!):

```
 1 withdraw[?h:HOST]:
 2 (and (exists USER ?u suchthat (linkto [?h.loc_user ?u]))
 3       (exists AGENT ?a suchthat (ancestor [?h ?a]))):
 4 (and        (?u.auth = True)
 5             (?a.status = Up)
 6       no_chain(?a.export = Exported))
 7 {EXPORTTOOL withdraw ?h.name}
 8 (and        (?a.todo = Check)
 9             (?u.auth = NotChecked)
10       no_forward(link [?a.chg_user ?u])
11       no_forward(?a.export = Withdrawn));
12 no_assertion;
```

Line 1 shows the name of the rule (*withdraw*) and the formal parameter object with its class. The statements in Line 2 and 3 identify the local user and the corresponding agent. In Marvel such objects are called derived parameters. They can be determined through existentially and all-quantified predicates inspecting the object graph along attribute links. From Line 3 to Line 6 list the precondition of the rule. The keyword no_chain in Line 6 ensures that the interpreter does not try to satisfy this precondition by firing other rules, as otherwise the interpreter could try to export an already withdrawn agent because this action would fulfill the precondition. If the precondition is satisfied, the activity in Line 7 will be activated, here the envelope withdraw in Figure 1 is executed. In this example the load-manager is instructed to withdraw the agent, given as parameter in the command. Two possible outcomes are then defined: In the case of a successful tool invocation, Line 8 to 11 will adjust the attributes of the involved objects, according to the new situation.

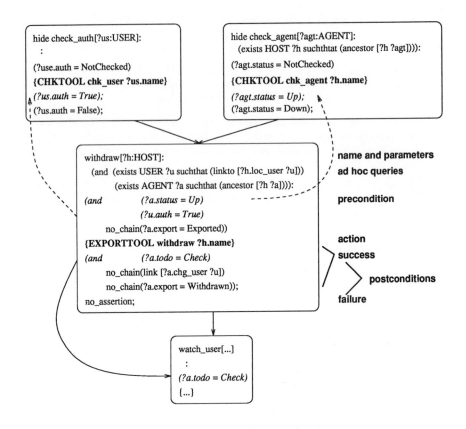

Figure 4: Rule Chaining

These changes may trigger other rules by the use of forward chaining. If the tool returns an error code, no attributes are changed according to Line 12.

Next we will assume that the policy has changed to:

> "Every **authorized** user is allowed to withdraw his host from the pool as long as the user is active."

The new situation is illustrated in Figure 4. The *withdraw* rule checks the user authentication, according to the policy. Rule *check_auth* is designed to activate the authorization mechanism and *check_agent* reads the current status of the corresponding agent. Both rules are hidden for a normal user and automatically fired through backward chaining as indicated with dashed lines.

If the *withdraw* rule was fired succesfully, forward chaining is automatically applied to initiate a watchdog mechanism. The box *watch_user* in Figure 4 indicates a set of rules, designed to check periodically if the user is still active. If the user becomes inactive, the previously withdrawn host is automatically exported.

Note that the example policy just gives authorization to withdraw the host, here the motivation comes from the fact that the user issued a command to the interpreter. As soon as the user becomes inactive, *watch_user* is motivated to export the host. The authorization is implicitly given.

This scenario shows another important advantage of our management environment: Marvel could be **asked** for all necessary preconditions as well as for all resulting post-conditions for a planned activity. These "What is necessary to..." or "What happens if..." questions are extremely useful for the management of complex distributed systems, because the system administrator can learn about the consequences of a planned activity without executing the command.

5 CONCLUSIONS AND FUTURE RESEARCH

We described the application of the rule-based software development environment Marvel as a management environment for distributed systems. The motivation behind this approach was to gain rapid feedback about our management model through the use of existing tools that showed sufficient potential to support this experiment. The encapsulation feature of Marvel allows an easy integration of separate management tools under a common policy. Our approach improves distributed systems management in several areas: The automation of simple management tasks is supported and the observance of policies is enforced. Another important feature of our approach lies in the fact that existing management tools can be embedded in the new environment. The possibility to ask the system for preconditions or consequences without actually executing the command reduces the potential for erroneous activities drastically. Currently this feature is not included in the Marvel environment, but an approach using an additional set of rules is under investigation.

Despite the encouraging results of our protoype implementation, further research is still needed, especially in the area of policy transformation into appropriate rules. The possibility for automated negotiations seems to be an indispensable feature for open distributed systems.

General policies, describing the management tasks on an abstract level, are more difficult to deal with than local policies. One approach currently under investigation translates abstract policies into special rules. They do not activate a management tool, but the postcondition of the rule changes the objectbase in such a way that the preconditions for other rules are affected.

Another problem with global policies is the potential for inconsistencies between global and local policies. A promising approach to this problem is described by Finkelstein et al. in [13]. Here a common logical representation of objects, policies and consistency rules is used to detect and identify inconsistencies. Meta-level rules may then be used to perform predefined actions on inconsistencies.

References

[1] A. Schill. *DCE Einführung und Grundlagen*. Springer Verlag, 1993. in German.

[2] The Common Object Request Broker: Architecture and Specification. Technical Report 91.12.1, OMG , December 1991.

[3] M. Bauer, N. Coburn, D. Erickson, P. Finnigan, J. Hong, P. Larson, J. Pachl, J. Slonim, D. Taylor, and T. Teorey. A distributed system architecture for a distributed application environment. *IBM SYSTEMS JOURNAL*, 33(3):399–425, 1994.

[4] Architecture Projects Management Limited, Castle Park, Cambridge. *Application Programming in ANSAware*, release 4.1 edition, February 1993.

[5] Reference Model - Open Distributed Processing - Part 2: Descriptive Model. Technical Report JTC 1/SC 21 N 8538, ISO/IEC, April 1994.

[6] Reference Model - Open Distributed Processing - Part 3: Prescriptive Model. Technical Report JTC 1/SC 21 N 8540, ISO/IEC, April 1994.

[7] Open Distributed Management Architecture - First Working Draft. Technical Report JTC 1/SC 21 N 8801, ISO/IEC, August 1994.

[8] A. Langsford and J.D. Moffett. *Distributed Systems Management*. Addison-Wesley, 1993.

[9] M. Sloman and K. Twidle. Domains: A Framework for Structuring Management Policy. In Morris Sloman, editor, *Network and Distributed Systems Management*, chapter 16, pages 433–453. Addison-Wesley, 1994.

[10] J. D. Moffet. Specification of Management Policies and Discretionary Access Control. In Morris Sloman, editor, *Network and Distributed Systems Management*, chapter 17, pages 455–480. Addison-Wesley, 1994.

[11] A. Goscinski and M. Bearman. Resource Management in Large Distributed Systems. *Operating Systems Review*, 24(4):7–25, October 1990.

[12] T. Koch, G. Rohde, and B. Krämer. Adaptive Load Balancing in a Distributed Environment. In *Proceedings of SDNE'94*, pages 115–121. IEEE, June 1994.

[13] A. Finkelstein, D. Gabbay, A. Hunter, J. Kramer, and B. Nuseibeh. Inconsistency Handling in Multi-Perspective Specifications. *IEEE Transactions on Software Engineering*, 20(8):569–578, August 1994.

[14] G.E. Kaiser; P.H. Feiler; S.S. Popovich. Intelligent Assistance for Software Development and Maintenance. *IEEE Software*, 40–49, May 1988.

[15] M.A. Gisi and G.E. Kaiser. Extending a tool integration language. In *First International Conference on the Software Process – Manufacturing Complex Systems*, pages 218–227, Redondo Beach, California, October 1991. Computer Society Press.

21

Flexible Management of ANSAware Applications[*]

B. Meyer[a] and C. Popien

[a]Aachen University of Technology, Computer Science IV,
Ahornstr. 55, 52056 Aachen, Germany, e-mail: bernd@i4.informatik.rwth-aachen.de

In contrast to monolithic application programms developed for large mainframe computers, applications for networks of interconnected workstations are nowadays developed using a client-server model of computer cooperation. Applications are still monolithic, but using certain servers for subtasks. The next step will be to devide an application into smaller subparts with common service and resource requirements in order to build units for concurrency and migration. These components of an application also need to be managed. We present a new approach to application management. For the distributed computing platforms ANSAware we developed monitoring and analysis tools. In addition we developed a formal notation for defining policies. Our approach to application management introduces policy controller objects, that enforces policies for policy target objects. Experiences gained with policy descriptions lead to a common policy handling function as it is required for Open Distributed Processing and management systems.

Keyword Codes: C.2.4; D.2.9;
Keywords: Distributed Systems, Management

1. INTRODUCTION

Applications running on large mainframe computers have a monolithic structure with respect to their resource utilisation. Ongoing with the evolution of computer and communication technology, modern computer systems are able to share their resources over different kinds computer networks. This enables applications to be distributed over a number of computer nodes with an expected performance gain. In order to cope with problems arising from distribution, a number of computing platforms have been developed, e.g. the Distributed Computing Environment (DCE) of the Open Software Foundation [DCE], the ANSAware [ANSA] and implementations of the Common Object Request Broker Archictecture (CORBA), that is a part of the Object Management Architecture (OMA) of the Object Management Group [OMG]. In addition, the ISO advanced standardization work on Open

---

[*] This work was partially supported by the Deutsche Forschungsgemeinschaft under Grant No. Sp 230/8-1

Distributed Processing (ODP) for defining a framework for classification, development and interworking of distributed processing platforms, see [ODP p1-p3]

Besides proceeding appropriate functions to support distributed processing, the managament of distributed systems becomes increasingly important. Most management concepts and systems have been developed for management of network components. In the last years a lot of work has been done in adopting these concepts for managing resources of a distributed system. Because of new requirements put up, network management concepts need to be extended. The notion of a service has been introduced as a further abstraction from underlying processing and networking resources. These services offer more functionality as those offered by an (distributed) operating system or a communication network. However, this distinction is a more logical one and is a bit fuzzy, e.g. mailing or database services are more complex services than a transport layer service or a file service. In this work are thinking of services in the former sense. Service management became an important area of research with the rising popularity of the client-server model for computer cooperation, see also [PK94], [PKM94]. In addition to services offered to a wide range of users, an application contains a huge amount of code specific to that application. Consistently, managing applications in a distributed system is a more complex task than service management. On the other side, application management has to fall back on service and resource management in order to not invent the same concepts again.

Currently distributed systems are becoming more and more large, containing hundreds of computer nodes. The configuration of these systems is likely to change dynamically due to a node or server failure. In addition, new services or components are introduced or removed from the system. To cope with changing configurations, components of distributed systems need to have flexible behaviour to react on changes. This behaviour has to be managed by human or computer manager agents. A reasonable approach to this goal have been introduced by so-called policies. There has been a lot of research on the topic of policies, e.g. see [MoS194], [Wie94].

The paper is structured as follows: the second section gives an introduction to basic management concepts and tasks of application management. A policy-based approach to application management is introduced in section three. Therefore a model for describing applications and policies is given. Section four shows, how these models can be realized using the ANSAware distributed platform as a base for an implementation of application management. Finally fields for future work will conclude this work.

2. MANAGEMENT OF APPLICATIONS IN DISTRIBUTED SYSTEMS

In the following, we introduce a number of notions and their interpretation according to the reference model of Open Distributed Processing (ODP), see [ODP p2]. Domains consists of a collection of managed objects, that are grouped together for a special purpose. One domain is managed by a single controller object. Policies can be defined an relation between one or more objects and some reference behaviour. The behaviour can be expressed in terms of obligations, permissions and prohibitions. An obligation behaviour must be fulfilled, whereas permitted behaviour is allowed to occure. A prohibition is in some way the opposed behaviour to an obligation, because it defines a behaviour, that must not occur.

Management concepts and tools can be classified using three orthogonal dimensions [HeAb93], see figure 2.1 . One dimension refers to the functional management areas as defined by OSI Management, namely accounting, configuration, fault, perfomance and security management [OSI Man]. Another is called time dimension and deals with phases in the life of a management application or a tool the description aims at. Some tools are only suited for planning an application, wheres others focus on its development or maintainace. The last dimension deals with the target object a management tool works on. There exists a large number of tools for network and system components. But also applications and enterprises need to be managed.

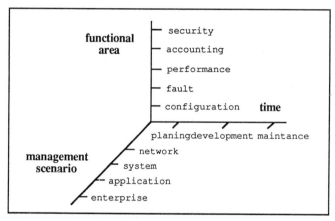

Figure 2.1. Three dimension of management

Management in ODP defines five different functional areas, namely quality of service, accounting and configuration management accompanied by monitoring and defining policies [Slo90], [ODP p1]. QoS Management encompasses fault and performance management, whereas security is classified to be an operational and not a managament function, same as e.g. communication.

Target of application management are components like applications, object cluster, application objects, server objects, interfaces and binding objects . According to the functional management areas, application management requires:

accounting management
- administration of payment methode for service usage
- administration of cost calculation methode for service usage

configuration management
- controlling the state of an application component
- storing properties of application components

- starting and stopping of application components
- modifying general configuration
- administration the migration of application components

fault management
- administration of application components checkpoints
- fault tolerant service usage
- collecting fault messages
- recognition of cause of fault messages

performance management
- monitoring performance characteristics of an interface, e.g. number of operation , invokations, volume of transferred data, response time
- monitoring performance characteristics of an application component, e.g. throughput, total delay, average service time
- determining performance bottlenecks
- avoiding performance bottlenecks
- monitoring and controlling reservation of network and system resources
- load balancing of application components

security management
- administration of authentication for multiple server access

3. POLICY-BASED APPLICATION MANAGEMENT

The following section introduces a policy-based approach to application management. First a model for applications and policies are presented, which are used by application management. Using these models a concept for an integrated application management tool will be developed.

3.1 The application model

In our approach an application consists of a number of cooperating objects. Each object is able to interact with other objects via operations it offers at its interfaces. In general, objects can have more than one interface; e.g. for management purposes there will be a separate management or control interface. The interfaces hides the internal object state and can only be manipulated by invoking interface operations or it forwards notifcations initiated by its internal objetc state fulfilling a given condition. Binding between objects can be established by an binding object, if it has to be managed by the application itself or a management application. A binding object is an abstraction from the communication channel established to enable interworking of two (or more) objects. The binding control interface enables users to change properties of a binding, e.g. quality of service parameter. New cooperating partners can be introduced or connected ones can be withdrawn. Applications of these capabilities are data stream handling as it is necessary for multimedia systems or control systems for technical production processes. Application objects with common service and resource requirements can be grouped into object clusters. Clustering is a means of configuration and an object cluster is intendend to be located at the same node as its resources or servers. In general not all functionality is provided by application objects, but a number of (external) server objects are also needed. These servers offer different services, e.g. operating system services like a file system, memory management, persistent data storage service offered by a database systems or a global name service according to the X.500 recommendations. All introduced application components can be mapped on ODP modelling concepts.

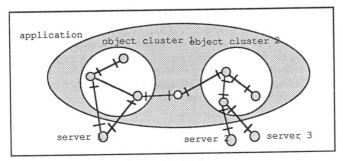

Figure 3.1. Components of the application model

In the following, we will give examples of information required by application management for each component of an application according to our model. These metrics can be useful either as current, or average value. For certain analysis tasks, a complete value history is required.

Table 3.1. Application management metrics

component	metric
application	required services (resources)
	contained object cluster / application objects
	application performance metrics (waiting time, throughput)
object cluster	required services (resources)
	used server objects
	contained application objects
	host computer node
	cluster performance metrics (response time, throughput)
application object	required services (resources)
	used server objects
	host computer node
	cluster performance metrics (response time, throughput)
server object	number of current users / interfaces
	waiting queue length
	server performance metric

component	metric
interface	number of invokations per operation
	number of failed operation invokations
binding object	protocol type
	communication channel peformance metrics (response time, throughput)
	bit error rate

3.2 The policy model

As mentioned above, policies in ODP are defined as an prescriptive relation between one or more objects and some reference behaviour. The behaviour can be expressed in terms of obligations, permissions and prohibitions. An obligation behaviour must be fulfilled, whereas permitted behaviour is allowed to occure. A prohibition is in some way the opposed behaviour to an obligation, because it defines a behaviour, that must not occur.

Our policy model has a certain view on objects it deals with. Objects are assumed to be black boxes hiding their internal state behind an interface. The policy state of an object is determined by the kinds of policy related to it and the current behaviour associated with each policy. In our model policy changes are caused by intervenition of a (human) system operator or by the management system itself. Policies in the presented approach are responsible for extending, reducing or modifying the behaviour of certain objects. It defines conditions under which a change of policy state appears, but it does not completely specify the object behaviour. Conditions for an automatic change can be external events like the reception of a notification or the occurance of an internal state. At a higher level of abstraction, policies only deal with event, behaviour or activity identifiers. They can be further refined by manipulating interfaces of objects. In the case of application management these objects are all above defined application components. For example, a certain behaviour can be achieved by restricting an operation parameter to a special value. In other cases, an activity defined in a policy definition correspondes to a certain operation (or a number of operations) at an interface. Prohibiting this activity can be done by disabling these operations at the corresponding interface. In addition it is possible, that the administrator role of enforcing a policy can be delegated from one object to another. An important design decision to be made for a policy-based management system is the question whether a policy should be handled at run-time or at compile-time. It is well-kown that interpreters are more flexible, but compiled code shows better performance. Due to promise of focussing on flexibility and scalability, an interpreter approach is chosen in this approach. Testing conformance of a given object against a behaviour determined in a policy is a very important task, but it is out of scope in our approach yet. Therefore, existing formal description techniques like SDL, LOTOS or Estelle, that allow conformance testing should be investigated for their usage as behaviour description for policy behaviour, see also [Tur93]. In our approach, it is assumed, that all objects are able to show the behaviour requested by a policy. Otherwise, a policy failure notification is to be forwarded. One difficulty in using existing techniques are the policy modalities, that are not directly expressable Modalities can be expressed in terms of model logic or temporal expressions over an state-transition model, see [HeMi85], [Got92]. For example, a forced behaviour can be interpreted by the model operator always. Therefore a state-transition model is constructed by policy states and changes of policy as transitions. It must be checked, that the forced behaviour is enabled in every state of the state-transition model. This procedure is called model checking and is a seperate field of research. Policies are a general purpose concept required for managing distributed systems. Their usefullness has been recognized in several projects and finally standarization is concerned about a general policy handling function. Although it is a typical management function, not only OSI Management tries to include policies, but also ODP is using policies for startegic modelling. In our point of view, a policy handling function will need to include at least activities like

- storing, naming and retrieving a policy,
- replacing or modifying a policy,
- merging two policies,
- solving conflicts between two policies or
- applying well-defined refinements on a policy.

3.3 Integrated model for application management

The environment of a policy-based application management system is shown in figure 3.2. Besides other objects, the application management contains a policy manager. Application components like whole applications, object cluster, objects, interfaces, binding objects and server objects are managed by the management system.

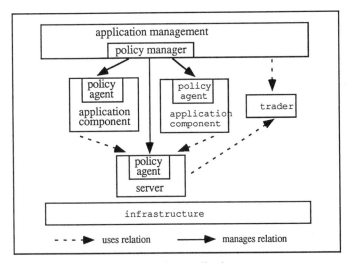

Figure 3.2. Global architecture of an application management system

Each application component and service object is equipped with a policy agent, which is in charge of controlling object behaviour with respect to its policy. This policy agent itself is part of a management agent, that is in charge of additional management task. For simplicity, the management agent is not integrated into figure 3.2. Of special importance in our application management model is the trader component [ODP TF]. It is used by the application management to get server interface references for services required by an application. In our approach it serves as a means to achieve a higher level of software dependability. Especially if a server failure occurs at run-time, it is necessary, that the application management is able to request a new server interface and to restart the failed operation call. Therefore it must be assured, that modifications already done by the failed server have to be cancled.

3.4 Related work

The META approach presented in [MCW+91] also uses policies to achieve flexible management behaviour. It also makes use of policies, but the only statement possible is a conditioned activity. Distributed applications are monitored by so-called sensors and operations on applications can be invoked by so-called actuators. Together, both concepts offer similar functionality that is given by the managed object concept defined in OSI Management. Another approach to distributed applications management has been developed at the universtity of Frankfurt within a project names PRISMA (a platform for integrated construction and management of distributed applications), see [ZBD+94], [ZFD93]. It supports an application model consisting of interfaces, communication contexts, application components and a global application configuration. Interfaces and components are related to either the application or the management. A new specification language has been introduced, but in addition these concepts are described using the guidelines for use of managed objects (GDMO). A related approach is established by the MELODY (management environment for large open distributed systems) project of the universtity of Stuttgart [KoWi94], [Kov93]. There an activity management has been proposed, which can be compared to transactions in database systems; except a complete rollback. An activity consists of a sequnce of service calls, where the actvity state is checkpointed after each service operation. On service failure, the activity is set to the last checkpoint and a new server is imported using a trader. The service call is then invoked again. The application model within MELODY assumes interacting components to as building blocks. A component itself is characterized by its type, its interfaces and the role it can play in an interaction.

4. IMPLEMENTATION USING ANSAWARE

ANSAware is an infrastructure for developing and running distributed applications. It is available for a number of operating systems like SunOS, HP/UX, VMS and MS DOS. Communication between computer nodes must be provided by a socket interface of a transport service using the TCP or UDP protocol. Besides services like remote procedure call (RPC) and thread support, ANSAware provides three server. One server is called the factory and is able to start server objects of a certain service type. An interface reference has to be obtained, before a client can use an object service. This reference can be obtained by making an service import request to the trader. The trader is the second server provided by ANSAware and delivers interfaces references of servers of a given service type. Before interface references to server objects can be forwarded, they have to be exported to the trader. In addition, cooperations between different trader server can be established.

Application management is based on management-relevant information about application components, binding and server objects. A means to this end is a monitor tool for distributed systems. We developed such a tool for ANSAware called ANSAmon [Hei94]. A special kind of these management information, performance metrics, can be calculated by analysing or simulating application models. They fall back on system parameters measured by a monitoring system. Another way to achieve information about performance characteristics is to use analytical techniques. Therefore we developed a method to evaluate fork-join queueing networks, see [MePo93], [PMS94].

In order to make our policy model understandable to computer, we developed a formal notation for policy description called the policy definition notation (PDN), see [MePo94]. It is based on the above defined object model. Here only a brief summary of the most important constructs will be given. Policies using PDN are named and consist of descritpion of a domain it is applied to and a behaviour description. The domain policy is enforced by a special controller object. Policy changes are initiated by the occurance of external events or internal state conditions. The prescribed policy behaviour can be an obligation, a permission or a prohibition.

```
<policy>::= "policy" <name> "for" <domain> "with" <behaviour_desc> "end policy".
<behaviour_change> ::= <event_triggered_behaviour> | <conditional_behaviour> .
<event_triggered_behaviour> ::= "on" <external_event> "=>" <behaviour>.
<conditioned_behaviour> ::= "if" <internal_state> "then" <behaviour>.
<policy_behaviour> ::= <modality> <behaviour>.
<modality> ::= "force" | "permit" | "prohibit" | ... | <empty>.
```

Figure 4.1 describes a policy for a trader domain, that is managed by a trader administrator. It determines where to place a new user depending on the size of the trader database (TDB) of each trader. It is assumed, that a user contacts its local trader first. If a database is changed, the corresponding trader sends a noticication to its administrator.

Policies can be generated by system administrators or directly by a management application. Refinement of policies likelywise needs to be done by human users. There is some research on developing automatic refinement tools, but we are convinced, that a certain amount of human activity will still remain. Changing current policy behaviour of an object can be initiated by a system administrator or by an analysis object. Analysis needs values for system and application parameters to be gathered by a monitoring system. Policies created or modified in the policy maker are defined using our notation and will be interpreted by a policy agent. It controls the behaviour of the distributed application and in some cases of the distributed computing environment, too.

```
policy UserPlacementPolicy
for TRADER enforced by trader_administrator with
behaviour AdaptivePlacement for trader_administrator is
        force
                on notification reception TDB_Updated from TRADER with TDB_Size =>
                    if (SizeTDB in [0, 320) )
                            then behaviour LocalPlacement for TRADER
                    if (SizeTDB in [320, 550) )
                            then behaviour PlaceAtLightestLoadTrader for TRADER
                    if (SizeTDB in [550, 600) )
                            then behaviour LocalPlacement for TRADER
where
        behaviour LocalPlacement is ...
        behaviour PlaceAtLightestLoadTrader is ...
end policy
```

Figure 4.1. Example policy for flexible placement of trader user requests

ANSA objects to be managed by policies are supported by an policy controller. This controller manages the access to the server object. Figure 4.2 shows a policy manageable trader object. Internally it consist of a policy controller and a server object. The trader has a management interface and a separate service interface. A system administrator or an management application can change policy via the managemebt interface, whereas the service interface is the same as the server interface, but interaction is controlled by the policy controller.

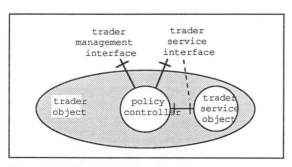

Figure 4.2. Policy-controlled trader object

There might be other contributing to the trader management interface, but for convinience we chose only one policy controller. Common operations at management interfaces of objects can devided into operations on one policy description and on the list of available policy descriptions. There are operations to add a new policy to a list, delete a certain policy from a list and to show all policies with the list. Policies can be read and modified by a new policy description. In addition, several notifications can be forwarded from the policy controller, e.g.

a notification announcing a policy behaviour change due to an external event or an internal condition.

5. CONCLUSIONS

In this paper we presented a policy-based approach to application management in distributed systems. Policies enable the application behaviour to be managed automatically. It consists of an application component model, that is based on concepts introduced by the RM ODP. Besides we developed an appropriate policy model. For the distributed computing environment ANSAware we present an application management approach. It makes use of a distributed montioring tool we developed. Policies are described formally using a new notation, which allows to be interpreted by computer. We propose to add an policy controller object to each object, whose behaviour is determined by a policy. This controller is able to enforce a policy by influencing interface activities. The policy controller object offers behaviour required by a general policy handling function. Future work will be done in the area of performance, fault and configuration management applications.

REFERENCES

[ANSA]   ANSAware Release 4.1, Manual Set, Document RM.102.02, February 1993

[DCE]   Open Software Foundation: Introduction to OSF DCE. Prentice Hall 1992

[Got92]   Gotzhein, R.: Temporal Logic and applications - a tutorial. Computer Networks and ISDN Systems, Vol. 24, 1992, pp. 203-218

[HeAb93]   Hegering, H.; Abeck, S. : Integrated Network- and Systemsmanagement (in german). Addison Wesley, 1993

[Hei94]   Heineken, M.: Performance Analysis of the Communication Components of the ODP Prototype ANSAware (in german). Diploma Thesis at Aachen University of Technology, Nobember 1994

[HeMi85]   Hennessey, M.; Milner, R.: Algebraic Laws for Non-determinism and Concurrency. Journal of the ACM, Vol. 32, 1985, pp. 137-161

[Kov93]   Kovács, E.: The MELODY management system for distributed applications (in german). Proceedings of EMVS´93, Frankfurt, 1993

[KoWi94]   Kovács, E.; Wirag, S.: Trading and Distributed Application Management: An Integrated Approach. Proceeding of 5th IEEE/IFIP International Workshop on Distributed Systems: Operation & Management, Toulouse, October 1994

[MCW+91]   Marzullo, K.; Cooper, R.; Wood, M. et al: Tools for Distributed Application Management. IEEE Computer, August 1991, pp. 42-51

[MePo94]   Meyer, B.; Popien, C.: Defining Policies for Performance Management in Open Distributed Systems. Proceeding of 5th IEEE/IFIP International Workshop on Distributed Systems: Operation & Management, Toulouse, October 1994

[MoSl93]   Moffet, J.; Sloman, M.: Policy Hierachies for Distributed Systems Management. IEEE Journal on Selected Areas in Communications, Vol. 11, No. 9, December 1993, pp. 1404-1414

[ODP p1]   ISO/IEC JTC1/SC21/N. (7S023) Information Technology - Open Distributed Processing - Basic Reference Model - Part 1: Overview. Meeting output for Committe Draft, July 1994

[ODP p2]   ISO/IEC JTC1/SC21 N7988 Information Technology - Open Distributed Processing - Basic Reference Model - Part 2: Descriptive Model. Committe Draft, December 1993

[ODP p3]   ISO/IEC JTC1/SC21 N8125 Information Technology - Open Distributed Processing - Basic Reference Model - Part 3: Prescriptive Model. Committe Draft, December 1993

[ODP TF]   ISO/IEC JTC1/SC21 N8409 Information Technology - Open Distributed Processing - ODP Trading Function. Working Document, January 1994

[OMG]   Object Management Group: The Common Object Request Broker Architecture: Architecture and Specification. Revision 1.2, December 1993

[OSI Man]   ISO/IEC IS 7498 Information Processing Systems - Open Systems Interconnection - Basic Reference Model - Part 4: Management Framework. International Standard, November 1989

[PK94]   Popien, C.; Kuepper, A.: A Concept for an ODP Service Management. Proceedings of IEEE/IFIP Network Operations and Management Symposium, (NOMS´94), Kissimmee/Florida 1994

[PKM94]   Popien, C.; Kuepper, A.; Meyer, B.: Service Management - The Answer of new ODP Requirements. In: Meer, J. de; Mahr, B.; Storp, S.: Open Distributed Processing II (ICODP´93), Berlin 1993, North Holland 1994, pp. 408-410

[PMS94]   Popien, C.; Meyer, B.; Sassenscheidt, F.: Efficient Modeling of ODP Trader Federations using P^2AM. (in german). In: Wolfinger, B. (ed.): Innovations for Computer and Communication Systems. Springer 1994, pp. 211-218

[Slo90]   Sloman, M.: Management for Open Distributed Processing. Proceedings of 2nd IEEE Workshop on Future Trends of Distributed Systems, 1990

[Tur93]   Turner, K. (ed.): Using Formal Description Techniques - An Introduction to Estelle, LOTOS, and SDL. Wiley 1993

[Wie94]   Wies, R.: Policy Definition and Classification: Aspects, Criteria, and Examples. Proceeding of 5th IEEE/IFIP International Workshop on Distributed Systems: Operation & Management, Toulouse, October 1994

[ZBD+94]   Zimmermann, M.; Berghoff, J.; Doemel, P. et al: Integration of Managed Objects into Distributed Applications. Proceeding of 5th IEEE/IFIP International Workshop on Distributed Systems: Operation & Management, Toulouse, October 1994

[ZFD93]   Zimmermann, M.; Feldhoffer, M.; Drobnik, O.: Distributed Applications: Design and Implmentation (in german). PiK, Vol. 16, No. 2, 1993, pp. 62-69

SESSION ON

Experience with Distributed Environments

22

DDTK Project: Analysis, Design and Implementation of an ODP Application

Spiros ARSENIS[a,b], Noëmie SIMONI[a], Philippe VIRIEUX[b]

[a] TELECOM Paris, Dept Réseaux, 46 rue Barrault, 75634 Paris Cedex 13, France. e-mail: arsenis@res.enst.fr
[b] DOXA INFORMATIQUE, 41-43 rue des Chantiers,78000 Versailles, France. e-mail: philippe.virieux@doxa.fr

In this paper, recently concluded work on the development of a communication support tool kit is reported. The analysis, design and implementation steps of this project follow closely the Open Distributed Processing reference model (RM-ODP). We have used the ODP concepts and design rules as the basis of our development. In this paper we present the results of ODP's application on a specific case along with all the additional elements resulting from our case analysis.

Keyword Codes : C.2.4.; D.2.10.; H.4.3.
Keywords: Distributed Applications, Object Orientation, Modelling

1. Introduction

In this paper we present the research activity carried on at both TELECOM Paris and Doxa Informatique for the realization of the Doxa Distribution Tool Kit (DDTK). Our research has been done within the framework of a pilot project for a telecom operator. DDTK provides communication support for distributed applications. Using portable communication tools as well as data presentation tools based on standard application interfaces, DDTK supports applications within distributed and heterogeneous environments. It administrates these tools in order to manage the communication from one application to an other and reuse the called treatments. The global objectives are:

- the provision of a client/server communication model supporting message exchanging through message queues,
- the treatment of the heterogeneity of a distributed environment as well as the interoperability between the different sub-sets constituting the information system,
- the securization of the system by considering the information distribution, the heterogeneity as well as the decentralisation of responsibilities,
- The administration of the transport network, the telecommunications infrastructure as well as of applications.

To cope with the complexity of distributed systems and to meet the applications requirements in terms of distribution, interworking, interoperability and portability we have followed the overall framework of ODP that provides us with design rules and guidance to meet the previously mentioned requirements. Following the five ODP viewpoints, we pass from the analysis, to the design and implementation of DDTK.

In this paper, considering the complexity of the subject, we focus our presentation on the development work concerning the DDTK communication module and its basic mechanism: the message queue. Following the five viewpoints' presentation order, we will present the modelling results of DDTK communication module development using ODP concepts as well as all the additional concepts resulting from our case study. The different viewpoints language models, basis of our analysis and design, are detailed. Starting with the requirements analysis of the enterprise viewpoint, we will arrive to DDTK implementation in a certain hardware and software context.

2 . Enterprise Viewpoint

The enterprise viewpoint should serve as the basis for specifying system or application goals on which all other viewpoint specifications will directly or indirectly depend. It explains and justifies the application functions by describing its global objectives as well as its integration in the information system concerning organisation aspects and in terms of member roles, actions, purposes, usage and administration policies.

The analysis of the enterprise viewpoint is recursively applicable to several levels: from the system level considering the set of system applications, to a specific application or a specific application component. For our case study, we begin with the tool kit analysis to focus progressively to the message queue communication model analysis, the basis of DDTK.

To estimate DDTK functional characteristics we analyse the enterprise information processing requirements. This analysis is guided by the following questions :

- who requires DDTK information processing?
- which activities and which purposes are concerned with DDTK information processing ?
- which are the different roles obligations (contracts) in a DDTK community ?

To respond to the first question, let us consider the different roles in the system :

- the *user* of DDTK requiring a service meeting his needs,
- the *system designer* who wants the system to be robust and easy to maintain, analyse and expand,
- the *system manager* whose objective is to maintain normal condition of the system;
- the *developer* of the applications that installs or/and develops the applications, tests them and makes them operational in the specific context of a system.

We can attempt the consideration of viewpoints inside the Enterprise viewpoint corresponding to each of the identified roles. Each of them has different functional requirements in order to reach its own objectives and considers the DDTK application from a different point of view.

There is no proper user's perception for DDTK in terms of human being. As users of DDTK, we consider any Application Process (AP) that at a given time requires DDTK service. The APs perception of DDTK service is the provision of a client/server communication model supporting message exchanges through message queues. For these APs, distribution exists only in terms of distribution requirements. They want to communicate with each other without any preoccupation for the localization of the service, for the possible existence of heterogeneity, for the security and the supervision of the service (QoS).

For the three other roles, the distribution exists and has to be considered and provided among a number of other system properties.

The system designer considers DDTK as a brick of his system and looks for the properties that will guide his system toward openness and integration and/or any other specific policy meeting the information system needs. To the system designer pursuit, DDTK leads toward reusability of application components by breaking the applications into functional units that interact with each other through well defined public interfaces. It also provides application portability to various platforms and distribution processing across heterogeneous environments (evolutivity). At last, it provides transparency (location, relocation, access and replication transparency) and security functions.

For the system manager's pursuit for manageability and QoS, DDTK provides functions for efficient configuration, monitoring and control of the tool behaviour and resources, in order to support evaluation, prediction, QoS, accounting and security policies.

For the developer of Client/Server applications, DDTK is perceived as a tool kit offering a message queue communication model but also interface specification and compilation, trading, data conversion, and transparency functions as well as installation and testing functions in order to facilitate applications conception, realisation and operation.

To summarise and also to answer the second question (DDTK, what for ?), concerning the roles we have the following *viewpoints of responsibility*:

- The AP that requires distribution. It will define its global requirements, usually in terms of service, distribution and transparency needs. Its viewpoint is limited to the DDTK basic function: communication between applications through a message queue mechanism and research of services provided by others applications (trading),
- The system designer's viewpoint, who sees in DDTK the service offered to the users but mostly the appropriate properties that cope with his system policy,
- The administrator's and developer's viewpoints, who have to satisfy the user's requirements and use DDTK tools in order to manage all the distribution problems and provide the required service (message queue communication model).

After having analysed the required basic function that DDTK has to provide to the user, as well as the functions provided to the other roles, we can proceed to the presentation of DDTK community contract in order to answer to the last question.

The set of applications using DDTK in a distributed system form one or more communities. DDTK introduces an additional object into the system: the message queue. It is an agent object providing APs service access to the system services as well as the management of this access. Normally, there are two basic actions performed for a AP : service request and service provision corresponding to a service request. The APs (DDTK users) maintain the objectives they had before joining the community but, in the DDTK context, they are always requesting a service search, sending or receiving messages. They post and retrieve messages in and from the message queue.

The DDTK vision of the system is through an organisation structure of services materialised by the queues. To cope with the different organisation structures of enterprises (topological, infrastructure, thematic structures, etc.), a trading mechanism is introduced that permits service selection and localization depending on the characteristics asked by the users and the characteristics of the available services. The administrator determines which trading information can be accessed by whom.

Thus, an AP service requester, after selecting the appropriate queue, posts to it a message containing his service request. The request will be retrieved from the message queue by the AP service provider, will be treated and the eventual result (response) will be placed into the queue waiting for the requester to retrieve it. The kind of action an AP can perform for his communication with another AP is fixed before the beginning of message exchange.

3. Information Viewpoint

The main purpose of the information viewpoint is to specify the entities that model the system and their relationships. We will pass from an abstract explicit definition of our application (*enterprise viewpoint*) to an abstract but well structured definition (*information viewpoint*), concerning the structure, the flows, the values and their interpretation of the application information as well as the constraints of time and coherence.

To pass from our application definition to its information modelling, a methodology must be provided. The basis of this methodology is an abstract, object oriented and recursive information model (figure 1) that helps us obtain a common basis for understanding the general behaviour of the application as well as of the system in which it is installed [1] [2]. It can model every component of the distributed system and can represent both system resources and services and also identify logical partitioning, composition and inference over information.

Using the *abstraction concept* we describe any system as a composition of interacting objects. Abstraction is a fundamental way of treating complexity. Abstraction is means of ignoring the minor differences between elements and focusing on their similarities. Thus, a node-link representation can be applied to any distributed system and its applications (e.g. telecommunication service, accounting application, inventory control) considered as a set of application objects (nodes) and relations (links) that support their communications. A link is a communication object with two access points as defined in [Inoue 90].

We represent the model in figure 1 by using a simple object notation where a rectangle represents an object class: the first part contains the name of the class and the second its attributes.

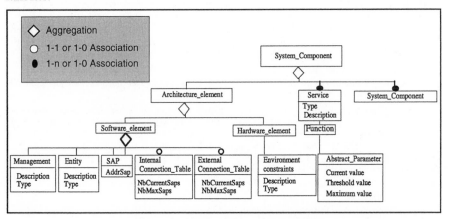

Figure 1 : Abstract information model

A *System Component* is a generic object. It may represent a domain, a node, a certain application, an application mechanism, a link, etc. Described according to the object concept, it is characterised by its own *architecture*, the *service* it provides (the system activity in which it participates) and the *SCs* it contains (*encapsulation*). Indeed, a SC carries the information of the SCs it contains since it depends on them in order to operate. Decomposition is applied recursively until all aspects of the information have been specified. Instantiation for a specific SC is realised through inheritance.

In terms of *architecture*, a SC is modelled as a composition of a software and a hardware part.

The software entity contains all the static, invariant and dynamic information of this SC. The entity class contains all the information describing the activity basic function (e.g., financial broker trading function). As the administration of any SC is one of our basic preoccupations, it has to be considered from the design phase. Thus, the *Management* contains all the necessary information (counters values, state values, etc.) permitting us to monitor and control the SC behaviour in order to reflect changes in the system but also to retrieve qualitative information on its behaviour.

The Service Access Point class, SAP, (to be compared to the Termination Point defined in [ITU92]) gives information on how users may access the entity and the Connection tables express the relations with other SCs. More precisely, the *external Connection Table* contains the access points of the (external) SCs that can be reached by this SC. The *internal Connection Table* contains the associated access points of the embodied objects.

The *service* part is modelled as a set of applicative and management functions, i.e., financial broker trading functions, fault-tolerance function, performance management function, etc. Every function is modelled as a set of abstract *parameters,* each one having current, threshold and conception values. As an example, the focus on fault tolerance may exhibit the availability, reliability, error rate and loss rate parameters.

The hardware entity contains all the information concerning the implementation or installation of this SC in a particular community (with a specific application policy), a particular station and operating system using a specific programming language (environmental constraints). It describes all the environmental constraints to be satisfied during the design and implementation steps (engineering and technology viewpoints).

The information model provides us with the image of the system concerning DDTK activity (figure 2). In a physical node one or more DDTK kernels are installed. Each one contains a number of messages queues associated with a number of Application Processes (AP) having the role of services providers. Other local (in the same Kernel) or distant APs establish accesses with the message queues in order to obtain the offered services.

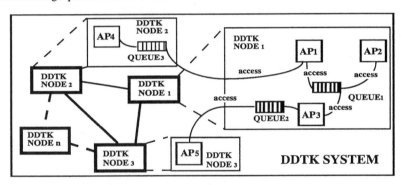

Figure 2 : Image of the information system concerning DDTK activity

Figure 2 represents DDTK application distributed over a system. Figure 3 illustrates the DDTK application using our model. The service in which this SC participates is DDTK activity: route the information through message queues connected by accesses. An access is conform to the link definition. Using the *decomposition concept* recursively enable us to specify all the aspects of the information concerning the APs, queues and accesses, that this SC contains as well as their relations.

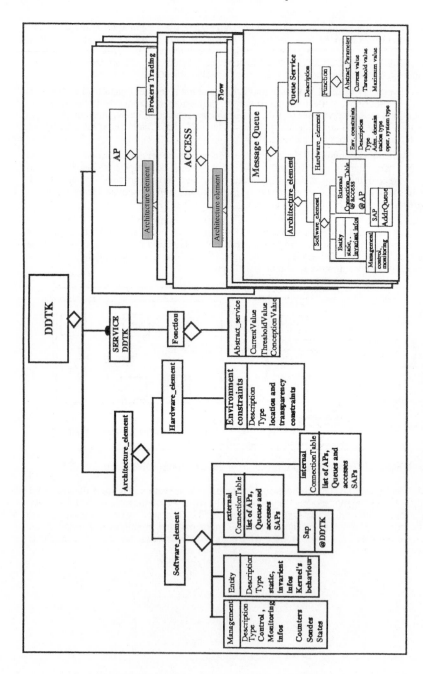

Figure 3 : Instantiation of the information model for DDTK application

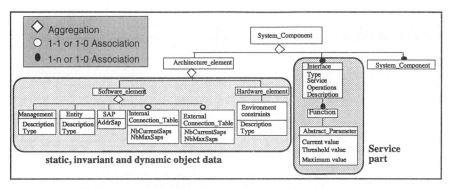

Figure 4 : Computational model

The information specification is completed with the use of state diagrams describing the states of the SCs concerning their usage, operation and administration as well as the rules that lead to the transitions from one state to an other (e.g., integrity rules imposed by the system policy). The model provides us information on the information partitioning within different administration domains. Using the diagrams, we can define the integrity rules permitting the co-operation of various sub-systems each of them having a different policy.

4. Computational viewpoint

The computation viewpoint focuses on the functional decomposition of system activities into computation interacting objects providing application specific functions. The basic aim of this viewpoint is to define the application objects configuration through a computational model describing the computational objects, their relations and specifying their interfaces as well as to define the object interaction model.

4.1. Computational model

We consider DDTK application as a set of cooperating objects that interact with each other through interfaces. Every object encapsulates internal states (dynamic object part), data and other objects (static part). It can modify their internal state by realizing treatments on their internal data. These treatments, resulting from the object interaction, are invoked by operations which are grouped into interfaces.

Let us consider once more the information model (figure 1). The architecture element contains all the static, invariant and dynamic information describing the object. The service part contains the services in which the object participates. These services correspond to a set of applicative and a set of management treatments (functions). For the computation viewpoint purposes, we express these services through the interfaces that are provided or requested (Figure 4).

We consider that an object has 3 types of Operational Interfaces: Service Provider interface (SP), through which it receives operations and provides the requested services; Service User interface, through which it requests services (SU); and the administration interface (ADM) for the administration purposes. Instantiation for a specific object is realized through inheritance, e.g., an agent has the three types of interfaces and can have more than one SP and SU interface types (he can have the client and server role at the same time for different interactions), an artefact has the SP and ADM interface types, etc.).

The functional decomposition of the abstract activity objects (information viewpoint) to the functional objects (computational viewpoint) is guided by the activity analysis results (in terms of roles, objectives, policies, etc.) of the enterprise viewpoint. These results permit us to identify the basic functions of each activity as well as its distribution requirements, each of them translatable in the objects operations and satisfied by a distinguished object function. Therefore, the user's role requirements of the enterprise viewpoint are reflected throught the SP interface, the administrator role throught the ADM interface and so on. With the computational model (figure 4) we capture all the information on the computation specification. As we notice, the model does not change from one viewpoint to the other. Its service part is refined in order to describe how the objects, that participate on specific services, interact with each other through interfaces of specific types (SP, SU) using operations selected for an operational interface signature. It also specifies which internal function (treatment) each operation invokes and describes it in terms of specific function parameters.

4.2. Interaction model

DDTK supports the applications break down into client/server modules. It proposes a C/S communication model based on message exchanges through message queues.

The Application Processes (APs) that use DDTK can be clients or servers or both. They have access to the DDTK service through an Application Programming Interface (API) that provides communication and trading functions. The APs interact using DDTK service primitives (operations) that we call messages. The primitive signature consist of an operation name and the number, names and types of argument parameters as well as the results of the primitive execution. The primitives convey to the message queue the information (messages) destinated to the server or to the client.

There are two types of AP operation mode : *synchronous mode*, where the AP is blocked waiting for the response of the execution results and *asynchronous*, where the AP does not wait for the results.

The message queue object is always collocated with the server. For the client APs the queue object provides the access to a system service (a server AP) with specific characteristics concerning the service type, the service context, the provided QoS but also specifies the type of interaction between the two objects in terms of synchronisation, QoS, etc.

A binding object supports the binding between the client APs and the queue and satisfies their environment contracts in terms of subtyping relationships between the environment contracts. The object binding is initiated by the APs using specific service primitives and specifying the queue name (id). As result they receive a binding identifier. It is also their responsibility to terminate the binding.

An AP (client or server) interacts with a queue by invoking operations at the queue interfaces. The DDTK interaction model supports two types of *interrogation* for an AP toward a queue : The posting of an AP message to the queue through a binding object and the retrieval of an AP message from the queue through a binding object. The clients post service requests and retrieve service replies. The servers retrieve services requests and post the replies.

Two queue types exist:
- *bi-directional* queues, support the emission of a client's request to the queue, the reception of this request by a server, the emission of the response to the queue and it's reception by the client.
- *Unidirectional* queues, support only the emission and reception request.

As we can see the selection of a queue "types" also the "dialogue" between the client and the server AP, imposing a type scheme that defines the APs interaction and assures type conformance. A client by choosing a bi-directional queue, chooses a synchronous "dialogue" mode as the queue guarantees, among other the presence of the server. The uni-directional queue supports an asynchronous "dialogue" that does not require the simultaneous server's presence. Moreover, the messages queues present other options as: messages save (on disk or on memory, that determines throughput performances and fault tolerance). The message queue manages simultaneous invocations of many clients to many servers. It takes care of access concurrency and also masks from the clients the presence of many servers. It transforms a simultaneous execution of operations to a sequential one.

The APs are also bound also to a trading function that permit them to select and localize a message queue by specifying the desired service characteristics as well as the queue characteristics mentioned above, and also to register a queue to the trader.

The modes of failure during interactions can be classified as binding failures (when the environment contracts cannot be satisfied, security failures (when the clients do not have access permission to a specific queue), communication failure and resource failure (due to a lack of resources).

5. Engineering viewpoint

The engineering viewpoint specifies the mechanisms supporting the distribution of the DDTK application in conformance to the computation viewpoint. It deals with the information, memory and communication proceedings necessary to support the application distribution in the context of transparency requirements, by defining their structure, their configuration, and their fine-grained placement onto the nodes.

5.1. Distribution infrastructure

In the computational viewpoint we have defined the functional objects of our application, their interfaces as well as the interaction model. The transition to the engineering viewpoint is showed in figure 5.

In this figure, we consider the interaction between two APs located in different nodes (stations). We consider a one to one correspondence between our computational objects and our basic engineering objects.

The DDTK cluster in fact represents the linking of the client or server AP with DDTK (DDTK API). The cluster configuration consists of the server APs providing a specific service (which does not forbid them from being clients of other services as well) and the message queues supporting the distribution of this service. If the client AP of the example also has the server role we should consider it as being clustered with a specific queue. The servers always have a direct binding with the message queue(s) that serves them. For the evolution of DDTK application toward servers and message queues location independence, this direct binding is also considered as a DDTK access (a link with no distribution needs for this time). The capsule represents the configuration of all the clusters and, in fact, of all the services supported by the DDTK application of a certain version. It corresponds to a unique address space. We consider these capsules to the nodes that represent the stations forming a single unit in the system in terms of location in space and embodies a set of processing, storage and communication functions.

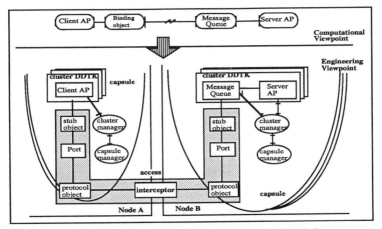

Figure 5 : Transition to the engineering viewpoint. Distribution infrastructure.

The DDTK accesses (links) correspond to the ODP channels. It is a set of objects that provide distribution mechanisms and transparency to the basic engineering objects. The access is a configuration of stub objects, binding objects and protocol objects. The stub object provides data conversion functions and supports access transparency between the engineering objects located in different stations with different operating systems, and written in different programming languages. The ports are binding objects that maintains the binding between the basic engineering objects, manage the end-to-end integrity of the access as well as the QoS of this access. They provide location transparency to the basic engineering objects. The protocol object provides the communication functions. When the protocol objects of an access are of different type they require an interceptor object to communicate.

Concerning the management functions we consider the cluster and capsule managers as part of the system global administration. We will only be interested in the management functions provided by DDTK. For the client or server AP no specific action is provided by DDTK in terms of deletion and management. The APs are clients of DDTK service, therefore is out of DDTK scope. DDTK is limited to the management of their communication. It applies controls and keep records on the AP only for security and accounting reasons (number of transactions, etc.). The queue message presents an interface with the cluster manager supporting the queue management function. Each DDTK cluster supports interactions leading to cluster creation, deletion, reactivation, disactivation, recovery and migration. It can be deleted after the desactivation or deletion of all the APs and queues it contains. The cluster management function is constrained by the management policy of it's cluster (of the APs activity and queue types). The capsule management function policy is guided by the activities that DDTK supports and mostly by the presence of DDTK communication and distribution support.

The AP, by interacting with the nucleus through the node management interface, can request an access creation that establishes a binding between the AP object and a message queue object. It can also request a message queue creation that creates the queue and generates an interface reference, to enable binding of other objects (clients) to the queue.

5.2. Engineering model

One of the basic aspects of the engineering viewpoint is the organization aspect. The engineering viewpoint imposes a structure of the basic engineering objects guided by ODP structure rules. By defining the node, capsule and cluster configuration, we specify the system organization policy. For the moment ODP basic engineering preoccupation, concerning this structure, is resource management and federation, but it can also support the others management aspects. The engineering structuring provides us with basic information on the application organization in terms of the programming language (cluster), process type (capsule), operating system (nucleus) and physical node (node). Using this information, we can organise the application operating and management in terms of activity policies in a first level and furthermore in terms of administration domain policies.

Figure 6 represents the instantiation of our model for the system SC. The system is composed of nodes that communicate with each other through links. Each node is structured as a set of capsules. The DDTK application is considered as a capsule of the node. To tell the difference between the break down concerning objects that communicate (nodes, APs, message queues, etc.), represented on the right side of the model, and the structure decomposition (capsule, cluster, etc.), we represent this decomposition on the left side of our model. The capsules are decomposed in clusters that contain the basic engineering objects (as we have already mentioned the grouping together of APs into clusters is made by activity). The basic engineering objects description (APs, accesses, stub, binding and protocol objects) is obtained using the computational model.

The engineering model contains all the necessary information for the objects and accesses realisation by considering the different environment and policy constraints of the system.

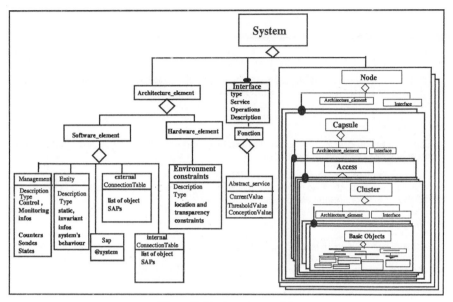

Figure 6 : Engineering model

6. Technology viewpoint

The technology specification defines the choice of software and hardware technology for the implementation of an ODP system. The figure 7 presents DDTK functions (thick line rectangular) as well as the environment that supports them.

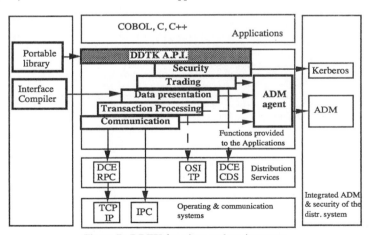

Figure 7 : DDTK functions and environment

Distributed Computing Environment (DCE) provides the distributed environment for DDTK. Although it does not fully conform to RM-ODP [5], DDTK uses it in order to build a support environment satisfying the ODP requirements.

Hence, DDTK uses: OSF/DCE Remote Procedure Call (RPC) for communication and IPC for local calls, Cell Directory Service for localization and Kerberos for security.

An Interface Definition Language (IDL) is used for describing the Application Service Interface (ISA) that links the APs with DDTK by describing the used data types and the permitted operations. The IDL is a extended version of IDL CORBA [OMG91]. It supports the construction of COBOL ISAs. A specific tool is provided, permitting the definition and description of interfaces.

The transaction processing function based on Open Distributed Systems - Transaction Processing (OSI TP) is not yet available.

DDTK is available for the moment on OS/2, AIX (BOS/X), Unix SCO and Solaris 2.3. The next stage is the porting to other environments, particularly workstation environments (Windows 3) and mainframe environments (OSF/DCE).

7. Conclusion

In this paper we have presented the development steps of an ODP application from analysis to design and implementation. ODP framework has served as the basic support providing us with basic distributed concepts, design rules and development guidance.

Following this approach, DDTK has been considered as an application developed on the ODP framework. Nevertheless, from DDTK user's viewpoint it can be considered as a middleware integrating communication and distribution tools and offering a platform solution.

8. References

[1]   J. Sclavos, N. Simoni, S. Znaty. "Information Model: From Abstraction to Application", IEEE NOMS'94.
[2]   S.Arsenis, N. Perdigues, N.Simoni. "Distributed Applications and Networks Integration : from modelling to implementation" TINA Fev.95.
[3]   J. Rumbaugh, M. Blaha, W. Premerlani, F. Eddy, W. Lorensen. "Object Oriented Modeling and Design". Prentice Hall, 1991.
[4]   OMG. "The Common Object Request Broker: Architecture and Specification". Dec 93.
[5]   Joint Xopen/NM Forum. Translation of GDMO/ASN.1 Specifications into CORBA-IDL. Draft 0.02, July 94.
[6]   A.D.Beitz, P.W.King, K.A.Raymond. "Is DCE a Support Environment for ODP?". ICODP 93
[7]   Y.Inoue, M.Hoshi, H.Uchinuma, Y.Hoshi : "Transport Network Architecture for the Information Modelling" TINA 90
[8]   ISO/IEC JTC1/SC 21/WG7. "Draft Recommendation X901, Basic Reference Model of Open Distributed Processing - Part1: Overview and Guide to Use", June 93.
[9]   ISO/IEC JTC1/SC 21/WG7. "Draft Recommendation X901, Basic Reference Model of Open Distributed Processing - Part2: Descriptive Model", June 93.
[10] ISO/IEC JTC1/SC 21/WG7. "Draft Recommendation X901, Basic Reference Model of Open Distributed Processing - Part3: Prescriptive Model", June 93.
[11] DOXA Informatique : "General Presentation of Doxa Distributed Tool Kit" June 1994

23

Experiences with Groupware Development under CORBA

Thilo Horstmann and Markus Wasserschaff

National Research Centre for Informatics and Information Technology (GMD),
CSCW Research Group, Postfach 1316, D-53731 St. Augustin, Germany

This paper summarizes our experiences of groupware application development based on the new CORBA technology. We designed and developed a system that implements the metaphor of circulation folders on computer systems. Like its real-world counterpart, an electronic circulation folder can contain any kind of object, e. g. documents, annotations, spreadsheets, or graphics. When a user has finished his work with a circulation folder it will automatically be sent to the next user according to an user defined control tag. The application is built upon the new Distributed Smalltalk, Version 2.1 of Hewlett-Packard. HPDST is an application framework for building distributed, cooperative applications. It provides a full implementation of the CORBA 1.1. standard and extensions that have been submitted to OMG for CORBA 2.0. Additionally, there are class libraries according to Hewlett-Packard's Distributed Application Architecture (DAA) which particularly simplify the implementation of groupware applications.

Keyword Codes: C.2.4, D.1.5, H.4.1
Keywords: CORBA, Smalltalk, CSCW, Groupware

1. INTRODUCTION

The CSCW group of the Institute of Applied Information Technology at the German Research Centre of Computer Science (GMD-FIT.CSCW) has been developing groupware applications for over 10 years. The institute's research was mainly focused on developing work flow and task management systems (Domino [5], The Task Manager [6]), e-mail systems, video conferencing tools and organization information systems [8]. Currently the research group is involved in the Politeam project which is intended to support the coordination of work in German Ministries. It was set up because of the decision of the German Bundestag to move main parts of the government and ministries from Bonn to Berlin [4].

Successful groupware development currently requires consideration of many different areas of computing, such as distributed computing, (data sharing and migration across networks, data consistency), database technology (long lasting transactions, data persistency), user interface design (multi-user interfaces, adaptable interfaces), and others. Although most of these points are not an immediate research subject to CSCW they require much conceptual and implementational effort during the groupware development cycle.

In particular, we were forced to focus much of our intention on the network layer of our applications. Although we designed and implemented our applications in an object-oriented fashion we have had to implement the communication level based on low-level protocols, e.g. TCP/IP, ISO/DE or RPC. This has been due to the fact that were no industrial-grade tools available for cross-platform, object-oriented programming.

The implementation effort of the network layer had often influenced the design process of our groupware applications. In some cases, we were even forced to drop important functional requirements to the application because the expected implementational effort would have exceeded our resources.[1]

To keep the overall design of the network layers simple, we have often chosen to provide the servers and clients with a functional interface, although this is contradictionary to the object-oriented programming paradigm.

As a result, one of the most important issues in CSCW, awareness, notification and eventing, was often implemented in a very rudimentary way. The classical (functional) client server approach does not allow the server to actively inform its clients of data changes. Instead, the server must first generate a representation of the changed data, which can be accessed by the clients by means of polling mechanisms. In turn, clients must filter and interpret the server's data to get the changes relevant for the corresponding user.

The classical client-server approach has turned out to be unacceptable for supporting many cooperative activities. In groupware applications supporting loosely coupled cooperation, in which changes occur rarely, the permanent client polling implies a disproportional load of both the network, and the clients' machines. This holds particularly in wide area networks in which the clients might be separated by long distances from the server. In tightly coupled systems, the generation and interpretation of data strongly decrease the server and client performance. These disadvantages become even more significant in applications in which cooperation changes from loosely to tightly coupled over time.

For these reasons our institute has made an evaluation of Distributed Smalltalk from Hewlett-Packard (HPDST). HPDST is a complete Smalltalk development environment based on VisualWorks from ParcPlace, Inc. extended with class libraries for distributed programming according to the CORBA 1.1 standard. Applications developed in HPDST will run without modifications to the source or binary code on all platforms supported by VisualWorks. In addition to the CORBA compliant class libraries, HPDST provides a framework for building cooperative applications. This framework includes libraries for building applications according to HP's Distributed Application Architecture (DAA) and libraries supporting the development of cooperative user interfaces.

In this paper we will first describe the functionality of the circulation folder application. After describing major concepts of the HPDST system in Section 3 and 4, we will present some details of our implementation. Finally, Section 6 analyses the experiences we made with the software development.

---

1. Grudin even argued that the lack of a common infrastructure has contributed to the failure of most existing cooperative systems to effectively support the work for which they were designed [2], [3].

2. THE EVALUATION APPLICATION

The HPDST application framework provides many promising features for the development of distributed, cooperative applications. To get an impression of the usefulness of these features, our institute decided to implement a typical groupware application based on HPDST. We developed a system that simulates *circulation folders* which are in common use in different kinds of organizations. It allows users to send arbitrary objects like documents, annotations, graphics, spreadsheets etc. to other users in a wide area network. An electronic circulation folder is routed across the net according to a user defined *control tag*. *Access rights* specify which users are allowed to modify the control tag and the contents of a circulation folder. A user should be kept informed about the state and current position of his circulation folder.

Fig. 1 shows the client start-up window. Once a user has successfully connected to the server he can access all circulation folders that have newly arrived ("In Tray"). To create and send new circulation folders, he has to open his personal workspace by double clicking on the workspace icon.

Fig. 2 shows a circulation folder which is ready to send to other users. The upper

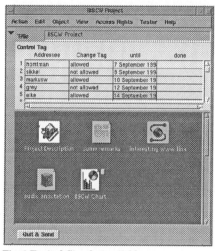

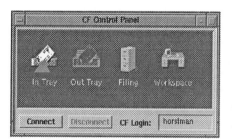

Fig. 1 Circulation Folder Login Window   Fig. 2 Typical Circulation Folder

part of the circulation folder represents its control tag. The creator stands in the first line followed by the recipients. The creator defines who is allowed to modify the control tag and the deadline each recipient should forward the folder. Similar, the creator specifies who is allowed to delete or add objects to the folder ("Access Rights").

The circulation folder may contain external objects, e.g. a standard Microsoft Word document. If the receiver of this folder clicks on this Icon the appropriate external application will be started with this file as argument. The application to start is user defined and does not necessarily have to be the same as the one with the file was created. This allows the use of documents of external applications platform independently. E.g. a WinWord document might be processed on Macintosh with MacWord and on Unix with FrameMaker.

To integrate the huge amount of information resources available in the World Wide Web, users can include Uniform Resource Locators (URLs) in a circulation folder. Clicking on this Icon starts Mosaic (or a different WWW-Browser) with this URL. Since a URL can point to files created by arbitrary applications, URL's provide another way to integrate external applications. E.g. if the URL points to a postscript file, clicking on the WWW-icon would start Mosaic which in turn starts a postscript viewer with the file depicted by the URL. This approach is more generic but require to start both Mosaic *and* the external application.

The other objects shown in the folder are objects shipped with the HPDST framework. These are a simple text object which can be used to include personal remarks concerning the folder, a graphic object showing for instance a business chart and an audio annotation.

A circulation folder is sent by either pressing the "Quit & Exit" button or by dragging the circulation folder icon onto the "Out Tray" icon.

When a circulation folder is sent the system creates a circulation folder adapter in the user's "Out Tray." The adapter contains the current position of the folder and is automatically updated each time the folder is sent to the next user.

Before we explain the implementation of the application in more detail, we give an overview of the HPDST system. Because of the numerous features of HPDST we will focus on those parts that mainly influenced our implementation.

3.   BRIEF OVERVIEW OF HPDST

HPDST extends the VisualWorks System with class libraries for distributed, object-oriented programming according to the CORBA 1.1 standard [7]. This includes a portable implementation of the Object Request Broker (ORB) and Object Services. Fig. 3 shows the architecture of the system. Because HPDST does

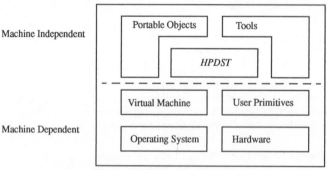

Fig. 3 HPDST Architecture.

not modify the hardware dependent parts of the VisualWorks Smalltalk system, it runs on all platforms supported by VisualWorks. The current implementation of HPDST relies on TCP/IP communication protocols. Support of OSF DCE RPC is planned for future versions.

With HPDST, distributed objects are accessed transparently. If a client requests a service, it is either sent directly to the service object, if it is local, or to the (local) object reference ("proxy"), if it is remote. Calls on proxies are automatically intercepted by the ORB and forwarded to the remote object. Results of the request returned to the client appear the same as results of local requests. A HPDST ORB can handle multiple requests of external calls. The calls are mapped to independent processes in the Smalltalk image.[1]

Standard core and standard object services

HPDST provides support for standard core and for standard object services. Standard core services include support for the Interface Definition Language (IDL), static and dynamic invocation interfaces, and queries to an Interface Repository. HPDST provides a language binding IDL-to-Smalltalk which is expected to become part of the CORBA 2.0 standard. Public interfaces are registered in the Interface Repository which can be dynamically accessed. It allows clients to determine how to make requests to distributed service objects at run time.

Standard object services are not defined in the 1.1. specification. Hewlett-Packard and SunSoft have jointly proposed an extension for object services which is implemented in HPDST. It consists of services for object persistency, properties, associations, naming and eventing.

Distributed Application Architecture (DAA)

In addition to CORBA services HPDST provides support for Hewlett-Packard's Distributed Application Architecture (DAA), an specification for building cooperative, distributed object-oriented systems. The DAA, which is built upon CORBA class libraries, has strongly influenced our application so that we want to describe briefly its main concepts here.

DAA *Object life cycle policies* specify standard ways for objects to implement creation, deletion, copying and moving. The *containment, link,* and *data view* services deal with relationships between objects. Containment establishes a hierarchical relationship between objects. A container can contain any kind of DAA *application object* which is a common desktop object like a text document, business chart or a folder. They are presented to the user as one single entity, for example as an icon on the desktop. A container may contain other containers. The resulting link structure is called a *containment tree*.

The standard HPDST user interface uses an office-building containment tree. In that, the root container is called a building which contains an arbitrary number of office containers. An office can in turn contain desks, file cabinets, folders, etc.

---

1. External calls are not mapped to autonomous threads of the operating system. One reason is the virtual machine of the VisualWorks systems which does not support multi-threading at operating system level.

HPDST provides many operations on containers and containment trees. The most important ones deal with object location management and object life cycle support. A container can be searched to locate its children, sub-trees can be moved to other containers by preserving the sub-tree structure and so on.

Using the containment model in combination with the CORBA class libraries allows the containers to be on different machines. This is a very powerful concept for building distributed, cooperative applications. Containers of objects can easily be moved to containers located on other machines. Remote containers can be traversed to locate, add, remove or modify objects.

Another central concept of the DAA used in our application is the *presentation / semantic split*, which splits each distributed object into two parts, a presentation and a semantic object. The semantic object can be compared to the model in the Model-View-Controller (MVC) concept, because it contains all the important information about the application domain. Multiple presentation objects may be connected to one semantic object to allow the users to have different view on the same, shared data. Both semantic and presentation object do not have to be on the same machine. When the shared data of a semantic object is changed, all presentation objects will be automatically updated. Semantic and presentation object communicate via IDL.

4.   CREATING DISTRIBUTED OBJECTS WITH HPDST

The creation of distributed objects with HPDST may start with the specification of their interfaces.[1] CORBA distinguishes strictly between the implementation and the interface of an object. The interface must be defined in the programming neutral Interface Definition Language (IDL) while the implementation can be made in any programming language.

Defining an object's interface

To allow local Smalltalk objects to communicate with remote objects via the ORB, the language neutral Interface Definition Language must be translated into a Smalltalk equivalent. The current CORBA 1.1 specification defines such a language binding for the C programming language only. HP's proposal for the Smalltalk binding used in HPDST is submitted to the OSF for the CORBA 2.0 specification.

IDL is a statically typed definition language. This makes a definition of a mapping from IDL to a dynamically typed language such as Smalltalk more difficult than for C++, for example.

Simple IDL data types are mapped to the corresponding Smalltalk objects in category Magnitude (IDL long integer, IDL short integer → Integer, IDL float → Float, etc.).[2] Smalltalk strings are mapped to IDL template type string. Because

---

1. Application development with HPDST does not necessarily have to start with the definition of the object interfaces. Alternatively, applications may be implemented by modifying existing (example) applications. IDL interfaces of existing Smalltalk classes might be defined later on.
2. In Smalltalk there is no notion of *types*. IDL types are all mapped to Smalltalk *classes*.

Smalltalk class `String` consists of several subclasses (`MacString`, `OS2String`, ...) which would be mapped to all the same IDL type string, return string values can be specified with the pragma `CLASS=<Smalltalk class name>`. Using IDL pragmas allow the Programmer to use Smalltalk more flexibly in the sense that he is not restricted to the IDL data types. But the use of pragmas will make an interface definition implementation dependent, because other IDL compilers will probably not be able to interpret these directives correctly.

HPDST supports multiple inheritance of IDL interfaces. To avoid conflicts where an interface inherits from two interfaces of the same name but different implementations, IDL does not allow name overloading. Inherited interfaces can override types, constants and exceptions but not operations and attributes.

Implementing the interface of an object

The Smalltalk implementation of an object's interface can be made as usual by means of the VisualWorks development tools with two exceptions: First, to uniquely identify the interface and its implementation, each class must implement the instance method `mostDerivedInterfaceID` which returns the unique id of its interface. The system wide unique identifiers, can easily be obtained from the HPDST system. Second, each class must implement the `abstractClassId` method to allow the CORBA factory finder[1] to uniquely identify the class to which instances can serve a request from a remote client.

Creating and accessing objects in a remote image

The way of accessing objects in a remote image depends on how the HPDST system is set up. According to the CORBA specification, HPDST can be used with at least one ORB and a unlimited numbers of Object Adapters (OA). To allow client objects to invoke methods of remote service objects, the server class must be registered in the server's factory finder. The interface of the server object must be registered in both the interface repository of the client and server image. A typical client invocation of a remote service looks like this (assuming a configuration with two ORBs running on remote machines):

1. get the naming service of the remote ORB:

   ```
   remoteNamingService := ORBObject namingService:'hera.gmd.de'.
   ```

2. get the name of the remote factory finder by means of using the remote ORB's naming service:

   ```
   remoteFactoryFinder :=
   
       remoteNamingService contextResolve: (DSTName onString: 'facto-
       ryFinder').
   ```

---

1. In CORBA terminology, an object which can be instantiated to create other objects is called a *factory*. Thus, all non-abstract Smalltalk classes can be used as factories. The *factory finder* is intended to find the appropriate factory whose instances can provide the service requested by an external client. The factory finder associates the object's abstract class id with its class name.

3. create a new object with class name <CLASSNAME> on the remote machine using the abstract class id:

```
remoteObject:= remoteFactoryFinder createObject: <CLASSNAME> get-
InstanceACL.
```

4. send messages to the remote object as usual. The arguments in <MESSAGE> must conform to the arguments of its IDL:

```
remoteObject <MESSAGE>.
```

In 4. the call to the remote object looks exactly the same as a call to a local object. For the programmer there is no difference between a local and a remote object even if the remote object is created on a different platform.

5.   IMPLEMENTATION OF THE EVALUATION APPLICATION

This section is concerned with the implementation details of our circulation folder application. It describes the basic concepts of the implementation and how we have used the concepts described above.

Architecture overview

The basic architecture of the circulation folder consists of a server and an unlimited number of clients (Fig. 4). Each client and the server are running Object

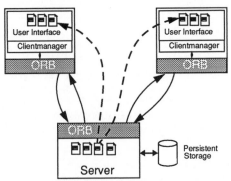

Fig. 4 Architecture of circulation folder application

Request Brokers in their Smalltalk images. In contrast to the classical client server architecture, the server can send messages to the currently connected clients at any time. Circulation folders are passed as physical copies from one client to another via the server. The server is responsible for the distribution of folders and manages a persistence storage of all circulation folders which can be used for recovery purposes after an system crash. The server keeps track of all currently connected clients. The solid arrows indicate method invocations from the clients to the server (e.g. send folder, get folder). The dotted arrows visualize the server-client communication which is invoked if the state (e.g. the position) of an circulation folder object changes. This concept allows a user to be aware of the current state of his circulation folders at any time.

In order to minimize network traffic and to enhance client autonomy we decided to pass circulation folders as physical copies. This allows a user to use a client without being connected to the server, e.g. on a mobile system. Eventing and information services about incoming circulation folders are handled by using CORBA mechanisms when the user reconnects to the server.

Implementation details

The current implementation makes intensive use of the DAA framework in combination with the HPDST CORBA class libraries.

In DAA terminology a *circulation folder* is simply a specialization of the DAA container classes. Since DAAContainer consists of a semantic and a presentation part (cf. Section 3) we subclassed both DAAContainerSO, and DAAContainerPO respectively. The contents of the control tag and the contents of the circulation folder itself belong to the view independent domain of the object. They are stored in the objects' semantic part. The presentation part of the circulation folder consists mainly of an adjustment of the views and the drag-and-drop protocol which allows a user to simply drag objects to and from the folder.

The implementation was straightforward. Almost all aspects concerning communication between the (distributed) parts of the objects were inherited from the superclasses. Our work was to define the IDL between both parts of the object, and to specify the persistent object data which have to be transmitted when the object migrates to a remote image.

The task of the *client manager* (Fig. 4) includes connecting and disconnecting to and from the server, requesting new circulation folders since a user's last login, transmitting folders to the server and propagating notifications about the progress of a particular folder to the user interface. The visual component of the client manager was shown in Fig. 1.

Connecting to the server is realized with the aid of CORBA's naming service (cf. also Section 4). The *server* coordinates all the circulation folders, passes them to the next user (in our terminology *station*), informs users about folders, that are currently in their *in-boxes* and actively notifies all users on a control tag list about the progress of the corresponding circulation folder (eventing service). Because these services are public to all (remote) clients, they have to be registered within CORBA's interface repository. HPDST's IDL differs in some points from standard IDL, but can be converted to standard syntax automatically. The most notable differences are the two pragmas CLASS, which were explained in Section 4, and SELECTOR. The second pragma should be added after every signature in an interface and specifies a mapping to a Smalltalk instance method.

Internally the server stores all circulation folders in a global dictionary keyed by a unique index. Actually, only small surrogates ("proxies") are kept in memory and point to regular files. The files serve as a persistency store. This workaround was necessary, because the persistent storage manager is still subject to OMG's standardization efforts and thus not yet available in the HPDST environment.

6.   SOFTWARE DEVELOPMENT WITH HPDST

This section describes our experiences with CORBA based software development under HPDST.

The use of CORBA technology has strongly affected many design and implementation decisions. CORBA allows implementation of the communication between objects at a high level. The low level protocol layer is completely encapsulated in Smalltalk classes. No work was necessary concerning data type conversion, serializing/de-serializing of objects, socket binding, etc. This encapsulation allowed us to design the whole application including communication layer fully in an object-oriented fashion independent of the platform.

However, using IDL as the interface language for Smalltalk objects restricts Smalltalk's flexible data type concept. Even simple Smalltalk data structures (e.g. Point, which contains two members each of arbitrary type) have to be awkwardly declared in IDL. Instances of more complex classes (e.g. dictionaries) can only be used in IDL if they are mapped by the programmer to IDL data types.

IDL is intended for definition of interfaces between objects which may not be implemented in the same programming language. However, IDL cannot cover the entire range of data types supported by all programming languages, so that there must be compromises in each particular IDL language binding. This means using IDL in a distributed Smalltalk application is reasonable when Smalltalk objects have to communicate with objects implemented in another programming language. However, there are currently no commercial ORB's available which support use of different implementation languages simultaneously. This is mainly due to the fact that the protocol between two ORB's is not yet defined (It's a major topic for the CORBA 2.0 standard).

Our implementation was simplified by the additional features supplied by HPDST, over and above those defined in CORBA 1.1. Many of these features, such as the naming service which provide a convenient way of locating remote objects, have been proposed as extensions to OMG and will probably be included with the CORBA 2.0 standard currently being defined.

On the other hand, there are many features still missing in CORBA 1.1 and HPDST. There are no means for implementing authentication, security and access rights which forced us to implement our own application dependent protection mechanisms. Some support for security is needed if CORBA is going to be used as the basis for industrial-grade applications.

As mentioned above, there is no object persistency store in HPDST. To make the circulation folders persistent we simply created a BOSS[1] stream on an object by calling the appropriate HPDST method. This stream wrote and read object details on the file system.

Again, this was a workaround not suitable for industrial applications. The use of a central database systems as an object store fails since it contradicts to the concept of distributed objects in the CORBA specification. The definition of a CORBA persistency store is an important future requirement.

The most powerful concept in HPDST for building the circulation folder applica-

---

1. Binary Object Streaming Service as provided in the VisualWorks System.

tion was the Distributed Application Architecture which in turn uses CORBA distribution mechanisms. Distributed objects can easily be created by simply subclassing pre-defined DAA classes. These classes are tailored for groupware applications. Particularly the concept of the semantic/presentation split strongly simplified our implementation. By using the presentation and semantic split, we were able to cleanly separate the user interface components from the underlying application components. This allows for tailoring of the interface in isolation by both developers and users. This separation is important for groupware development, as cooperating users may work better with interface representations they can tailor individually [1]. Moreover, it allows reuse of the interface and application independently, and to easily the porting of the application to other platforms.

Outside the scope of CORBA but very useful in the development cycle was the HPDST remote debugger. It allows transparent inspection and debugging of objects which reside on remote machines. It allows access of objects in remote images as well as easy testing of the marshalling and unmarshalling process.

The HPDST application framework provides very short edit-compile-debug cycles even for distributed applications. The main reasons for this are the use of VisualWorks for the programming environment and the powerful concepts of the Distributed Application Framework which is based on the CORBA class libraries. The development of the evaluation application was finished in approximately 2 months with a team of 4 developers. Compared to our experiences with the development of similar applications, this is an enormous enhancement of productivity.

The performance of the evaluation application was sufficient[1]. A circulation folder containing up to 10 small-sized objects was transferred from one client to another almost immediately. The performance was dependent on the performance of the network instead on the performance on the application itself. This holds for many groupware applications.

7. CONCLUSION

We have shown how the use of the CORBA compliant class libraries and the Distributed Application Architecture (DAA) has simplified the design and implementation of the evaluation application. The use of CORBA and the DAA has led to a fully object-oriented design of the application which allows us to reuse parts of the application independently. In the development process we were able to separately design the user interface, the application, and the network layer.

However, CORBA 1.1 lacks important features. There is no way to connect ORB of different vendors since the protocol between ORB is not yet defined. This makes the use of different implementation languages currently impossible, although this is a major goal of OMG. Further requirements from a groupware developer point of view are the definition of object dependent standards concerning security, authentication and access rights.

HPDST provides support for concepts that are not defined in CORBA 1.1 but are submitted to OMG to be included in the 2.0 standard. These are the object naming

---

1. We used a cluster of Sun workstations. The server ran on a Sparc Station 10, the clients on Sparc IPX.

service, a specification for information insertion into the repository (e.g. compiling IDL), a concept of asynchronous messaging (event channels), the IDL-to-Smalltalk language binding, and others. Most of them are advantageously used in our application.

We found HPDST is very powerful tool for building distributed, cooperative applications across different platforms. The network layer is completely encapsulated in Smalltalk classes. Together with the encapsulation of the operating system dependencies by the VisualWorks system, HPDST allows development of distributed applications which run without modification to the source or binary code on all major platforms. Due to the incremental compilation in the VisualWorks system, HPDST allows to rapid prototype distributed applications which strongly increases the productivity of system development.

ACKNOWLEDGEMENTS

The evaluation application was developed and implemented by Ulrich Frank, Rüdiger Grey, and the authors. Richard Bentley and Uwe Busbach read numerous drafts of the paper and made valuable comments.

REFERENCES

[1] Bentley, R., Rodden, T., Sawyer, P., and Sommerville, I. "Architectural Support for Cooperative Multiuser Interfaces", *IEEE Computer*, Vol. 27, No. 5, May 1994, pp. 37-46.

[2] Grudin, J. "Why CSCW systems fail: problems in the design and evaluation of organizational interfaces" In Proc. CSCW'88, Portland, Sept. 1988, pp. 85-93.

[3] Grudin, J. "Groupware and cooperative work: Problems and prospects" *In: Laurel, B. (Ed.): The art of Human-Computer Interface Design*, Addison Wesley, 1990, pp. 171-185.

[4] Hoschka, P., Kreifelts, T., and Prinz, W. "Gruppenkoordination und Vorgangsbearbeitung" *In: Kirn, S., and Klöckner, K. (Eds.): Betrieblicher Einsatz von CSCW-Systemen (in German)*, GMD-Studien Nr. 230, St. Augustin, 1994, pp. 91-112.

[5] Kreifelts, T., Hinrichs, E., Klein, K.-H., Seuffert, P., and Woetzel, G. "Experiences with the DOMINO office procedure system" *In: Bannon, L., Robinson, M., and Schmidt, K.(Eds): Proc. of the Second European Conference on Computer Supported Work-ESCW'93*, Kluver, Dordrecht, 1991, pp. 117-130.

[6] Kreifelts, T., Hinrichs, E., and Woetzel, G. "Sharing To-Do Lists with a Distributed Task Manager" *In: De Michelis, G., Schmidt, K., and Simone, C. (Eds.): Proc. of the Third European Conference on Computer Supported Work-ESCW'93*, Kluver, Dordrecht, 1993, pp. 31-46

[7] Object Management Group "The Common Object Request Broker: Architecture and Specification", Revision 1.1, OMG Press, 1991.

[8] Prinz, W. "Providing organizational information to CSCW applications" *In: De Michelis, G., Schmidt, K., and Simone, C. (Eds.): Proc. of the Third European Conference on Computer Supported Work-ESCW'93*, Kluver, Dordrecht, 1993, pp. 139-154.

24

Performance Analysis of Distributed Applications with ANSAmon*

B. Meyer[a], M. Heineken and C. Popien

[a]Aachen University of Technology, Computer Science IV,
Ahornstr. 55, 52056 Aachen, Germany, e-mail: bernd@i4.informatik.rwth-aachen.de

Distributed computing platforms like DCE or ANSAware offer a number of services to cope with distribution in application programming. With these services some aspects of distribution can be made transparent to programmers. In return programmers partially abandom control over performance of their application. To avoid performance bottlenecks, a distributed applications behaviour needs to be monitored. In this paper we present a distributed monitor called ANSAmon, that we developed for the ANSAware platform. It is realized according to a master-slave architecture and is a distributed application itself. Monitoring experiments can be configured in advance, whereas analysis is done after collecting all events. Global times will be estimated by a new method improving existing ones. Applying ANSAmon we investigated the ANSA REX protocol and the ANSA trader. Important conclusions for the efficient use of ANSAware for distributed applications can be deduced from the results obtained.

Keyword Codes: C.2.4; C.4;D.4.4
Keywords: Distributed Systems, Performance of Systems, Communications Management

1. INTRODUCTION

The ongoing migration from large mainframe computers to networks of workstations derives new requirements to application programmers. Whereas the former systems allow monolitic programs to be developed, modern systems are organized in a client-server fashion. As a consequence, a means for programming inter-workstation cooperation is necessary. The most common method is the remote procedure call (RPC), see [BiNe84]. Several distributed computing platforms offering appropriate services have been developed in the last years. Examples are the distributed computing environment (DCE) of the Open Software Foundation [DCE], the common object request broker architecture (CORBA) of the Object Management Group [OMG] or the ANSAware (Advanced Network Systems Architecture) of the AMP [ANSA]. Besides these commericial products, ISO developes a reference model for Open Distributed Processing (ODP) [ODP p2], [ODP p3]. It consists of concepts and their

---

* This work was partially supported by the Deutsche Forschungsgemeinschaft under Grant No. Sp 230/8-1

formal interpretation for distributed system design, general functions for building an infrastructure and means for bridging heterogenity between interfaces.

A lot of services and distribution transparency are offered by distributed computing platforms, that makes the developement of distributed applications more convinient. In return application programers partially abandom control over the performance of their application. Often, unconvenient use of offered services leads to enormous perfomance loss. This is partially caused e.g. by communication or syncronisation delay. First step is to find the performance bottlenecks before improvement can be realized. Therefore, performance characteristics have to be ascertained for subparts of an application and for infrastructure services. Performance evaluation of communication and computer system have been studied for a longer time, see [Jai91], [Kob78]. These techniques can be classified into analytic methods, simulations and measurement tools. First of all, a model of the system under study has to developed. Queueing systems and Petri Net models are common means for analytical tools and simulations, see [Kle75], [ICT93]. We developed an analytic tool for evaluating fork-join queueing nets, that supports performance management of service and applications presented in [MePo93], [PMS94], but we are not going into detail here. Whereas simulation tools are able to handle more general kinds of nets, analytical techniques are much faster in calculation. Often simulations still produce results, where analytic tools fail because of an state explosion of the corresponding models. Measurement can be further subdiveded into monitoring and benchmarking. In this work we will focus on monitoring, see also [ChSu92] for monitoring in network management systems.

The second section introduces the topic of distributed monitoring, especially focussing on differences to monitoring a single computer. A distributed monitor for ANSAware will be introduced in section three, whereas section four shows an example monitor session using ANSAmon and performance results obtained by monitoring the RPC protocol of ANSAware and its trader. A brief summary and fields for future work will conclude this article.

2. MONITORING OF DISTRIBUTED SYSTEMS

In the area of monitoring, an object system denotes the system to be mointored, whereas all components used for monitoring are named the monitoring system. These are logical concepts, that does not necessarily be physically separate, e.g. the instrumentation statements in an application program logically belong to the monitoring system, but physically to the object system. The complete monitoring process can be divided into several subtasks, see figure 2.1. First of all, events producted by the object system need to be made accessible to the monitor system. In general, there are hardware and software monitors. The difference depends on the kind of components used for monitoring. Hardware monitors employ special hardware devices for recording system events. Software monitors need event submitting statements to be introduced into application program code. This activity is called instrumentation of the application. Each event statement forwards a message to the monitor system containing event type and additional data in a special data structure called event record. This is called the recording of events. The monitor system checks incoming event records if they contain a relevant event type. All event records containing relevant event types are stored in a so-called event trace, which gives a temporal ordering to event records induced by their arrival time. After the recognition of events it is possible to filter certain events. Filtering checks if data in event records satisfy certain (filter) constraints, which are specific to each experiment. Even in central computers it is possible to have more than one event recording unit, e.g. if software and hardware monitoring is combined. Before starting analysis, event traces of all event recording components are collected. After collection, all monitoring data resides in a single place, where analysis takes place. In general it has to merge all event traces into a single temporal ordering. This requires a global time for all events [Lam78], which can be provided by a global time service or can be estimated after finishing an experiment. The result is usually stored in a program activity graph, that defines a partial order on events of a program. From this model, more complex metric values can be calculated. These data can be presented to human users by different kinds of graphical techniques like curves or diagrams. This completes to monitoring process. There have been several approaches proposed, see [LKG92], [JLS+87], [Mil89], but most of them are bound to a special kind of computer or operating system.

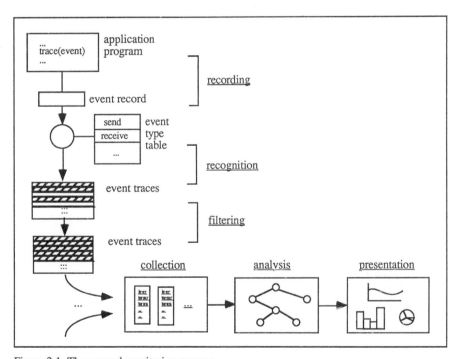

Figure 2.1. The general monitoring process

The monitoring results can be further interpreted and finally leading to instructions determining how to change the behaviour of the object system. In the best case, an improvement can be achieved. These activities are related to network and system management and commonly divided into functional areas like accounting, configuration, fault, performance and security management, see [OSI Man]. Before starting monitoring experiments, relevant performance metrics have to be determined. Target performance metrics for monitoring distributed applications can be e.g.

- the time spent in the communication system for executing remote procedure calls,
- the time spent for transport service,
- the throughput of the communication channel,
- the time spent in server waiting queues,
- the current length of server queue,
- critical paths of an application execution,
- the time spent with joining spawned threads.

All performance metrics can be useful as current or average values or as an value trace, that is combined with a time stamp, see also [Rol94].

3. CONCEPTS AND REALISATION OF ANSAMON

After a brief introduction to ANSAware we are going to describe concepts and design decisions made for the developement of ANSAmon.

3.1 Basic concepts of ANSAware

ANSAware is an infrastructure for developing and running distributed applications. It is available for a number of operating systems like SunOS, HP/UX, VMS and MS DOS. Communication between computer nodes is provided by socket interface of a transport services according to the TCP or UDP protocol. Heterogenity of programming language and operating system is bridged by a RPC mechanism and a common language for abstract interface description called the interface definition language (IDL).

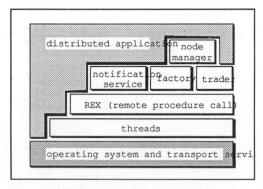

Figure 3.1. Global architecture of ANSAware

Besides RPC ANSAware provides a thread mechanism and three server types, see figure 3.1. A factory server is able to create server processes called capsules, and interfaces for a certain service type dynamicaly on the local node. References to these interfaces are mediated by a server called trader. It stores all public service offers and delivers appropriate references on request. A more comfortable way to deal with services and servers is provided by the node manager. It enables creating and deleting server on any node in the distributed system. In addition, cooperations between different trader server can be established.

3.2 The design of ANSAmon

Main goal for the development of ANSAmon is to achieve a minimal pertubation of the running application to be monitored. We decided to develope a software monitor, because we want the monitor system to be portable. Hardware monitors requires certain devices to be available for different kind of systems and vendors, but this assumption is not reasonable in our case. Although we used functions of the ANSAware platform, design has been done carefully to encapsulate ANSA-specific code in a few modules. Since ANSAware does not provide a global time service, we chose to estimate global timings. This choice was only possible, because we decided to perform analysis post-mortem, that means after finishing an experiment. It allows us to compute time-consuming performance characteristics as well as it contributes to the goal of interferring regular program execution as less as possible.

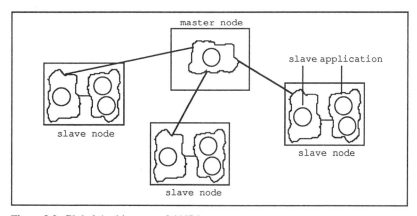

Figure 3.2. Global Architecture of ANSAmon

ANSAmon is designed as a distributed monitor with master-slave architecture, see figure 3.2. This choice contributes to the fact, that almost all of todays distributed computing is done in a client-server manner. Consistenly, ANSAware applications cooperate via RPC. A complete monitoring system consists of one master monitor (or master) and an arbitrary number of slave monitors (slaves). For each node participating in a distributed application there will be one slave. The master starts a monitor session at all its slaves. In order to have a minimal time for recording events and forwarding event records, each application manages two buffers. One is currently used for storing occuring events until it is full. Afterwards, the complete buffer is sent to the slave residing on the same node as the application, while the next events are store in the other buffer. The slaves collect incoming events from the application, stores and forward them to the master after the end of an experiment. While monitoring an application ANSAmon traces the occurence of certain events. These events can be predefined or defined by users. ANSAmon supports five kinds of predefined events, see table 3.1. These events occur while performing an RPC, as it is common in client-server-style programming.

Table 3.1. Predefined ANSAmon events

event type	description
send	transmission of data from client buffer, e.g. data for a RPC
receive	arrival of data in server buffer
dispatch	start of RPC operation at server
fork	spawning a new thread
join	waiting for a spawned thread to terminate

Within ANSAmon events are stored in a data structure called event record, see figure 3.3. It consists of an event type denoted by an identifier, some common event properties and some event-specific data. All events are described by a time stamp using local time and identifiers for monitored thread and capusule. For example an event record of a dispatch event additionally contains beside the event type the current number of threads in the server capsule. With this information the distribution of clients served for a certain server can be calculated for the monitored interval of time. Some analysis already can be done on local event traces, but this option is not supported yet.

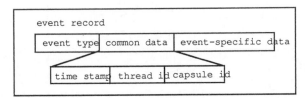

Figure 3.3. Structure of an event record

Before starting analysis, the master has to collect event traces of all slaves, which have to be merged. The basis for analysing traced events is a so-called program activity graph (PAG), see figure 3.4. It consists of a number of nodes representing traced events and directed arcs connecting two or more nodes giving the events a temporal ordering. Some events e.g. fork nodes have more than one outgoing arcs, whereas others e.g. join nodes can have more than one incoming arcs. A send node is always followed by at least one receive node. It represents an ordinary directed point-to-point communication. For the case of more than one successing receive nodes, a multicast communication is modelled. First, all event traces have to be merged and ordered. This requires finding a corresponding `receive` event for each `send` event. Before this can be done, all local time stamps have to be transformed to an estimated global time. Afterwards, the correspondence between `send` and `receive` events can be determined using global time stamps and an unique sequence number for each communication. Therefore it is assumed, that each message sending causes a reception on the other side and sending order is preserved. The sequence number is specific to `send` and `receive` events. Once a pair is detected, both nodes are replaced by a single node including the delay for communication.

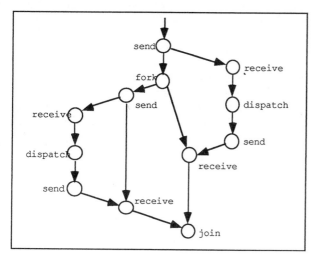

Figure 3.4. Example of a program activity graph

For estimating global time, an improved procedure for ANSAmon has been developed in [Hei94]. Like [DHH+87] it assumes the divergence between local times to develope in a linear manner. Global time estimation is consistently done by linear regression. Precision of global time approximation depends on the number of communication events occuring in an experiment. In contrast to [HKL+90] it does not only use extreme values for sending and receiving for computing regression parameter, but follows an adaptive approach using all available communication events.

4. MONITORING ANSAWARE USING ANSAMON

This section contains results obtained by sample monitor runs analysing the average response time of the ANSA RPC and of import requests to the ANSA trader. The underlying network for all experiments is a LAN.

4.1 Monitoring the ANSAware REX protocol for inter-capsule cooperation

First task of a RPC is to store a compressed version of an operation name and its parameters in a buffer. This is called marshalling, whereas the reverse operation is named unmarshalling. The first time a cooperation via RPC is set up between two capsules, an entry is added to both, channel and session table. Whereas the channel table holds information about all connections between capsules, the session table holds only information about currently active connections; that means, the entries of the session table are a subset of those of the channel table. The remote execution protocol (REX) handles the RPC. If the buffer is too large to be transported by a single packet of the underlying transport service, REX divides it into a number of appropriate fragments. These fragements are handed to the message passing system, that sends them using one of the offered transport services, e.g. TCP or UDP. All following experiments make use of UDP transport service. On the receiving side each incoming call is stored in a thread queue. Threads on the other side need a task to be processed. ANSAware performs non-preemptive scheduling for assigning free tasks to waiting threads. The RPC response time is calculated adding times of its most significant subtasks. It consists of the time for calling, performing a remote operation and returning results. In the following it is shown how the calling time is calculated within our experiments

time for call = marshalling + access channel table + access session table + REX
+ MPS + transport service delay+ MPS + REX + access session
table + access channel table + unmarshalling + thread queue +
task queue

In the following, we present some results obtained by monitor runs investigating ANSAware performance. Figure 4.1 shows average response time of a sequence of RPCs being sent to the same server depending on the size of the client REX buffer.

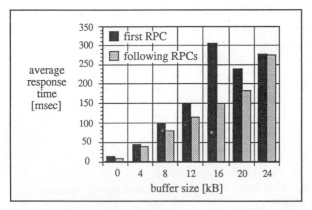

Figure 4.1. Average response time depending on REX buffer size

It can be recognized, that the first RPC for each cooperation lasts longer due to session and channel management within ANSAware. Values for the average response time grow almost constantly with buffer size. It is assumed for all RPCs that call processing at the server takes no time and only one client is attached to a server. Resulting values are comparable to those obtained for DCE RPC in [RaSc93].

4.2 Monitoring import requests to the ANSAware trader

The following three experiments measure the mean response time of import requests to the ANSAware trader. Figure 4.2 shows portions of the average response time of significant subtasks while processing a trader import request.

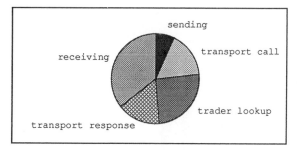

Figure 4.2: Portions of time for a trader import

For a small number of service offers (here for 9) only a fourth part is needed for looking up the trader service database. The time for database lookup will be certainly larger for bigger service databases, as we will show in the next experiment. Even more interesting is the fact, that time required receiving a call until it is performed is longer than the time required for transport service usage.

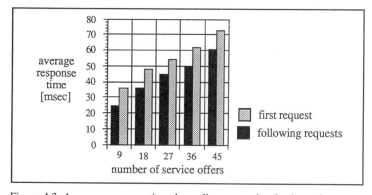

Figure 4.3. Average response time depending on service database size

Further investigation have shown, that REX processing and thread scheduling are responsible for this delay. The major time is used for REX processing, especially when fragementation of the REX buffer is required. We have investigated the influence of service database size in more detail. Figure 4.3 shows the average response time of a trader import request depending of the number of service offers in the service database. It can be concluded from the results obtained, that response time linearly increases with the size of the trader database. This phenomen is not very surprising, because the ANSAware trader always checks all service offers for matching. In addition, a constant amount of time has to be added due to administration purposes. In order to investigate an dependency between the average response time of a trader import request and the number of clients attached to a trader, we did several experiments. The results are shown in figure 4.4. For more than three objects the average response time increases strongly until the converges for 9 and 10 clients.

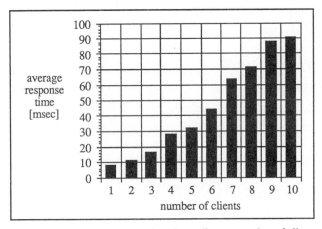

Figure 4.4. Average response time depending on number of clients

5. CONCLUSIONS

Available distributed computing platforms offer a rich functionality to provide distribution transparency to application programmers. In return programmers abandom some control over their application performance. In order to find performance bottlenecks, we developed a distributed monitor for the ANSAware platform. It has been designed to be used for general monitoring tasks, but it also supports special features for performance monitoring. Because analysis is carried out post-mortem, global times can be estimated. Therefore we improved existing methods.

Using ANSAmon we investigated several aspects of the ANSAware platform inter-connect via LAN. At first we measured performance of the REX protocol for inter-process cooperation. The resulting response time values are comparable to those measured for DCE

RPC. In addition, we checked the performance of import requests processed by the ANSAware trader. For traders managing a large number of service offers, looking them up takes most of the response time. In order to achieve more efficient service offer database lookup, service offers should be clustered reasonably, e.g. with respect to service types. In addition, clustering service offers into contexts allow a finer granularity for trader request processing than it is implemented right now. Another performance bottleneck is the REX protocol execution. Usually REX parameter values are fixed. For making efficient use of REX, parameters need to be adapted at run-time according to application requirements.

Until now we only investigated trader cooperation via LAN, but we are preparing monitoring trader cooperation over WAN. Since ANSAmon only offers post-mortem analysis at present, it will be investigated, which performance metrics might be obtained at run-time in order to achieve reasonable on-line analysis. With on-line analysis it would be possible to to efficiently manage applications and services by automatic managers, see [MePo94], [PKM94], [PK94].

REFERENCES

[ANSA]   ANSAware Release 4.1, Manual Set, Document RM.102.02, February 1993

[BiNe84]   Birell, A.; Nelson, B.: Implementing Remote Procedure Calls. ACM Transactions on Computer Systems, Vol. 2, No. 2, Februar 1984, pp. 39-59

[ChSu92]   Chiu, D.; Sudama, R.: Network Monitoring Explained - Design and Application. Ellis Horwood 1992

[DCE]   Open Software Foundation: Introduction to OSF DCE. Prentice Hall 1992

[DHH+87]   Duda, A.; Harrus, G.; Haddad, Y. et al: Estimating Global Time in Distributed Systems. Proceedings of 7th International Conference on Distributed Computing Systems, Berlin 1987, pp. 299-306

[HKL+90]   Hofmann, R.; Klar, R.; Luttenberger, N. et al: Integrating monitoring and modelling of performance evaluation methodology. In: Haerder, T.; Wedekind, H.; Zimmermann, G. (eds.): Design and Operation of Distributed Systems, Springer 1990, pp. 122-149

[Hei94]   Heineken, M.: Performance Analysis of the Communication Components of the ODP Prototype ANSAware (in german). Diploma Thesis at Aachen University of Technology, Nobember 1994

[ICT93]   Ibe, O.; Choi, H.; Trivedi, K.: Performance Evaluation of Client-Server Systems. IEEE Transactions on Parallel and Distributed Systems, Vol. 4, No. 11, November 1993, pp. 1217-1229

[JLS+87]   Joyce, J.; Lomov, G.; Slind, K. et al: Monitoring Distributed Systems. ACM Transactions on Computer Systems, Vol. 5, No. 2, May 1987, pp. 121-150

[Jai91]   Jain, R.: The Art of Computer Systems Performance Analysis - Techniques for Experimental Design, Measurement, Simulation and Modelling. Wiley 1991

[Kle75]   Kleinrock, L.: Queueing Systems - Volume 1: Theory. Wiley 1975

[Kob78]   Kobayashi, H.: Modeling and Analysis: An Introduction to System Performance Evaluation Methodoloy. Addison Wesley 1978

[Lam78]   Lamport, L.: Time, Clocks, and the Ordering of Events in a Distributed System. Communications of the ACM, Vol. 21, No. 7, July 1978, pp. 558-565

[LKG92]   Lange, F.; Kroeger, R.; Gergeleit, M.: JEWEL: Design and Implementation of a Distributed Measurement System. IEEE Transactions on Parallel and Distrbuted Systems, Vol. 3, No. 6, November 1992, pp. 657-672

[MePo93]   Meyer, B.; Popien, C.: Concepts for Modeling and Evaluation of ODP architectures (in german). In: Walke, B.; Spaniol, O. (eds.): Measurement, Modeling and Evaluation of Computing and Communication Systems (MMB'93), Aachen 1993, Springer 1993, pp. 77-89

[MePo94]   Meyer, B.; Popien, C.: Defining Policies for Performance Management in Open Distributed Systems. Proceedings of 5th IFIP/IEEE International Workshop on Distributed Systems: Operations and Management (DSOM'94), Toulouse 1994

[Mil89]   Miller, B.: Performance Characterization of Distributed Programs. Ph. D. Dissertation, University of California, Berkeley, 1984

[OMG]   Object Management Group: The Common Object Request Broker Architecture: Architecture and Specification. Revision 1.2, December 1993

[ODP p2]   ISO/IEC JTC1/SC21 N7988 Information Technology - Open Distributed Processing - Basic Reference Model - Part 2: Descriptive Model. Committe Draft, December 1993

[ODP p3]   ISO/IEC JTC1/SC21 N8125 Information Technology - Open Distributed Processing - Basic Reference Model - Part 3: Prescriptive Model. Committe Draft, December 1993

[OSI Man]   ISO/IEC IS 7498 Information Processing Systems - Open Systems Interconnection - Basic Reference Model - Part 4: Management Framework. International Standard, November 1989

[PK94]   Popien, C.; Kuepper, A.: A Concept for an ODP Service Management. Proceedings of IEEE/IFIP Network Operations and Management Symposium, (NOMS'94), Kissimmee/Florida 1994

[PKM94]   Popien, C.; Kuepper, A.; Meyer, B.: Service Management - The Answer of new ODP Requirements. In: Meer, J. de; Mahr, B.; Storp, S.: Open Distributed Processing II (ICODP'93), Berlin 1993, North Holland 1994, pp. 408-410

[PMS94]   Popien, C.; Meyer, B.; Sassenscheidt, F.: Efficient Modeling of ODP Trader Federations using P^2AM. (in german). In: Wolfinger, B. (ed.): Innovations for Computer and Communication Systems. Springer 1994, pp. 211-218

[RaSc93]   Rabenseifner, R.; Schuch, A.: Comparison of DCE RPC, DFN-RPC, ONC and PVM. In: Schill, A. (ed.): DCE - The OSF Distributed Computing Environment, Karlsruhe 1993, LNCS 731, Springer 1993, pp. 39-46

[Rol94]   Rolia, J.: Distributed Application Performance, Metrics and Management. In: Meer, J. de; Mahr, B.; Storp, S.: Open Distributed Processing II (ICODP'93), Berlin 1993, North Holland 1994, pp. 235-246

SESSION ON

Quality of Service

25

Class of Service in the High Performance Storage System

S. Louis[a] and D. Teaff[b]

[a]Lawrence Livermore National Laboratory, Livermore CA 94551-9900 USA louis@nersc.gov

[b]IBM U.S. Federal, Houston TX 77058-1199 USA teaff@vnet.ibm.com

Quality of service capabilities are commonly deployed in archival mass storage systems as one or more client-specified parameters to influence physical location of data in multi-level device hierarchies for performance or cost reasons. The capabilities of new high-performance storage architectures and the needs of data-intensive applications require better quality of service models for modern storage systems. HPSS, a new distributed, high-performance, scalable storage system, uses a Class of Service (COS) structure to influence system behavior. We summarize the design objectives and functionality of HPSS and describe how COS defines a set of performance, media, and usage attributes assigned to storage objects managed by HPSS servers. COS definitions are used to motivate appropriate behavior and service levels as requested (or demanded) by storage system clients. We compare the HPSS COS approach with other quality of service concepts and discuss alignment possibilities.

Keyword Codes: C.4; H.3.4
Keywords: Performance of Systems; Information Storage and Retrieval, Systems and Software

1. INTRODUCTION

A mass storage system is that portion of a computing facility responsible for long-term storage of information. These systems are shared among users and organized around specialized hardware devices. The complexity of storage systems has undergone rapid advancement over the past twenty years as modern computers placed increasing demands on support services. The evolution of storage architectures has been shaped by ever-larger capacities and the rapid growth of interactive processing, networks, and distributed computing. Storage systems grew from simple, large peripheral disk and tape devices and utilities, through centralized, but shared service nodes, and finally to large, complex, often highly distributed, systems supporting powerful supercomputers and parallel processors.

The sheer size of some storage problems meant that the largest systems were developed at organizations such as large government research laboratories and scientific supercomputer centers, using in-house systems engineering expertise. These individual efforts brought about systems that were heavily dependent on unique elements at each site. Unfortunately, developers usually made assumptions about who users were and how storage capabilities would be used, forcing users to interact in prescribed ways to use archival services. Varying levels of transparency were provided to reduce the complexity of system interaction, but the disadvantage of some transparencies is that efficiency may be lost in resource utilization or performance. Different levels of service quality were generally not offered. While developers of early storage systems were certainly aware of service level issues, the term *quality of service* (QoS) sometimes became a catch-all bucket into which were deposited all manner of long-term, difficult implementation issues concerning successful administration and operation of a high-performance storage system.

Client-server and consumer-provider models have been examined for many years within the mass storage community. The IEEE Mass Storage Systems Reference Models (MSSRMv4 and MSSRMv5) [1,2] identified high-level abstractions that underlie modern storage systems. The IEEE view of a storage system is one or more storage device hierarchies, implemented using an architecture that allows storage services to be distributed throughout the system. Consumers of storage service interact with standardized providers through well-defined Application Programming Interfaces (APIs). These interactions may be subject to several environmental constraints, including *storage system management* policies, administrative requirements, and operational procedures. Storage system management is discussed in MSSRMv5, but QoS is not explored.

During the time that the IEEE MSSRMs were developed, system managed storage initiatives were also launched by the GUIDE and SHARE user groups of IBM equipment. These initiatives stemmed from growing concern about use of manual techniques to allocate and manage large data center storage, and resulted in new technologies associated with the IBM Data Facility Storage Management Subsystem product. Automation functions were applied to the management of storage space, performance, availability, and configuration, but these functions did not address heterogeneous, distributed environments.

In distributed computing efforts, such as the Reference Model for Open Distributed Processing (RM-ODP) [3] and several emerging enterprise management technologies, concepts of QoS, service level agreements, environment rules and contracts play key roles in determining whether users receive services that meet their needs. Although the IEEE storage models are not identical to the RM-ODP storage function, many issues surrounding QoS are common. Alignment of future storage system standards with RM-ODP and QoS may prove beneficial for management of complex, multi-level device hierarchies in highly distributed computing and storage infrastructures. We describe below how HPSS [4,5], a newly developed high-performance storage system, uses a Class of Service (COS) capability allowing users to observe and specify differing service levels within the storage system.

HPSS is a high-performance storage system for highly parallel computers, as well as traditional vector supercomputers and workstation clusters, and is a major development project of the National Storage Laboratory (NSL). The NSL is an industry and U.S. Department of Energy collaborative project organized to investigate, develop, and commercialize new hardware and software technologies for high-performance distributed storage [6]. The principal development partners for HPSS are the U.S. Department of Energy's Lawrence Livermore, Los Alamos, Oak Ridge and Sandia National Laboratories, and IBM U.S. Federal. Other development partners include Cornell University, NASA Lewis, and NASA Langley Research Centers. A major driver for HPSS was to develop a distributed, high-speed storage system that provides scalability to meet demands of new high performance computers and applications where vast amounts of data are generated [7,8], such as those under development in the Department of Energy's Grand Challenges science program, and to also meet the needs of a national information infrastructure [9].

COS in HPSS is not based on the RM-ODP QoS model, but incorporates similar ideas. In RM-ODP, QoS is viewed as a set of *user-perceived* attributes expressed in *consumer-understood* language that describes an available service. A *service-boundary* is defined separating provider and consumer. Consumers see QoS but not necessarily service performance. Similarly, providers see service performance but not necessarily QoS. In contrast, HPSS COS is a set of *system-defined* attributes, expressed in a *provider-understood* language that describes storage capabilities. Both QoS and COS help describe the collective behavior of distributed system objects that may be subject to contractual agreements. The COS design attempts to separate consumer requirements from actual storage device characteristics, but COS and related data structures in HPSS are biased toward the service provider. This is because COS was initially implemented to provide single or parallel data transfer capabilities over possibly striped storage devices. Providing a consumer view was of secondary concern during early HPSS design. Enlarging COS beyond its current use to incorporate more of the common QoS parameters (e.g., delay, availability, reliability, accuracy, security) is under consideration.

Extending HPSS COS toward a consumer-oriented view to better serve new non-traditional clients of mass storage suggests aligning future enhancements to COS with RM-ODP standards and QoS. In addition, the RM-ODP Trading Function [10] may also prove useful in implementing *middleware* software solutions to communication, service offer, and service discovery problems between existing storage systems and new types of clients. These problems occur frequently in mass storage applications because services requested from applications do not necessarily coincide with a storage system's internal view of its offered services. Examples of middleware tasks might include deciding which of several replicated copies of data to select based on load or cost optimization schemes, and abilities to screen out or set aside data requests that may result in hundreds or thousands of random tape mounts and tape read requests. Location-independent operations and interfaces may be necessary. Third-party processes to translate consumer-oriented requests to provider-oriented services in a highly distributed storage system would be beneficial.

2. HPSS OVERVIEW AND DESIGN OBJECTIVES

The HPSS software architecture is based on the IEEE MSSRMv5, and is network-centered. The architecture includes a high-speed network for data transfer and a separate network for control (see Figure 1). The control network uses the Open Software Foundation's (OSF) Distributed Computing Environment (DCE) Remote Procedure Call (RPC) technology. In actual implementations, the control and data transfer networks may be physically separate or shared [11]. Another feature of HPSS is its support for both parallel and sequential input/output (I/O) and standard interfaces for communication between processors (parallel or otherwise) and storage devices.

In typical use, clients direct a request to store or retrieve data to an HPSS server. The HPSS server directs the network-attached storage devices to transfer data directly, sequentially or in parallel, to or from the client node(s) through the high-speed data transfer network. Local devices can also transfer data through the HPSS server. HPSS currently supports a TCP/IP socket programming interface and IPI-3 over HIPPI. Future plans include support for Fibre Channel Standard (FCS) and Asynchronous Transfer Mode (ATM) networks. COS specifications can be included with file creation requests to influence behavior of the HPSS servers regarding initial data placement and subsequent data migration operations. The COS identifier becomes part of the persistent HPSS metadata for a new file.

The HPSS I/O architecture is designed to scale as technology improves by using data striping as a parallel I/O mechanism. The system is designed to support application data transfers from hundreds of megabytes up to a gigabyte per second. File size scalability must meet the needs of billions of data sets, each potentially terabytes in size, for total storage capacities in petabytes. The system must also scale geographically to support distributed systems with hierarchies of distinct storage systems. Multiple systems located in different areas must integrate into a single logical system accessible by personal computers, workstations, and supercomputers. HPSS design was also driven by modularity of software components. Each software component is responsible for a well-defined set of storage objects, and acts as a service provider for those objects.

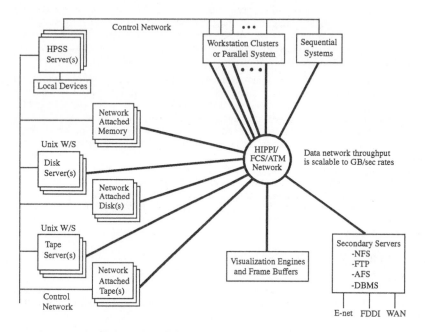

Figure 1. An Example HPSS Configuration

Current applications access HPSS and specify file-related COS characteristics at the file interface level. A COS identifier for a new file can be passed to HPSS using the quote command in FTP or through a Client API call. The Client API also provides an ability to pass prioritized hints that can force the assignment of an appropriate COS for a new file. Files in HPSS are composed of lower-level objects at both a logical and physical level. The management of these lower-level objects and their individual or collective behavior is also controlled through appropriately defined storage class identifiers related to the COS. The importance of new COS capabilities for lower-level objects will grow as the HPSS architecture is used to accommodate applications that may not be file-based, such as digital libraries, object stores, and large data management systems.

3. HPSS SOFTWARE ARCHITECTURE AND INFRASTRUCTURE

A simplified view of major HPSS software components is shown in Figure 2. Servers are shown together with their basic communication paths (thin lines). The thicker lines show data movement. Infrastructure components (the *glue* holding servers together) are shown at the top. Where multiple boxes of a particular server appear, it indicates that more than one of those servers may be running in a specific site implementation.

3.1 Servers

The *Name Server* maps a file name to an HPSS *bitfile* object. This Name Server provides a POSIX view of a hierarchical name space structure consisting of directories, files, and links. File names are human readable ASCII strings. In addition to mapping names to objects, the Name Server provides access verification to objects.

The *Bitfile Server* provides an abstraction of logical bitfiles to its clients. A logical bitfile is an uninterpreted bit string and is identified by a *bitfile id*. Mapping of a human readable name to the bitfile id is provided by the Name Server. Clients may reference byte-addressable portions of a bitfile by specifying the bitfile id, a starting address, and length. Using one or more Storage Servers, the Bitfile Server maps logical portions of bitfiles onto physical storage devices using *storage segments*. COS is primarily used to support this mapping of logical to physical storage and thus assist the Bitfile Server in choosing appropriate physical storage.

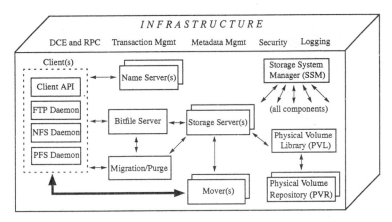

Figure 2. HPSS Software Architecture Diagram

The *Storage Server* provides a three-layer hierarchy of storage objects: storage segments, *virtual volumes* and *physical volumes*. All layers of the Storage Server can be accessed by its clients. The Storage Server translates references to storage segments into references to virtual volumes and finally to physical volumes. It also schedules the mounting and dismounting of removable media. Clients of the Storage Server are typically the Bitfile Server at the segment interface and the Storage System Manager at the virtual and physical volume interface.

The *Mover* is responsible for transferring data from a *source* device to a *sink* device. A device can be a standard I/O device with geometry (e.g., a tape or disk), or a device without geometry (e.g., a network or memory). The Mover also performs a set of device control operations.

The *Physical Volume Library* (PVL) manages all HPSS physical volumes. Clients can ask the PVL to atomically mount and dismount sets of physical volumes. Clients can also query status and characteristics of physical volumes. The PVL maintains mappings of physical volumes to *cartridges*, and cartridges to PVRs. The PVL also controls allocation of drives. When the PVL accepts client requests for volume mounts, the PVL allocates resources to satisfy the request.

The *Physical Volume Repository* (PVR) manages HPSS cartridges. Clients ask the PVR to mount, dismount, inject and eject cartridges. Every cartridge in HPSS must be managed by exactly one PVR. Clients can also query the status and characteristics of cartridges.

The *Migration and Purge Server* provides storage management facilities for HPSS. This server moves (or copies) bitfiles (or storage segments) from one storage level down to the next as specified in a hierarchy data structure to allow space on the original level to become free. Disk migration is used to free disk space. Tape migration is used to free tape volumes.

The *Storage System Manager* (SSM) monitors and controls resources of the storage system according to site policies. Monitoring includes querying values of managed object attributes representing storage system resources, and receiving notification of fault alarms and significant events. Resource control includes abilities to set managed object attribute values and storage system policy parameters. SSM may also request specific operations be performed on resources within the system (e.g., adding and deleting logical or physical resources). HPSS managed objects are based on OSI management model concepts.

3.2 Infrastructure

HPSS design uses a DCE service infrastructure, including DCE RPCs for control messages and DCE threads for multitasking. HPSS uses DCE Security, Cell Directory, and Time services as well. A library of DCE convenience functions was also developed for HPSS to facilitate server communication and to detect failing components.

Requests to HPSS to perform actions, such as creating bitfiles or accessing data, result in client-server interactions between multiple HPSS components. Transactional integrity to guarantee consistency of server state and metadata is required if a component should fail. Encina, a Transarc product, was selected by the HPSS project as its transaction manager and provides distributed commit-abort semantics, transactional RPCs, and nested transactions. Each HPSS software component has metadata associated with the objects it manages, and each server requires an ability to reliably store its metadata. The Structured File Server, another Encina product, is used by HPSS as a metadata manager and is integrated with the transaction manager.

The security components of HPSS provide authentication, authorization, enforcement, and audit capabilities for the HPSS components. HPSS developed security libraries that utilize DCE security. The authentication service, which is part of DCE, is based on Kerberos v5. A logging service records alarms, events, requests, security audit records, accounting records, and trace information from system components. A central log and local-node logs are supported. A *delog* function is provided to extract, format, and display log records. Delog options support filtering by time interval, record type, server, and user.

3.3 Interfaces

HPSS provides several data transfer interfaces. The Client API provides an interface that mirrors POSIX.1 specifications. Extensions to the POSIX interface are also provided to utilize HPSS parallel data transfer capabilities, and to allow applications to take advantage of COS hint and priority structures that can be passed during file creation.

HPSS also provides standard and parallel FTP server interfaces to transfer files from HPSS to a local file system. Parallel FTP, an extension of standard FTP, was implemented to provide high performance data transfers and provides high performance FTP transfers to the client while still supporting standard FTP commands. Use of Parallel FTP requires additional FTP client code.

The NFS Server interface provides transparent access to HPSS name space objects and bitfile data for client systems through the industry-standard Network File System interface. HPSS also can act as an external file system to the IBM SPx Parallel File System (PFS). The user may issue a command from an application to import or export files directly to or from HPSS Movers to PFS. COS specifications may be provided in the PFS import/export request to HPSS to facilitate parallel data transfers between systems.

4. HPSS USE OF COS

COS in HPSS defines a set of performance, media, and usage attributes related to the behaviors of a bitfile and its underlying physical storage. Every bitfile must have a COS identifier associated with it. The attributes of a COS are implicitly or explicitly linked with one or more *device hierarchies* and *storage classes* within the storage system. Device hierarchies in HPSS represent particular combinations of storage devices with policies controlling caching and migration of data between the devices. A storage class identifies the storage *type* (e.g., disk, tape) of a particular device, together with COS-related characteristics of the device. COS definitions, associated hierarchy identifiers, and storage classes are used by the Bitfile Server to select appropriate devices and servers for space allocation and new storage segment creation. Each COS definition used by the Bitfile Server is stored in Encina as non-volatile metadata. A simplified COS metadata structure kept by the Bitfile Server is shown below:

```
struct bfs_cos_md {
       Version;           /* HPSS version number       */
       COSId;             /* Class of Service identifier*/
       OpsSupported;      /* I/O Operations supported  */
       MaxSize;           /* Max size of bitfiles in COS*/
       MinSize;           /* Min size of bitfiles in COS*/
       Activity;          /* Amount of expected access  */
       Reliability;       /* Expected reliability level */
       XferRate;          /* Expected transfer rate     */
       Latency;           /* Expected transfer delay    */
       HierId;            /* Associated hierarchy id    */
       } bfs_cos_md_t;
```

In comparison, a storage class metadata structure in HPSS will contain many device-dependent attributes as shown in the following example:

```
struct hpss_sclass_md {
       SClassId;             /* Storage Class identifier    */
       SClassType;           /* Device type for this class */
       TransferRate;         /* Transfer rate in kilobytes */
       StripeSize;           /* No. of elements in a stripe*/
       StripeWidth;          /* Size of a stripe in bytes  */
       BlockSize;            /* Blocksize used on device   */
       OptimalAccessSize;    /* Size for best data transfer*/
       StorageSegmSize;      /* Segment size used by client*/
       MaxFileSize;          /* Max size in bytes for class*/
       MinFileSize;          /* Min size of bytes for class*/
       MigrPolicyId;         /* Migration policy to use    */
       PurgePolicyId;        /* Purge policy to use        */
       MPSId;                /* Migration/Purge Server Id  */
       MediaType;            /* General type (e.g., tape)  */
       MediaSubType;         /* Specific type (e.g., 3490E)*/
       AvgLatency;           /* Delay before start of xfer */
       WriteOps;             /* Valid write I/O operations */
       ReadOps;              /* Valid read I/O operations  */
       } hpss_sclass_md_t;
```

Access activity is typically daily, weekly, monthly, or archival. Latency is the delay in seconds between the time a request is received by a Storage Server and the time data begins to be transmitted. This will normally be non-zero for tape devices due to mount delays. Valid I/O operations for a storage class may be RANDOM, PARALLEL, WRITE, WRITE_MANY, APPEND, and READ. A COS hints structure and a COS priorities structure, both roughly equivalent to the COS definition described above, also exist to assist a client in selecting a suitable COS definition for a newly created bitfile. The priorities structure allows an HPSS client to specify a weighting for each attribute supplied in the hints structure. Priorities currently represent one of the following values: NONE, LOW, DESIRABLE, HIGHLY_DESIRABLE, or REQUIRED. Pointers to the hints and priorities structures are two of the input parameters to the Bitfile Server *bfs_Create* API, which is used to create a new bitfile, allocate space, and save relevant metadata. A pointer to the COS definition structure *actually* used by the Bitfile Server when creating the new bitfile is returned to the client as an output parameter of the call.

Using the HPSS Client API library, applications can specify an existing COS identifier for a file, or fill in the COS hints and priorities structures to describe desired/required service attributes for the file. The FTP quote command can also be used to specify a COS identifier when using HPSS's FTP Daemon. Creating a new file through the Client API is performed through an *hpss_Open* call whose input parameters include (possibly null) pointers to COS hints and priorities structures. When null pointers are passed, the Bitfile Server is free to use a default COS definition for the new bitfile. The hpss_Open call returns a pointer to the COS definition used by the Bitfile Server. The COSId of a bitfile can be obtained or modified by Bitfile Server *bfs_BitfileGetAttrs* and *bfs_BitfileSetAttrs* calls. Changing the COSId may be subject to administrative or operational constraints. The Bitfile Server also maintains the necessary device hierarchy and storage class information that describe where and how the storage segments that comprise the bitfile were physically stored. Bitfiles may reside on multiple devices simultaneously depending on the migration and caching schemes employed in a specific hierarchy.

In initial versions of HPSS, COS use is preliminary and some attributes supplied in a client-generated hints structure will not affect system behavior. Currently, only the transfer rate attribute has significant effect. The Bitfile Server either uses the specific COS identifier supplied by the client, or finds an *appropriately close* COS identifier based primarily on the supplied transfer rate value. If the client specifies the transfer rate's priority as REQUIRED, and the Bitfile Server does not have an existing COS definition that can satisfy the desired rate, the request fails and a *NO_SUPPORT* error is returned to the client. Similarly, if an invalid COS identifier is requested by a client, an error will be returned. Valid COSs are those that have been previously defined in an SSM administrative procedure to create new Storage Server virtual volumes and storage maps (the entities that actually provide storage space in HPSS). Virtual volumes and storage maps are identified by storage class, and are used to provide storage segments of that class to clients of the Storage Server.

The storage segment service is the mechanism used to obtain and access internal storage resources. Clients of the Storage Server are presented with a storage segment address space from 0 to N-1 where N is the byte length of the segment. The Bitfile Server provides a storage class identifier and an allocation length during creation of new storage segments. To ensure locating free space of appropriate type, the storage class must represent storage service conforming to any client-specified COS hints and priorities. During the creation of new space, only storage maps that have proper storage class are searched. If no storage map exists to fit the requirements, a *NO_SPACE_FOUND* error is returned.

The COS structure was designed to be extensible, and additional attributes are planned to more heavily influence server actions during data placement, data transfer, and file/fragment migration operations. A goal for future releases is better integration with large data management systems, whose needs will require COS attributes for objects other than files. In particular, I/O operations on data fragments necessary for resolving complex database queries will require new

COS capabilities. COS attributes are planned for controlling placement or collocation of related files and data fragments on physical media to enable better use of HPSS by new data management applications.

In the current COS implementation, an administrator must be responsible for creating COS and storage class structures at the time HPSS servers are configured. This is necessary because the storage resource objects managed by the Storage Server (i.e., virtual volumes, storage segments, and storage maps) are all identified by the kind of storage they support. At least one COS must be created for the Bitfile Server to use as a default for client requests that do not specify a COS identifier or COS hint and priority structures. Using HPSS Storage System Management facilities, which are based on an X-Windows graphical user interface environment, administrators create new physical volumes, then virtual volumes, and finally, storage maps for the virtual volumes. These must all exist before the creation of any storage segments. When an administrator creates a COS for the Bitfile Server, an accurate determination must be made whether the attribute combination for the COS is sound. Definition of these structures might be based on *a priori* knowledge of devices. Specifying a COS needing a stripe width of four to meet a high data rate when HPSS has only two drives at its disposal for parallel transfers would not work. Administrative creation, modification, or deletion of metadata representing COS and storage class is accomplished through the SSM management windows.

5. RELATED WORK

In previous hierarchical storage management systems used at Lawrence Livermore National Laboratory (LLNL), QoS and COS capabilities were rare. A storage system called FILEM, in use between 1976 and 1986 at LLNL's National Energy Research Supercomputer Center (NERSC), restricted users to specifying a life-span code of *archival, long-life,* or *medium-life*. Archival kept the file forever, but forced its migration to the lowest level device, at that time a manual shelf operation. In exchange for long delays on retrieval, the user was charged a lower cost. Long-life also kept the file in the system indefinitely, but an attempt was made to maintain the file in a robotic archive for faster data retrieval than shelf. Medium-life caused the file to be deleted after a time period determined by local site policies. Medium-life also tended to keep the file on disk for faster access, but at substantially higher cost. No other attributes were available to influence the level of service received or corresponding cost accrued.

The Common File System (CFS) [12], developed by Los Alamos National Laboratory (LANL) in 1980 and still running at LANL, NERSC, and other DOE laboratories, provided additional QoS mechanisms that were somewhat improved but still limited. CFS allowed users to specify a usage characteristic for new files (or to change that attribute for existing ones). Users could tell the storage system that the file was to be active daily, weekly, monthly, infrequently, or for only for a few days and then never again. The system used this access hint to place the file at an appropriate initial level in the storage hierarchy, later migrating it to lower levels accordingly. Users also could specify that a file be written to sets of mutually exclusive devices. If a user wanted to write a file twice, and ensure each copy ended up on separate groups of disks or tapes for the life of the file, it could be done with one command during initial storage.

The ability to determine disposition of data improved with NSL-UniTree, an early software development project of the NSL. In NSL-UniTree, dynamic storage hierarchies [13] were implemented to let clients define into what hierarchy their file would be placed. The placement determined how caching and migration to different storage devices would be performed and affected access time to the file, as well as data rates. Clients of NSL-UniTree were able to specify a hierarchy identifier when creating and storing a new file in the same manner as specifying COS identifiers in HPSS. Dynamically managed hierarchies eased the insertion of new technologies and allowed existing files to automatically take advantage of new devices.

New kinds of metadata, resource attributes, and other abstractions have been proposed or implemented to help optimize use of archival storage systems by non-traditional clients, including relational or object-oriented databases and scientific data management systems [14,15,16,17]. Non-traditional clients do not necessarily use a file as the data entity that is stored and retrieved. In many of these applications, specifying values for overall data transfer rate or parallel transfer stripe widths, as is done in HPSS, may not be meaningful. Other performance issues such as overall latency reduction, close clustering of related data chunks, deadline and continuity requirements, data compression, and redundancy may take precedence.

A recent IEEE-sponsored effort [18] to investigate metadata issues for access to large scientific and technical databases explored problems of storage and archive. Metadata requirements are driven by applications, but also affect storage and software system performance. Since metadata is used to improve the understanding of data content, but also to describe data access concerns, system-level metadata addressing accessibility is closely related to HPSS COS. Attributes are needed that address files or data fragments related by application usage and how these fragments should be stored and migrated. This becomes important in applications that need to manage and query specific, but possibly widely scattered, pieces of information in large data collections. There are several efforts underway to better understand requirements to effectively manage large volumes of scientific data stored on mass storage devices.

One such project, Optimass [19], has ties to the developers of HPSS, and concentrated on multi-dimensional climate modelling data. In Optimass, large datasets are passed through a partitioning engine driven by several query prediction tools that help estimate data usage patterns. Data fragments, related by application use, are then stored appropriately in the archival system. Fragments are also re-assembled after retrieval from the storage system, based on actual application queries. In Optimass, the partitioner constructs and stores partitioning information in an external metadata database for subsequent use by the reassembler as necessary. This project designed a COS-like interface between the data partitioning/reassembly engines and high-performance mass storage systems such as NSL-UniTree and HPSS. The interface provides an ability to influence or control allocation of space and physical placement of data by defining several key COS attributes associated with *data clusters* (fragments of data related by application use), and to provide these attributes to the storage system through modified client interfaces. This permits the storage system to intelligently bundle the data clusters for a targeted tertiary storage device (usually a slow, sequential-access tape).

The Sequoia 2000 project [20] and the related Mariposa effort [21] are also investigating integration of storage and large data management systems. These projects are working on extending database management system optimizations to deal effectively with tertiary devices and the movement of data between storage systems. Mariposa proposes using several QoS and Trading Function concepts, including subcontracts between subsystems, and open competition for services in a free-market economic model, to explore service guarantees among distributed, cooperating servers. Some of these servers will be performing hierarchical storage management tasks. A typical application might be to provide guaranteed delivery of frames (at a fixed rate) to a high-resolution rendering engine and display. A storage system may need to decide whether or not to accept or decline a subcontract for data movement out of tertiary storage at a specific transfer rate as part of an overall contract guarantee made by a networking service.

6. CONCLUDING REMARKS

Most current work on QoS concepts for high-performance storage, including HPSS, concentrate on attributes related to the cost and performance concerns of initial hierarchical data placement and subsequent data transfer speeds. Emphasis on QoS attributes for other issues such as guaranteed delivery, reliability, and continuity is also needed. Providing storage systems with negotiating capabilities in free-market network environments may also be required. Understanding guaranteed services and third-party brokers using RM-ODP Trading Function ideas would be a beneficial addition to new mass storage system implementations and could provide better communication with new types of storage service consumers.

A significant problem is that archival mass storage systems and new consumers of storage, such as large data management systems, do not communicate well. This is an ongoing research area, but continues to present problems for non-traditional clients of storage service. For example, storage systems and database systems can both provide various request optimization, data replication, and parallel execution capabilities, but integration of these two types of systems is difficult. An ability for customers to negotiate and receive adequate QoS means high-performance mass storage developers must address how to communicate available services to a consumer-oriented world, possibly through brokers. This involves more than development of standard or extended file transfer interfaces.

Properly applying metadata to manage data storage and access has also not yet been addressed in a systematic manner. An issue for HPSS is how to decide where application-related metadata and COS information belongs. Does this information always belong in an external database? How should the information be translated into HPSS COS attributes? Existing Encina metadata management capabilities in HPSS are not infinitely scalable. HPSS is investigating increasing and decreasing the total metadata associated with storage objects under server control and effects on system efficiency.

As requirements grow for high-performance storage systems to support application-specific views instead of traditional file-system views, the need for a richer set of COS and QoS attributes for storage is obvious. As storage systems and computing environments become more distributed, the need to provide better alignment between high-performance storage and open distributed processing standards is also clear. We have incorporated several open systems concepts into the HPSS hierarchical storage management design, but new extensions to COS are required to become aligned with RM-ODP QoS concepts. We believe this to be worth further investigation.

ACKNOWLEDGMENTS

This work was, in part, performed by the Lawrence Livermore National Laboratory under contract number W-7405-Eng-48 and Cooperative Research and Development Agreements under auspices of the U.S. Department of Energy, and by IBM U.S. Federal under High Performance Data Systems Independent Research and Development and other internal funding. For more information about the National Storage Laboratory and HPSS contact:

Dick Watson, LLNL   (or)   Bob Coyne, IBM U.S. Federal
+1 510 422 9216   +1 713 282 8039
dwatson@llnl.gov   coyne@vnet.ibm.com

Access the HPSS tutorial on World Wide Web at URL http://www.ornl.gov/HPSS/HPSS.html

REFERENCES

1. S. Coleman and S. Miller (eds.), *Mass Storage System Reference Model: Version 4*, IEEE Technical Committee on Mass Storage Systems and Technology, May 1990.

2. R. Garrison, et al. (eds.), *Reference Model for Open Storage Systems Interconnection: Mass Storage Reference Model Version 5*, IEEE Storage System Standards Working Group, September 1994.

3. ISO/IEC JTC1/SC 21/WG7, *The Basic Reference Model of Open Distributed Processing*, ITU-TS Recs. X.901 to X. 904 | ISO/IEC 10746, 1994.

4. R. Coyne, H. Hulen, and R. Watson, The High Performance Storage System, *Proc. Supercomputing '93*, Portland, OR, November 15-19, 1993.

5. D. Teaff, R. Coyne, and R. Watson, The Architecture of the High Performance Storage System, *Fourth NASA GSFC Conference on Mass Storage Systems and Technologies*, College Park, MD, March 1995.

6. R. Coyne, and R. Watson, The National Storage Laboratory: Overview and Status, *Proc. Thirteenth IEEE Symposium on Mass Storage Systems*, Annecy, France, June 13-16, 1994.

7. S. Coleman and R. Watson, The Emerging Paradigm Shift in Storage System Architectures, *Special Issue on Storage, Proc. of the IEEE*, April 1993.

8. S. Coleman and R. Watson, New Architectures to Reduce I/O Bottlenecks in High Performance Systems, *Proc. 26th Hawaii International Conference on System Sciences*, Maui, HI, January 5-8, 1993.

9. S. Howe (ed.), *High Performance Computing and Communications: Toward a National Information Infrastructure*, A Report by the Committee on Physical, Mathematical, and Engineering Sciences; Federal Coordinating Council for Science, Engineering, and Technology; Office of Science and Technology Policy, 1994.

10. ISO/IEC JTC1/SC21/WG7, *Draft ODP Trading Function*, ITU-TS Rec. X.9tr | ISO/IEC 13235, 1994.

11. R. Hyer, R. Ruef, and R. Watson, High Performance Direct Network Data Transfers at the National Storage Laboratory, *Proc.Twelfth IEEE Symposium on Mass Storage Systems*, Monterey, CA, April 26-29, 1993.

12. T. McLarty, B. Collins and M. Devaney, A Functional View of the Los Alamos Central File System, *Digest of Papers, Sixth IEEE Symposium on Mass Storage Systems*, Vail, CO, June 1984.

13. L. Buck and R. Coyne, Dynamic Hierarchies and Optimization in Distributed Storage Systems, *Digest of Papers, Eleventh IEEE Symposium on Mass Storage Systems*, October 7-10, 1991.

14. L. Roelofs and W. Campbell, Applying Semantic Data Modeling Techniques to Large Mass Storage System Designs, *Digest of Papers, Tenth IEEE Symposium on Mass Storage Systems*, Monterey, CA, May 7-10, 1990.

15. R. Grossman, et al., A Proof-of-Concept Implementation Interfacing an Object Manager with a Hierarchical Storage System, *Proc. Twelfth IEEE Symposium on Mass Storage Systems*, Monterey, CA, April 26-29, 1993.

16. M. Tankenson and S. Wright, A Data Distribution Strategy for the 90s (Files Are Not Enough), *Compilation of Papers, Third NASA GSFC Conference on Mass Storage Systems and Technologies*, College Park, MD, Oct. 1993.

17. J. Shiers, Data Management Requirements for High Energy Physics in the Year 2000, *Proc. Twelfth IEEE Symposium on Mass Storage Systems*, Monterey, CA, April 26-29, 1993.

18. S. Louis and M. Gary, Storage and Archive Group Summary Report, IEEE Computer Society Technical Committee on Mass Storage Systems Workshop on Metadata for Scientific and Technical Databases, May 16-18, 1994.

19. L. Chen, et al., Efficient Organization and Access of Multi-Dimensional Datasets on Tertiary Storage Systems, submitted to *Information Systems Journal, Special Issue on Scientific Databases*, to be published 1995.

20. M. Stonebraker, J. Frew and J. Dozier, *The Sequoia 2000 Architecture and Implementation Strategy*, Sequoia 2000 Technical Report 93/23, University of California, Berkeley, March 1993.

21. M. Stonebraker, P. Aoki, R. Devine, W. Litwin and M. Olson, Mariposa: A New Architecture for Distributed Data, *Proc. Tenth Int. Conference on Data Engineering*, Houston, TX, February 1994.

26

AN APPROACH TO QUALITY OF SERVICE MANAGEMENT FOR DISTRIBUTED MULTIMEDIA APPLICATIONS*

A. Hafid and G.v.Bochmann

Université de Montréal, Département d'Informatique et de Recherche Opérationnelle,
Montréal, H3C 3J7, Canada
{hafid,bochmann}@iro.umontreal.ca

Emerging high-speed networks and powerful end-systems give rise to a new class of applications such as video-on-demand and teleconferencing. Such applications are very demanding on Quality of Service (QoS) because of the isochronous nature of media they are using. To provide QoS support on an end-to-end basis, the need for the integration of network, transport, and operating services arises. Thus to support the new emerging services, an end-to-end QoS management is required. This paper deals with QoS management for multimedia applications by taking remote access to multimedia database as a case study. The example application is introduced and the entities involved in QoS provision are identified. QoS management activities are defined and a basic QoS management architecture for multimedia applications is presented. A general framework for QoS (re)negotiation is defined and an instantiation of this framework in the context of the example application is presented.

Key words: multimedia applications, quality of service, quality of service management

1. INTRODUCTION

Currently new emerging services, particularly distributed multimedia (MM) applications, e.g. video conferencing [1,2,3,4] and video-on-demand [5], based on broadband communications, are of great interest in industry, academic research and standardization. The introduction of these new services provides a new quality of communications. The new quality of distributed MM applications is characterized by handling continuous media and by managing various media at the same time. Different types of continuous media require different levels of quality of service (QoS) and they require guarantees for the level of service to be maintained. This implies stringent requirements for the communication systems and the end-systems to support the requirements of MM applications. Hence the new types of application need end-to-end *QoS management* to ensure that the requirements of the users are satisfied.

Several approaches and protocols have been proposed to support QoS at the communication level [6,7]. At the end-system level, system software operating systems, based on earliest deadline scheduling [8] enhanced with priority parameters [9] have been defined to manage the time con-

---

*. This work was supported by a grant from the Canadian Institute for Telecommunication Research (CITR), under the Networks of Centres of Excellence Program of the Canadian Government

straints for the new services. All those approaches provide schemes to support resource reservation in order to guarantee the requested QoS.

A key issue in QoS support is QoS negotiation activity which makes use of most of QoS management functions. However, to date, no framework has been defined to support end-to-end (user-
to-user) QoS (re)negotiation for distributed multimedia applications. This paper proposes a basic
architecture for QoS management on which a general framework for QoS (re)negotiation is
defined.

The paper is structured as follows. Section 2 defines a QoS management architecture for
remote access to MM database applications. Section 3 defines different QoS management activities and presents a basic QoS management architecture for MM applications. A general framework for QoS (re)negotiation is presented in Section 4. Section 5 concludes the paper.

2. REMOTE ACCESS TO MULTIMEDIA DATABASES

Remote access to multimedia (MM) database systems enables users to browse, search, request
and display MM documents which are digitally stored in one or more high-capacity storage
devices [10]. To support such an application, stringent requirements in terms of QoS must be provided by the underlying system.

In the following we give a short description of the components of a simplified architecture of a
prototype system for remote access to multimedia databases which we implemented (Figure 1).

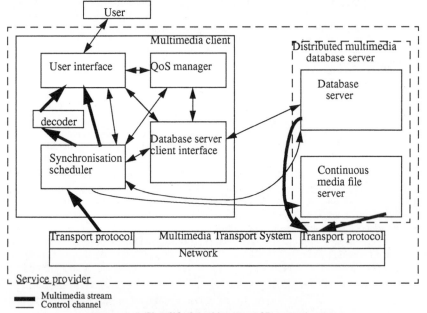

Figure 1. A Simplified Architecture of Remote Access
to Multimedia Database Application

- *User interface* enables the user to select a document, to negotiate the QoS of the document to be played, to display a selected document, and to renegotiate the QoS of the current document.

- *QoS manager* performs a QoS negotiation protocol to find an agreement which is compatible with the client's constraints, e.g. devices characteristics, the user's desires, and the database server constraints, e.g. access delay [11]. Typically the QoS manager must support the QoS negotiation, renegotiation, monitoring, adaptation, etc. by managing the configuration of the concerned objects.

- *Database server* provides reliable and coherent storage of multimedia objects as well as concurrent access to these objects and to their components. Information stored in the database is classified in to two categories: (1) MM information, (2) control information such as synchronization scenarios or QoS parameters.

- *Continuous media file server* provides support for the storage and retrieval of time sensitive continuous media, e.g. video and audio [12].

- *Synchronization scheduler* computes a time flow graph (TFG) and derives the object delivery schedule in order to compensate for variable network delays. The TFG is derived from the information stored in the database server [13].

- *MM transport system* is used for MM data that may have to be transferred between the physically distributed components of the system [14]. It is composed from high-speed transport protocol(s) and a high-speed underlying network.

- *Decoder* is used to decode the coded information, e.g. MPEG video, for presentation to the user [15].

In the context of remote access to MM database applications (Figure 1), we have identified six main objects that are involved in the QoS provision of a given activity: the user, the MM document, the file system, the transport objects, the network, the decoder, the synchronization object and the presentation device (Figure 2).

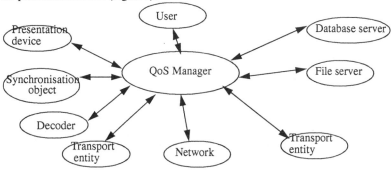

Figure 2. A QoS Management Architecture for Remote Access
to Multimedia Database Application

In the following we will give a short description of each object identified, excepted the user, and define the main QoS parameters that must be supported by each entity.

(1) MM document: Multimedia object components of a given MM document, e.g. video and audio, may be stored in a number of file servers at different locations. To handle these documents, MM applications require the specification of document structures which integrate the different components of MM objects such as texts, graphics audio or video. These structures

must be defined in order to represent the control information used during consultation or display. During the consultation phase, information describing the document localization is used for query processing, while information concerning the media type, the size or the frame rate is used for the negotiation of the QoS. During the display phase, the system uses information describing the spatial and temporal relationships to guarantee the synchronization between components of the multimedia documents. More details concerning the integration of QoS parameters in our MM document structure can be found in [11]

(2) *File server:* The file server(s) provides access to MM documents with a requested quality of service. When several streams are being retrieved, guarantee of a desired QoS for a given stream becomes non trivial since all streams are competing for bandwidth, e.g. storage access, workstation bus, and CPU slots. Hence real-time server resources management is required [10]. This means that before accepting a new connection, the server must check if there are enough resources to support this connection without affecting the existing ones. Given the huge quantity of information to be accessed, e.g. 30 Mbits a second for uncompressed video, and the severe temporal guarantees required by continuous media, the file system(s) must guarantee a minimum throughput and a maximum delay to retrieve MM data [16].

(3) *Transport entity:* Protocol stack processing may introduce a non-desirable delay and jitter, e.g. through error control and segmentation/reassembling. Protocol architectural performance issues have been well studied in [17]. In [18] it was reported that parallel protocol processing is required for end-systems to keep with the improvements in network performance. Some new architectures for protocol development have been proposed, such as the horizontally oriented protocol (HOPS) architecture [19], or the high speed protocol development (HIPOD) architecture [20]. Guarantees for the speed of protocol processing are required to support the desired QoS. Thus, for a requested throughput, the delay and the jitter must be bounded at the transport service level.

(4) *Network:* The network must be able to transfer various media types with complex and variable QoS. The new characteristics of such a network can be summarized as follows: The support of constant bit-rate and variable bit-rate, the suitability of the transport for services with bit-rates varying from tens of bits to tens of millions of bits per second (e.g. full motion compressed video requires at least 2 Mb/s even using the most sophisticated compression schemes), the support of multicast and broadcast services, no distinction among the nature of information, e.g. video and audio are transmitted in the same way, and the support of a rich set of mechanisms to cope with bandwidth management. To support a requested QoS, the MM networks control the flow of new connections and the flow of the existing connections in order to guarantee a maximum transport delay and jitter, a minimum throughput, and a maximum loss rate [21].

(5) *Decoder:* Due to the huge memory size required to store voice and video data directly, compressions schemes are used to compress the data before storage or transport. To play a compressed multimedia document, it must be decompressed by a decoder before presentation to the end-user. Such an activity may introduce a non-desirable delay and jitter. To provide end-to-end QoS guarantees, the latter must be bounded.

(6) *Synchronisation object:* Temporal relationships may exist among the media of a given multimedia document. Each of the related media passes through several components, such as network links, and nodes, accumulating some delay at each component. Furthermore the paths used by the related media may be different. Thus the differences in the transit delay which are encountered by related streams may be important and possibly not acceptable to the applica-

tions. The aim of the synchronisation object is to smooth the jitter and skew introduced by the underlying system [14]. Such an entity must be able, by using some buffer management support, to restitute the temporal relationships between the related media of the document, typically by introducing some additional suitable buffering delay.

(7) *Presentation device:* After decompression, the multimedia data is sent to the presentation device, e.g. monitor and speaker, for presentation to the end-user. Such an activity may introduce variable jitter and delay depending on the used device. Thus the latter must be able to provide guarantees to support the requested delay and jitter.

3. QOS MANAGEMENT

Generally, there are three steps during the lifetime of a session, and particularly of a multimedia session: the establishment phase, the active phase and the clearing phase. Hence, to provide a requested service, the service provider has to establish a MM session with the desired QoS, to control the session QoS during the active phase and to clear the session when requested by the user or because of service provider problems, e.g. network congestion. In other words, to support multimedia applications requirements, QoS management functions are needed during the three phases (Figure 3). A brief description of these functions is given below.

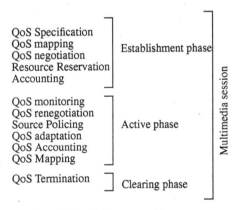

Figure 3. QoS Management Functions

(1) QoS specification and mapping: In the OSI Reference Model, the notion of QoS is defined as a set of qualities related to the provision of an (N)-service as perceived by an (N)-service-user [22]. Hence the QoS concept is associated with parameters to quantify the characteristics of the transfer of data between service access points of an OSI layer.

Mapping functions are required to map the QoS parameters of (N) layer to QoS parameters of (N-1) layer. The layer model of QoS for OSI is only concerned with those aspects of QoS requirements and the flows of QoS data across service boundaries and inside (N)-subsystems that are related to the operation of layer protocols.

More generally, for each component of the system, e.g. human user or the transport system, a set of QoS parameters must be defined [16]. Such parameters must be specified in a language understandable to the corresponding components. As an example, it is not acceptable to present to the human user the parameter jitter, since it is not meaningful to him/her. Consequently mapping functions are required to translate the user QoS parameters values to the provider ones. Thus the service provider can use such QoS parameters values, results of the translation, to reserve the resources required to support the requested QoS.

(2) *QoS negotiation and resource reservation:* The QoS negotiation activity permits to find a system configuration which might support the requested QoS. Thus a QoS negotiation is required with all components of the configuration. The aim of this activity is to obtain a commitment from the service provider to support the requested QoS. Thus each involved entity is asked about the level of QoS it is able to support. At the end of the negotiation, if the 'sum' of the levels of QoS of each entity corresponds to the QoS required by the user, each entity must reserve the corresponding resources to support its level of QoS. In the case that the negotiation fails, a notification, preferably indicating the failure cause, is sent to the user. The latter has the choice to try another QoS negotiation, or simply to abandon. In the case that the negotiation succeeds (each entity committed to support a specific level of QoS), each entity must dedicate resources to support the requested QoS. In other words, each entity must reserve a certain number of CPU slots and buffers to meet its commitment.

(3) *QoS monitoring and source policing:* Monitoring mechanisms permit to perform a continuous measurement of the QoS actually provided by the underlying system. They have mainly two tasks: (1) To detect and notify any QoS violation (notification task): When the measured value of a QoS parameter does not meet the agreed one, a notification indicating the violation, and preferably the cause, is sent to the QoS manager, and (2) to store information (collection task): a description of the information to be selected when monitored and a description of any computations to be performed of the retained information are required. QoS monitoring requires QoS measurements, measurement procedures and methods must be defined; only measurable parameters must be considered, points where measurements can be affected must be identified, the periodicity of measurements must be determined [23].

Source policing is defined [24] as the set of actions taken by the network to monitor and control users' traffic to guarantee the QoS for any source which keeps within the parameter values agreed upon during the establishment phase, and hence to protect the network resources from user misbehaviour (malicious or not) which can affect the QoS of active connections. An obvious action to take when a user goes beyond the negotiated frame rate, is to discard data to meet the negotiated value. For example, if the negotiated frame rate is 20 frames/s and the user is sending data at 30 frames/s, the network may discard 10 frames/s.

(4) *QoS adaptation:* Overload, e.g. network congestion, will be a common occurrence in communication systems and workstations [9] because of the stringent requirements of the new emerging services. Thus QoS adaptation is required to react to such occurrences. QoS adaptation activity must be able to exhibit graceful degradation reacting adaptively to changes in the environment. Indeed, it may be more desirable to degrade the quality of the affected service (violation of its agreed QoS) rather than to abort it.

(5) *QoS renegotiation:* A renegotiation may be initiated by the user or the underlying system (e.g.communication system). The user initiated renegotiation allows a user to request a better quality, e.g. a user asks for colour quality while the current quality is black&white, or to reduce his requirements from the service provider to reduce, for example, the cost of the current session. On the other hand, renegotiation initiated by the underlying system is generally due to lack of resources (e.g. network congestion) and aims to reduce the provided quality to avoid a service interruption.

(6) QoS accounting: QoS accounting concerns the determination of the cost relative to a service requested by a user. Since distributed multimedia applications provide a vast range of QoS, QoS accounting is a complex activity. It must be based at least on the requested QoS to limit the user's greediness, e.g. if the cost does not depend on the requested QoS, all users will ask for the best QoS.

(7) QoS termination: When the service is terminated, notifications are sent to all entities involved in the QoS provision relative to this service to free the reserved resources.

Toward a general QoS management architecture

We can generalize the QoS management architecture for remote access to MM database application (Figure 2) to more general MM applications. Given an application, the following steps are required to build a framework that supports QoS management functions such as QoS (re)negotiation: (1) to identify the entities involved in the QoS provision relative to this application, and (2) to define their characteristics and functionalities, such as the attributes and the methods supported. Our QoS management framework consists of a QoS manager and a collection of entities required to support the requested QoS. When a QoS manager receives a request from the user, it determines an optimal configuration, whenever possible, that should support the requested QoS and sends requests to each entity to reserve the necessary resources to support the required QoS. Each entity sends a response to the QoS manager. The response may be yes, no, or an alternate proposal. If all parties respond yes, the configuration is established and the QoS management activities of the active phase, such as QoS monitoring and adaptation, can start. If one or more entities respond no, the QoS manager rejects the user request. If any other proposal was sent by one of the entities, the QoS manager will try to find another configuration. Close cooperation between the QoS manager and network managers is required to optimally support the QoS management functions.

4. A GENERAL FRAMEWORK FOR QOS NEGOTIATION

Distributed MM applications provide a vast range of QoS. Hence a user must be able to specify a desired QoS depending on his needs, his end-system characteristics and his financial capacity. In other terms he must be able to negotiate the QoS with the service provider. Furthermore it is not acceptable that the negotiated QoS at the establishment phase applies for the whole session lifetime; a user must be able to decrease or to increase the QoS currently provided, e.g. to receive a less expensive service or a better quality. Consequently a QoS negotiation and renegotiation protocol is required. This protocol must be designed on an end-to-end basis and must not be restricted to the communication systems, as is the case for most existing QoS negotiation protocols [25, 26, 27, 28]. Based on the QoS management framework defined above, we present a general framework for QoS negotiation. For sake of clarity, we will instantiate the framework by considering the application of remote access to MM databases.

Facilities to support QoS (re)negotiation

Upon the receipt of the user request, e.g. user_QoS request, for a specific service with a desired QoS, the QoS manager makes use of the following facilities to answer the user request.

(1) The information facility permits to collect information pertinent for a given user request. Such information concerns QoS parameters relative to the entities involved in providing the QoS. A management information database (MIB) [29] is a candidate to store such information. Two types of information may be considered: (1) static information which does not depend on the

actual system load, e.g. maximum packet size supported by a given network or a compression scheme supported by an end-system, and (2) dynamic information which changes with the system state, e.g. available bandwidth.

The QoS manager communicates with the MIB(s) to get QoS information within the system using the *QoS_Information* service (confirmed).

In the context of our case study, the information collected concerns only the QoS related to the selected multimedia document, the database server and the transport user system. Such information is stored in the database server and concerns the document media-type, format, and size. Additional information are available such as the access time and copyright cost [16]. When the user asks to play a specific document, the client gets the required information from the database server.

(2) *The configuration selection facility* permits to determine an acceptable configuration that supports the requested QoS. Such a configuration is determined based on the information obtained from the information facility. Optimization schemes may be used to find the optimal configuration. Some of the tasks to be performed to construct the configuration are:

(a) Choice of compression schemes: The compression scheme to be used must satisfy the required delay and reliability.

(b) Choice of networks: Such a choice may be based on the cost, resources availability, etc.

(c) Choice of transport protocols: Such choice may be based on error control functions, performance, etc.

(d) Choice of resource locations: If the user's end-system does not have enough resources, e.g. it does not support the compression scheme with which the selected document is stored, then another machine with available resources may be used.

In the context of our case study, this facility permits to get the final QoS parameters values taking into account the user QoS, the local end-system characteristics and the QoS information obtained from the server using the information facility. However the algorithm used to find a configuration that supports the QoS is very simple [11], given that we are using mainly static QoS parameters.

(3) *The resources reservation facility* permits to gain a commitment from the components of the configuration, identified by the configuration selection facility, indicating their agreement to support the requested QoS. Each component must formally commit resources for this purpose. Mapping schemes to transform QoS parameters values to resources, e.g. CPU slots and buffers, are used to provide this facility.

The QoS manager asks the components of the selected configuration to reserve the resources to support the requested QoS using *QoS_reserve* service (confirmed).

Several solutions have been proposed to support guaranteed performance communication. They all adopt a connection-oriented and reservation-oriented approach [6, 7, 30, 31]. At the system level, several systems have been developed to support MM applications requirements such as TRDM [9], SUN OS 5.0 [32], real-time (RT) MACH [33], DASH [34] and extended CHORUS [35]. Most of them use earliest deadline [8] as the scheduling policy.

(4) *The data transfer facility* enables the control entities, e.g. QoS manager, to communicate with the different system components, e.g. server and MIB. Such a facility is used by almost all the negotiation facilities. In the context of our prototype we used remote procedure calls (RPC) to realize this facility.

(5) *The monitoring facility* enables to detect QoS violation because of resources shortage of one or more components involved in the QoS provision. When a QoS violation is detected an indi-

cation is sent to the QoS manager. Depending on the established policies, the session is aborted, the violation is ignored, or a renegotiation is initiated.

The QoS manager sends a QoS violation notification to the user using the *QoS_violation* service (non confirmed)

(6) *The renegotiation facility* supports a QoS renegotiation initiated by the user or by the underlying system (e.g.communication system). The user initiated renegotiation allows a user to request a better quality or to reduce his requirements, for example, to reduce the cost of the current session. On the other hand, the underlying system initiated renegotiation is generally due to lack of resources and aims to reduce the provided quality to avoid a service interruption. The renegotiation facility uses all the other facilities to provide its service.

The user can renegotiate, with the QoS manager, the current QoS at any moment during the session lifetime using *QoS_renegotiate* service (confirmed).

(7) *The termination facility* enables a user to terminate a session. All the resources reserved at the components of the configuration are deallocated.

At any moment the QoS manager can terminate the session using the *QoS_terminate* service (non confirmed).

States transitions for the QoS manager

Based on the facilities defined above, the QoS manager always goes through the states shown in Figure 4. The model shown in Figure 4 is a general and abstract model which may be refined depending on the system architecture and the management protocols used. A refinement of this model for the case of remote access to MM database application can be found in [11].

The transitions between the six states of Figure 4 are based on service primitives. Initially the QoS manager is in the *Idle* state. When the user asks for a specific service with a desired QoS (user_QoS_request), e.g. play a specific multimedia document, the QoS manager sends the QoS_information_request primitive(s) to the appropriate MIB(s) and moves to the *waiting to get information* state (transition T1).

When the QoS manager gets the information, related to QoS, from the MIB(s) (QoS_information_confirm) it enters the *selecting configuration* state (T2). At the end of a successful configuration selection the QoS manager asks for a reservation from the involved entities (QoS_resource_request), e.g. database server and video file servers, to support the QoS required by the selected configuration and moves to the *resource reservation* state (T3). In the case where no acceptable configuration is found within a specific interval, the QoS manager moves to the *idle* state (T5).

After receiving resource reservation commitment from the different entities (QoS_reserve_confirm), the QoS manager enters the *active* state (T4). In this state a renegotiation event may be initiated by the user, or the underlying system if a QoS violation occurs. In the case where one or more entities do not commit themselves to support the required QoS, the QoS manager reacts by a transition to the *idle* state (T6).

Depending on the policy used to deal with QoS violation, on receipt of some QoS violation (QoS_violation_indication) the QoS manager moves to *waiting for user response* if the "renegotiation" policy is used (T7), moves to *idle* if the "abort" policy is used (T8), or remains in the same state if the "ignore" policy is used (T9). The QoS manager reacts to the receipt of a renegotiation request from the user (renegotiate_QoS_indication) by a transition to *configuration selection* state (T10).

On the receipt of a *terminate_session_indication* () primitive from the user, in any state, the

QoS manager enters the *idle* state (T11). When the QoS manager is in *waiting to get information, selecting configuration, resource reservation* or *waiting for user response* states, a timer T is initialized and started. If no transition occurs from the current state before the timer expires, the QoS manager moves (T12, T13, T14) to the *idle* state or (T15) to the *active* state.

The parameters relative to each primitive, e.g. QoS_information request, have been defined for remote access to database application in [16, 11]

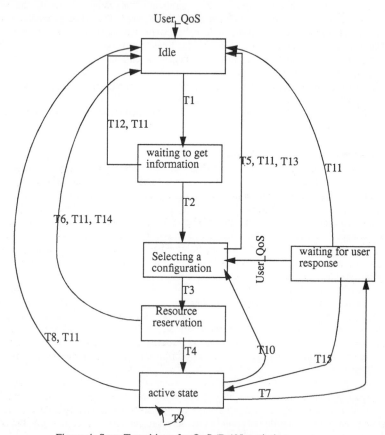

Figure 4. State Transitions for QoS (Re)Negotiation

5. CONCLUSION

The paper describes a remote access to MM database application. A QoS management architecture for this type of applications is defined. The QoS management activities are described and a QoS management architecture for MM applications is presented. A general framework for QoS negotiation is presented. An instantiation of this framework has been built for remote database access applications in order to show its feasibility by an example. Currently, we are working on elaborating the framework for QoS negotiation, particularly the information and configuration selection facilities. Thus we are developing a performance model that enables us to find an optimal configuration for a requested service. We also study the required MIB(s) and their distribution. More generally we are aiming at specifying QoS management functions such as negotiation, monitoring and adaptation, for remote access to MM database and other applications.

ACHNOWLEDGEMENT: We would like to thank R.Velthuys from Waterloo University for fruitful discussions on a first draft of the paper, and A.Ezust from University of Montreal for helping in the presentation of the last version of the paper.

REFERENCES

1. G.Dermler, T.Gutekunst, B.Plattner, E.Ostrowski, F.Ruge and M.Weber, Constructing a Distributed Multimedia Joint Viewing and Tele-Operation Service for Heterogeneous Workstation Environments, IEEE Workshop on Future Trends of Distributed Computing Systems, Lisboa, 93
2. T.Gutekunst and B.Plattner, Sharing Multimedia Among Heterogeneous Workstations, 2 Inter. Confer. on Broadband Islands, Greece, 93
3. T.Gutekunst, T.Schmidt, G.Schulze, J.Schweitzer and M.Weber, A Distributed Multimedia Joint Viewing and Tele-Operation Service for Heterogeneous Workstation Environments, Workshop on Distributed Multimedia Systems, Stuttgart, 93
4. IEEE Communications magazine, Special Issue: Multimedia communications, May 1992
5. A.Rowe and B.Smith, A continuous Media Player, 3rd International Workshop on Network and Operating System Support for Digital Audio and Video, San Diego 1992
6. D.Ferrari, A.Banerjea and H.Zhang, Network Support for Multimedia, Technical report 92-072, International Computer Science Institute, Berkeley, November 1992
7. C.Topolcic, Experimental Internet Stream Protocol, Version 2 (ST-II), Internet RFC 1190, 90
8. C.Liu and J.Layland, Scheduling Algorithms of Multiprogramming in a Hard Real-Time Environment, Journal ACM, Volume 20, Number 1, 1973
9. J.Hanko, E.Kuerner, D.Northcut and G.Wall, Workstation Support for Time-Critical Applications, Second International Workshop Heidelberg, Germany, November 1991
10. V.Rangan, H.Vin and S.Ramanathan, Designing an on-demand multimedia service, IEEE Communications Magazine, July 1992
11. A.Hafid, A.Bibal, G.Bochmann, T.Burdin, R.Dssouli, J.Gecsei, B.Kerherve and Q.Vu, On News-on-Demand Service Implementation, Technical Report, Université de Montréal, Montréal, Canada
12. P.Lougher and D.Sheperd, The Design of a Storage Server for Continuous Media, The Computer Journal, February 1993
13. L.Lamont, Component Interactions and Messaging System for Multimedia Synchronisation, Technical Report, MCRLab University of Ottawa, May 1994and Hypermedia Information
14. B.Metzler, I.Miloucheva and K.Rebensburg, Multimedia Communication Platform: Specification of the Broadband Transport Protocol XTPX, CIO, RACE Project 2060, 60/TUB/CIO/DS/A/002/b2, 92
15. D.le Gall, A video Compression Standard for Multimedia Applications, Communications of the ACM, April 1991

16. B.Kerherve, A.Vogel, G.Bochmann, R.Dssouli, J.Gecsei and A.Hafid, On Distributed Multimedia Presentational Applications: Functional and Computational Architecture and QoS negotiation, High Speed Networks Conference, Vancouver, Canada, August 1994

17. D.Feldmeir, A framework of Architectural Concepts for High Speed Communications Systems, IEEE JSAC, May 1993

18. G.Neufeld, M.Ito, M.Goldberg, M.McCutcheon and S.Ritchie, Parallel Host Interface For an ATM Network, IEEE Network Magazine, July 1993

19. Z.Haas, A Communication Architecture for High Speed Networking, IEEE Network Magazine, January 1991.

20. A.Krishnakumar, J.Knener and A.Shaw, HIPOD: An Architecture for High Speed Protocol Implementation, High Performance Networking IV, 1993

21. D.Towsley, Providing Quality of Service Packet Switched Networks, Joint Tutorial Papers of Performance 93 and Sigmetrics 93, Lecture Notes in Computer Science

22. ISO/IEC JTC1/SC21, Quality of Service Framework, Project JTC1.21.57, Working Draft #3

23. A.Hafid, J. de Meer and A.Rennoch, On JVTOS QoS Experiments, Technical Report, GMD-FOKUS, Berlin, Germany, 1994

24. CCITT I.311, B-ISDN General Network Aspects, 1990

25. ISO/IEC JTC1/SC6/WG4 N831, Multimedia Communication Platform: Specification of the Enhanced Broadband Transport Service, 1993

26. ISO/IEC JTC1/SC6, High Speed Transport Service Definition (HSTS), Preliminary Draft, 6 July, 1992

27. A.Danthine, Y.Baguette, G.Leduc and L.Leonard, The OSI 95 Connection-Mode Transport Service: The enhanced QoS, High Performance Networking IV, Liege, Belgium, 1993

28. I.Miloucheva, XTP and ST-II Protocol Facilities for Providing the QoS Parameters of Connection-Mode Transport Services, Research Note TUB-PRZ-W-1029, Berlin, 92

29. M.Rose, The Simple Book: An Introduction to Internet Management, Prentice Hall, 1994

30. A.Cramer, M.Farber, B.McKellar and R.Steinmetz, Experiences with the Heidelberg Multimedia Communication System: Multicast, Rate enforcement and Performance, High Networking Performance IV, 1993

31. T.LaPorta and M.Schwartz, The Multistream Protocol: A highly Flexible High Speed Transport Protocol, IEEE JSAC, 1993

32. S.Khanna, M.Sebree and J.Zolonowsky, Real-Time Scheduling in SUN OS 5.0, USENIX Winter Conference, SAN Francisco, January 1992

33. H.Tokuda, T.Nakajima and P.Rao, Real-Time Mach: Towards a Predictable Real-time Systems, USENIX Association Mach Workshop, October 1990

34. D.P.Anderson, Support for Continuous Media in the DASH Systems, Proceedings of the Tenth International Conference on Distributed Computing Systems, 1990

35. G.Coulson, G.Blair, B.Stefani, F.Horn and L.Hazard, Supporting the Real-Time Requirements of Continuous Media in Open Distributed Processing, Technical Report 1993, Lancaster University, UK

36. CCITT recommendation G.106, 1984

27

Integration of Performance Measurement and Modeling for Open Distributed Processing

Richard Friedrich, Joseph Martinka, Tracy Sienknecht, Steve Saunders

Hewlett-Packard Company, HP Laboratories, Palo Alto, California 94304 USA
{richf, martinka, tracy, saunders}@hpl.hp.com

Successful deployment of open distributed processing requires integrated performance management facilities. This paper describes measurement and modeling technologies that provide quality of service (QoS) measures and projections for distributed applications. The vital role of performance instrumentation and modeling is applied to the Reference Model for Open Distributed Processing. We discuss an architecture and prototype for an efficient measurement infrastructure for heterogeneous distributed environments. We present an application model useful for application design, deployment and capacity planning. We demonstrate that integrated measurement and modeling yields the QoS measures that guide application deployment and increase management capability.

Keyword Codes: C.4, I.6.3, C.2.4
Keywords: Performance of systems; Simulation; Distributed systems

1 INTRODUCTION

Open Distributed Processing (ODP) offers advantages in performance, availability and resource sharing. However, managing applications in a distributed environment is a complex task and the lack of integrated performance management facilities is an impediment to large-scale deployment. The performance management tasks of application design, deployment, bottleneck analysis, and capacity planning require the collection, modeling and analysis of workload data. Previous techniques used to design and manage high performance, monolithic applications are inadequate for ODP. A systematic approach based on performance engineering is required, supplemented by stronger support of performance metrics in the Reference Model for ODP (RM-ODP) [11].

This paper describes the architecture of an efficient, scalable Distributed Measurement System (DMS). The DMS is a software-based measurement infrastructure that defines standard performance metrics, instrumentation and interfaces. DMS provides correlated performance metrics across objects and their channels, integrates disparate measurement interfaces from a node's nucleus object (operating system) and channels (networking), and efficiently transports collected data. The architecture is realized in an object-oriented prototype based on the OSF Distributed Computing Environment (DCE).

Strong interdependence exists between measurement and modeling. The DMS architecture was designed in concert with distributed application modeling requirements. We demonstrate the benefit of this integrated methodology on the QoS specification and measurement of a distributed application. The models ease application deployment by estimating expected resource demands and QoS for various designs and network topologies. DMS enables performance management by measuring application QoS and reporting exceptions.

Section 2 motivates our research and summarizes related research. We describe how performance management maps to the ODP framework in section 3. The DMS architecture and prototype are discussed in section 4. Derived metrics are provided to performance models of two distributed applications in section 5. Conclusions and future work are summarized in the last section.

2   MOTIVATION

The complexity of distributed applications bewilder application designers and system managers. We illustrate this complexity in Figure 1 with the RPC-flow diagram of a simple query transaction of a client-server application using a distributed transaction processing monitor. Each arrow indicates an RPC request/response. The thick arrows indicate the logical RPC operation from the application developer's perspective and the thin arrows represent supporting RPCs necessary for explicit binding from a transaction monitor. Note that RPCs are nested such that the primary RPC will not return until the secondary RPC is complete (e.g., RPC 13 and its nested RPC 14), thus further complicating analysis. Arrows that cross a dotted line indicate that a network communication occurs with the potential of adding tens or hundreds of milliseconds to the transaction's latency. *How is the user's QoS estimated or measured in this complex and dynamic environment?*

Our research focuses on measurement and modeling solutions that decrease the risk of ODP application deployment. The application of these techniques in a user environment results in *performance management*. Specifically, our objective is to ensure realizable application deployment by:

- estimating the cost of an object's invocation as a function of resource consumption, target hardware capacity, and channel latency and contention, prior to deployment;
- creating an efficient, pervasive measurement infrastructure that collects, transports, correlates and analyzes the performance metrics of monitored applications;
- providing effective performance management that supplements the ODP architecture with abstractions of performance metrics and measurement interfaces, and integrates measurements and models to address system management *what-if* questions before initiating an expensive, risky course of action.

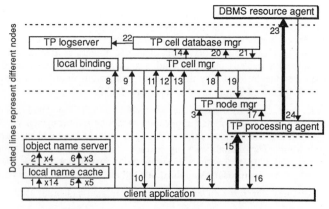

Figure 1   RPC flow for a distributed TP monitor database query.

2.1 Related research

Our work extends previous research in measurement and modeling and integrates them to provide distributed application performance management. Several trends exist in the performance measurement of distributed systems. A hardware monitor approach [20] has the advantage of minimal perturbation effects on the system under test. However, it cannot support application measurements in an enterprise environment and deployment cost is excessive. A second approach is a software monitor [17] that collects application and distribution infrastructure metrics, but requires careful design to minimize perturbation of the system under test. A variation is hybrid software and hardware monitors [5][12]. These suffer from the enterprise deployment issues of hardware monitors. A growing trend is to use the Internet SNMP protocols [3] for more than network management. Software subsystems from operating systems to databases are implementing SNMP access for distributed management. We believe that SNMP's polling nature makes it unscalable for large distributed application environments and that its trap function can flood the network with data. Our research uses a software monitor to provide an efficient, scalable infrastructure for operational enterprise environments.

Researchers and practitioners have recognized the critical need for model and instrumentation integration [19]. There has been work in developing requirements and standards for performance measurement and management [1][3][8] but others note the current paucity of instrumentation and tools in distributed client/server systems [4][16][21]. With the increasing role that middleware components play in ODP applications, Software Performance Engineering (SPE) methods [19] are dependent on pervasive performance instrumentation. Advocates of performance modeling early in the design cycle support the notion of decompositional techniques [21]. Impressive efforts by ESPRIT to create this performance design environment have been mounted for ANSAware systems [10]. Franken and Haverkort [7] describe a performance model-based *Performability Manager* that uses SPN techniques to analyze QoS in a distributed environment, guaranteeing user-requested QoS and reliability. They demonstrate its use in an ANSAware-based environment for performability management, but recognize the complexity involved in the mapping of SPN components to alternative configurations.

3   PERFORMANCE MANAGEMENT IN ODP

We map the aspects of performance engineering as applied to the RM-ODP framework in Table 1. ODP transparencies are subject to application constraints specified in the enterprise viewpoint. Transparencies mask the difficult programming decisions about distribution semantics yet their performance determines if the application can achieve the enterprise QoS requirements.

We believe that ODP objects that manage the enterprise viewpoint's QoS requirements are essential to knit together the dynamic needs of an application across the RM-ODP viewpoints, and negotiate the channel-object relationships as necessary. This need is highlighted and argued forcefully for multimedia applications by Campbell and Fédaoui [2][6]. Indeed, we believe that the QoS architecture described in [2] is extensible to on-line transaction processing (OLTP). Their focus is on the multimedia *streams* interactions between objects in the computational viewpoint of the RM-ODP and specifies a QoS manager object using its own channel. Our emphasis is similar but applied to the *operational* interactions between objects using a client-server model. Our research on performance methodologies has focused on OLTP and its version of QoS, the service level agreement.

Table 1   Performance Engineering Requirements for the RM-ODP.

RM-ODP Viewpoint	Applicable Aspects of Performance Engineering
Enterprise	Establish end-to-end user QoS requirements. These specifications guide design decisions and dynamic binding and are compared to measurements or performance models.
Information	The information schemas of this viewpoint define performance metrics that describe application behavior. This schema describes logical groupings of metrics that allows user-level QoS analysis and isolation.
Computational	An object's Interface Description Language (IDL) serves to map object functions to other objects, independent from distributional concerns. Since objects interact through their IDL, this enables performance management objects to collect and aggregate performance data measured at an object's interface. The *environment contract* for each object is specified here which includes QoS constraints. These object contracts must satisfy mappings to the engineering viewpoint.
Engineering	This viewpoint is expressed in terms of nodes and channel objects that are mapped from the computational viewpoint using transparencies. These mappings require QoS agreements comprised of performance measures that are dynamic and change in real-time. A measurement architecture is needed to implement transparencies efficiently and provide performance knowledge of an application's distributed domain.
Technology	The distribution infrastructure must have a low overhead, pervasive performance instrumentation facility that is complete and provides metrics to monitor and manage QoS.

4   DISTRIBUTED MEASUREMENT SYSTEM

This section discusses the DMS architecture and prototype implementation.

4.1 DMS architecture

DMS is a software architecture that facilitates the routine measurement of performance metrics on distributed objects. Furthermore, it provides a measurement infrastructure that collects and transports data independent of the underlying distribution mechanism.

DMS provides information for application design, deployment, QoS monitoring, load balancing and capacity planning. DMS specifies a common set of performance metrics and instrumentation to ensure consistent collection and reporting across heterogeneous nodes [8]. It defines standard application programming interfaces (APIs) to ensure pervasive support for performance metric measurement. The DMS objects shown in Figure 2 are capable of monitoring distributed technologies such as DCE and CORBA. DMS also supports integrating performance measurement interfaces from external sources, such as the host operating system and networking, into a single unified measurement infrastructure. This results in a seamless, integrated view of the behavior of a distributed application in a heterogeneous environment.

Instrumentation is specialized software incorporated into programs or libraries to calculate performance metrics. DMS *sensor* objects are instances of a general performance metric for a specific function. The sensors calculate and buffer primitive statistical quantities such as counts, summations, or interval times, but defer the computation of more complex statistics such as moments to higher-level DMS objects. Individual sensors are uniquely named within the enterprise with a string name or an *object identifier* (OID).

A sensor's collection granularity, and thus overhead, is controlled by specifying an *information level*. The information level controls the statistical detail of the collected data. The lowest information level corresponds to instantiated but inactive sensors and incurs nearly zero overhead. The *threshold* information level is the lowest overhead setting for active sensors and is used to monitor QoS. This level reports data only when user specified threshold values are exceeded. Higher levels provide moments and histograms for analyzing distributions.

Three categories of *standard sensors* are defined: *timers* provide interval times, *counters* record the number of occurrences or state of an event, and *composers* return an arbitrary structure to higher level DMS objects.

Application objects may define *custom sensors* that supplement the standard sensors by recording application specific behavior. These sensors support extensible measurement of business or organizational units of work that are not available in the distribution infrastructure.

An *observer* object resides within each instrumented process and provides a sensor control point. It minimizes in-line overhead by allowing the sensors to defer some computation and off-loads sensors of the need to manage and transmit data. The observer exports a sensor access and control interface named the *Performance Measurement Interface* (PMI). The observer supports sensor registration and unregistration and transmits intervalized data generated by sensors within the local process address space to the collector object. Multiple sensors are transferred simultaneously to minimize the overhead.

A *collector* is a network-node level object that controls sensors via the PMI and performs local node data management. There is one collector per node. It provides network transparent sensor access and control to higher levels of the architecture using the *Collector Measurement Interface* (CMI). The collectors accumulate sensor data from all observers on the node using the *Collector Data Interface* (CDI). The observer summarizes sensor data and periodically "pushes" it to the collector using the CDI. The CDI eliminates the need for polling of sensor data by providing an asynchronous data transport channel. The collector provides a *sensor registry* that contains the state of all registered sensors on the node.

An *analyzer* object analyzes the data gathered by collectors within a specific domain. It applies statistical routines to compute the distributional characteristics of the collected data,

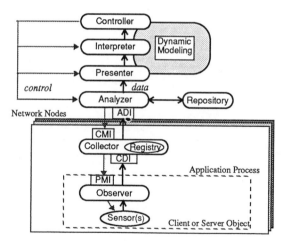

Figure 2 Distributed Measurement System Architecture.

correlates data from application elements residing in different processes and on different nodes, and prepares data for expert system or human analysis. Simple operations, such as counting events, are most efficiently provided by the sensor, but an analyzer performs complex statistical calculations such as computing moments. One analyzer can request subsets of sensor data from multiple collectors; multiple analyzers can access a specific collector. The analyzer provides data to a presentation service for visualization or an interpretation service for autonomous agent operation. An analyzer provides the basis for dynamic end-to-end QoS negotiation and monitoring.

The CMI is a network communication interface exported by a collector. An analyzer uses this interface to communicate with collectors anywhere in the network to request sensor data and specify sensor configurations. Multiple sensor values are batched by the collector and returned in bulk using the *Analyzer Data Interface* (ADI). This minimizes the number of network packets between the collector and analyzer. The collector periodically "pushes" data to the ADI which eliminates the overhead of an analyzer polling for its next update. This technique improves scalability and reduces overhead. The collector must use self-describing data techniques to facilitate interpretation of data formats since collector and analyzer may reside on heterogeneous nodes.

Additional DMS objects complement the measurement infrastructure objects. The *presenter* object supports the human user interface and interactively produces reports, displays, and alarms. It presents a logically integrated view that resembles that of an application executing on a single, centralized node. Visualization techniques are necessary to provide efficient on-line analysis of the collected data. The *interpreter* object is an intelligent entity, human or expert system, that evaluates and identifies complex relationships between the data and the environment and then draws meaningful interpretations. It relies on *dynamic models* to estimate and compare measured data with QoS requirements. The *controller* object provides control of the application and system parameters and states. The controller decides to set or change system parameters or configurations based on the interpretation of monitored performance data. Together with the measurement system and dynamic models, the controller provides the closed feedback loop necessary for providing *self-adapting* systems that manage themselves with less human intervention.

Collecting performance data must not significantly impact the applications and systems under measurement. We have used optimization techniques to minimize network bandwidth utilization and to improve scalability. We used thresholds that report data only when QoS levels are violated, sensors that report summarized data periodically but do not report unchanged data, and bulk transport of aggregated sensor data at the observer and collector level.

4.2 DMS prototype implementation

The DMS prototype provided a research tool to evaluate functional partitioning of the architecture. It implemented the sensor, observer, collector, and presenter objects, and was based on OSF DCE [15] because of availability and commercial interest. The use of DCE as the prototype's distribution infrastructure impacts only the implementation of sensors and observers but not the interface.

Sensors were developed and placed within the DCE runtime library (RTL). A copy of the RTL is linked with all clients and servers. Thus standard sensors are available without modifying client and server application source. The observer was implemented within the RTL. The collector was implemented as a daemon process that communicated with all the observers on the node via IPC using the PMI and CDI interfaces. The analyzer was implemented as a daemon and it communicated with all the collectors in the network via RPC.

Figure 3 illustrates the prototype run time environment. The RTL supports a pool of application call threads, represented by vertical arrows, that execute within a single application address space. Curved arrows represent data transfer paths. The RTL and application RPC requests are executed on an available thread. Sensors are reentrant and use locks to access the global sensor data since they execute in a threaded environment. The analyzer periodically retrieves the data temporarily held for it from each collector. DCE RPC marshalling/ unmarshalling handles data translation.

We integrated DMS into a network management infrastructure provided by current SNMP-based tools. These tools provide a possible foundation for QoS monitoring and we wanted to demonstrate coexistence. In Figure 4 we display the *Performance Management Information Screen* (PMIS) that includes DMS integration into the Hewlett-Packard OpenView network management framework [14].

Effectively displaying an application's performance is necessary after collecting, correlating and analyzing its data. The presentation of performance management information is shown in Figure 4 for the application *PhoneDB*. The row of icons at the top of this Motif-based screen represents the integrated view of the application's performance metrics. The data available to this display is independent of the location of the application's objects. Selecting one of these icons displays a graph of the corresponding performance metrics.

DMS objects measure, map, and isolate any violations of the application's specified QoS. During normal operation, a particular ODP node or channel becomes visible to the administrator only if a QoS service level exception occurs. The inset of Figure 4 graphs a client and server object's response time. It illustrates that client perceived performance has a server latency that is only one component of its response time (solid line in graph). Channel latency and binding, stub, and protocol objects that participate in the distribution transparencies contribute an additional response time component (difference between dashed and solid lines). In this example, network loading markedly degrades the client's response time but does not affect server response.

The lower portion of the PMIS screen shows two nodes, *boom* and *winch*, with their corresponding performance metrics and node-centric tools. Other node based tools, such as SNMP or PerfView (a Hewlett-Packard performance monitor), are used as necessary for further analysis.

We learned valuable architectural and implementation lessons from the prototype. First,

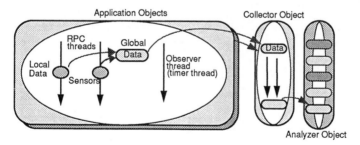

Figure 3 DMS implementation run time environment.

measurement infrastructure activities that have bursty behavior, especially sensor registration and data transportation, require bulk functions in the CDI and ADI interfaces to minimize the use of expensive communication mechanisms. Second, obtaining timestamps is an expensive operation if using standard library functions such as *gettimeofday*. We were motivated to create a custom, low overhead timer mechanism, portable to other operating systems. Finally, the viability of the DMS architecture depends on overhead incurred and channel bandwidth utilized. Our prototype's primary focus was on flexibility and adaptability, yet the overhead of the prototype was negligible (on the order of a few percent).

5   MODELING AND MEASUREMENT INTEGRATION

This section demonstrates the benefit of modeling distributed applications with data provided by DMS measurements.

Modeling eases the design, deployment, and management of ODP applications. Models provide the basis for evaluating the partitioning of application functionality, upgrading performance, developing new designs, and planning capacity. Capacity planning, for example, requires modeling of various application environments and network topologies since few large distributed environments are built solely to benchmark application design alternatives. Furthermore, the dynamic nature of ODP applications makes transparency mapping of location, migration, resource, and transaction semantics impossible to predict and manage without modeling techniques. Models answer *what-if* questions regarding load balancing, capacity planning, or QoS violation causes.

DMS plays a crucial role in modeling by providing data for workload characterization, and

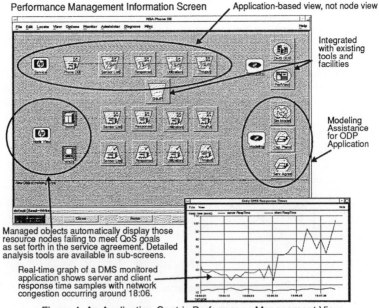

Figure 4   An Application-Centric Performance Management View.

model construction, parameterization, and validation. Models have little value without effective parameterization based on instrumentation. As further illustration of the synergy between measurement and modeling, the placement of DMS sensors was driven by modeling requirements.

5.1 Model description

Based on the complexity of distributed systems and the uncertainty in treating them analytically we approached this research using a general discrete simulation engine. We used SES/Workbench from Scientific and Engineering Software, Inc. [18]. Our prototype model was designed to meet the functional modeling requirements of client/server applications in an OLTP environment [13]. The model was designed to satisfy requirements for flexible topology specifications, nested and asynchronous RPC's, statistical model termination, and a network abstraction with realistic latencies. The simulation model is based on a resource-centered event paradigm that lends significant flexibility to the specification of the model. A simple ASCII configuration file specifies a large variety of distributed applications and topologies. This file input includes number and type of compute nodes, application transactions, users, networks, routers and routing technology. Nested and asynchronous RPCs were accommodated.

5.2 Benefits of integrated measurement and modeling

We illustrate the benefit of modeling and measurement by describing their use for two distributed applications.

PhoneDB is a client-server application that provides a database of telephone numbers for a user community. The *PhoneDB* server supports several remote functions that includes adding database entries, deleting entries, and searching for an entry using either a regular expression or a binary search algorithm.

We used DMS to measure application transactions using no-load single class techniques. This provided an estimate of the service time for both classes of transactions: regular expression and binary search (77 msec and 14.5 msec respectively). The network delay was measured at 1 msec. The client contribution to the service time was 3 msec. We assumed one packet exchange per RPC based on our knowledge of the application. The CPU times were converted to instruction pathlength using an assumed MIPs rate for the nodes on which this test was run.

Clients in the sample application were simulated to initiate transactions with a uniform inter-arrival time distribution between 0.25 and 0.75 seconds with a mean of 0.5 seconds, yielding an average arrival rate of 2 TPS. We modeled this application for a combined transaction rate load that varied from 4 transactions to 36 transactions per second. The model simulated the contention at the server for the two transaction types and generated expected mean service times for each transaction type. Up to the knee of the response time curve (around 10 TPS) the model results agreed to within 10% of the actual system measured with DMS. In Figure 5 we plot the estimate of two workload classes' average transaction response time and DMS measurements as a function of increasing load.

Using this validated model provides the benefit of estimating QoS applications and ODP topologies while in the design phase. For example, we used the model to predict the impact on QoS for a deployment environment that placed the PhoneDB server object on a node 1000 miles from the client object [13]. We also used it to demonstrate the improvement in QoS provided by upgrading the Phone DB server node to a 2-way multiprocessor. Furthermore, the model improves performance management effectiveness by estimating QoS impacts for load balancing and capacity planning before a system manager decides on a course of action. An extension to the model and the addition of a QoS manager mechanism in the distribution infrastructure would provide dynamic QoS negotiations among objects, and quantitative evaluation before binding client and server objects.

In the second use of this model, we applied it to an OLTP application built on a DCE-based middleware product that provides transactional-RPC semantics. A distributed order processing application, *Telshop,* was measured using event-tracing and then modeled. In Figure 6 we plot the modeled and measured results of a transaction consisting of five read queries to an inventory database for a range of workload intensity. The agreement with the measured results was within the normal modeling goals of 15% accuracy.

In summary, the integration of measurement and modeling results in two major benefits. First, the validated models provide insight into application resource consumption and expected QoS behavior prior to full-scale deployment. Second, measurements supply the data necessary to comprehend dynamic application behavior, manage QoS, and parameterize models.

6   CONCLUSIONS AND CONTRIBUTIONS

This paper highlighted the benefits of integrating modeling and measurement for application design, deployment and management in ODP.

RM-ODP provides a framework that requires extension to address the performance engineering problems of building and managing distributed applications. We described a distributed measurement architecture that provides an integrated view of application resource consumption across heterogeneous nodes and network channels that node-based tools cannot. DMS provides an extensible architecture for integrating disparate performance measurement interfaces from operating system and networking software with the distribution infrastructure. It provides efficient mechanisms for controlling, collecting and transporting performance data. We have used techniques to minimize the amount of processing and network bandwidth required.

Distributed client/server application models are crucial to ensure efficient design, deployment, and scalability. RM-ODP transparencies need modeling to help guide the binding, replication, location, and migration activities specified by the application's environmental contracts. The necessary modeling technologies are not generally available. We developed a simulation model to address these requirements and validated it with the DMS prototype.

Many problems remain in providing performance management of large-scale ODP

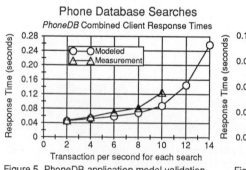

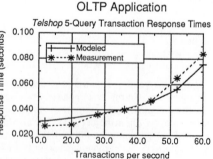

Figure 5 PhoneDB application model validation.   Figure 6 OLTP application model validation.

applications. A few of the most important include:

• Include standardized performance instrumentation in the RM-ODP and implement the standard interfaces proposed by the OSF DCE community [9].
• Provide a methodology to decompose the activities of middleware dependent applications so that a compound object's behavior is more easily characterized by simpler objects, preferably through the RM-ODP language specifications.
• Extend the models to include QoS negotiation for dynamic control of object selection based on service requested, service pricing, and current service QoS levels.

Acknowledgments

We wish to acknowledge the consistent support of HP's management. We are grateful to the referees who provided constructive feedback. We are indebted to the HP-Chelmsford development laboratory for providing access to the OSF DCE RTL source and development environment. We also thank Muthusamy Chelliah and Aravindan Ranganathan for contributions to the design and development of the DMS prototype.

References

1 ANSA, *ANSA Architecture Report: Monitoring in Distributed Systems*, Report Number AR.010.00, Feb 1993.
2 A. Campbell, et al, *Resource Management in Multimedia Communication Stacks*, Telecommunications, 1993, IEEE Conference Publication 371, pp 287-295.
3 Jeffery Case, et al, *A Simple Network Management Protocol (SNMP)*, Internet RFC 1157, May 1990.
4 Peter Dauphin, et al, *ZM4/Simple: A General Approach to Performance Measurement and Evaluation of Distributed Systems,* Readings in Distributed Computing Systems, IEEE Computer Society Press, Los Alamitos, CA, 1994, pp 288-309.
5 Oliver Endris, et al, *NETMON-II a Monitoring Tool for Distributed and Multiprocessor Systems*, Performance Evaluation 12 (1991) 191-202, North Holland.
6 Linda Fédaoui, Wassim Tawbi, Eric Horlait, *Distributed Multimedia Systems Quality of Service in ODP Framework of Abstraction: A First Study*, 2nd International Conference on Open Distributed Processing '93, Elsevier Science B.V. (North Holland), pp 265-273.
7 Leonard Franken and Boudewigm Haverkort, *The Performability Manager*, IEEE Network, Jan/Feb 1994, pp 24-32.
8 Richard Friedrich, *The Requirements for the Performance Instrumentation of the DCE RPC and CDS Services*, OSF DCE RFC 32.0, June 1993.
9 Richard Friedrich, Steve Saunders and Dave Bachmann, *Standardized Performance Instrumentation and Interface Specification for Monitoring DCE Based Applications*, OSF DCE RFC 33.0, November 1994.
10 Peter Hughes and Dominique Potier, *The Integrated Modelling Support Environment*, Esprit 89 Conference Proceedings, Document IMSE R-1.2-4, STC plc and Thomson CSF, 1989.
11 ISO/IEC JTC1/SC21/WG7 N885, *Reference Model for Open Distributed Processing Part 1*, November 1993.
12 Frank Lange, et al, *JEWEL: Design and Implementation of a Distributed Measurement System*, IEEE Transactions on Parallel and Distributed Systems, Volume 3 Number 6, November 1992, pp 657-671.
13 Joseph Martinka, *Requirements for Client/Server Application Performance Modeling- An Implementation using Discrete Event Simulation*, submitted for publication.
14 *HP Open View Distributed Management Platform - Administrator's Reference*, Manual J2322-90003, Hewlett-Packard, Dec 1992.
15 Open Software Foundation, *Introduction to OSF DCE*, Cambridge, MA, USA, 1992.

16 Jerome A. Rolia, *Distributed Application Performance Metrics and Management*, Proceedings from 2nd International Conference on Open Distributed Processing '93, Elsevier Science B.V. (North Holland), pp 235-246.

17 Hemant Rotithor, *Embedded Instrumentation for Evaluating Task Sharing Performance in a Distributed Computing System*, IEEE Transactions on Instrumentation and Measurement, Volume 41 Number 2, April 1992, pp 316-321.

18 *SES/workbench User's Manual*, Scientific and Engineering Software, Inc., Austin, Texas, 1992.

19 Connie Smith, *Software Performance Engineering*, in the Tutorial Proceedings of PERFORMANCE '93, Rome, Italy, September 1993.

20 Ram Sudama, et al, *Experiences of Designing a Sophisticated Network Monitor*, Software Practices and Experience, Volume 20 (6), June 1990, pp 555-570.

21 Vidar Vetland, *Measurement-Based Composite Computational Work Modelling of Software*, Ph.D. Thesis, Norwegian Institute of Technology, University of Trondhem, 1993.

28

A Quality of Service Abstraction Tool for Advanced Distributed Applications

A. Schill, C. Mittasch, T. Hutschenreuther, F. Wildenhain

Dresden University of Technology, Faculty of Computer Science, 01062 Dresden, Germany
e-Mail: [schill|chris|tino|wildhai]@ibc.inf.tu-dresden.de

Abstract

High performance communication protocols are being used in our lab as an experimental base for advanced distributed applications with a special focus on quality of service (QoS) characteristics. During our recent work, we recognized that quality of service abstractions would prove very useful for facilitating distributed application development. As a first base, we present a ma-nagement tool for inspecting and controlling quality-of-service related characteristics on top of XTPX (eXpress Transfer Protocol Extended) and TCP/IP. Based on a classification of QoS requirements of various multimedia data streams,the design and implementation of a higher-level QoS manager tool is described then. We finally illustrate how such an approach will be integra-ted with traditional client/server platforms such as the OSF Distributed Computing Environment.

Keyword Codes: C.2, C.2.3, C.2.4
Keywords: Computer Communication Networks, Network Operations, Distributed Systems

1. Introduction

High performance transport systems have to support flexible QoS characteristics in order to enable advanced distributed applications, including multimedia communication. XTPX /MIB94/ as an extension of XTP /STR92/ allows the specification, usage, and monitoring of such QoS related parameters. It is therefore being used as a base for experimenting with high performance communication in our lab. TCP/IP is also used in parallel.

However, during our initial experimentation and analysis, we also identified several advanced requirements. These are mainly concerned with network management support, with the level of abstraction of QoS specifications, and with the integration of high performance communication facilities into existing distributed systems frameworks:

- *Management support:* Based on the existing management specification for XTPX /MIT94/, advanced runtime and tool support for QoS-based management is required. An appropriate management protocol with additional management policies is desirable. This will also enable dedicated QoS management applications or at least a QoS supervision in case of TCP/IP.

- *QoS specification:* The level of QoS specification of existing protocols is relatively low; it is required that applications directly specify parameters such as data rate, jitter, error rate etc. It is desirable to introduce facilities for higher-level, application-related QoS specifications. The task of an appropriate support tool is to map these specifications onto low-level QoS parameters automatically while providing sufficient flexibility, too.

• *Framework integration:* Distributed applications are widely supported by standardized distributed systems frameworks with services such as remote procedure call (RPC), naming, and security. It will provide significant benefits to integrate QoS-related high performance communication facilities with the services of such distributed platforms.

The paper addresses these requirements by the following contributions: After a brief discussion of related work in section 2, section 3 presents the architecture and implementation of a new management framework for QoS support. Advanced concepts for a semi-automatic threshold management as a base for QoS related control mechanisms are also discussed. Section 4 first presents an analysis of distributed multimedia application requirements concerning QoS. The relationship of the most important communication facilities and media encoding standards with QoS parameters is discussed furthermore. Based on the results, a higher level QoS management tool is described. It supports an abstract, application-related QoS specification and maps it onto concrete QoS parameters. Section 5 illustrates how these concepts can be integrated into a distributed systems architecture such as the OSF Distributed Computing Environment (DCE) /OSF94, SCM93/. Finally, section 6 concludes with an outlook to future work.

2. Related Work

Similar QoS management approaches and distributed multimedia systems have been developed recently. The *Cinema* architecture /BDH93/ provides programming abstractions for distributed multimedia applications with explicit QoS mapping functionality. Abstract components of an application are partitioned into more detailed components that are finally mapped onto basic resources (such as CPU, communication bandwith or buffer storage). Future research within Cinema will address the details of this mapping functionality. Within the Touring Machine Project /TMS93/, an infrastructure for advanced multimedia communication is developed, too. Abstract ports and connections enable the representation of point-to-point and point-to-multipoint interactions for audio, video and other data streams. Connections are embedded in logical sessions with an explicit session management component. An effective abstraction for multimedia programming is offered, based on a high-level API. However, details of QoS management are not addressed yet.

Such QoS issues are addressed by /VOG94/ to a QoS negotiation approach, for example. An extended QoS architecture that also addresses flow control of media streams is presented in /CCH94/; it is similar to our approach but is more oriented towards the transport layer. /MBA94/ discusses a QoS monitoring approach top identify QoS problems dynamically, similar to the threshold management facility discussed below. In /ADB94/, an extension of the ANSA architecture with continuous media communication facilities is presented. Based on multimedia device and stream abstractions, distributed multimedia applications can be configured. Moreover, media synchronization requirements can be specified explicitly in a script language. An integration with OSF DCE has also been outlined. Considering these and other related approaches discussed below, we present our QoS abstraction facilities in the following sections.

3. Management Architecture

Our design and implementation of a QoS-related management architecture is based on the following requirements: (1) It shall be possible to analyze and modify QoS parameters of the transport protocol in a flexible, decentralized way based on managed object (MO) specifications. (2) Management shall be supported by a graphical interface, and MOs shall be grouped into logical categories. (3) The implementation shall be based on SNMP (Simple

Network Management Protocol) and on public domain components. (4) A major focus shall be put on layer management and on mid-term management tasks.

General model:
The approach is based on the traditional SNMP interaction model /CFS90/. A management app-lication on the manager's site observes, analyzes and controls several management agents representing protocol instances. Each protocol instance comprises a MIB that is specified in a standardized way and is mapped onto internal protocol parameters by kernel-level access functions. The interaction between a manager and an agent is based on SNMP Get, Next, and Set requests for accessing single MOs, for reading groups of MOs, and for modifying MOs, respectively. Moreover, it is possible for an agent to issue a trap for immediately notifying a manager of a significant status change concerning MO characteristics.

Management architecture and implementation:
Based on this management model, we designed and implemented a management architecture for the XTPX protocol (see fig. 1) and also partially applied it to a TCP/IP environment. It specifically addresses layer-3 and -4 management. On the site of the manager, a MIB catalog has been defined in ASN.1. It comprises the major MOs for configuration and performance management. The specification has been transformed into an internal representation using the MOSY compiler of the ISO Development Environment (ISODE).

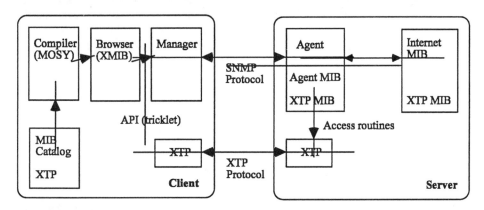

Figure 1: Management architecture

The manager provides access to the remote agents based on the existing MO specifications. It has been implemented by using the tricklet v1.4 tool of Delft University. In particular, it offers an API that has been used for integrating a MIB browser. This XMIB browser is based on the xmib1.0b implementation of Hebrew University in Jerusalem. The agent implementation uses snmp1.5 of Carnegie Mellon University. It accesses a uniform agent MIB that is eventually being mapped onto the internal XTPX MIB (or the simpler TCP/IP MIB) via access routines.

All components have been integrated and have been ported onto the OSF/1 operating system. The relevant MOs have been defined and have been integrated into the architecture. Examples of typical MOs according to the management areas of layer 4 of XTPX are:

Configuration management:
- ConfigMaximumSizeReceiveWindow (receiver's default window size)

- ConfigReceivingBufferSize (receiver's buffer size)
- ConfigSendingBufferSize (sender's buffer size)
- ConfigMaximumNumberCNTLRetries (retries until acknowledgement is delivered)
...

Performance management:
- PerfConnectionsAttempt (number of attempted connection establishments)
- PerfConnectionsEstablished (number of successfully established connections)
- PerfNumberOfSentPackets
- PerfNumberOfReceivedPackets
...

Some of these parameters are for monitoring only (such as the number of sent packets), and can be used for long-term decisions concerning the adaptation of QoS related parameters. Others can be directly adapted (such as the buffer sizes) and can therefore be used to implement QoS management strategies (for example, addressing jitter control features).

Support for the MIB-II specification has been enabled by defining various MIB-II MOs at the agent site. MIB-II support is already being provided by the manager and the browser. The manager can also be configured in a flexible way by adapting the timeout interval and retry counter for agent interactions. The tools initially did not support write access to MOs. Therefore, several extensions for enabling the set routines were made. Moreover, the Next operation was also implemented by supporting flexible browsing of tables within the manager.

In summary, the approach enables a flexible interactive management of XTPX. It provides a comfortable graphical interface with an explicit MO tree representation, enables standardized access to potentially heterogeneous agents via SNMP, and is based on highly portable public domain tools. In the case of TCP/IP, it is simpler due to the lack of specific QoS related parameters but still provides a base for the threshold management as discussed below.

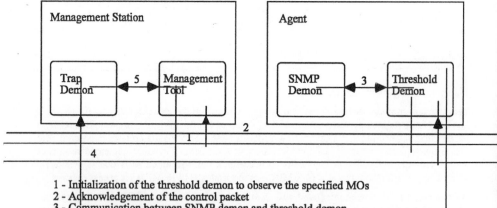

1 - Initialization of the threshold demon to observe the specified MOs
2 - Acknowledgement of the control packet
3 - Communication between SNMP demon and threshold demon
4 - SNMP trap PDUs
5 - Communication between management tool and trap demon

Figure 2: Architecture of threshold management

Extension: Threshold management:

A recent extension includes the definition of a management API at the manager's site, as well as the design and implementation of management applications. In particular, dedicated threshold management is supported based on the implemented layer management functionality (see fig. 2). A separate threshold demon observes the status of the respective protocol instance and notifies a trap demon on the manager's site in case of threshold violations. After initializing an appropriate threshold (1), an acknowledgement is sent to the manager (2). The threshold monitoring is based on a local interactions between the threshold demon and the SNMP demon (3). Threshold violations are propagated via SNMP traps (4), and the trap demon may inform the management tool about it (5).

The thresholds themselves can in principle be modeled and implemented via different alternatives:

- *Explicit MOs:* In this case, a threshold is represented as an MO explicitly. The advantage is relative simplicity, the disadvantage is the overhead associated with the various MOs.
- *Global table:* With this alternative, only one MO per agent is used. It represents a global table with references to all other MOs that are to be monitored, together with the respective threshold values. This alternative is more flexible and can be implemented more efficiently.
- *Logical names:* With this approach, the manager specifies the MOs to be monitored by logical names. This way, a flexible naming approach is introduced that is easy to use. However, a mapping to the internal MOs must be provided.
- *Semantic specification:* This alternative is even more far-reaching. MOs are specified by names or by semantic constructs, and the threshold calculation functions can also be given by the manager based on semantic expressions. Ths approach is most flexible but difficult to implement.

We selected the second alternative due to its relative efficiency and implementability. The approach also requires the definition of reasonable thresholds for QoS related parameters such as roundtrip time, throughput, or dropped packets. Based on these definitions, the agents perform a decentralized threshold monitoring by explicit threshold checks, and can notify the manager using the existing trap feature of SNMP within our architecture. This also requires new facilities for threshold initialization. In particular, the threshold values have to be specified in advance, also providing reasonable defaults. These issues are addressed in the next section.

4. Quality of Service Abstractions

The management facilities discussed above support monitoring and manipulation of explicit QoS parameters. However, for distributed applications, a higher level of abstraction concerning QoS specification is desirable. An application should not be required to explicitly deal with para-meters such as buffer sizes, delay or jitter. It should rather specify its requirements in terms of the data streams or media to be communicated. This should be supported by reasonable defaults and should cover a wide range of communication scenarios, not limited to multimedia only.

Analysis of QoS requirements:

Based on these considerations, we performed a detailed analysis of QoS requirements of existing distributed communication scenarios and media encoding schemes. The major goal was to specify a set of predefined QoS classes that can automatically be mapped onto QoS parameters. The overall classification is illustrated in fig. 3. According to /SCZ93/, we defined 4 major classes, dividing application requirements into reliable and unreliable communication and time-dependent and time-independent interactions, respectively. The resulting classes may partially overlap; for example, the transfer of a pixel image may either be time-constrained or not although no isochronous traffic characteristics are required.

isochronous	non isochronous		
unreliable	reliable		unreliable
time-dependent		time-independent	
I	II	III	IV
speech, music, video, HDTV realtime text, graphics, images	realtime text, graphics, compressed images, data	text, data graphics, images	text, graphics, compressed images

Figure 3: Classification of distributed interaction scenarios

For each kind of interaction, the basic characteristics and the associated QoS parameters have been analyzed. Results for audio and video are summarized in figure 4. For low-quality interactive audio communication, relatively high residual error rates can be tolerated, and relatively low throughput is sufficient. This changes significantly for high-quality audio based on CD or MPEG encoding. The delay and jitter values have to be constrained in any case in order to guarantee reasonable quality of service characteristics. This is similar for video, however, much higher throughput must be supported. The residual error rate, on the other hand, can be significantly higher without compromising quality of service.

For conventional file transfer and RPC interactions, there are no jitter constraints, of course. However, the delay should be constrained in the range of several milliseconds up to one second for data-intensive interactions. The residual error rates must be close to zero. Analysis of various other kinds of communication scenarios has been performed, too. As an example for mapping the results onto QoS parameters in more detail, a sample QoS specification for JPEG-based binary image transfer is shown below:

	audio				video						
encoding formats	IMA 8 kHz	IMA 11,025 kHz	IMA 22,05 kHz	IMA 44,1 kHz	CIF	SIF	CCIR 601	HDTV Europa	H.261	MPEG	MPEG II
throughput (Mbit)	0,064	0,0861	0,1722	1,378	37	30,5	206	2373	0,064 - 1,92	1,5	4 - 10
residual error rate	10^{-2}	10^{-8}	10^{-8}	10^{-8}	10^{-4}	10^{-4}	10^{-4}	10^{-4}	10^{-6}	10^{-6}	10^{-5}
delay (ms)	50 - 100										
jitter (ms)	5 - 10										

Figure 4: QoS characteristics of audio and video

```
jpeg.throughput.min = 1000; /* bit/s, default minimum */
jpeg.throughput.target = FileSize / 100000; /* for a transfer in typically less than a second */
jpeg.throughput.max = 5000000; /* assumption */
```

```
jpeg.throughput.lower_threshold = jpeg.throughput.min;
jpeg.throughput.upper_threshold = jpeg.throughput.max;
jpeg.transit_delay.min = 5; /* ms */
jpeg.transit_delay.target = 50;
jpeg.transit_delay.max = 100;
jpeg.max_delay_jitter = NO; /* no jitter requirement */
```

jpeg.max_error_rate = 10^{-6};

jpeg.max_establishment_delay = 200;jpeg.max_establishment_failure_probability = 10^{-3};

jpeg.max_transfer_failure_probability = 10^{-6};

```
jpeg.max_resilience = max;
jpeg.max_release_delay = 500;
```

jpeg.max_release_failure_probability = 10^{-3};

```
jpeg.target_error_handling = 4; /* error correction for lost packets and bits */
jpeg.target_guarantee_class = 1; /* reliable service, therefore bandwidth reservation desired */
jpeg.target_protection = 0; /* no protection */
jpeg.target_priority = 100; /* medium priority */
jpeg.max_cost = min; /* no realtime service */
```

QoS manager tool:
Based on these results, we designed and implemented a QoS manager tool in order to provide the desired higher level of QoS abstraction. The integration of the tool is shown in fig. 5. An application requests to establish a connection for a specific data stream. Rather than specifying the QoS requirements in detail at the transport layer interface, it sends a request to the QoS manager tool, specifying the desired QoS class and the selected kind of media encoding (the tool is typically co-located with the application or even residing within the same process). The tool analyzes the input and issues a complete low-level QoS data structure according the given requirements. The result is passed back to the application and is used for connection establishment. Should the connection fail due to limited resources, the application contacts the tool once more, requesting a reduced set of default QoS parameter values. This way, the connection estab-lishment may afterwards succeed, providing some reduced but still acceptable QoS character-istics. In the case of TCP/IP, QoS characteristics can not be specified directly at the protocol level but can at least be monitored by our threshold management, leading to a similar approach.

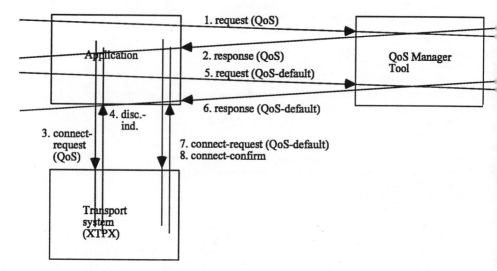

Figure 5: Integration of QoS manager tool

In summary, the application is relieved from specifying QoS details at the transport system interface. It can make use of predefined QoS characteristics; these definitions can also be easily extended and adapted according to specific application requirements. They are structured as a class library of predefined media communication classes; the inheritance structure of the library is shown in fig. 6. The abstract class *channel* provides basic mechanisms for connection management. Basic, time-independent communication facilities are offered by *BulkDataChannel* (for regular mass data transfer) and *COControlDataChannel* (for connectionless transfer of the control data of an application). As opposed to that, *TimeDependentChannel* enables the dynamic supervision of time-related QoS parameters based on our threshold management approach. Further distinctions are made concerning the error handling at the protocol level; for various compressed data streams with less redundancy (such as MPEG or adaptive PCM), reliable channels are required while unreliable transfer is sufficient for other encoding formats. The inheritance tree is extensible in order to introduce other kinds of media encoding formats (such as MPEG2, for example).

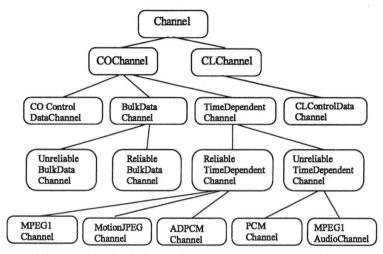

Figure 6: Media object class library

At instantiation time, predefined QoS characteristics can be overridden for a specific object of one of these classes. For example, a high- or low-quality video according to the application requirements and the expected available bandwidth can be selected. The QoS manager tool can provide additional adjustments based on the described two-phase approach as discussed above. The detailed QoS specification is performed by using the (overloaded) constructors of the media channel classes. Various levels of abstraction are available as illustrated by the following example:

```
MPEG1Channel (...);
MPEG1Channel (..., int min_throughput);
MPEG1Channel (..., QoS *user_qos);
MPEG1Channel (..., int resolution, int frame_rate, QoS *user_qos);
```

In addition to some regular control parameters (not shown here), QoS-specific parameters can be specified. In the first case, however, they are defaulted by the system. The second constructot allows the specification of throughput requirements with defaults for other QoS parameters, still at a rather low level.The third constructor enables a direct layer-4-level-specification of QoS with all details. The last contructor given above provides for application-level QoS specification (for example, a resolution of 768x576, a frame rate of 30 per second and a maximum delay of 90 milliseconds). Even higher levels of abstraction have been considered (namely, abstract media quality classes), but have not been implemented yet. The corresponding QoS data structure used in two of the constructors has the following outline (based on QoS in XTPX, but generalized and now protocol-independent):

```
struct QoS (
        long min_throughput;
        long lower_threshold_throughput;
        long target_throughput;
        long upper_threshold_throughput;
        long max_throughput;
```

```
long target_delay;
long upper_threshold_delay;
long max_delay;
long max_jitter;
long max_establishment_delay;
long max_release_delay;
long target_error_rate;
long max_error_rate;
long target_priority,
long min_protection;
long max_cost;
);
```

The structure comprises absolute limits, target values, and threshold values. Absolute limits specify the intervals of QoS values that have to be guaranteed in any case. Target values are recommended and should be achieved if possible. Threshold values are within the absolute intervals and are used for threshold management. This way, the system can react and take appropriate measures before the limits are reached, using our threshold management approach. For all fields of the QoS structure, explicit values can be given. However, the application can also request that the system should use an appropriate default (indicated by *REQ_EMPTY*) or can state that the parameter should not be used for QoS management at all (*REQ_NO*).

Several alternatives and extensions have also been considered. It would also be possible to implement the tool as a complete intermediate layer between the application and the transport system. This way, the tool would be responsible for connection management, too. We initially rejected this solution as we wanted to keep the QoS tool independent from the regular connection management. This way, mapping of QoS details onto various different protocols is enabled in a flexible way. In particular, with the emerging IPng (next generation) protocol in the Internet, this will become very important. It will also be interesting to use various underlying reservation protocols, namely RSVP. Such alternatives will be investigated in more detail in the near future.

5. Integration into Client/Server Programming Models

Motivation:
The presented approach provides QoS abstractions for relatively straightforward communication scenarios. However, general distributed applications require more advanced support based on standardized distributed platforms /NWM93/. Most approaches of this kind are based on the client/server model and are using an RPC-style interaction mechanism. In particular, hetero-geneous environments are supported by canonical data transfer formats. These facilities are typi-cally being enhanced with directory services for mapping servers to client requests by logical na-mes, and by security services for guaranteeing authorization and privacy in distributed environments.

However, such approaches are not yet using high performance communication protocols nor considering QoS aspects at all. On the other hand, protocols such as XTPX do not provide any facilities above the transport layer and therefore lack the already well-established functionality of distributed platforms. For these reasons, an integration of both approaches is most desirable. As a base for our integration work, we are using the OSF DCE /OSF94/.

Integration approach:
The basic integration architecture is outlined in fig. 7. A distributed application consists of seve-ral decentralized components that regularly communicate via remote procedure call. Initially, components locate each other by querying the directory service for peer components based on their name and on their static characteristics. Moreover, RPC communication can be protected by encryption and access control. Even in advanced multimedia applications, the control part of all interactions (such as the management of dynamic participation in a videoconference) be im-plemented in a comfortable way using RPC. Major advantages are similarity with local proce-dure calls, masking of operating system and hardware heterogeneity, and relative reliability.

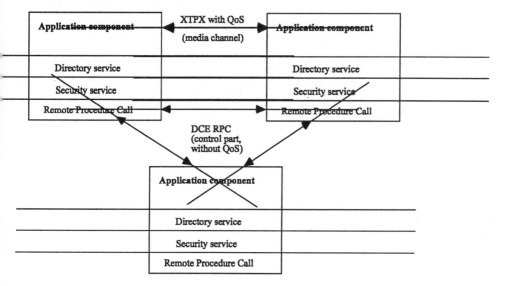

Figure 7: Integration with OSF DCE

However, it is also possible to pass control objects via RPC that initiate the establishment of an media channel in addition to existing RPC interaction paths. For example, an RPC parameter may comprise a reference to a document with audio and video annotations. The server of an RPC interaction does not receive the actual media data but just the media control information this way. Thereafter, it may initiate additional calls to the received parameter objects in order to access and display the media information. Due to the discussed QoS requirements, this cannot be implemented effectively by conventional RPC; rather, it is reasonable to bypass RPC, open a connection in a transparent way, and transfer the required data while guaranteeing QoS characteristics. This will be achieved by furnishing RPC control objects with an interface to the QoS manager tool and to the transport system.

This initial integration will present a first step towards high performance client/server applications. However, a more seamless integration is desirable. Therefore, we are currently designing a QoS-based class library that offers high performance RPC interaction channels (to be mapped onto the transport protocol), and various media object classes imbedded into an inheritance hierarchy. Instances of these classes represent specific media with given encoding formats and individual characteristics (such as frame rates of video objects). By parameterizing channels with media objects, it will be possible to automatically select QoS characteristics for applications in the traditional world of client/server interactions.

6. Conclusions and Future Work

This paper presented a management framework with special respect to QoS characteristics. Based on SNMP, a management architecture has been designed and implemented; it allows for interactive, graphical access to QoS parameters via managed objects. Moreover, it supports interactive modifications of such parameters and enables functional extensions towards threshold management. We also presented a classification of application-level QoS requirements and derived a QoS abstraction architecture and tool. This approach supports a high-level specification of QoS requirements, relieving the application from QoS details. It has been outlined how the approach can be integrated with traditional client/server platforms for distributed systems.

Future work will focus on the refinement of QoS requirements and characteristics. It will be investigated how the QoS requirements of combined media streams can be specified and how such specifications can be imbedded into an overall application framework. The results will be used as a base for implementing a more advanced QoS manager tool. The implementation will also address a seamless integration with OSF DCE and shall be validated by several examples. Moreover, the installation and usage of a local ATM network /BOU92/ is also planned for the near future in our lab. This will raise the question how the discussed QoS characteristics can be mapped onto ATM and how applications can be supported in a mixed environment with ATM networks and conventional Ethernet-based communication.

Acknowledgements

We would like to thank all colleagues and students who contributed to the design and implementation of the presented approach. In particular, we are most grateful to Stephan Albrecht, Frank Breiter, Samer Habib, Ulf Kieber, Olaf Kiese, Jörg Kretschmar, Sabine Kühn and Jens Watzek for their important contributions. An earlier version of the paper was presented at the Workshop on new protocols for multimedia systems, Berlin, June 1994.

References

/ADB94/   Adcock, P., Davies, N., Blair, G.: Supporting Continuous Media in Open
            Distri-   buted Systems Architectures; Lancaster University, Dep. of Computing,
1994
/BDH93/   Barth, I., Dermler, G., Helbig, T., Rothermel, K. Sembach, F., Wahl, T.:
            CINEMA: Eine konfigurierbare, integrierte Multimedia-Architektur;
            GI/ITG Workshop "Verteilte Multimedia-Systeme", Stuttgart, 1993
/BES94/   Besse, L., Dairaine, L. et.al.: Towards an Architecture for Distributed Multi-
            media Applications Sopport; Proceedings of the International Conference on
            Multimedia Computing and Systems, Boston, Massachusetts, USA, May 1994
/BOU92/   Boudec, J.Y.: The Asynchronous Transfer Mode: A Tutorial; Computer
            Networks and ISDN Systems, Vol. 24, 1992, pp. 279-309
/CCH94/   Campbell, A., Coulson, G., Hutchison, D.: A Quality of Service Architecture;
            ACM SIGCOMM Computer Communications Review, Vol. 24, No. 2,

April 1994, pp. 6-27

/CFS90/ Case, J.D., Fedor, M., Schoffstall, M.L., Davin, C.: Simple Network
         Management Protocol; RFC 1157, May 1990

/MBA94/ Mourelatou, K., Boulloutas, A., Anagnostou, M.: An Approach to Identifying
         QoS Problems; Computer Communications, Vol. 17, No. 8, Aug. 1994,
         pp. 563-570

/MIB94/ Miloucheva, I., Bonnesz, O.: XTPX - Version 1.3 Technical Documentation;
         Technical University of Berlin, 1994

/MIT94/ Miloucheva, I., To, K.: XTPX Management Information Base (MIB); Design
         and Implementation; Technical University of Berlin, PRZ Document, 1994

/NWM93/ Nicol, J.R., Wilkes, C.T., Manola, F.A.: Object-Orientation in Heterogeneous
         Distributed Computing Systems; IEEE Computer, Vol. 26, No. 6, 1993, p. 57-
67

/OSF94/ Distributed Computing Environment - An Overview; Open Software
Foundation,   1994

/SCM93/ Schill, A., Mock, M.: DC++: Distributed Object-Oriented System Support on
         top of OSF DCE; Distributed Systems Engineering Journal, Vol. 1, No. 2, 1993

/SCZ93/ Schill, A., Zitterbart, M.: A System Framework for Open Distributed
Processing;
         Journal of Network and Systems Management, Vol. 1, No. 1, 1993, 71-93

/STR92/ Strayer, W.T., Dempsey, B.J., Weaver, A.C.: XTP: The Xpress Transfer
         Protocol. Addison-Wesley, 1992

/TMS93/ Touring Machine System Project: Touring Machine System; Comm. of the
         ACM, Jan.1993, Vol.36, No. 1

/VOG94/ Vogel, A., Kerherve, B. et.al.: QoS-Negotiation for Distributed Multimedia
         Presentational Applications; Proceedings of Workshop on Distributed Multi-
         media Applications and Quality of Service Verification, Montreal, Quebec,
         Canada, May /June, 1994

29

Quality-of-Service Directed Targeting Based on the ODP Engineering Model

Georg Raeder and Shahrzade Mazaher

NR – Norwegian Computing Center
P.O.Box 114 Blindern, 0314 Oslo, Norway
Georg.Raeder@nr.no, Shahrzade.Mazaher@nr.no

When developing large distributed systems, the final transformation steps toward executable components (targeting) are usually complex and platform dependent. The ODP Engineering Model provides a framework for alleviating this situation, but methods and tools that realize its potential are still lacking. This paper gives an example of a *distribution configuration language* and shows how it can be used, together with SDL, to achieve platform-independent distributed targeting. Furthermore, targeting decisions should be made on the basis of quality-of-service constraints, such as performance and reliability requirements, and the paper shows how our language can serve as a basis for stating and checking such constraints. Finally, the relation of our work to dynamic reconfiguration is discussed.

Keywords: Distribution configuration, configuration languages, distributed targeting, quality of service, open distributed processing.

1. INTRODUCTION

Targeting is a complex issue in distributed systems. Targeting involves all the steps taken to transform platform-independent application code to software components that can be executed in a concrete environment. While this is a relatively trivial task for a monolithic program, targeting distributed systems involves non-trivial choices and trade-offs as to how the system is to be split into distributed components, achieving the desired performance, reliability, etc.

Thus, methods and tools have to be developed for *distribution configuration*, allowing system implementors to express how a system is to be partitioned and assigned to platform entities, such as computers and operating system processes.

There are many degrees of freedom in such a configuration. To control it, a framework of *quality of service* (QoS) must be developed so that end-user requirements can be readily translated to technical constraints that will guide the distribution configuration.

To make these steps possible in more than an ad hoc manner, we need a *model* of the target environment, i.e., an abstract framework that 1) will help us tie together the various concepts of applications, target platforms, and end-user requirements, and 2) will do so in a manner independent from concrete platform implementations.

This paper presents a language for distribution configuration, and it also describes an approach to using the language for checking that QoS constraints are satisfied. The language is based on the Engineering Model of the Open Distributed Processing (ODP) draft international standard [6] as its abstract framework.

The paper assumes familiarity with ODP. Knowledge of SDL, the ITU-T Specification and Description Language, would also be helpful, but is not essential.

1.1. Configuration of distributed systems

Most contemporary work on configuration of distributed systems (see [7] for a collection of recent reports on this topic) deals with *logical* configuration of system components. In terms of ODP, this concerns configuration at the Computational Model level. Typically, these approaches are based on a modules/ports model, where individual components (*modules*) are encapsulated and send and receive messages to and from local *ports*. Configuration consists of creating and deleting instances of these modules and connecting their ports to each other. This is an example of programming-in-the-large, as distinct from the programming-in-the-small of each module.

Configuration must also handle *engineering-level* concerns, such as the specification of physical distribution of components in the system. This aspect becomes more prominent in ODP-based systems, since the ODP Engineering Model defines comprehensive structuring concepts. The components of a distributed application have to be mapped to the physical nodes of the infrastructure and they have to be structured according to these engineering concepts.

Configuration of distributed systems may be *static* (done "off-line"). Executables are typically generated statically, and an initial distribution configuration may also be specified statically. Dedicated systems that do not evolve may in fact have their engineering configuration statically determined.

In many cases, however, the system needs to change *dynamically*. Functional (logical) changes, in terms of new modules being instantiated and connected, or parts being shut down, can often be pre-specified in terms of change scripts (see, e.g., [1, 3, 8]). Non-functional changes, such as reconfigurations to counter load imbalances, usually cannot be anticipated in this way. *Management* objects could be supplied for this purpose, with each application object providing a corresponding management interface (a form of this is presented in [5]). A major challenge is how one can keep a system consistent according to some criteria in the face of ad hoc changes. One would like to describe an initial, consistent configuration, and then make sure that every reconfiguration operation takes the system from one consistent state to another, for example by means of checking *constraints* that always must hold [2, 3].

Engineering configuration in ODP belongs to the dynamic category, since the state of the network in an open system typically will vary in a non-deterministic manner. Achieving a desired level of quality of service must therefore be subject to continual evaluation and reconfiguration. Introducing this type of management in ODP seems to imply more centralization (via management objects), but it is nevertheless very important for achieving performance and reliability.

1.2. Quality of service

Quality of service (QoS) encompasses a wide range of essentially non-functional requirements that users may have on a system, such as timeliness (delay bounds, jitter bounds), throughput (bandwidth), cost, dependability, security, etc. ([12]; [10] Vol. 5). As the system design is refined into components, so will the QoS requirements be broken down into constraints on individual components, such as ODP Computational Model objects, and on the engineering infrastructure that supports them. There may also be a translation from user-level requirements, such as the quality perception of moving images, to technical constraints, such as a minimal number of frames per second.

The current ODP standards documents refer to QoS, but without detailing the subject. In [12] a framework is proposed for how QoS can be integrated in the ODP Computational Model

(CM). Briefly, interface definitions are extended with

- QoS clauses that express the quality of service *required* by this interface type from its environment,
- QoS clauses that express the quality of service *provided* by this interface type to its environment, given that its requirements on the environment are fulfilled, and
- a list of the types of interfaces that implementations of this interface type may invoke operations on.

The QoS clauses are expressed in a language based on first-order logic. The last item above is necessary when performing QoS analyses, which essentially consist of checking whether the QoS provided by servers match that required by their clients. This gives rise to an extended type conformance relation.

These CM-level QoS constraints must be further interpreted in terms of the supporting infrastructure. Whether an object can provide the level of QoS specified depends on its engineering and technology support. Section 2.3 of this paper describes our approach to engineering-level QoS analysis and its role in guiding distribution configuration.

1.3. The context

The RACE II project SCORE,[1] in which the authors are involved, aims to establish environments for the creation of new generation telecom services. These services are distributed applications, and the ODP framework is a prime target architecture.

One of the main formalisms used in SCORE for system design is SDL, the ITU-T Specification and Description Language. In particular, the SDL-92 definition, with its object-oriented extensions, is the focus of much of the novel work in the project. Since SDL allows detailed specification of system behavior, it is possible to generate code automatically from SDL diagrams. SCORE is developing a translator from SDL-92 to C++ [11]. The SDL input to this translator has to adhere to certain guidelines to narrow the semantic gap between SDL and ODP ([10] Vol. 5). The output code conforms to ODS, a pilot ODP implementation.[2]

Thus, distribution configuration support developed in SCORE will have to deal with how to map SDL processes (the SDL "objects") to ODP engineering concepts. We have designed a *distribution configuration language* (Sdcl) for expressing this type of information. We are currently prototyping tools for mapping an application onto an ODP-based platform (such as ODS) according to a Sdcl specification. Figure 1 illustrates the tool chain.

SCORE aims at developing support for multimedia services. This type of service implies strict performance and reliability requirements, which are expressed mainly in terms of different kinds of QoS constraints [4]. An ideal tool environment would contain means for automatically deriving Sdcl configuration descriptions from QoS requirements. Similarly, to deliver the required performance, whenever a system is subjected to dynamic reconfiguration, or when the QoS constraints change, the configuration should be validated against the QoS constraints.

Translating QoS constraints into an engineering specification is a complex problem, however, calling for heuristics and knowledge-based techniques. This is outside the scope of our work.

---

[1] RACE Project 2017, Service Creation in an Object-oriented Reuse Environment.
[2] The Open Distributed Systems platform from BNR Europe Ltd. Another platform experimented with in SCORE is ANSAware from APM Ltd.

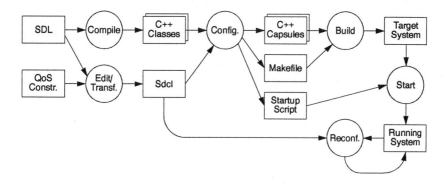

Figure 1. The distribution configuration tool chain.

We have focused on developing a specification language that can be used for expressing a distribution configuration, be it generated by humans or an intelligent front-end tool.

1.4. The approach

Our goals are 1) to build practical tools to automate the task of configuring distributed systems, and 2) to include QoS concerns in this tool support. We can split the task of configuring a particular system as follows:

1. Specify the QoS requirements and provisions of the interfaces of each object class in terms of engineering-level constraints.
2. Specify capsule files, i.e., the executables that should be generated and which object classes they should have compiled in.
3. Specify an initial configuration, indicating the capsules to start, where to start them, and which object instances they should initially contain, and check it against the QoS constraints.
4. Create a management object that maintains the QoS specifications of the system, and that can check requests for reconfigurations against these constraints.

Steps 1–3 involve *static* specifications that can be processed off-line. This paper mainly concerns these aspects. The management object in step 4 can also be statically generated, but the object itself handles requests for *dynamic* reconfigurations, and must make decisions based on QoS requirements, current system configuration, and the current platform status (node loads, network traffic, etc.). The following sections detail our contributions covering these steps.

2. AN ODP-BASED DISTRIBUTION CONFIGURATION LANGUAGE

Distribution configuration comprises several kinds of information. In this section, we will list the main issues of concern, and we will show via examples how our language, Sdcl, expresses these aspects.

2.1. Specifying capsule files

A logical system structure must in some way be mapped to a corresponding engineering structure. For example, in SCORE we would like to map a SDL system specification to the ODP engineering concepts of nodes, capsules, clusters, and objects. This is the core task of distribution configuration, and it should therefore form the central feature of a notation for this purpose.

To this end, our language supports the definition of types based on the engineering concepts, and that of an initial configuration in terms of the assignment of instances of the defined types to explicit physical nodes:

- We interpret ODP capsule and cluster definitions as *templates* used for code generation. The modules generated from these templates can be instantiated several times and in several different places in the target system.
- ODP nodes, on the other hand, correspond to physical system entities. A node description therefore gives a snapshot of part of the system, in terms of the capsules and clusters instantiated on it. In particular, a set of *initial* node configurations can be used for starting up the system.

Assume that we have the SDL-92 design shown in Figure 2.[3] The figure depicts a system[4] consisting of five process types—*Customer, Subscriber, Terminal, Session,* and *Database*— along with their instances *cu, sb, tr, sn,* and *db*. There is one instance each of *Customer* and *Database*, whereas the other three types may have any number of instances. The communication patterns are as indicated by the arrows (dotted arrow means process creation).

Figure 3 (a) presents the capsule and cluster types used in the distribution configuration. It also specifies the interfaces that each object class (i.e., SDL process type) supports and uses (**provides** and **requires**). These **class** statements provide the link to the SDL specification. A cluster type is defined in terms of the types of its constituent objects (SDL process types). A capsule is in turn defined in terms of the types of the clusters and objects[5] that it comprises, defining the object classes that the capsule can support at run-time. For example, objects of type *Session* can be instantiated in two different capsule types.

2.2. Specifying an initial configuration

Figure 3 (b) shows an initial configuration of the system in terms of the structure of the nodes comprising it. For exposition, we have instantiated two subscribers and their terminals. We have distributed the system over one central computing node (containing one capsule) plus one node for each terminal (containing a capsule with the terminal object plus an empty session capsule).

---

[3]The example is derived from a Virtual Private Network (VPN) test service elaborated in SCORE. Briefly, the *Customer* represents the service, aided by a *Database* of system information. *Subscriber* handles subscriptions and sets up calls. A *Terminal* is instantiated for each connected terminal. A *Session* is instantiated for each call.
[4]Since SDL *system* and *block* concepts do not map to engineering structures, we have omitted them from the figure.
[5]An object class directly within a capsule is a shorthand for a cluster containing only that class.

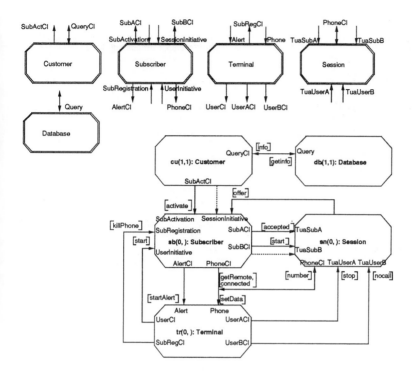

Figure 2. The example SDL-92 system.

Those instances that are to be created dynamically are not part of the initial configuration of the system.

The client-server relationships (**uses** clauses) are included, since configuration tools may need them. These may actually be generated from the SDL specification. In general, Sdcl code may either be written by hand, or it may be generated by various tools, such as a visual programming tool or a constraint resolution tool.

In addition to the basic structure requirements, there may be many other engineering aspects we wish to express. We have focused on the following aspects, also depicted in Figure 3 (b).

Location

For each node in the distributed system, it may be necessary to explicitly express its location, the concrete computer in the network on which it should reside. This is taken care of by the **on** *<node>* construct. All capsules specified within node *vpn* will run on that machine.

```
class Customer                                  initial
    requires SubActivation, Query;                  node vpn on "vpn.nr.no" is
class Subscriber                                        CustSub: CustSubCap[
    requires Phone, Alert, TuaSubA, TuaSubB;                Cs: CustSubCls[
    provides SubActivation, SubRegistration,                    cu: Customer; sb(1:2): Subscriber]]
        UserInitiative, SessionInitiative;          end
class Terminal
    requires TuaUserA, TuaUserB                      node ola on "ola.nr.no" is
        SubRegistration, UserInitiative;                Term: TerminalCap[tr: Terminal];
    provides Phone, Alert;                              SessionMgr: SessionCap[]
class Session                                       end
    requires Phone, SessionInitiative;
    provides TuaSubA, TuaSubB,                        node per on "per.nr.no" is
        TuaUserA, TuaUserB;                             Term: TerminalCap[tr: Terminal];
class Database                                          SessionMgr: SessionCap[]
    provides Query;                                  end

cluster CustSubCls is                                exist
    class Customer;                                      db: Database with "C=no; O=nr"
    class Subscriber                                 end
end                                              end

capsule CustomerCap is                          vpn.CustSub.Cs.cu
    cluster CustSubCls;                             uses vpn.CustSub.Cs.sb(1).SubActivation;
    class Session                               vpn.CustSub.Cs.cu
end                                                 uses vpn.CustSub.Cs.sb(2).SubActivation;
                                                vpn.CustSub.Cs.cu uses db.Query;
capsule TerminalCap is                          vpn.CustSub.Cs.sb(1) uses ola.Term.tr.Alert;
    class Terminal;                             vpn.CustSub.Cs.sb(1) uses ola.Term.tr.Phone;
    class Session                               ola.Term.tr uses vpn.CustSub.Cs.sb(1).SubRegistration;
end                                             ola.Term.tr uses vpn.CustSub.Cs.sb(1).UserInitiative;
                                                (Analogous clauses for "per" as for "ola")
```

         (a)   (b)

Figure 3. Sdcl code for type definition (a) and initial distribution (b).

Instantiation vs. binding

When introducing a new system, it might be the case that instances of the same types as some of its components are already running in the network. An issue is thus whether to create a new instance or to use an existing one. If an existing object is used, there is also a question of whether a specific instance is desired or any available instance will do.

In Figure 3 (b), the *db* object is specified to exist already, and no instance of this object will therefore be created. Instead, its client *cu* will be bound to an instance already running in the distributed environment. If there is more than one appropriate instance, one is chosen (by the trader) based on the additional information given in the optional **with** clause, as shown.

2.3. QoS constraints and dynamic reconfiguration

The distribution configuration of an application must satisfy the QoS (non-functional) requirements imposed on it. As the system is elaborated through design and implementation, the original user-level requirements must be broken down into more specific constraints. At the computational level, QoS requirements are stated by putting constraints on the computational interfaces of objects ([12]; [10] Vol. 5). During implementation, these constraints should be further translated into *engineering-level constraints*.

Engineering constraints would handle aspects such as the location of objects, whether to replicate objects for added reliability or performance, performance of channels connecting the objects of the application, etc. Checking configurations against these constraints, expressed in terms of the supporting infrastructure, may be expected to be relatively simple.

Engineering constraints may conveniently be included as an *invariant* part of the Sdcl specification. An initial configuration of the system should satisfy these constraints. Furthermore, a (dynamic) reconfiguration should essentially update the Sdcl description (the initial configuration part) that could then be checked against the invariant. The capsule/cluster class part of the Sdcl description is used for capsule generation to *facilitate* that appropriate reconfiguration is possible (capsules with the proper contents must be available).

The constraints would take the form of first order logic formulae, and could contain functions such as:

node(x), capsule(x), cluster(x) – The node, capsule, or cluster of an object x.

numcaps(x), numclus(x), numobjs(x) – The number of capsules, clusters, or objects in a node or capsule x.

type(x) – The type of an object x.

proc(x) – A performance factor, given as the time taken to complete a specific computation on the infrastructure of object x.

comm(x,y) – A performance factor, given as the time taken to complete a specific communication between the objects x and y.

A set of constraints for our example system could be:

node(cu) = "vpn.nr.no"
$\forall$ x $\in$ Subscriber : cluster(x) = cluster(cu)
$\forall$ y, z $\in$ Session, y $\neq$ z : capsule(y) $\neq$ capsule(z)
10 $\geq$ 4·proc(tr) + proc(sb) + 2·proc(sn) + comm(tr,sb) + comm(tr,sn)

The first constraint ties the object *cu* to a physical node, i.e., the object has to permanently be placed on that node (due to security reasons, for example). The second constraint implies collocation of objects (which is directly reflected by the *cluster* construct of Sdcl). The third constraint expresses the condition that no two objects of type *Session* should be placed in the same capsule (e.g., for reliability reasons).

The last constraint expresses a relation between the performance factors of the potential nodes and capsules where the objects would be placed and of the communication channels between them. Consider Figure 4, which is a partial ODP realization of the system in Figure 2. Assume that there is a requirement from the environment (via an interface *Call* not shown in Figure 2) on *tr*, stating a maximum response time of 10ms for invocation of a certain operation. Assume further that executing this operation uses 4 processing units in *tr* plus one invocation on each of *sb* (delay *x*) and *sn* (delay *y*). The executions of the latter two operations use 1 and 2 processing units, respectively. Adding communication delays, we get:

$$10 \geq 4 \cdot proc(tr) + x + y \wedge x \geq proc(sb) + comm(tr,sb)$$
$$\wedge \; y \geq 2 \cdot proc(sn) + comm(tr,sn)$$

or:

$$10 \geq 4 \cdot proc(tr) + proc(sb) + 2 \cdot proc(sn) + comm(tr,sb) + comm(tr,sn)$$

All the *proc* and *comm* parameters should be obtained from a database of network information and from the Sdcl description that states the clusters, capsules, and nodes within which the objects are contained. The objects *tr*, *sb*, and *sn* should be distributed over the network such that the corresponding values for *proc* and *comm* parameters satisfy the above formula. For example, this analysis will uncover whether *tr* can fulfill its requirements when using an *sn* in another capsule, or whether only the one in its own capsule is fast enough.

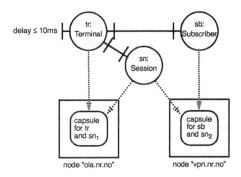

Figure 4. Engineering constraints.

Thus, the expression of the QoS of the system will involve a system of equations which cannot be solved until a concrete platform is chosen. The external system requirements (the 10ms in the example) provide boundary conditions on the equations. In theory, one could automatically derive a configuration satisfying the equations, but in practice, one will manually propose a configuration (using Sdcl) and check that it satisfies the equations (possibly having to go through some iterations).

Many QoS characteristics can be expressed in a manner similar to this response-time example, but the combination rules for the equations for, e.g., a probabilistic dependability measure, would vary. The analysis is crude, and it must be based on average or worst case figures, but we conjecture that automated tools supporting even this level of analysis would be quite useful in complex systems.

The need for constraints checking implies the existence of management objects that can query the objects in the system to verify the system state, and access a database of network information for relevant data. (The ODP node manager could serve this purpose.) It also implies that the objects must provide a management interface. One interesting approach would be to generate automatically the management interface and operations for application objects, and to generate a management object that could maintain and update a Sdcl description of the system and periodically check the configuration part of it against the specified engineering constraints.

3. CONCLUSIONS

This paper has presented a language for specifying the distribution configuration of a system, for targeting onto a platform conformant to ODP. Realizing that the major factor influencing this kind of configuration is the QoS requirements of the system, we also described how the language can be used as a basis for QoS-directed targeting. Finally, we outlined an approach to managing the dynamic reconfiguration of the system based on the initial configuration and a set of engineering-level constraints derived from QoS requirements.

3.1. The role of the ODP Engineering Model

As we have seen, the distribution configuration specification defines how the target system is to be structured in terms of ODP concepts, such as nodes, capsules, and clusters. It does not, however, have to handle the mapping of these concepts to a concrete platform, such as ODS or ANSAware. That step is taken care of by the translation tools. The ODP Engineering Model provides the abstract model that makes this *platform-independent targeting* possible. Without such an abstract target model, platform independence extends only to the computational model (SDL) level.

3.2. The role of SDL

In the SCORE tool chain (Figure 1), systems are specified and designed in SDL-92, and C++ code is automatically derived from the specification. Defining the application at such a high level is essential for platform independence. Code may be generated from the same SDL description for many different target platforms. Writing object code files in C/C++, on the other hand, makes platform details visible.

There are other benefits from using a well-defined high-level language as well. For SDL there exist powerful tools for analysis and simulation of the system.

Note that SDL has many similarities with the modules/ports model mentioned in Section 1.1. In fact, SDL block diagrams (such as Figure 2) can be thought of as a programming-in-the-large configuration language, with process diagrams supplying programming-in-the-small functionality. Thus, in our case, SDL takes care of the logical configuration, leaving engineering configuration for Sdcl.

The fact that SDL offers another computational model than ODP, does, however, result in some "impedance mismatch" [9]. (One way to model ODP in SDL is documented in [10] Vol. 5.) The ideal situation would be to have a high-level language embodying the ODP Computational Model, strengthened with logical composition facilities. Since SDL here is used as an ersatz ODP CM language, we see that SDL process types and instances actually show up in the Sdcl language, where only ODP interface types and objects should have been present.

3.3. Status

As of this writing, a tool for a previous version of Sdcl has been implemented that transforms a set of individual object files written for ANSAware (which is C-based) to capsule files corresponding to the Sdcl code. It also generates a makefile and startup script. We will move our work to the ODS platform (C++-based) and the new version of Sdcl, and integrate it with the other tools of SCORE. The work will continue through 1995.

3.4. Further work

In order to achieve predictable, consistent, and appropriate quality of next-generation systems and services in open distributed environments, a QoS-directed development approach is needed. To this end, an ODP-based framework (methods and tools) for stating QoS constraints and carrying them through analysis, design, and implementation into engineering-level properties should be developed. These properties should guide the generation of service implementations so that they meet the QoS constraints (e.g., by selecting proper library components and choosing an appropriate distribution of components in the network). The properties could also be used in the management domain to control dynamic reconfigurations of services.

The design of a distribution configuration language, and its use in conjunction with engineering-level QoS constraints, as presented in this paper, is one step in this direction.

Acknowledgements

This work has been supported by the Research Council of Norway and the CEC through the RACE II project 2017 SCORE (Service Creation in an Object-oriented Reuse Environment). We particularly thank our SCORE colleagues from PTT Research (The Netherlands) and Tele Danmark Research (Denmark) for the stimulating collaboration on ODP targeting. This paper represents the view of its authors.

REFERENCES

1. M.R. Barbacci, C.B. Weinstock, D.L. Doubleday, M.J. Gardner, and R.W. Lichota. Durra: a Structure Description Language for Developing Distributed Applications. *Software Engineering Journal*, 8(2), March 1993.
2. T. Coatta and G. Neufeld. Configuration Management via Constraint Programming. In *Proc. Int'l Workshop on Configurable Distributed Systems*. IEE, U.K., 1992.
3. M. Endler and J. Wei. Programming Generic Dynamic Reconfigurations for Distributed Applications. In *Proc. Int'l Workshop on Configurable Distributed Systems*. IEE, U.K., 1992.
4. L. Hazard, F. Horn, and J.-B. Stefani. Notes on Architectural Support for Distributed Multimedia Applications. Technical Report CNET/RC.W01.LHFH.001, ISA Project, March 1991.
5. C. Hofmeister, E. White, and J. Purtilo. Surgeon: a Packager for Dynamically Reconfigurable Distributed Applications. *Software Engineering Journal*, 8(2), March 1993.
6. ISO/IEC JTC1/SC21/WG7. Basic Reference Model of Open Distributed Processing – Part 3: Prescriptive Model. ISO/IEC 10746-3.
7. J. Kramer, editor. *Proc. Int'l Workshop on Configurable Distributed Systems*. The Institution of Electrical Engineers, U.K., 1992.

8. J. Magee, N. Dulay, and J. Kramer. Structuring Parallel and Distributed Programs. *Software Engineering Journal*, 8(2), March 1993.

9. S. Mazaher and G. Raeder. SDL and Distributed Systems—a Comparison with ANSA. In O. Faergemand and A. Sarma, editors, *SDL '93, Using Objects. Proc. 6th SDL Forum.* North-Holland, 1993.

10. SCORE-METHODS AND TOOLS. Report on Methods and Tools for Service Creation (First Version). Deliverable D203—R2017/SCO/WP2/DS/P/027/b2, RACE project 2017 (SCORE), December 1993.

11. SCORE-METHODS AND TOOLS. Report on Methods and Tools for Service Creation (Final Version). Deliverable D204—R2017/SCO/WP2/DS/P/028/b2, RACE project 2017 (SCORE), December 1994.

12. J.-B. Stefani. Computational Aspects of QoS in an Object-based Distributed Architecture. In *Proc. 3rd International Workshop on Responsive Computer Systems*, 1993.

30

Quality of Service Management in Distributed Systems using
Dynamic Routation

Leonard J.N. Franken[a], Peter Janssen[a],
Boudewijn R.H.M. Haverkort[b] and Gidi van Liempd[a]

[a] PTT Research, P.O. Box 15000, 9700 CD Groningen, the Netherlands
[b] University of Twente, P.O. Box 217, 7500 AE Enschede, the Netherlands

With the advent of multimedia in computing and communication systems, a new set of
requirements has been imposed. Because multimedia applications demand a guaranteed
Quality of Service (QoS), the sharing of resources, by means of allocating applications and
their communications on computing and communication resources, is an important issue.
In this paper, we address the issues of communication routing and process allocation in an
integrated way, the so-called *routations*. It is only in such an integrated way that *end-to-end
Quality of Service* requirements of multimedia end-users can be fulfilled.

For the determination of a routation, we propose to use the A^* algorithm, an intelligent
tree-search algorithm, in combination with a heuristic, $H3''$. The A^* algorithm always finds
an optimal solution if one exists, whereas the heuristic algorithm comes up with (sub)optimal
solutions. The heuristic, however, accomplishes the same task as the A^* algorithm with
far reduced computational effort. With the combination, denoted as $A^*/H3''(x)$, the A^*
algorithm will, when time is available, find an optimal solution, and in case of limited time
the heuristic will calculate a suboptimal solution. This combination has been found suitable
for a real time multimedia environment by a thorough statistical analysis after testing the
heuristics on 1750 randomly generated test cases.

Finally, we present an experimental multimedia, ANSAware-based, distributed system
in which the routations take place using the heuristic we developed. To demonstrate the
capabilities of the routation algorithms the *xallocator* has been implemented. The *xallocator*
guarantees the QoS using heuristic routation algorithms and dynamic reconfiguration whenever a change in the distributed environment occurs.

Keyword Codes: C.2.4; D.4.4; D.4.5; G.1.6; I.2.0; I.2.8
Keywords: Quality of Service; Performance of Systems; Communications Management; Reliability; Optimization; AI General; Problem Solving, Control Methods and Search

1. INTRODUCTION

In large current-day distributed systems, such as e.g., multimedia telecommunication systems like videophones, the sharing of resources, by means of allocating a wide variety of applications and their communications on a pool of computing and communication resources, is an important issue for Quality of Service guarantees. Because computing and communications are both important for current-day discrete and continuous end-to-end applications, we advocate an integrated approach towards the allocation of processes and the routing of communications as well. In this paper we present this integrated approach by the use of a so-called *routation algorithm* that can be used in the context of the *performability manager* introduced by Franken *et al.* [1, 2].

The performability manager, a distributed system component, guarantees and maintains the quality of service (QoS) requirements of an application in a distributed system [3]. On the basis of a model-based procedure, the performability manager decides which configuration is put into effect, thereby using dynamic reconfiguration facilities of the underlying supporting computing platform. The ultimate goal in using a model-based reconfiguration approach is to be sure that the reconfigurations performed do result in a desired change of the end-user perceived QoS. This goal, however, is in general too ambitious to be reached in an environment in which (mild) real-time requirements are present as well. It is simply infeasible to evaluate all possible alternative configurations. Therefore, we split the generation process of possible alternative configurations in two steps. We first generate a number of reasonable alternatives, based on static system and application properties and constraints, i.e., in this step we do not address issues related to queueing. The intend is, however, to create the alternatives as reasonably as possible. Then, in a second and more time consuming step, we automatically create queueing models for the candidates selected in the first step, evaluate them, and decide then which alternative configuration has to be put into effect. This second step has already partly been addrssed in [2].

This important issue of routing and allocation has been dealt with before [4, 3, 5, 6, 7]. However, we address the issues of routing and process allocation in an *integrated* way. This is a desirable approach as it is often unpractical to separate these highly connected issues. For the configuration creation we present routation algorithms based on an intelligent tree search algorithm, the A^* algorithm, known from artificial intelligence theory [8] as well as some heuristics.

This paper is further organized as follows. In Section 2 we present and evaluate (test) the A^* algorithm and the derived heuristics. In Section 3, we discuss a prototype implementation of a routation component in an ANSAware-based distributed environment using a videophone application. Section 4 concludes the paper.

2. COMBINED ROUTING AND ALLOCATION: ROUTATION

We start this section with the explanation of the basic terminology for routation algorithms in Section 2.1. After that, we discuss the algorithms we used and tested their performance on the 1750 test cases in Sections 2.2 and 2.3 respectively.

2.1. Notation and terminology

When determining a valid routation for a distributed environment, properties and constraints of the elements of the distributed environment are to be used. The elements are the application components, dataflows, devices, communication links and communication paths. In Figure 1, we graphically present a small distributed environment. The rectangles full of lines represent the application components, and the arrows between them the logical interaction between application components. Application components are executed on devices, which are interconnected via a transport network.

The properties and constraints of the elements of the distributed environment provide the algorithms with sufficient information for the search to a routation. When trying to find an optimal routation, an objective function is needed. We define these here as follows[1]:

property A property is a statement about a characteristic of an element, which can be identified by examining the element.

constraint A constraint is a statement about an element that restricts the routation of one set of elements (e.g. applications) onto another set of elements (e.g. computing and communication resources)). If a certain routation violates a constraint, that routation is an invalid mapping.

objective function An objective function assigns a value to each routation in such a way that a more preferred routation is assigned a higher (or lower) value. Objective functions thus rank routations.

To each element in Figure 1 a *descriptor* is attached which contains the name of the element, the properties of the element (given in the left half of the descriptor) and the possible constraints on it (given in the right half of the descriptor). The following properties and constraints are used (the subscript x denotes the element to the which property/constraint belongs):

Properties: et_x: execution time, er_x: execution rate, l_x: load, s_x: size (memory required), pr_x: precedence relations, v_x: volume, pol_x: processing overhead local communications, por_x: processing overhead remote communications, t_x: type (certain device type required), m_x: memory available, pc_x: processing capacity, fr_x: failure rate, rr_x: repair rate, c_x: cost, tr_x: transmission rate, td_x: transit delay, cl_x: number of communication links.

Constraints: d_x: device (specific device required), rt_x: response time required, rel_x: reliability required, av_x: availability required, c_x: max cost allowed, cl_x: number of communication links required, lb_x: lower bound load, ub_x: upper bound load, m'_x: available memory on a device, ac_x: upper bound on number of application components, df_x: upper bound on number of dataflows.

The objective of the optimization algorithms is to maximize the probability that a failure does not occur while the devices and communication paths are active. For every device the probability that it does not fail when it is processing the application components assigned to it is computed. For example, the time that an application component ac is using a device d is defined by $et_{AC}(ac, d) = \frac{et_{AC}(ac)er_{AC}(ac)}{pc_D(d)}$. For device d under routation $\mathcal{RA}$ this probability $R_d(\mathcal{RA})$ is equal to (with $x_{ac,d}$ equal to 1 if application component ac is assigned to device d, and 0 otherwise):

$$R_d(\mathcal{RA}) = \exp\left(-\sum_{ac=1}^{|AC|} x_{ac,d}et_{AC}(ac,d)fr_D(d)\right) \qquad (1)$$

---

[1]In the literature, many different terms are used to indicate characteristics of an element (e.g., parameter [9, 10]), limitations on assignments (e.g., constraint [5, 11] or requirement [12]) and functions that need to be optimized (e.g., cost function [4, 13] or performance goal [4]).

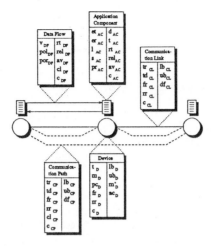

Figure 1: View on distributed environments with respect to routations

For the processing overhead for (local and remote) communications, expressed by R_{por}, and communication paths transporting data, expressed by R_{cp}, a similar expression can be derived [6, 7, 14, 15]. Since for an infinite time horizon these probability tend to zero, they are calculated for a certain time period. Using these probabilities, $R(\mathcal{RA})$ represents the probability that a failure does not occur during the period of length t in which the devices and communication paths are processing or transmitting the application components and dataflows assigned to them by routation $\mathcal{RA}$. The total reliability $R(\mathcal{RA})$ can then be calculated as follows (under the assumption these probabilities are independent and the failure times are exponentailly distributed):

$$R(\mathcal{RA}) = (\prod_{d=1}^{|D|} R_d(\mathcal{RA}))(\prod_{d=1}^{|D|} R_{por}(\mathcal{RA}))(\prod_{cp=1}^{|CP|} R_{cp}(\mathcal{RA})),$$
$$= \exp(-F_R(\mathcal{RA})). \qquad (2)$$

Maximizing the probability $R(\mathcal{RA})$ is equivalent to minimizing the function $F_R(\mathcal{RA})$. The function $F_R(\mathcal{RA})$ is the objective function our algorithms try to minimize in order to find an optimal routation.

2.2. The algorithms and their performance

For an efficient and effective solution of the routation problem we present a smart combination of a heuristic and the A^* algorithm. We first preseent the A^* algorithm, then a smart heuristic and then a combination of these two.

The A^* algorithm The A^* algorithm searches for an optimal routation by representing it as the (intelligent) search for a minimal-cost path in a tree [8]. Each node in the search tree represents a different, *partial* (not all application components and dataflows are yet assigned), routation. An edge along the path represents the assignment of an application component or dataflow.

In the partial routation the first components and the first dataflows are assigned. This implies that the set of applications and dataflows are ordered. We assume that not every mapping exists, but that only those exist in which all dataflows that are assigned are used by already assigned application components.

The root node represents an empty routation, i.e. no application components nor dataflows assigned, all leaf nodes represent complete routations, i.e. all application components and dataflows have been assigned. The basic cycle of operations of the A^* algorithm starts with the selection of a node from a set of not yet expanded nodes, called the *OPEN* set. Initially, only the root node is in *OPEN*.

If the node selected is a leaf node, we have found a complete routation. Under a certain condition on the selection mechanism (see below), we can guarantee that the first leaf node encountered represents an optimal routation, at which point the A^* algorithm may terminate.

If the selected node is not a leaf node, A^* will *expand* the node, i.e., generate all children of that node. A child node represents the partial routation of the parent node, with an extra application component or dataflow assigned. We take care to generate only child nodes that do not violate constraints. The children are then inserted into the *OPEN* set, and A^* will select the next node to expand.

Choosing the node to expand is done using a heuristic evaluation (objective) function $f(n)$ (our $F_R(\mathcal{RA})$), which is calculated for every node n upon generation. The node in *OPEN* that has currently the lowest value for $f(n)$ is called the *most promising node*, and this is the node that A^* selects for expansion next. Note that the nodes in the set *OPEN* do not have to be all of the same level, and that the most promising node is not always one of the nodes of the deepest level in *OPEN*. This implies that backtracking is allowed. In our application, $f(n)$ represents an estimate of the cost of a complete optimal routation.[6, 14]

The heuristic $H3''$ algorithm The $H3''$ algorithm is based on two other algorithms, the $H1''$ and the $H2''$ algorithm. $H3''$ first determines two routations, one in the way algorithm $H1''$ does, the other in the way $H2''$ does. From these two routations the best is chosen. These algorithms themselves are derived from the $H1'$ and the $H2'$ algorithm which again are derived from $H1$ and the $H2$. In this section we start with the $H1$ and the $H2$ algorithm, which are based on algorithm 4 of Shatz [6] and describe the succesive improvement made whichs lead to the $H3''$ algorithm. The details of this improved versions are presented by Franken *et al.* [14].

The $H1$ routation algorithm first sorts the application components in decreasing order of total volume of communication with other application components. The second heuristic routation algorithm, $H2$, sorts the application components in decreasing order of execution time per second. Hereby, we treat both communication and processing-oriented applications in a fair way. Algorithms $H1$, $H2$ sort the application components and assign the application components and the dataflows one by one to the devices and communication paths. Backtracking is not allowed, so once a decision has been made, it cannot be turned back. At each decision point that pair of device and set of communication paths is chosen that has the minimum increase in the objective function and that does not violate any constraint. But it could be that in this way a device d is already occupied, by a number of application components that do not *have* to be assigned to d, at the moment an application component *ac* having a *device constraint* $d_{AC}(ac) = d$ will be assigned. In such a case, the algorithms can not determine a valid routation although one might exist.

For application components having a device constraint we can use *advance reservations*. Adding these reservations to algorithms $H1$, $H2$ and $H3$ results in the enhanced heuristic algorithms $H1'$, $H2'$ and $H3'$ for which many of the device-constraint related problems in the basic algorithms can be circumvented. Running the enhanced heuristic algorithms on a number of cases showed that using reservations for application components having a device constraint resulted indeed in a higher number of solved cases. But in cases that an application component *ac* has a *device type constraint* $t_{AC}(ac)$ and it was assigned as one of the last application components, it could be that all devices of the required type or types were already occupied by other application components, not depending on these devices. Using reservations will not solve this completely, since an application component could still be assigned to more than one device.

A solution is the use of a *possible set* for each application component. This set contains all devices to which an application component can still be assigned. First for each application component the possible set is determined, using all constraints on the application components. Then, at each decision point all possible sets are checked and reservations are made for those application components having a possible set containing just one device. After each decision, all possible sets are adjusted according to the assignment made. This is another form of *constraint propagation* [16, 17]. Adding possible sets to algorithms $H1'$, $H2'$ results in the re-enhanced heuristic algorithms $H1''$, $H2''$ and $H3''$.

The combined algorithm: $A^*/H3''(x)$ The A^* algorithm determines optimal routations, but it has an exponential worst case time complexity and sometimes it runs out of memory. Algorithm $H3''$ has a polynomial time complexity, but it determines (sub)optimal routations and it does not always find a routation, although one exists. If a (sub)optimal routation does not have to be determined as fast as possible, but an algorithm is allowed to spend a certain time searching for an optimal routation, a combination of algorithms A^* and $H3''$, denoted as $A^*/H3''(x)$, can be used to find a routation. If the A^* algorithm finds a routation within x seconds, this routation will be the result of algorithm $A^*/H3''(x)$. If after x seconds, however, the A^* algorithm does not find a routation, it is stopped and a routation will be determined by algorithm $H3''$.

2.3. Results

The algorithms have been tested on 1750 cases. These cases were divided into seven classes of 250 cases, each class having a specific number of application components (4, 6 and 8) and a specific number of devices (3, 4 and 5). The properties and constraints were, within a range of allowed values, completely randomly generated for each case. Giving every application component and dataflow every possible constraint would result in a high number of cases for which a routation does not exist. Therefore we used probabilities for the existence of every ⟨application component , constraint⟩-pair and ⟨dataflow , constraint⟩-pair during the generation of the cases. For the same reason the cost constraints c_{AC} and c_{DF} were not used. In [14] the ranges of allowed values for all properties and constraints, together with the mentioned probabilities, are given. For the comparison of the algorithms we use the following criteria ; the *number* of found routations, the *quality* of the determind routations, and the *time* needed to determine routations. For the comparison of the quality we use the relative error $d(H)$. Thie relative error for algorithm H is equal to

$$d(H) = \begin{cases} \dfrac{R(\mathcal{RA}_{A^*}) - R(\mathcal{RA}_H)}{R(\mathcal{RA}_{A^*}) - R(\mathcal{RA}_{\overline{A^*}})}, & \text{if } R(\mathcal{RA}_{A^*}) \neq R(\mathcal{RA}_{\overline{A^*}}), \\ 0, & \text{if } R(\mathcal{RA}_{A^*}) = R(\mathcal{RA}_{A^*}), \end{cases} \tag{3}$$

where A^* is an adjusted version of the A^* algorithm that determines the valid routation with the *highest* instead of the lowest value for the objective function. The relative error $d(H)$ ranges from 0 to 1, and is equal to 0 if algorithm H determined an optimal routation, and equal to 1 if algorithm H determined the worst case routation.

The number of routations found by the algorithms A^*, $H3''$ and $A^*/H3''(x)$ for the various classes of cases are given in Table 1. For $A^*/H3''(x)$ we vary x over to 1, 2, 5, 10, and 15.

Table 1: Number of found routations by A^*, $H3''$ and $A^*/H3''(x)$

	A^*	$H3''$	$A^*/H3''(x)$				
			$x=1$	$x=2$	$x=5$	$x=10$	$x=15$
$\lvert AC \rvert = 4, \lvert D \rvert = 3$	194	188	193	193	193	193	193
$\lvert AC \rvert = 6, \lvert D \rvert = 3$	135	111	125	127	129	130	130
$\lvert AC \rvert = 8, \lvert D \rvert = 3$	45	36	38	39	40	41	41
$\lvert AC \rvert = 6, \lvert D \rvert = 4$	180	166	173	175	177	178	180
$\lvert AC \rvert = 8, \lvert D \rvert = 4$	132	107	108	109	112	116	117
$\lvert AC \rvert = 6, \lvert D \rvert = 5$	194	189	189	190	192	194	194
$\lvert AC \rvert = 8, \lvert D \rvert = 5$	162	155	155	155	155	156	158
Total	1042	952	981	988	998	1008	1013

The relative errors $d(H)$ of the routations found by algorithm $H3''$ and $A^*/H3''(x)$ are given in Table 2 as well as the percentage of the found routations that have a relative error $d(H)$ of at most 10%.

Table 2: Average $d(H)$ and percentage $d(H) \leq 0.1$ of algorithms $H3''$ and $A^*/H3''(x)$

	$H3''$		$A^*/H3''(x)$									
			$x=1$		$x=2$		$x=5$		$x=10$		$x=15$	
	$d(H)$	≤ 0.1	$d(H)$	≤ 0.1	$d(H)$	≤ 0.1	$d(H)$	≤ 0.1	$d(H)$	≤ 0.1	$d(H)$	≤ 0.1
$\lvert AC \rvert = 4, \lvert D \rvert = 3$	0.0168	96%	0.0023	99%	0.0005	100%	0.0002	100%	0.0000	100%	0.0000	100%
$\lvert AC \rvert = 6, \lvert D \rvert = 3$	0.0602	79%	0.0144	95%	0.0082	97%	0.0045	98%	0.0029	99%	0.0027	99%
$\lvert AC \rvert = 8, \lvert D \rvert = 3$	0.0963	64%	0.0613	80%	0.0592	81%	0.0505	84%	0.0229	91%	0.0176	94%
$\lvert AC \rvert = 6, \lvert D \rvert = 4$	0.0704	78%	0.0459	87%	0.0320	90%	0.0200	93%	0.0106	96%	0.0054	98%
$\lvert AC \rvert = 8, \lvert D \rvert = 4$	0.1144	62%	0.1031	64%	0.0962	65%	0.0656	76%	0.0410	85%	0.0291	89%
$\lvert AC \rvert = 6, \lvert D \rvert = 5$	0.0469	83%	0.0364	87%	0.0320	88%	0.0222	91%	0.0097	97%	0.0055	98%
$\lvert AC \rvert = 8, \lvert D \rvert = 5$	0.0849	71%	0.0832	71%	0.0821	71%	0.0724	73%	0.0541	78%	0.0374	85%
Total	0.0594	80%	0.0408	86%	0.0352	88%	0.0261	91%	0.0160	94%	0.0107	96%

From these tables it follows that for larger distributed environments, the increase in the quality of the determined routations is less than for smaller distributed environments, when x increases. This could be expected, since the larger the distributed environments, the more time the A^* algorithm needs to find a routation, i.e., the A^* algorithm exceeds in more cases its time limit. This means that in more cases the routation determined by algorithm $H3''$ is used, which determines suboptimal routations.

Table 3: Average time in seconds needed to find a routation by A^*, $H3''$ and $A^*/H3''(x)$

	A^*	$H3''$	$A^*/H3''(x)$				
			$x = 1$	$x = 2$	$x = 5$	$x = 10$	$x = 15$
$|AC| = 4, |D| = 3$	1.193	0.029	0.129	0.160	0.214	0.271	0.322
$|AC| = 6, |D| = 3$	37.369	0.045	0.537	0.795	1.281	1.891	2.350
$|AC| = 8, |D| = 3$	120.010	0.056	0.887	1.519	3.167	5.236	6.836
$|AC| = 6, |D| = 4$	14.754	0.086	0.773	1.202	2.023	2.810	3.364
$|AC| = 8, |D| = 4$	276.780	0.113	1.086	1.987	4.149	6.734	8.617
$|AC| = 6, |D| = 5$	13.399	0.161	0.968	1.560	2.826	4.140	4.967
$|AC| = 8, |D| = 5$	128.626	0.200	1.185	2.132	4.761	8.551	11.678
Total	70.349	0.105	0.758	1.260	2.441	3.900	5.006

The times needed to find a routation by algorithms A^*, $H3''$ and $A^*/H3''(x)$ are given in Table 3 (in seconds). From this table it follows that the larger the distributed environments become, the more algorithm $A^*/H3''(x)$, on average, approaches or even exceeds its time limit allowed for algorithm A^*. If, on average, the time limit is exceeded, this indicates that the routation determined by algorithm $H3''$ is often the result. A careful choice of the time limit is therefore necessary in order to use the advantages of algorithm $A^*/H3''(x)$.

3. ROUTATION IN A MULTI-PARTY VIDEO-PHONE

To demonstrate some of the capabilities of the heuristic routation algorithm $H3''$ the *xallocator* has been implemented in an experimental distributed system providing a multi-party videophone service. The experimental distributed system is described in Section 3.1 and the *xallocator* is described in Section 3.2.

3.1. The distributed environment of the multi-party video-phone

The multimedia application used in the demonstration is a *multi-party video-phone*. This application has been implemented using *ANSAware*. Video recorded at one computer system, using a video camera and a special video card, is sent to two other computer systems, which perform a similar task. For this test we equipped three computer systems. At each computer all three video streams are displayed.

The application consists of a number of application components: three video-phones, three duplicators, a stream manager and a stream factory. These application components and the dataflows between them are shown in Figure 2.

At each computer to which a video camera is attached, a video-phone resides. Each video-phone sends the video signal it records to a duplicator, which sends this data to the other two video-phones. The stream manager and the stream factory are used for managing

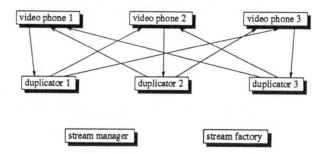

Figure 2: The application components and the dataflows of the multi-party video-phone

the video streams between the video-phones and the duplicators, but have, after starting up the application, no interaction with the other application components.

The properties of the application components are shown in Table 4. These properties are theoretically determined approximations of the real properties; actual monitoring of the application components has not been performed. The three video-phones have equal properties, just like the three duplicators.

Table 4: Properties of the application components

	et_{AC} (ms)	$er_{AC}(/s)$	$l_{AC}(/s)$	s_{AC}(Kbytes)
video-phone	20	20	0.4	3072
duplicator	10	20	0.2	1024
stream manager	1	1	0.001	1024
stream factory	1	1	0.001	1024

Except for the video-phones, the application components originally do not have any constraints. The video-phones all have a device constraint d_{AC}, since they interact with the user of a specific device. Reconfiguring the distributed environment according to a new routation means migrating one or more application components and/or dataflows. Since the only application components that can be migrated are the duplicators, we assign a device constraint d_{AC} to the stream manager and the stream factory. The dataflows all have identical properties and no constraints. Since each frame has a size of at most 8 Kbytes, the volume v_{DF} of each dataflow is equal to 160 Kbyte/s. The processing overhead por_{DF} for each dataflow is assumed to be 40 ms/s.

The distributed system used for the demonstration consists of a number of different Sun workstations connected by an Ethernet. The properties of and the constraints on these devices as used in the demonstration are given in Table 5. Note that the devices do not have a cost property c_D. The memory constraint m'_D is set to 1, meaning that all memory present is available.

The values presented in these table do not intend to represent the actual properties of and constraints on the devices. The devices are fully connected, i.e., a communication path exists between each pair of devices. The properties of and constraints on these communication paths allow all possible routings for the dataflows of the multimedia application.

Table 5: Properties of and constraints on the devices

	t_D	m_D (Kbytes)	pc_D	fr_D(/s)	ub_D	ac_D
sun012/015/039	Sparc10/Parallax	16384	5.0	5^{-5}	0.8	16
sun027/029	Sparc IPC	8192	1.0	5^{-5}	0.8	16
sun035	Sparc ELC	8192	0.75	5^{-5}	0.8	16

3.2. Overview of and experiences with xallocator

The *xallocator* is an application that allows the user to view the current allocation of the application components of the multimedia application, change the properties of and the constraints on the devices and simulate the presence of other applications in the distributed environment, by changing the usage of the devices. After one or more changes to the distributed environment have been made, a routation is determined using algorithm $H3''$. If the current routation is no longer valid or the just determined routation has a higher reliability, the distributed environment is reconfigured according to this new routation. In Figure 3, an example view of the main screen of *xallocator* is shown. This screen contains the current allocation of the application components and a number of buttons. A description, in terms of properties and constraints, of an application components or a device will be shown after selecting it. Note that the allocation of two other application components is shown as well: that of the trader and the *xallocator* itself.

After using *xallocator* some time, we noticed that the possible assignment of the duplicators to other devices after changing one or more values very well visualizes what a routation algorithm is capable off. The time needed for a migration is considerably larger than the times needed to determine a new routation. In all cases the time needed for a reconfiguration should be minimized and un-noticable by the users. However, there will always be a trade-off between serious QoS degradation and a short unavailability (because of a reconfiguration). Futhermore, despite the fact that, considering the changes in the properties and constraints, only one duplicator had to migrate, sometimes a new routation was determined for which all three duplicators had to migrate. The reason for this is that when there is more than one device that increases the objective the least, the algorithm chooses the first of these devices to assign an application component to.

One possible solution to this problem is to exchange the first device in the set of devices with the device to which application component ac is currently assigned, just before assigning application component ac when determining the new routation.

4. CONCLUSIONS AND FURTHER RESEARCH

In this paper we have addressed the issues of routing and allocation in distributed systems. In particular, we addressed these issues in an integrated way, the so-called routations, in order to ensure an optimal realization of the end-to-end quality of service. We then proposed an optimal routation algorithm based on the A^* algorithm and a heuristic, $H3''$, derived from the A^* algorithm. A smart combination of the A^* algorithm with the $H3''$ heuristic gives us a very good routation algorithm, called $A^*/H3''(x)$. This algorithm can be used in an

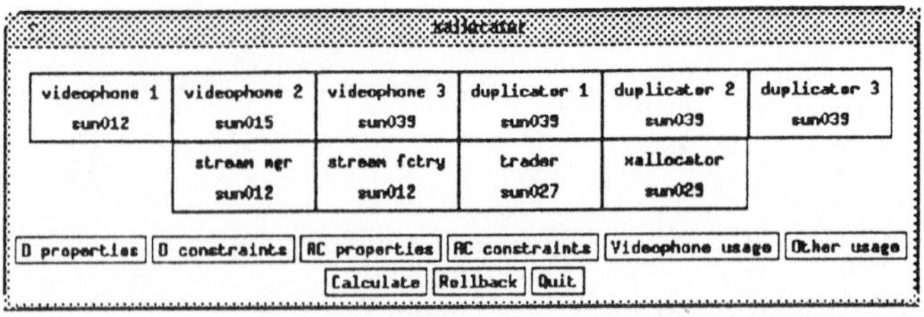

videophone 1	videophone 2	videophone 3	duplicator 1	duplicator 2	duplicator 3
sun012	sun015	sun039	sun039	sun039	sun039
	stream mgr	stream fctry	trader	xallocator	
	sun012	sun012	sun027	sun029	

D properties	D constraints	AC properties	AC constraints	Videophone usage	Other usage

Calculate	Rollback	Quit

Pressing a button results in the following actions: **D properties**: An overview of the properties of the devices will be shown. Selecting one of these properties allows the user to modify this property; **D constraints**: An overview of the constraints on the devices will be shown. Selecting one of these constraints allows the user to modify this constraint; **AC properties**: An overview of the properties of the application components will be shown. These properties cannot be modified; **AC constraints**: An overview of the constraints on the application components will be shown. These constraints cannot be modified; **Video-phone usage**: An overview of the usage of the devices by the application components of the multi-party video-phone is shown; **Other usage**: An overview of the usage of the devices by the other, simulated, applications in the distributed environment is shown. Selecting one of these usages allows the user to modify this usage; **Calculate** A routation for the current settings of the distributed environment is determined. If no routation can be determined, or if the current routation is still valid and has a reliability equal to the reliability of the just determined new routation, this is reported. If a new routation is determined that is closer to optimal than the current routation or the current routation is no longer valid, the distributed environment is reconfigured according to this new routation. The dynamic reconfiguration is performed; **Rollback** The changes made to the distributed environment since the last determination of a valid routation are discarded. **Quit** Terminates *xallocator.*

Figure 3: An example view of the main screen of xallocator

environment with real-time requirements. This claim is proven to be true by a statistical analysis of 1750 randomly selected test cases. The lower the timing requirements, the more this algorithm operates as a normal A^* algorithm, i.e., the better it becomes. Also, the higher the timing requirements, the more it acts as a pure heuristic algorithm. In this way, this algorithm combines the best of both worlds, in an adaptable fashion.

We applied the $H3''$ algorithm on an ANSAware-based videophone application using the *xallocator*. It appeared that routation is useful but more work is needed on the algorithms as well as on the migration aspect of application components.

As topics for future research, we envisage the following. The current routation algorithms optimize only one objective function. Different applications, however, will have different QoS requirements. Therefore, research is needed to allow for the use of different objective functions for different applications during the creation of a single routation.

Another topic of interest is the use of incremental or adaptable routations. The aim of these is to allow for the (stepwise) addition or removal of new applications without re-routating, i.e., reallocating and rerouting, the existing environment, thereby still satisfying all QoS requirements.

With respect to the dependability and reliability aspects of the QoS requirements, research is needed towards strategies for the routation of replicated applications [18].

A final topic we would like to mention is research towards distributed or decentralised routation algorithms. The algorithms described in this paper are fully centralised. In order to achieve acceptable performance in larger systems, decentralisation might be a good way to go.

References

[1] L.J.N. Franken and B.R.H.M. Haverkort. The Performability Manager. *IEEE Network: The Magazine of Computer Communications, Special Issue on Distributed Systems for Telecommunications,* 8(1):24–32, Januari 1994.

[2] L.J.N. Franken, R.H. Pijpers, and B.R. Haverkort. Modelling Aspects of Model Based Dynamic QoS Management by the Performability Manager. In G. Haring and G. Kotsis, editors, *Computer Performance Evaluation. Modelling Techniques and Tools. Proceedings of the 7th International Conference, Vienna, Austria,* pages 89–110. Lecture Notes in Computer Science, Springer-Verlag, Volume 794, May 1994.

[3] T.C.K. Chou and J.A. Abraham. Load Balancing in Distributed Systems. *IEEE Transactions on Software Engineering,* SE-8(4):401–412, July 1982.

[4] N.S. Bowen, C.N. Nikolaou, and A. Ghafoor. On the Assignment Problem of Arbitrary Process Systems to Hetrogeneous Distributed Computer Systems. *IEEE Transactions on Computers,* 41(3):257–273, March 1992.

[5] W.W. Chu, L.J. Holloway, M. Lan, and K. Efe. Task Allocation in Distributed Processing. *IEEE Computer,* 13(11):57–69, November 1980.

[6] S. M. Shatz, J. Wang, and M. Goto. Task Allocation for Maximizing Reliability of Distributed Computer Systems. *IEEE Transactions on Computer,* 41(9):1156–1168, December 1992.

[7] C.M. Woodside and G.G. Monforton. Fast Allocation of Processes in Distributed and Parallel Systems. *IEEE Transactions on Parallel and Distributed Systems,* 4(2):164–174, February 1993.

[8] T.H. Cormen, C.E. Leiserson, and R.L. Rivest. *Introduction to Algorithms.* MIT Press, Massachusetts Institute of Technology, Cambridge, Massachusetts, 1990.

[9] D.P. Anderson. Metascheduling for Continuous Media. *ACM Transactions on Computing,* 11(3):226–252, August 1993.

[10] K.G. Shin, C.M. Krishna, and Y. Lee. Optimal Dynamic Control of Resources in a Distributed System. *IEEE Transactions on Software Engineering,* 15(10):1188–1197, October 1989.

[11] K. Efe. Heuristic Models of Task Assignment Scheduling in Distributed Systems. *IEEE Computer,* 15(6):50–56, June 1982.

[12] J. P. Huang. Modeling of Software Partition for Distributed Real-Time Applications. *IEEE Transactions on Software Engineering,* 11(10):1113–1126, 1985.

[13] S.M. Shatz and J-P. Wang, editors. *Tutorial: Distributed Software Engineering.* IEEE Computer Society Press, 1989.

[14] L.J.N. Franken, P. Janssens, B.R.H.M. Haverkort, and E.P.M. Van Liempd. Dynamic Routation in Distributed Environments. Submitted for publication, 1994.

[15] J. Laprie and K. Kanoun. X-Ware Reliability and Availability Modeling. *IEEE Transactions on Software Engineering,* 18(2):130–147, February 1992.

[16] V. Kumar. Algorithms for Constraint-Satisfaction Problems: A Survey. *AI Magazine,* 13:32–44, Spring 1992.

[17] P. Meseguer. Constraint Satisfaction Problems: An Overview. *AI Communications,* 2(1):3–17, March 1989.

[18] L.J.M. Nieuwenhuis. *Fault Tolerance Through Program Transformation.* PhD thesis, University of Twente, 1990.

SESSION ON

Using Formal Semantics

31

Some Results on Cross Viewpoint Consistency Checking

Howard Bowman, John Derrick and Maarten Steen
University of Kent at Canterbury, U.K. (hb5,jd1,mwas)@ukc.ac.uk.

The ODP multiple viewpoints model prompts the very challenging issue of cross viewpoint consistency. This paper considers definitions of consistency arising from the RM-ODP and relates these in a mathematical framework for consistency checking. We place existing FDTs, in particular LOTOS, into this framework. Then we consider the prospects for viewpoint translation. Our conclusions centre on the relationship between the different definitions of consistency and on the requirements for realistic consistency checking.

Keyword Codes: D.2.1, D.2.10
Keywords: Requirements/Specifications, Design

1. INTRODUCTION

Multiple viewpoints are a cornerstone of the Open Distributed Processing (ODP) model [12]; they enable a different perspective of a system to be presented to different observers. Each viewpoint is a partial view of the complete system specification. It is through this separation of concerns that the inherent complexity of a complete distributed system is decomposed. ODP supports five viewpoints: *enterprise, information, computational, engineering* and *technology*.

However, the subdivision of a system specification raises the issue of *consistency*. Descriptions of the same or related entities will appear in different viewpoints and it must be shown that the multiple specifications are not in conflict with one another. The development of tools and techniques to check the consistency of viewpoint specifications is of great importance, however, it is also extremely challenging. In particular, in its most general form, consistency checking requires specifications in different notations to be related. This is because it has been recognised that different notations are appropriate for different viewpoints. Relating model based specification notations, such as Z, to languages which explicitly model the 'temporal ordering' of abstract events, such as LOTOS or SDL, is particularly challenging.

This paper addresses the question: what is an appropriate definition for consistency? The RM-ODP is ambiguous in this respect. We will clarify the relationship between a number of possible consistency definitions and we will consider how different FDTs, in particular LOTOS, can be integrated into a consistency checking framework and then we will discuss the different options for translation. The results of the paper centre on the relative strengths of definitions and the information that needs to be made available in

---

[0]The work presented in this paper was partially funded by British Telecom Labs., U.K. and partially by the U.K. Engineering and Physical Sciences Research Council (grant number GR/K13035.)

order that an appropriate consistency check can be applied.

We consider consistency in very general terms. In particular, we do not consider specific instances of consistency, such as between the information and computational viewpoints. This reflects our adopted strategy, which is to clarify the general form of consistency as a relationship between arbitrary specifications before considering specific instances of consistency. This paper is reporting results of the initial, general, phase of our work.

The paper begins by exploring the extent of consistency relationships in ODP (in section 2). Section 3 discusses appropriate definitions of consistency arising from the RM-ODP and then section 4 relates these to a mathematical framework for consistency checking. Section 5 places existing FDTs into this framework. Then we outline a number of possible approaches to translation in section 6. Finally, we present concluding remarks in section 7.

2. THE EXTENT OF CROSS VIEWPOINT RELATIONSHIPS

Due to the central role viewpoints play, consistency relationships are extremely pervasive in ODP. Consistency arises in the following situations:-

Conformance Assessment. Conformance assessment for ODP is extremely broad. In particular, it encompasses both *conformance testing* (i.e. relating real implementations to specifications) and *specification checking* (i.e. specification to specification relationships), this distinction was particularly emphasised in PROST [7]. Verification of cross viewpoint consistency is an important example of specification checking.

System Development. The RM-ODP does not prescribe a particular system development methodology and a number of development methodologies could be envisaged. However, each viewpoint specification is, at least potentially, at the same level of abstraction; suggesting that viewpoints are related horizontally relative to a vertical system development. This is in contrast to classic waterfall development methodologies. PROST [7] has investigated such a, fully general, system development methodology for ODP. This is depicted in figure 1 and uses a number of specification to specification transformations, such as *translation*, *refinement* and *unification*, in order to generate a composite 'implementation' specification. Translation maps specifications into new languages, refinement has the usual meaning and unification is a transformation which enables specifications in the same language to be combined. Consistency is implicit in such a system development methodology. For example, two specifications would be viewed as inconsistent if a common unified specification did not exist. Thus, consistency arises during unification of specifications in models of ODP system development.

Architectural Semantics. The use of different FDTs in defining the ODP architectural semantics and the fact that the architectural semantics (when complete) will span a number of the viewpoint languages suggests consistency relationships will have relevance in this domain as well. Two forms of consistency relationship can arise. Firstly, there is a need to relate the architectural semantics of different viewpoints in order to determine that the FDT interpretations are consistent. Secondly, there is a need to demonstrate that descriptions in different FDTs of particular architectural semantics entities are consistent.

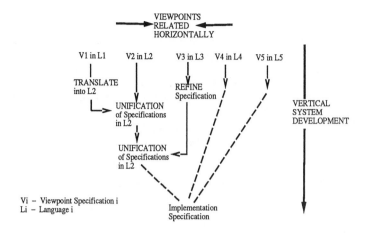

Figure 1: PROST System Development Scenario

We strongly believe that a formal approach to consistency checking should be employed. In particular, the ability to reason rigorously about the specifications under consideration is of vital importance. We will assume the use of formal description techniques as viewpoint languages in the remainder of this paper.

3. CONSISTENCY DEFINITION

This section highlights three possible interpretations of consistency that appear in the RM-ODP, the first two appear in part 1 (clause 12.2) and the third apears in part 3 [12] (clause 10). Although, the first of these definitions is only alluded to; it is not formally proposed as a definition.

Definition 1
(1.1) Two specifications are consistent iff they do not impose contradictory requirements.
(1.2) Two specifications are consistent iff it is possible for at least one example of a product (or implementation) to exist that can conform to both of the specifications.
(1.3) Two specifications are consistent iff they are both behaviourally compatible with the other.

This last interpretation is a rewording of the RM-ODP definition. This is because the RM-ODP definition is expressed in terms of relating specific viewpoints. We are considering more generalised notions of consistency, thus, we have brought the definition into line with the other definitions in order to facilitate a direct comparison. In addition note, that all these definitions are symmetric, i.e. if a specification S is consistent with a specification R then R is consistent with S. This is a reasonable intuitive requirement for consistency.

Behavioural compatibility is defined as follows:

Definition 2 (Behavioural Compatibility) *An object is behaviourally compatible with a second object, with respect to a set of criteria, if the first object can replace the second*

object without the environment being able to notice the difference in the objects behaviour on the basis of the set of criteria.

These three consistency interpretations blur over the fact that specifications may be in different FDTs and that it may not be possible to relate specifications directly without some element of translation. In fact, in the RM-ODP the third of these definitions includes a notion of translation which is described in terms of 'information preserving' transformations between languages. Translation will be discussed in section 6.

Each of these notions of consistency is intuitively reasonable. However, the question arises: what is the relationship between the interpretations and, in particular, are these definitions of consistency themselves consistent? In fact, the different interpretations are likely to be applicable in different settings. For example, definition 1 is relevant to consistency checking in a logical setting, e.g. in an FDT such as Z which is based on first order logic.

We seek to reconcile these interpretations through formalisation. We formalise the first notion of consistency as follows,

Definition 3 $S_1\ C_1\ S_2\ \textit{iff}\ \neg(\exists \psi\ s.t.\ S_1 \models \psi\ \wedge\ S_2 \models \neg\psi)$

where $\models$ is the satisfaction relation of the specification's logic. This definition states that two specifications are consistent if and only if there is no property that holds over one of the specifications and its negation holds over the other specification.

To interprete consistency 1.2 we need a formal interpretation of conformance. There is a difficulty here because conformance relates implementations to specifications and implementations are not amenable to formal interpretation. The classical approach to handling this difficulty is to only consider conformance up to a, so called, *implementation specification*. This is a specification that describes a real implementation in as much detail that a direct mapping from the implementation specification to the real implementation can be found. Thus, it is normal just to consider conformance relations between specifications, see [4] [5] [14] for typical approaches. However, implementation specifications relate to real implementations in different ways for different FDTs and, in particular, for some FDTs not all implementation specifications are implementable. For example, a Z specification that contains an operation $[n! : N | n! = 5 \wedge n! = 3]$ has no real implementation.

Our approach then is to divide conformance testing into two parts. Firstly, we consider conformance up to implementation specifications, using a relation $conf \subseteq SPEC \times SPEC$, and then we consider conformance of implementation specifications to real implementations, using a relation $\mathbf{conf} \subseteq SPEC \times IMP$ [1]. Where $SPEC$ is the set of possible ODP specifications and IMP is the set of possible ODP implementations.

By way of clarification, $S_1 conf S_2$ expresses the property that specification S_2 conforms to specification S_1, i.e. according to tests derived from S_1, S_2 cannot be distinguished from S_1. It should be noted that we have not specified how and what form of tests are derived from S_1; there are many options for such derivation [4] [5]. In a similar way $S\mathbf{conf}I$ expresses the property that I conforms to S. Interpretation 1.2 is now formalized as:-

---

[1] The order of our relations is in accordance with Z conventions and is opposed to LOTOS conventions

Definition 4 $S_1 \, C_{2.1} \, S_2$ *iff* $\exists S \in SPEC, I \in IMP \, s.t. (S_1 conf S \wedge S_2 conf S) \wedge S$ **conf** I.

i.e., two specifications are consistent iff an implementation specification which conforms to both and a real implementation of the implementation specification can be found. This definition is correct, but is not very useful since it uses **conf**, which is not subject to formal interpretation. In order to resolve this difficulty we introduce the concept of *internal validity* which holds whenever a specification is implementable:-

Definition 5 S *is internally valid, denoted* $\Psi(S)$, *iff* $\exists I \in IMP \, s.t. \, S conf I$

Ψ acts as a receptacle for properties of particular FDTs that make specifications in that FDT unimplementable. For example, a Z specification which contains contradictions would not be internally valid. Now we can redefine C_2 in a more usable way:

Definition 6 $S_1 \, C_{2.2} \, S_2$ *iff* $\exists S \in SPEC \, s.t. \, (S_1 conf S \wedge S_2 \, conf S) \wedge \Psi(S)$.

The third and final consistency interpretation hinges on the notion of behavioural compatibility which is defined in terms of an environment and unspecified criteria. We will consider specific instantiations of behavioural compatibility when we look at specific FDTs; at this stage we formulate the interpretation completely generally, for bc a particular instantiation of behavioural compatibility.

Definition 7 $S_1 \, C_3 \, S_2$ *iff* $S_1 \, bc \, S_2 \wedge S_2 \, bc \, S_1$.

Since consistency checking will occur at the specification checking stage of conformance assessment we actually need a mechanism to assess consistency that uses only specification checking relationships, i.e. refinement, unification and equivalence. We will seek to define natural interpretations of refinement, unification and translation and then consider how the different definitions of consistency can be related to the above three consistency interpretations.

4. A SPECIFICATION CHECKING FRAMEWORK

Translation. It seems natural to require that translation enforces equivalence, i.e. a translation of a specification should be equivalent to the original specification. The actual notion of equivalence required will be FDT dependent. However, we would certainly want translation to preserve equivalence due to conformance, which we denote $\equiv_{cf}$:

Definition 8 $S_1 \equiv_{cf} S_2$ *iff* $\{S : S_1 conf S\} = \{S : S_2 conf S\}$

Intuitively, two specifications are equivalent iff they determine exactly the same set of valid implementation specifications through $conf$. It should be pointed out that $\equiv_{cf}$ does not imply standard semantic equivalence; the equivalences of FDTs (such as observational and testing equivalences of process algebra) are likely to be stronger than $\equiv_{cf}$.

Refinement. Following [14] we define that S_2 is a refinement of S_1 as:-

Definition 9 $S_1 \sqsubseteq S_2$ *iff* $\{S : S_2 conf S\} \subseteq \{S : S_1 conf S\}$

i.e. refinement restricts the set of conformant implementation specifications. But, importantly, the implementations of a refinement are also implementations of the original specification.

Unification. Unification takes two specifications in the same language and produces a unified version which is a combination of the two specifications. By *combination* of specifications, we mean that unification should satisfy the property of common refinement, i.e. that $T_1, T_2 \sqsubseteq U(T_1, T_2)$, since an implementation that conforms to $U(T_1, T_2)$ should also conform to the original specifications T_1, T_2. In fact, we characterize unification as the least refinement of two specifications, with the following construction: $U(T_1, T_2) \in \{T : T_1, T_2 \sqsubseteq T \text{ and if } T_1, T_2 \sqsubseteq S \text{ then } T \sqsubseteq S\}$, see [8] for a discussion.

Consistency. A natural specification checking definition of consistency is that two specifications are consistent if their unification can be implemented.

Definition 10 *Given S_1 in language L_1 and S_2 in language L_2. Then $S_1 C_4 S_2$ iff there exists a specification language L_3 such that $S_1 \equiv_{cf} T_1, S_2 \equiv_{cf} T_2$ and there exists a $U(T_1, T_2)$ in L_3 such that $\Psi(U(T_1, T_2))$ for some T_1, T_2 in L_3.*

Notice in particular that the internal validity condition guarantees that a conformant implementation of the unification exists. In addition, this is our first interpretation of consistency that embraces translation. Properties of refinement, equivalence, unification and consistency can be found in appendix (ii).

Discussion. We now have four definitions of consistency C_1, $C_{2.2}$, C_3 and C_4. The first three of these arise from the ODP reference model and the third is a natural specification checking definition, which links notions of conformance to specification checking relationships such as refinement, unification and equivalence. We would clearly like to relate these definitions. However, a number of aspects of these definitions are FDT dependent. We will make the required FDT dependent comparison in the next two sections. We can, though, clarify our general approach, which is the following. Firstly, we view C_1 as a specialised form of consistency which is relevant to consistency checking in a logical setting and it will be captured by the internal validity property where it is relevant. The main focus of this paper, though, will be the relationship between $C_{2.2}$, C_3 and C_4 which are clearly in the same domain of reference.

The specification checking relationships of a particular FDT will not be equivalent to the corresponding definitions in our framework. However, our interpretation in this respect is that FDT relations that are stronger or equal to the framework definitions are appropriate, but relations that are either weaker or only partially intersect with the corresponding framework definition are not appropriate. Our intuition behind this interpretation is that consistency checking occurs during specification checking and that the specifier has knowledge about the nature of the specifications under consideration that is relevant to consistency, thus, at this stage of system development we can be more discriminating than is implicit in the framework. For example, the specifier may know that a specification is a functionality extension of another specification; that two specifications are strictly equivalent or that two specifications are related by reduction of non-determinism. This extra information should be used at the specification checking phase as long as it does not contradict the weaker conformance oriented definitions.

5. INSTANTIATING PARTICULAR FDTs

5.1 LOTOS Consistency Checking Relationships

Existing LOTOS relations can be instantiated into the consistency framework as follows:-

Conformance. A natural instantiation of our $conf$ relation is the LOTOS conformance relation, which we denote $\underline{conf}$ (a definition of $\underline{conf}$ can be found in appendix i).

Internal Validity. The internal validity concept is targetted at FDTs such as Z where specifications can exist which do not have implementations. All LOTOS specifications can, at least 'theoretically', be implemented (and we apologize for the circularity here). Thus, we view all LOTOS specifications as internally valid.

C_3. Consistency definition 1.3 is dependent upon the interpretation of behavioural compatibility, which in turn hinges on the interpretation of a specification's environment and the criteria imposed on that environment. The looseness of the definition of behavioural compatibility implies that one of a number of interpretations of C_3 could be made. It is our view that C_3 could be interpreted as any of the following:-

Definition 11
(i) $S_1 C_3^{\sim} S_2$ iff $S_1 \sim S_2$ - *Strong Bisimulation*
(ii) $S_1 C_3^{\approx} S_2$ iff $S_1 \approx S_2$ - *Weak Bisimulation*
(iii) $S_1 C_3^{te} S_2$ iff $S_1 te S_2$ - *Testing Equivalence*
(iv) $S_1 C_3^{cs} S_2$ iff $S_1 \underline{conf} S_2 \wedge S_2 \underline{conf} S_1$ - *$\underline{conf}$ symmetric*

Definitions (11.i) and (11.ii) view the environment as an unconstrained observer, in the sense of standard observational equivalences. In contrast, (11.iii) and (11.iv) view the environment as a tester for the specifications. The distinction between (11.iii) and (11.iv) is that (11.iii) implies robustness testing and (11.iv) implies restricted testing, see [4] [5] for a discussion of these alternatives. In the remainder of this paper we will concentrate on C_3^{cs}. Our reasons for this choice are two fold. Firstly, this interpretation agrees with the LOTOS definition of behavioural compatibility in Part IV of [12] and, secondly, we will show that, in comparison with $C_{2.2}$ and C_4, C_3^{cs} is a strong interpretation of consistency. Furthermore, C_3^{cs} is the weakest behavioural compatibility definition. Thus, since $C_3^{\sim} \implies C_3^{\approx} \implies C_3^{te} \implies C_3^{cs}$, from process algebra theory, C_3^{cs} bounds the relationship between C_3 and the other consistency definitions.

Refinement. We will focus on two of the most important LOTOS refinement relations, *extension* (which we denote $\underline{ext}$) and *reduction* (which we denote $\underline{red}$), see appendix 1 for definitions. Intuitively, the former of these characterizes when a specification validly extends the behaviour of another specification and the latter relation characterizes refinement through reduction of non-determinism. In order to accept $\underline{ext}$ and $\underline{red}$ as suitable refinement relations we must show that both imply $\sqsubseteq$. Extrapolating from the results of [14] we get that $\underline{ext} \Rightarrow \sqsubseteq$, but $\underline{red} \not\Rightarrow \sqsubseteq$ and $\underline{red} \not\Leftarrow \sqsubseteq$. Thus, $\underline{ext}$ can be instantiated without any difficult, but $\underline{red}$ causes problems. We resolve this problem by considering a relation $\underline{red*}$ which we define as follows: $\underline{red*} = \underline{red} \cap \sqsubseteq$.

    We will denote the instantiation of $\underline{ext}$ as the refinement relation in C_4 as $C_4^{\underline{ext}}$ and, similarly, the instantiation of $\underline{red*}$ in C_4 as $C_4^{\underline{red*}}$

Results. The following results arise from applying LOTOS relations to consistency:-

Proposition 1 *For $\underline{conf}$ all pairs of LOTOS processes are consistent by $C_{2.2}$*

Proof This follows from [13] which provides an algorithm that determines a common extension (i.e. $\underline{ext}$) for any pair of LOTOS processes and since $\underline{ext} \implies \underline{conf}$.

Proposition 2 *For $\underline{conf}$, $C_3^{cs} \subset C_{2.2}$*

Proof All we have to do is to demonstrate a pair of processes that are not related by $\underline{conf}$. This is straightforward. For example, for the processes, $S_1 := b; stop[]i; a; stop$ and $S_2 := b; c; stop[]i; a; stop$, $\neg(S_2 conf S_1)$. This is because $Ref(S_1, b) \not\subseteq Ref(S_2, b)$, e.g. $c \in Ref(S_1, b)$ but $c \notin Ref(S_2, b)$.

Proposition 3 $C_4^{ext} = C_{2.2}$

Proof This follows from the results of [13].

Proposition 4 *(i) $C_3^{cs} \cap C_4^{red*} \neq \emptyset$, (ii) $C_3^{cs} \not\subseteq C_4^{red*}$ and (iii) $C_4^{red*} \not\subseteq C_3^{cs}$.*

Proof We provide example LOTOS processes to demonstrate each of the properties.
(i) Consider the following trivial example. Take $S_1 = S_2 := a; b; stop$. Clearly, $S_1 \; C_3^{cs} \; S_2$. In order to show that also $S_1 \; C_4^{red*} \; S_2$, we choose their common refinement to be $S = S_1 = S_2 := a; b; stop$. Obviously, $S_1 red* S$ and $S_2 red* S$.
(ii) Take $S_1 := a; stop$ and $S_2 := i; a; stop [] b; c; stop$. Now $S_1 C_3^{cs} S_2$, but we will show that $\neg(S_1 C_4^{red*} S_2)$. Firstly, the only possible reduction of both S_1 and S_2 is the process $S := a; stop$. Now, take the implementation $T := a; stop [] b; stop$. This is a valid implementation with respect to S, i.e. $S conf T$. However, we can see that $\neg(S_2 conf S)$, because S refuses action c after the trace b. Therefore, $S_2 red* S$ does not hold.
(iii) Take $S_1 := a; (b; stop [] i; stop)$ and $S_2 := a; b; stop [] i; stop$. We can easily check that $\neg(S_2 conf S_1)$ and $\neg(S_1 conf S_2)$. Therefore, we have $\neg(S_1 \; C_3^{cs} \; S_2)$. However, $S_1 \; C_4^{red*} \; S_2$, which can be shown by taking $S := a; b; stop$ as the common refinement of S_1 and S_2. This is because $S_1 red S$ and $S_2 red S$, since all non-determinism in S_1 and S_2 has been resolved in S. In addition, as $Tr(S) = Tr(S_1) = Tr(S_2)$ we know that $S_1 ext S$ and $S_2 ext S$. Moreover, since $\underline{ext} \Rightarrow \sqsubseteq$, from [14], we know that $S_1 \sqsubseteq S$ and $S_2 \sqsubseteq S$.

These results are depicted in figure 2. Interestingly, though unification construction algorithms can be given which demonstrate that $C_3 \subset C_4^{red}$ and $C_3 \subseteq C_4$, these algorithms will not always yield the same unification, thus $C_4^{red} \cap C_4 \neq C_4^{red*}$. For further discussion of these relations see [16]. The following implications can be drawn from these results.

1. For LOTOS $C_{2.2}$ is very weak. In fact, it does not distinguish any processes.

2. In contrast, C_3 is a strong relation for LOTOS. In particular, none of the specification checking consistency relationships, i.e. C_4^{red*}, C_4^{red}, C_4^{ext}, imply C_3.

3. The relationship between C_4^{red*} and C_3^{cs} is not very satisfactory and contrasts with the more natural relationship of C_4^{red} and C_4 with C_3^{cs}.

4. Under C_4^{ext} all pairs of LOTOS specifications are consistent. This may seem a surprising result at first, but it reflects the fact that extension of functionality across pairs of specifications can always be reconciled.

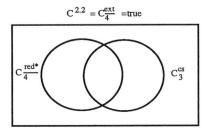

Firgure 2: LOTOS Consistency Relations

Probably the most important implication of these results is that consistency checking must be performed selectively. In particular, it is inappropriate to view consistency checking as a single mechanism which can be applied to any pair of specifications. For example, it would be inappropriate to check two specifications which express exactly corresponding functionality with C_4^{ext}. Thus, in order to apply suitable consistency checks the relationship of the specifications being checked must be made available. The RM-ODP has no provision for the communication of such information. The correspondence rule concept is used in the reference model as a means to locate portions of viewpoint specifications that should be compared. However, there is no means to define how these portions of specifications should be related.

5.2 Z Consistency Checking Relationships

A conformance relation for Z does not exist, but refinement has been extensively investigated. Thus, our work on consistency checking in Z has focussed on instantiating the C_4 definition of consistency. As indicated earlier internal validity is a central issue with Z, specifically, we define:-

Definition 12 *For S, a Z specification, $\Psi(S)$ iff $\neg\exists\psi$ s.t. $S \models \psi, \neg\psi$.*

An algorithm can be given which will unify two Z specifications [8]. This algorithm is divided into three stages: normalization, common refinement (which we usually term unification itself), and re-structuring. Normalization identifies commonality between two specifications, and re-writes the specifications into normal forms suitable for unification. Unification itself takes two normal forms and produces the least refinement of both. Restructuring is performed to re-introduce the specification structure that is lost during normalization.

The major issue with Z consistency checking is not demonstrating that a unification exists, rather it is showing that the unification is internally valid. This is in obvious contrast to LOTOS where finding a unification with respect to a refinement relation is the central task. Demonstrating internal validity of Z specifications using theorem proving tools is a central area of our current research. A companion paper [8] contains a full discussion of consistency checking for Z.

6. TRANSLATION - THE OPTIONS

There has been some success in relating FDTs that have similar underlying semantics, e.g. [15] [2], although, it should be pointed out that the common semantic form underlying these approaches is typically very ugly and significant research is required before usable translations can be generated. ODP consistency checking though, requires translation across FDT families. There are very few positive results on this topic, although a number of approaches could be considered, the following are the most likely:-

Syntactic Translation. Translation based upon a direct relating of syntactic terms in one FDT to terms in another FDT is a possible approach. However, it is difficult to envisage that such an approach could offer a general solution. In particular, a lot of semantic meaning will certainly be lost in such a crude relating of FDTs. Partial syntactic translations may though be feasible.

Common Semantic Model. Translation into a common semantic model is a more realistic approach. Such translation could either use the semantics of one of the FDTs as the intermediate semantics or use a third semantics. The former of these is not fully general, for example, Z and LOTOS are so fundamentally different that relating one to the others semantic model is very difficult to envisage. Relating FDTs using a third intermediate form is a more likely approach.

- There is a link between model based action systems (and thereby Z) and CSP made by showing that refinements (forwards and backwards simulation) in an action system are sound and jointly complete with respect to the notion of refinement in CSP [18].

- The requirement for highly expressive intermediate semantics suggests that logical notations may be appropriate. [10] and [3] consider logical characterisations of LOTOS in temporal logic. However, relating temporal logic to the Z first order logic remains an open issue. Categorical approaches and the theory of institutions offer a possible solution [3].

- An alternative logical approach is that by [19]. This work uses first order logic to express relationships between states and events. Thus, they offer a single notational link between model based specification and formal descriptions based on transition systems. The approach uses logical conjunction as composition and sketches how consistency checking can be performed in this framework. The pragmatic nature of this work reflects the compromises that will have to be made when performing translation in the ODP setting. Specifically, [19] acknowledge that their approach does not preserve the semantic equivalences of particular FDTs.

- A final alternative which has the benefit of being ODP specific is suggested by the work of [6]. This work offers a denotational semantics for the computational viewpoint language. These semantics could, theoretically, be used to relate different FDT interpretations of the computational viewpoint language. Clearly, this work does not give a complete solution to consistency as the semantics are restricted to a single viewpoint. However, it may be possible to extrapolate this approach to a general solution.

A further issue affecting translation is the role of the ODP architectural semantics. Specifically, Part 4 should provide a basis for relating FDTs. ODP concepts, in particular viewpoint languages, are defined in different FDTs in the architectural semantics. Thus, when relating complete viewpoint specifications in different FDTs these definitions can be used as components of a consistency check. However, it is important to note that the architectural semantics will only provide a framework for consistency checking. Actual viewpoint language specifications will extend the ODP architectural semantics, which are non prescriptive by nature, with FDT specific behaviour. There is then a need to combine the framework provided by the architectural semantics with actual consistency checking relationships arising from FDTs.

It is clear though that a usable translation mechanism is likely to represent a pragmatic, compromise solution. In particular, complete preservation of semantic meaning during translation will not be possible.

7. CONCLUDING REMARKS

We have described how consistency arises in ODP. We have formalized a number of possible definitions of consistency, three of which are presented in the RM-ODP. We have considered instantiations of these consistency definitions with particular FDTs, viz, LOTOS and Z and finally we have discussed the thorny issue of translation between FDTs.

We believe that consideration of consistency is timely, not just from an ODP perspective. In particular, a number of recent software engineering methodologies consider relating multiple specifications of a single system, e.g. [19] [1]. The interest in such approaches reflects a general move away from classical single threaded waterfall system development scenarios. Furthermore, OO methodologies, require specifications to be related horizontally. Related issues can be found in OSI [9].

There are very few published results on consistency checking for Open Distributed Processing, [17] and [11] are exceptions to this. Both of these consider strong notions of consistency based on process algebra equivalences and in this sense take a quite different approach to us. The work presented in this paper suggests the following concrete results:-

1. The consistency interpretations arising in the RM-ODP have very different meanings. In particular, for LOTOS, all pairs of specifications are consistent by $C_{2.2}$, while C_3 is significantly stronger. In addition, by defining suitable conditions on the relationship between *conf* and $\models$ we can use C_1 'consistently' with our conformance definitions. We can guarantee that $C_4 \Rightarrow C_{2.2}$ and $C_4 \Rightarrow C_1$, thus, C_4 provides an important link between logical notions of consistency and conformance notions.

2. It is appropriate to determine consistency using stronger relationships than the basic conformance definitions, since the extra knowledge available during specification checking enables system developers to apply consistency more discriminatingly.

3. With LOTOS all instantiations of C_4 with LOTOS refinement relations (trivially) imply $C_{2.2}$, while none of the instantiations imply C_3.

4. Consistency checking in Z and in LOTOS have a very different character. With LOTOS the central issue is finding a unification, while with Z the central issue is demonstrating that a unification does not contain any contradictions and can thus be implemented.

5. Pragmatic approaches to translation, in which some semantic information is lost, will have to be accepted.

We make the following recommendations; these are all required if realistic cross viewpoint consistency checking is to be undertaken:-

1. More specification to specification information must be made available to the consistency checking process. The nature of the consistency relationship to be checked must be made known. In addition, knowledge of the specification style used will be of value in performing consistency checking. It may even be necessary for specifiers to highlight particular cross viewpoint assertions that need to be tested.

2. Work on Part 4 of the RM-ODP must be undertaken as a priority. The architectural semantics provide an essential basis for consistency checking. In addition, the architectural semantics must themselves be shown to be 'consistent'. i.e. different FDT interpretations must not conflict.

3. Examples of multiple viewpoint specifications must be undertaken and be made available to the ODP community. Without realistic examples, consistency checking research will be poorly focussed.

In conclusion then, our inital results suggest that reasonable intra language consistency relationships can be found, however, inter language consistency checking remains a very challenging proposition. It is likely that this will only be possible with considerable prescriptive help from viewpoint language specifiers and in a pragmatic manner. However, this challenge must be met since without a realistic approach to maintaining the consistency of specifications across multiple viewpoints the potential of the existing and ongoing work on the ODP model cannot be fully realised.

References

[1] M Ainsworth, AH Cruickshank, LJ Groves, and PJL Wallis. Viewpoint specification and Z. *Information and Software Technology*, 36(1):43–51, February 1994.

[2] D. Bert, M. Bidoit, C. Choppy, R. Echahed, J.-M. Hufflen, J.-P. Jacquot, M. Lemoine, N. Lévy, J.-C. Reynaud, C. Roques, F. Voisin, J.-P. Finance, and M.-C. Gaudel. Opération SALSA: Structure d'AccueiL pour Spécifications Algébriques. Rapport final, PRC Programmation et Outils pour l'intelligence Artificielle, 1993.

[3] H. Bowman and J. Derrick. Towards a formal model of consistency in ODP. Technical Report 3-94, Computing Laboratory, University of Kent at Canterbury, 1994.

[4] E. Brinksma. A theory for the derivation of tests. In S. Aggarwal and K. Sabnani, editors, *Protocol Specification, Testing and Verification, VIII*, pages 63–74, Atlantic City, USA, June 1988. North-Holland.

[5] E. Brinksma, G. Scollo, and C. Steenbergen. Process specification, their implementation and their tests. In B. Sarikaya and G. V. Bochmann, editors, *Protocol Specification, Testing and Verification, VI*, pages 349–360, Montreal, Canada, June 1986. North-Holland.

[6] AFNOR cont. *A direct computational language semantics for Part 4 of the RM-ODP*. ISO/IEC JTC1/SC21/WG7 approved AFNOR contribution, July 1994.

[7] G. Cowen, J. Derrick, M. Gill, G. Girling (editor), A. Herbert, P. F. Linington, D. Rayner, F. Schulz, and R. Soley. *Prost Report of the Study on Testing for Open Distributed Processing*. APM Ltd, 1993.

[8] J. Derrick, H. Bowman, and M. Steen. Maintaining cross viewpoint consistency using Z. In *ICODP'95*, Brisbane, Australia, February 1995.

[9] A. Fantechi, S. Gnesi, and C. Laneve. Two standards means problems : A case study on formal protocol descriptions. *Computer Standards and Interfaces*, 9:11–19, 1989.

[10] A. Fantechi, S. Gnesi, and G. Ristori. Compositional logic semantics and LOTOS. In L. Logrippo, R.L. Probert, and H. Ural, editors, *Protocol Specification, Testing and Verification, X*, Ottawa, Canada, June 1990. North-Holland.

[11] K. Farooqui and L. Logrippo. Viewpoint transformations. In J. de Meer, B. Mahr, and O. Spaniol, editors, *2nd ICODP*, pages 352–362, Berlin, September 1993.

[12] ISO/IEC JTC1/SC21/WG7. *Basic reference model of Open Distributed Processing - Parts 1-4*, July 1993.

[13] F. Khendek and G. v. Bochmann. Merging behaviour specifications. Technical Report 856, University of Montreal, Department of Computing, 1993.

[14] G. Leduc. A framework based on implementation relations for implementing LOTOS specifications. *Computer Networks and ISDN Systems*, 25:23–41, 1992.

[15] R. Reed, W. Bouma, J.D. Evans, M. Dauphin, and M. Michel. *Specification and Programming Environment for Communication Software*. North-Holland, 1993. ISBN 0 444 89923 5.

[16] M. Steen, H. Bowman, and J. Derrick. Consistency in LOTOS. Technical Report in preparation, Computing Laboratory, University of Kent at Canterbury, 1995.

[17] A. Vogel. *Entwurf, Realisierung und Test von ODP-Systemen auf der Grundlage formaler Beschreibungstechniken.* PhD thesis, Humboldt-Universität zu Berlin, 1993. submitted.

[18] J.C.P. Woodcock and C.C. Morgan. Refinement of state-based concurent systems. In D. Bjorner, C.A.R. Hoare, and H. Langmaack, editors, *VDM '90 VDM and Z - Formal Methods in Software Development*, LNCS 428, pages 340–351, Kiel, FRG, April 1990. Springer-Verlag.

[19] P. Zave and M. Jackson. Conjunction as composition. *ACM Trans. on Soft. Eng. and Method.*, 2:379–411, 1993.

APPENDIX (i): LOTOS Relations. P, P_1 and P_2 are processes; $\mathcal{L}$ is the alphabet of observable actions; $\mathcal{L}^*$ denotes strings over $\mathcal{L}$; $Tr(P)$ denotes the set of traces of P and $Ref(P, \sigma)$ denotes the refusal set of P after the trace σ.

Definition 13
(i) $P_2 \underline{conf} P_1$ iff $\forall \sigma \in Tr(P_2) : Ref(P_1, \sigma) \subseteq Ref(P_2, \sigma)$.
(ii) $P_2 \underline{red} P_1$ iff $Tr(P_1) \subseteq Tr(P_2) \wedge P_1 \underline{conf} P_2$.
(iii) $P_2 \underline{ext} P_1$ iff $Tr(P_1) \supseteq Tr(P_2) \wedge P_1 \underline{conf} P_2$.
(iv) $P_1 \underline{te} P_2$ iff: $Tr(P_1) = Tr(P_2) \wedge \forall \sigma \in \mathcal{L}^* : Ref(P_1, \sigma) = Ref(P_2, \sigma)$.

APPENDIX (ii): Further Results. Proofs of these results can be found in [3].

Proposition 5 *Properties of* $\sqsubseteq$
(i) $\sqsubseteq$ *is a pre-order (i.e. reflexive and transitive)*
(ii) $S_1 \equiv_{cf} S_2$ *iff* $S_1 \sqsubseteq S_2$ *and* $S_2 \sqsubseteq S_1$ *(i.e.,* $\sqsubseteq$ *is a partial order with respect to equivalence)*
(iii) $(\sqsubseteq \circ conf) = conf$
(iv) *For all R, we have* $R \subseteq \sqsubseteq$ *iff* $(R \circ conf) \subseteq conf$
(v) *For all R, we have* $Id \subseteq R$ *implies that* $(R \subseteq \sqsubseteq)$ *iff* $(R \circ conf = conf)$
(vi) $\sqsubseteq$ *is the least relation R such that* $R \circ conf = conf$.

Proposition 6 *Unification satisfies the following properties:*
(i) $U(T_1, T_2) = U(T_2, T_1)$ *- commutativity*
(ii) $U(T_1, U(T_2, T_3)) = U(U(T_1, T_2), T_3)$ *- associativity*
(iii) $T_1, T_2 \sqsubseteq U(T_1, T_2)$ *- common refinement*
(iv) *If* $T_1 \sqsubseteq T_2$ *then* $U(T_1, T_2) = T_2$

Proposition 7 *Properties of consistency:*
(1) *Consistency is a symmetric relation, but it is neither reflexive nor transitive.*
(ii) $S_1 C_4 U(S_2, S_3)$ *iff* $S_2 C_4 U(S_1, S_3)$ *iff* $S_3 C_4 U(S_1, S_2)$.
(iii) *Global consistency of three or more specifications implies pairwise consistency.*
(iv) *Pairwise consistency does not imply global consistency.*

32

Maintaining Cross Viewpoint Consistency using Z

John Derrick, Howard Bowman and Maarten Steen †
University of Kent at Canterbury, U.K. {jd1,hb5,mwas}@ukc.ac.uk.

This paper discusses the use and integration of formal techniques, in particular Z, into the Open Distributed Processing (ODP) standardization initiative.

One of the cornerstones of the ODP framework is a model of multiple viewpoints. During the development process it is important to maintain the consistency of different viewpoints of the same ODP specification. In addition, there must be some way to combine specifications from different viewpoints into a single implementation specification. The process of combining two specifications is known as unification. Unification can be used as a method by which to check consistency. This paper describes a mechanism to unify two Z specifications, and hence provide a consistency checking strategy for viewpoints written in Z.

Keyword Codes: C.2.4, D.2.1, D.2.2.
Keywords: Distributed Systems, Specification, Tools and Techniques.

1   INTRODUCTION

This paper discusses the implications and integration of formal techniques, in particular Z, into the Open Distributed Processing (ODP) standard initiative.

The ODP standardization initiative is a natural progression from OSI, broadening the target of standardization from the point of interconnection to the end-to-end system behaviour. The objective of ODP [9] is to enable the construction of distributed systems in a multi-vendor environment through the provision of a general architectural framework that such systems must conform to. One of the cornerstones of this framework is a model of multiple viewpoints which enables different participants to observe a system from a suitable perspective and at a suitable level of abstraction [11, 14]. There are five separate viewpoints presented by the ODP model: Enterprise, Information, Computational, Engineering and Technology. Requirements and specifications of an ODP system can be made from any of these viewpoints.

Formal methods are playing an increasing role within ODP, and we aim to provide a mechanism by which specific techniques can be used within ODP. The suitability of a wide spectrum of FDTs is currently being assessed. Amongst these Z is likely to be used for at least the information, and possibly the enterprise and computational, viewpoint. The first compliant ODP specification, the Trader, is being written using Z for the information and computational viewpoint.

† This work was partially funded by British Telecom Labs., Martlesham, Ipswich, U.K; the Engineering and Physical Sciences Research Council under grant number GR/K13035 and the Royal Society.

Whilst it has been accepted that the viewpoint model greatly simplifies the development of system specifications and offers a powerful mechanism for handling diversity within ODP, the practicalities of how to make the approach work are only beginning to be explored. In particular, one of the consequences of adopting a multiple viewpoint approach to development is that descriptions of the same or related entities can appear in different viewpoints and must co-exist. *Consistency* of specifications across viewpoints thus becomes a central issue. Similar consistency properties arise outside ODP. For example, within OSI two formal descriptions of communication protocols can co-exist and there is no guarantee that, when the two protocols are implemented on the basis of these specifications, processes which use these two protocols can communicate correctly, [5]. However, the actual mechanism by which consistency can be checked and maintained is only just being addressed [8, 7, 6]. In particular, although Z is being used as a viewpoint specification language in ODP, there is as yet no mechanism to describe the combination of different Z viewpoint specifications, or the consistency of them.

In Section 2 we develop a unification mechanism for Z specifications. In Section 3 we present an example of the technique by specifying the dining philosophers problem using viewpoints. Section 4 discusses consistency checking of viewpoint specifications, and we make some concluding remarks in Section 5.

2   UNIFICATION IN Z

One of the cornerstones of the ODP framework is a model of multiple viewpoints. Clearly the different viewpoints of the same ODP specification must be consistent, i.e. the properties of one viewpoint specification do not contradict those of another. In addition, during the development process there must be some way to combine specifications from different viewpoints into a single implementation specification. This process of combining two specifications is known as *unification*. Furthermore, the unification of two specifications must be a refinement of both, see [3]. Unification can also be used, because of this common refinement, as a method by which to check consistency. To check the consistency of two specifications, we check for contradictions within the unified specification.

The mechanism we describe is a general strategy for unifying two Z specifications. As such it is not specific to any particular ODP viewpoint, nor is it tied to any particular instantiation of the architectural semantics. However, this generality does not reduce its applicability, indeed it is possible that unification can be used to describe an interaction mechanism between descriptions in Z of objects in such a way that is currently not supported by Part 4 of the reference model.

Given a refinement relation, $\sqsubseteq$, defined in a formal specification techniques, we can characterize the unification of two specifications as the least refinement of both, ie:

$$U(T_1, T_2) = \{ T : T_1, T_2 \sqsubseteq T \text{ and if } T_1, T_2 \sqsubseteq S \text{ then } T \sqsubseteq S \}$$

Unification of Z specifications will therefore depend upon the Z refinement relation, which is given in terms of two separate components - data refinement and operation refinement, [12]. Two specifications will thus be consistent if their unification can be implemented [1]. The ability for the unification to be implemented is known as *internal validity*, and for Z specifications this holds when the specification is free from contradictions.

Z is a state based FDT, and a Z specification describes the abstract state of the system (including a description of the initial state of the system), together with the collection of available operations, which manipulate the state. One Z specification refines another if the state schemas are data refinements and the operation schemas are operation refinements of the original specification's state and operation schemas. We assume the reader is familiar with the language and refinement relation, introductionary texts include [12, 13, 16].

The unification algorithm we describe is divided into three stages: normalization, common refinement (which we usually term unification itself), and re-structuring. Normalization identifies commonality between two specifications, and re-writes the specifications into normal forms suitable for unification. Unification itself takes two normal forms and produces the least refinement of both. Because normalization will hide some of the specification structure introduced via the schema calculus, it is necessary to perform some re-structuring after unification to re-introduce the structure chosen by the specifier. We do not discuss re-structuring here.

2.1 Normalization

Given two different viewpoint specifications of the same (ODP) system, the commonality between the specifications needs to be identified. Clearly, the two specifications that are to be unified have to represent the world in the same way within them (eg if an operation is represented by a schema in one viewpoint, then the other viewpoint has to use the same name for its (possibly more complex) schema too), and that the correspondences between the specifications have to have been identified by the specifiers involved. These will be given by co-viewpoint mappings that describe the naming, and other, conventions in force. Once the commonality has been identified, the appropriate elements of the specifications are re-named.

Normalization will also expand data-type and schema definitions into a normal form. The purpose of normalization is to hide the structuring of schemas (which needs to be hidden in order to provide automatic unification techniques) and expand declarations into maximal type plus predicate declarations. For example, normalization of a declaration part of a schema involves replacing every set X which occurs in a declaration $x : X$, with its corresponding maximal type and adding predicates to the predicate part of the schema involved to constrain the variable appropriately.

Normalization also expands schemas defined via the schema calculus into their full form. All schema expressions involving operations from the schema calculus can be expanded to a single equivalent vertical schema. Examples of normalization appear in [12].

2.2 State Unification

The purpose of state unification is to find a common state to represent both viewpoints. The state of the unification must be a data refinement of the state of both viewpoints, since viewpoints represent partial views of an overall system description. Furthermore, it should be the least refinement whenever possible. This is needed to ensure we do not add too much detail during unification because additional detail might add inconsistencies that were not due to inconsistencies in the original viewpoint specifications. Clearly, unification as a consistency checking strategy is more useful if it is also true that an inconsistent unification implies inconsistent viewpoint specifications, rather than just consistent unifications implying consistent viewpoints.

The essence of all constructions will be as follows. If an element x is declared in both viewpoints as $x : T_1$ and $x : T_2$ respectively, then the unification will include a declaration $x : T$ where T is the least refinement of T_1 and T_2. The type T will be the smallest type which contains a copy of both T_1 and T_2. For example, if T_1 and T_2 can be embedded in some maximal type then T is just the union of $T_1 \cup T_2$. The proof of correctness of this unification is given in [2]. If T_1 and T_2 cannot be embedded in a single type then the unification will declare x to be a member of the disjoint union of T_1 and T_2. In these circumstances we again achieve the least refinement of both viewpoints. Lack of space precludes a discussion of this construction here.

Given two viewpoint specifications both containing the following fragment of state description given by a schemas D_1 and D_2, then D represents the unification of the two:

$$
\begin{array}{|l}
\hline
D_1 \rule{3cm}{0pt} \\
x : S \\
\hline
pred_S \\
\hline
\end{array}
\qquad
\begin{array}{|l}
\hline
D_2 \rule{3cm}{0pt} \\
x : T \\
\hline
pred_T \\
\hline
\end{array}
\qquad
\begin{array}{|l}
\hline
D \rule{3cm}{0pt} \\
x : S \cup T \\
\hline
x \in S \implies pred_S \\
x \in T \implies pred_T \\
\hline
\end{array}
$$

whenever $S \cup T$ is well founded. (Axiomatic descriptions are unified in exactly the same manner.) This representation is needed in order to preserve the widest range of possible behaviours.

2.3   Operation Unification

Once the data descriptions have been unified, the operations from each viewpoint need to be defined in the unified specification. Unification of schemas then depends upon whether there are duplicate names. For operations defined in just one of the viewpoint specifications, these are included in the unification with appropriate adjustments to take account of the unified state.

For operations which are defined in both viewpoint specifications, the unified specification should contain an operation which is the least refinement of both, wrt the unified representation of state. The unification algorithm first adjusts each operation to take account of the unified state in the obvious manner, then combines the two operations to produce an operation which is a refinement of both viewpoint operations.

The unification of two operations is defined via their pre- and post-conditions. Given a schema it is always possible to derive its pre- and post-conditions, [10]. Given two schemas A and B representing operations, both applicable on some unified state, then

$$
\begin{array}{|l}
\hline
\mathcal{U}(A, B) \rule{6cm}{0pt} \\
\vdots \\
\hline
pre\ A \vee pre\ B \\
pre\ A \implies post\ A \\
pre\ B \implies post\ B \\
\hline
\end{array}
$$

represents the unification of A and B, where the declarations are unified in the manner of the preceding subsection. This definition ensures that if both pre-conditions are true, then the unification will satisfy both post-conditions. Whereas if just one pre-condition is true, only the relevant post-condition has to be satisfied. This provides the basis of the consistency checking method for object behaviour which we discuss below.

2.3.1 Example

As an illustrative example we perform state and operation unification on a simple specification of a classroom. The example consists of the state represented by the schema *Class*, and operation *Leave*. The two viewpoint specifications to be unified are:

$Max : \mathbf{N}$

$\underline{\quad Class\quad\quad\quad\quad\quad\quad\quad\quad\quad}$
$d : \mathbf{P}\{1,2\}$

$\#d \leq Max$

$\underline{\quad Leave\quad\quad\quad\quad\quad\quad\quad\quad}$
$\Delta Class$
$p? : \{1,2\}$

$p? \in d$
$d' = d \setminus \{p?\}$

$Min : \mathbf{N}$

$\underline{\quad Class\quad\quad\quad\quad\quad\quad\quad\quad\quad}$
$d : \mathbf{P}\{2,3,4\}$

$\#d \geq Min$

$\underline{\quad Leave\quad\quad\quad\quad\quad\quad\quad\quad}$
$\Delta Class$
$p? : \{2,3,4\}$

$\#d > Min + 1$
$p? \in d$
$d' = d \setminus \{p?, 2\}$

As described above, we first unify the state model, i.e. the schema *Class* in this example, which becomes:

$\underline{\quad Class\quad\quad\quad\quad\quad\quad\quad\quad\quad\quad\quad\quad\quad\quad\quad}$
$d : \mathbf{P}\{1,2\} \cup \mathbf{P}\{2,3,4\}$

$d \in \mathbf{P}\{1,2\} \Longrightarrow \#d \leq Max$
$d \in \mathbf{P}\{2,3,4\} \Longrightarrow \#d \geq Min$

With this unified state model we can unify the operation *Leave* on this state. To do so we calculate the pre and post-conditions in the usual manner, and for this we need to expand the schema *Leave* into normal form in each viewpoint. This will involve, for example, declaring $p? : \mathbf{N}$ and adding $p? \in \{1,2\}$ as part of the predicate for the description of *Leave* in the first viewpoint. The pre-condition of *Leave* in the first viewpoint is then $p? \in d \cap \{1,2\}$ (in fact this is the part of the pre-condition which is distinct from the pre-condition in the second viewpoint, the rest acting as a state invariant). Hence, the unified *Leave* becomes:

$\underline{\quad Leave\quad\quad\quad\quad\quad\quad\quad\quad\quad\quad\quad\quad\quad\quad}$
$\Delta Class$
$p? : \mathbf{N}$

$(p? \in d \cap \{1,2\}) \vee (p? \in d \cap \{2,3,4\} \wedge \#d > Min + 1)$
$(p? \in d \cap \{1,2\}) \Longrightarrow d' = d \setminus \{p?\}$
$(p? \in d \cap \{2,3,4\} \wedge \#d > Min + 1) \Longrightarrow d' = d \setminus \{p?, 2\}$

To show that the unified *Leave* is indeed a refinement of *Leave* in viewpoint one we will decorate elements in viewpoint one with a subscript one. We use the retrieve relation

$$\begin{array}{|l}
\hline
R_1 \rule{3cm}{0pt}\\
\quad Class\\
\quad Class_1\\
\hline
\quad d_1 \in \{d\} \cap \mathbf{P}\{1,2\}\\
\hline
\end{array}$$

to describe the refinement between the unified state and the state in the first viewpoint. To demonstrate the refinement is correct, we make the following deductions. Suppose *pre Leave*$_1$ $\wedge$ ΔR_1 $\wedge$ *Leave*, we have to show the result of this schema is compatible with *post Leave*$_1$. Now if *pre Leave*$_1$, then $p? \in d_1 \in \{d\} \cap \mathbf{P}\{1,2\}$, and hence $d' = d \setminus \{p?\}$. Then $d'_1 \in \{d'\} \cap \mathbf{P}\{1,2\} = \{d \setminus \{p?\}\} \cap \mathbf{P}\{1,2\}$. So $d'_1 = d' \cap \{1,2\} = (d \setminus \{p?\}) \cap \{1,2\} = d_1 \setminus \{p?\}$, since by *pre Leave*$_1$, $p? \in \{1,2\}$. The deduction that *pre Leave*$_1$ $\wedge$ $R_1 \implies$ *pre Leave* is similar. These two deductions complete the proof that the unification is a refinement of viewpoint one. The case for viewpoint two is symmetrical.

3   EXAMPLE - DINING PHILOSOPHERS

To illustrate unification with Z, we shall consider the following viewpoint specifications of the dining philosophers problem. In the dining philosophers problem, [4], a group of N philosophers sit round a table, laid with N forks. There is one fork between each adjacent pair of philosophers. Each philosopher alternates between thinking and eating. To eat, a philosopher must pick up its right-hand fork and then the left-hand fork. A philosopher cannot pick up a fork if its neighbour already holds it. To resume thinking, the philosopher returns both forks to the table.

The three viewpoint specifications we define are the philosophers, forks and tables viewpoints. The philosophers and forks describe individual philosopher and fork objects and the operations available on those objects. The table viewpoint describes a system constructed from those objects and the synchronisation mechanism between operations upon them. We shall then describe the unification of the three viewpoints.

Although this example is not one of an ODP system, it provides a suitable illustration of the issues involved in viewpoint specification and consistency checking.

3.1   The Philosophers Viewpoint

This viewpoint considers the specification from the point of view of a philosopher. A philosopher either thinks, eats or holds her right fork. Note that since the latter is just a state of mind there is no need to describe the operations from a forks point of view at all in this viewpoint. A philosopher object is just defined by the state of the philosopher, and initially a philosopher is thinking.

$$PhilStatus ::= Thinking \mid HasRightFork \mid Eating$$

$$\begin{array}{|l}
\hline
PHIL \rule{2.5cm}{0pt}\\
\quad status : PhilStatus\\
\hline
\end{array}
\qquad
\begin{array}{|l}
\hline
InitPHIL \rule{2cm}{0pt}\\
\quad PHIL'\\
\hline
\quad status' = Thinking\\
\hline
\end{array}$$

We can now describe the operations available. A thinking philosopher can pick up its right-hand fork. Philosophers who hold their right fork can begin eating upon picking up their left-hand fork. Finally to resume thinking, a philosopher releases both forks.

```
┌─ GetRightFork ─────────
│ ΔPHIL
├─────────────────────
│ status = Thinking
│ status' = HasRightFork
```

```
┌─ GetLeftFork ─────────
│ ΔPHIL
├─────────────────────
│ status = HasRightFork
│ status' = Eating
```

```
┌─ DropForks ─────────
│ ΔPHIL
├─────────────────────
│ status = Eating
│ status' = Thinking
```

3.2 The Forks Viewpoint

This viewpoint specifies a fork object. Each fork is either free or busy. The fact that the philosopher might change state when a fork is picked up or dropped does not concern forks. The state of the fork is given by a *FORK* schema, and initially a fork is free.

$ForkStatus ::= Free \mid Busy$

```
┌─ FORK ──────────────────────
│ fstatus : ForkStatus
```

```
┌─ InitFORK ──────────────────
│ FORK'
├──────────────────────────
│ fstatus' = Free
```

The operations available allow a free fork can be picked up, and both forks can be released.

```
┌─ Acquire ──────────────────
│ ΔFORK
├──────────────────────────
│ fstatus = Free
│ fstatus' = Busy
```

```
┌─ Release ──────────────────
│ ΔFORK
├──────────────────────────
│ fstatus = Busy
│ fstatus' = Free
```

3.3 The Tables Viewpoint

This viewpoint has a number of schemas from the other viewpoints as parameters, these are given as empty schema definitions. Upon unification the non-determinism in this viewpoint will be resolved by the other viewpoint specifications, and thus unification will allow functionality extension of these parameters. The parameters we require are:

```
┌─ PHIL ──────────
```

```
┌─ InitPHIL ──────────
```

```
┌─ GetRightFork ──────
│ ΔPHIL
```

```
┌─ GetLeftFork ──────
│ ΔPHIL
```

```
┌─ DropForks ──────────
│ ΔPHIL
```

```
┌─ FORK ──────────────
```

```
┌─ InitFORK ──────────
```

```
┌─ Acquire ──────────
│ ΔFORK
```

```
┌─ Release ──────────
│ ΔFORK
```

The system from the table viewpoint is defined by a collection of fork and philosopher objects:

$$| \quad N : \mathbb{N}$$

```
┌─ Table ─────────────────────────────
│ forks : 1..N → FORK
│ phils : 1..N → PHIL
└─────────────────────────────────────
```

Initially the table consists of forks and philosophers all in their respective initial states.

```
┌─ InitTable ─────────────────────────
│ Table'
├─────────────────────────────────────
│ ∃ InitFORK, InitPHIL • ran forks' = {θInitFORK} ∧ ran phils' = {θInitPHIL}
└─────────────────────────────────────
```

Here we use promotion (ie the θ operator) in the structuring of viewpoints, which allows an operation defined on an object in one viewpoint to be *promoted* up to an operation defined over that object in another viewpoint. As we can see, this can be used effectively to reference schemas in different viewpoints without their full definition.

In order to define operations on the table, we define a schema $\Phi\, Table$ which will allow individual object operations to be defined in this viewpoint. See [13] for a discussion of the use of promotion.

```
┌─ Φ Table ───────────────────────────
│ Δ Table
│ ΔPHIL
│ ΔFORK
│ m? : 1..N
│ n? : 1..N
├─────────────────────────────────────
│ phils(n?) = θPHIL ∧ phils' = phils ⊕ {phils(n?) = θPHIL'}
│ forks(m?) = θFORK ∧ forks' = forks ⊕ {forks(m?) = θFORK'}
└─────────────────────────────────────
```

Note that we use two inputs $m?, n?$, because we want to control later the synchronisation between operations on forks and those on philosophers. System operations to get the left and right forks, and to drop both forks can now be defined.

$$GLF \cong (\Phi\, Table \wedge GetLeftFork \wedge Acquire \wedge [\, n?, m? : 1..N \mid m? = n?\,]) \setminus (\Delta FORK, \Delta PHIL)$$
$$GRF \cong (\Phi\, Table \wedge GetRightFork \wedge Acquire \wedge [\, n?, m? : 1..N \mid m? = (n? \bmod N + 1)\,]) \setminus (\Delta FORK, \Delta PHIL)$$
$$DF \cong (\Phi\, Table \wedge DropForks \wedge Release \wedge [\, n?, m? : 1..N \mid m? = n?\,]) \setminus (\Delta FORK, \Delta PHIL)$$

The last schema in each conjunction performs the correct synchronisation between the individual object operations.

3.4   Unifying the Viewpoints

Since the fork and philosopher object descriptions are independent, ie there are no state or operation schemas in common, the unification of these two viewpoints is just the concatenation of the two specifications. We do not re-write that concatenation here.

The Table specification does have commonality with the other two viewpoints. For each state or operation schema defined in two viewpoints (ie the Table and one other), we build one schema in the unification. In fact, the separation and object-based nature (in a loose sense) of this example means that we will not make extensive use of unification by pre- and post-conditions. This is desirable, since it reduces the search for contradictions in the consistency checking phase. In fact, our experiences with viewpoint specifications confirms that such a viewpoint methodology is really only feasible if one adopts this object-based approach.

For example, the schema *FORK* defined in the Table viewpoint is just a parameter from the fork viewpoint, and consequently its unification will just be:

FORK
---
fstatus : *ForkStatus*

Similarly the unification of *GetLeftFork* from the Table and Philosophers viewpoint is

---
GetLeftFork
$\Delta PHIL$
---
status = *HasRightFork*
status' = *Eating*
---

since the pre-condition of *GetLeftFork* in Table is just false. Notice that this provides a mechanism in Z by which to achieve functionality extension across viewpoints in a manner previously not supported.

4   CONSISTENCY CHECKING OF VIEWPOINT SPECIFICATIONS

The unification mechanism can be applied to yield a consistency checking process. In terms of the ODP viewpoint model, consistency checking consists of checking both the consistency of the state model and the consistency of all the operations. Consistency checking of the state model ensures there exists at least one possible set of bindings that satisfies the state invariant; and the Initialization Theorem (see below) ensures that we can find one such set of bindings initially.

In addition, we require operation consistency. This is because a conformance statement in Z corresponds to an operation schema(s), [15]. Thus a given behaviour (ie occurrence of an operation schema) conforms if the post-conditions and invariant predicates are satisfied in the associated Z schema. Hence, operations in a unification will be implementable whenever each operation has consistent post-conditions on the conjunction of their pre-conditions.

Thus a consistency check in Z involves checking the unified specification for contradictions, and has three components: State Consistency, Operation Consistency and the Initialization Theorem.

State Consistency : From the general form of state unification given in Section 2.2, it follows that the state model is consistent as long as both $preds_S$ and $pred_T$ can be satisfied for $x \in S \cap T$.

Operation Consistency : Consistency checking also needs to be carried out on each operation in the unified specification. The definition of operation unification means that we have to check for consistency when both pre-conditions apply. That is, if the unification of A and B is denoted $\mathcal{U}(A, B)$, we have:

$$pre\ \mathcal{U}(A, B) = pre\ A \vee pre\ B, \quad post\ \mathcal{U}(A, B) = (pre\ A \Rightarrow post\ A) \wedge (pre\ B \Rightarrow post\ B)$$

So the unification is consistent as long as $(pre\ A \wedge pre\ B) \Rightarrow (post\ A = post\ B)$.

Initialization Theorem : The Initialization Theorem is a consistency requirement of all Z specifications. It asserts that there exists a state of the general model that satisfies the initial state description, formally it takes the form:

$$\vdash \exists\, State' \bullet InitState$$

For the unification of two viewpoints to be consistent, clearly the Initialization Theorem must also be established for the unification.

The following result can simplify this requirement: Let $State$ be the unification of $State_1$ and $State_2$, and $InitState$ be the unification of $InitState_1$ and $InitState_2$. If the Initialization Theorem holds for $State_1$ and $State_2$, then state consistency of $Initstate$ implies the Initialization Theorem for $State$. In other words, it suffices to look at the standard state consistency of $Initstate$.

If, however, $Initstate$ is a more complex description of initiality (possibly still in terms of $InitState_1$ and $InitState_2$), the Initialization Theorem expresses more than state consistency of $Initstate$, and hence will need validating from scratch. An example of this is given below.

Example 1 : The classroom

State Consistency : The unified state in this example was given by

```
┌─ Class ──────────────────────────────
│  d : P{1, 2} ∪ P{2, 3, 4}
├──────────────────────────────────────
│  d ∈ P{1, 2} ⟹ #d ≤ Max
│  d ∈ P{2, 3, 4} ⟹ #d ≥ Min
└──────────────────────────────────────
```

To show consistency, we need to show that if $d \in P\{1, 2\} \cap P\{2, 3, 4\}$, then both $\#d \leq Max$ and $\#d \geq Min$ hold. Suppose the class consisted of just the element 2, i.e. $d = \{2\}$. Both pre-conditions in the unified state, $d \in P\{1, 2\}$ and $d \in P\{2, 3, 4\}$, now hold giving the state invariant $Min \leq \#d \leq Max$. Thus the consistency of the viewpoint specifications of the classroom requires that $Min \leq Max$. This type of consistency condition should probably fall under the heading of a *correspondence rule* in ODP, [9], that is a condition which is necessary but not necessarily sufficient to guarantee consistency.

Operation Consistency : In the classroom example, this amounts to checking the operation *Leave* when

$$(p? \in d \cap \{1, 2\}) \wedge (p? \in d \cap \{2, 3, 4\} \wedge \#d > Min + 1)$$

In these circumstances, the two post-conditions are $d' = d \setminus \{p?\}$ and $d' = d \setminus \{p?, 2\}$. These two pre-conditions apply when $p? = 2$ and $2 \in d$. A consistency check has to be applied for all possible values of d. For example, let $d = \{1, 2\}$, then $d' = d \setminus \{p?\}$. If further $\#d > Min + 1$, then in addition we have $d' = d \setminus \{p?, 2\}$. These two conditions are consistent (since $p? = 2$) regardless of Max or Min.

Let $d = \{2\}$, then both pre-conditions apply iff $Min < 0$, in which case the post-conditions are $d' = d \setminus \{2\}$ and $d' = d \setminus \{2\}$, and thus consistent.

Hence the two viewpoint specifications are consistent whenever the correspondence rule $Min \leq Max$ holds.

Example 2 : Dining Philosophers

Inspection of the unification in the Dining Philosophers example shows that both state and operation consistency is straightforward (note, however, that with non-object based viewpoint descriptions of this example, consistency checking is a non-trivial task, this points the need for further work on specification styles to support consistency checks). Hence, consistency will follow once we establish the Initialization Theorem for the unification.

The Initialization Theorem for the unification is: $\vdash \exists\, Table' \bullet InitTable$, which upon expansion and simplification becomes

$$\vdash \exists\, forks' : 1..N \rightarrow FORK, phils' : 1..N \rightarrow PHIL \bullet \text{ran } forks' = \{Free\} \wedge \text{ran } phils' = \{Thinking\}$$

which clearly can be satisfied. Hence the viewpoint descriptions given for the dining philosophers are indeed consistent.

5   CONCLUSIONS

The use of viewpoints to enable separation of concerns to be undertaken at the specification stage is a cornerstone of the ODP model. However, the practicalities of how to make the approach work are only beginning to be explored. Two issues of importance are unification and consistency checking. Our work attempts to provide a methodology to undertake unification and consistency checking for Z specifications.

There are still many issues to be resolved, not least the relation to the architectural semantics work. Currently the architectural semantics associates an ODP object with a complete Z specification. Thus the configuration and interactions of objects is then outside the scope of a single Z specification. The architectural semantics comments upon the lack of support for combining Z specifications; we are currently investigating the extent to which unification can provide that support and hence model interaction and communication between Z specifications which represent ODP objects.

Not withstanding this, consistency checking of two Z specifications is still important. It provides a mechanism by which to assess different descriptions of the same object, and will be needed if consistency checking of specifications written in different FDTs is to be achieved. For example, one method would involve translating a LOTOS object into a Z specification (and this type of translation is the extremely challenging part), which could then be checked for consistency via unifying the two Z specifications. Thus the solutions presented in this paper are only part of the whole consistency problem, and much work remains including application to a larger case study.

We are currently funded by the EPSRC and British Telecom to extend our approaches to unification and consistency checking to other formal languages, in particular LOTOS, and to develop tools to support the process.

References

[1] H. Bowman and J. Derrick. Towards a formal model of consistency in ODP. Technical Report 3-94, Computing Laboratory, University of Kent at Canterbury, 1994.

[2] H. Bowman and J. Derrick. Modelling distributed systems using Z. In K. M. George, editor, *ACM Symposium on Applied Computing*, pages 147–151, Nashville, February 1995. ACM Press.

[3] G. Cowen, J. Derrick, M. Gill, G. Girling (editor), A. Herbert, P. F. Linington, D. Rayner, F. Schulz, and R. Soley. *Prost Report of the Study on Testing for Open Distributed Processing*. APM Ltd, 1993.

[4] E. W. Dijkstra. Cooperating sequential processes. In F. Genuys, editor, *Programming Languages*. Academic Press, 1968.

[5] A. Fantechi, S. Gnesi, and C. Laneve. Two standards means problems : A case study on formal protocol descriptions. *Computer Standards and Interfaces*, 9:11–19, 1989.

[6] K. Farooqui and L. Logrippo. Viewpoint transformations. In J. de Meer, B. Mahr, and O. Spaniol, editors, *2nd International IFIP TC6 Conference on Open Distributed Processing*, pages 352–362, Berlin, Germany, September 1993.

[7] J. Fischer, A. Prinz, and A. Vogel. Different FDT's confronted with different ODP-viewpoints of the trader. In J.C.P. Woodcock and P.G. Larsen, editors, *FME'93: Industrial Strength Formal Methods*, LNCS 670, pages 332–350. Springer-Verlag, 1993.

[8] K. Geihs and A. Mann. ODP viewpoints of IBCN service management. *Computer Communications*, 16(11):695–705, 1993.

[9] ISO/IEC JTC1/SC21/WG7. *Basic reference model of Open Distributed Processing - Parts 1-4*, July 1993.

[10] S. King. Z and the refinement calculus. In D. Bjorner, C.A.R. Hoare, and H. Langmaack, editors, *VDM '90 VDM and Z - Formal Methods in Software Development*, LNCS 428, pages 164–188, Kiel, FRG, April 1990. Springer-Verlag.

[11] P. F. Linington. Introduction to the Open Distributed Processing Basic Reference Model. In J. de Meer, V. Heymer, and R. Roth, editors, *IFIP TC6 International Workshop on Open Distributed Processing*, pages 3–13, Berlin, Germany, September 1991. North-Holland.

[12] B. Potter, J. Sinclair, and D. Till. *An introduction to formal specification and Z*. Prentice Hall, 1991.

[13] B. Ratcliff. *Introducing specification using Z*. McGraw-Hill, 1994.

[14] K. A. Raymond. Reference Model of Open Distributed Processing: a Tutorial. In J. de Meer, B. Mahr, and O. Spaniol, editors, *2nd International IFIP TC6 Conference on Open Distributed Processing*, pages 3–14, Berlin, Germany, September 1993.

[15] R. Sinnott. *An Initial Architectural Semantics in Z of the Information Viewpoint Language of Part 3 of the ODP-RM*, 1994. Input to ISO/JTC1/WG7 Southampton Meeting.

[16] J.M. Spivey. *The Z notation: A reference manual*. Prentice Hall, 1989.

33

ODP Types and Their Management: an Object-Z Specification

W. Brookes, J. Indulska

Department of Computer Science, CRC for Distributed Systems Technology,
The University of Queensland, Brisbane 4072, Australia

Defining a type model and maintaining a persistent type repository is an approach taken by a number of distributed platforms for managing the heterogeneity present in distributed systems. In particular, the Reference Model of Open Distributed Processing (RM-ODP) defines a basic type model and type repository function for supporting application interoperability in an open, heterogeneous and autonomous environment. This paper presents the specification of one approach to type description and type management that is compliant with the current version of the RM-ODP standard. The specification extends the scope of the RM-ODP type model by introducing relationship types. It shows their description and role in type matching. A summary of the important features of this specification is also presented.

Keyword Codes: C.2.4; D.1.3; D.2.1
Keywords: Distributed Systems, Concurrent Programming, Requirements/Specifications

1. INTRODUCTION

The Reference Model of Open Distributed Processing (RM-ODP) is an emerging ISO standard that recognises the need for applications and services to be able to interwork in an open, heterogeneous and autonomous environment [14]. In an open system little commonality can be assumed, as systems can be developed independently. Therefore, in order to use services globally, there must be a common understanding of the nature of those services, independent of their representation or provision by a particular network. Some distributed computing platforms have addressed this issue by building type management systems [2, 3, 18]. They differ in their type models and the functionality of those systems. RM-ODP introduces a Type Repository function which provides a framework for describing and relating the types of information, services and entities in an open distributed system [14].

With such a repository of type information, end-users and components of a distributed system infrastructure can use the type information to support *interoperability* of applications

---

The work reported in this paper has been funded in part by the Cooperative Research Centres Program through the Department of the Prime Minister and Cabinet of the Commonwealth Government of Australia. It was also partially supported by an Australian Government Postgraduate Research Scholarship (APRA).

(since there is a common agreement on the types), *resource discovery* (by providing a repository of information about types of services that exist), and *system evolution* (by recording information about different types which provide compatible functionality). ODP infrastructure components requiring support from the type repository are the ODP Trader (which selects services on behalf of users) and functions responsible for binding (to bind interfaces in a type-safe manner).

We refer to the type repository function as a *Type Manager*, however its functionality need not be implemented by a single ODP object. This paper presents an overview of a formal specification of types which are basic for interoperability (and therefore have to be described in the Type Manager) and of type management functions. The latter includes two components: the type description repository and the relationship repository.

The issue of formal specification of the ODP type model and management of types has already been addressed in other research. Najm & Stefani [17] and RM-ODP Annex A [14] present basic specifications of the ODP type model assuming that it is a first order type system with the structural ODP subtyping relationship defined in the form of inference rules on judgements. It addresses operational interfaces only. Z and Object-Z specifications of an ODP trader [9, 16] include some specification of type management functions. They are simplified and focus on the trading function only; they do not specify types and specification of type management is limited to adding and deleting types as well as a high level specification of service type matching.

Our goal is to provide a formal specification for aspects of the whole type management system including both the type model and management of types, to precisely and unambiguously describe ODP type management concepts. We use an object-oriented approach to provide a definition of types and a set of management operations. The type model presented is compliant with the ODP type model, but provides a more general approach to relationship types. As ODP defines a built-in subtyping relationship with one particular semantics, we extend this by allowing a definition of various kinds of relationships (*e.g.* different kinds of subtyping relationships) to be introduced. This creates a basis for mapping between subtyping relationships from different domains (*e.g.* mapping between an ODP-like system and OMG CORBA [19] or DCE [20]).

Several candidate specification techniques exist for the type model including ACT.ONE [8], and the lambda calculus [6]. There are also various languages which could be used to specify the whole Type Manager including LOTOS [11] and Z [21]. However, Object-Z was chosen for the following reasons: (a) the language Z (as developed by Spivey, [21]) is able to readily accommodate object-oriented concepts including inheritance [10, 15], (b) some work involving Z specifications already exists in ODP [9, 16], and (c) Object-Z [22] is an object-oriented extension of the Z specification language and it "displays sufficient expressive power for use in modelling ODP systems" [23]. Other factors which influenced the decision in favour of Object-Z were the difficulties in interpreting inheritance and subtyping in LOTOS [7, 15] and the lack of facilities for semantic specification in the other techniques mentioned (specification of the semantics of ODP types, not the semantics of the language in which the types are described). Semantic specification is not exploited in this paper as the specification presented considers only the syntax of types, but is necessary for future research.

The remainder of this paper is organised as follows. Section 2 presents a specification and discussion of types and the type description repository component of a Type Manager. Section 3 takes a similar approach in presenting the relationship repository. Section 4 contains a short description of the specification of a complete ODP-based Type Manager. Finally, section 5 highlights the most important sections of the specification, and presents areas for future work.

2. THE TYPE DESCRIPTION REPOSITORY

The Type Manager keeps type descriptions of types which are basic for ODP interoperability: Datatypes, Operations, Flows, Signals, Interfaces, Objects and Relationships.

Datatypes are essential as they are the building blocks for other types (for example, the arguments and results of operations). Descriptions of operations, flows, signals, interfaces and objects are necessary in an ODP-based type model since they are fundamental concepts for ODP, and provide the basic units of interaction in an ODP system. Our work extends the RM-ODP type model to include relationship types. Since the Type Manager should support many different relationships, it is necessary to store the description of each relationship. As defined by RM-ODP [13], a type is described by a predicate. Relationship descriptions can be treated as types because they may easily be described as a predicate (defining the relationship).

The Type Manager must be able to learn about new kinds of types, *e.g.* service types required by a Trader, therefore the set of kinds of types is extensible to accommodate definitions of new kinds. Types should be uniquely identified. Having type identifiers allows applications to reference types by names, and allows a relationship repository to be built in a more flexible manner since the relationships may be stored separately from the descriptions [1]. For the purposes of this paper a type is identified by a *TypeId* (which is unique in its domain). The issues of context-based naming (which is necessary to reduce the scope of unique names) and name versioning are not addressed in this paper.

$$[TypeId, Name]$$

The type description repository maintains type descriptions (each with an associated name). From the type descriptions together with type identifiers a repository is constructed with operations to manipulate the contents (*e.g.* Add, Delete, Lookup). In the following specification, bold subscripted numbers associated with schema names represent the RM-ODP clause on which the schema is based. The specification of object types has been omitted.

2.1. Type descriptions

We start by defining the notion of a type in general. The only thing the different kinds of types have in common is that it is possible to identify to which kind a particular type belongs.

```
_ TypeDefn _____
|  _____
| | kind : TypeKind
| |_____
```

As discussed previously, the type kind is one of the following:

$$TypeKind ::= Datatype \mid Operation \mid Flow \mid Signal$$
$$\mid Interface \mid Object \mid Relationship \mid \ldots (other\ types)$$

2.1.1. Operation type description

There are two basic kinds of operation in RM-ODP: interrogation and announcement. An interrogation is what is widely known as a remote procedure call. It consists of two interactions: an invocation followed by a termination. An announcement consists only of the invocation, and no termination action is expected (or allowed).

An invocation action consists of the operation name, and the number, names and types of the argument parameters. The following specification constrains the argument parameter names to be unique and that each parameter name has exactly one type.

```
_ OperationInvocationTemplate 7.1.26 _____
opName : Name
nrArguments : N
arguments : Name ↛ TypeId
_____
#(dom arguments) = nrArguments
```

A termination action is similar to an invocation. Each termination has a name, and the number, names and types of result parameters. This specification assumes that each termination is uniquely identified by a name.

```
_ TerminationTemplate 7.1.26 _____
termName : Name
nrResults : N
resultTypes : Name ↛ TypeId
_____
#(dom resultTypes) = nrResults
```

Having defined invocation and termination actions, we can now define interrogations and announcements. An interrogation consists of one invocation, plus possibly many terminations.

```
_ InterrogationSignature 7.1.26 _____
TypeDefn
  _____
  invocation : OperationInvocationTemplate
  responses : F₁ TerminationTemplate
  _____
  kind = Operation
```

An announcement contains only a single invocation action.

```
_ AnnouncementSignature 7.1.27 _____
TypeDefn
  _____
  invocation : OperationInvocationTemplate
  _____
  kind = Operation
```

Now we can define a more general notion of an operation. An operation type signature is either an interrogation signature or an announcement signature.

$$OperationSignature_{7.1.28} ::= Interrogation⟪InterrogationSignature⟫$$
$$| \quad Announcement⟪AnnouncementSignature⟫$$

2.1.2. Flow type description

RM-ODP defines stream interfaces to support modelling of applications which manage continuous flows of data (*e.g.* multimedia applications). Each stream interface consists of smaller building blocks called "flows". Each flow contains the type of information which will pass along the flow, as well as an indication of the direction of data flow.

```
┌─ FlowSignature 7.1.16 ──────────────────────────────
│ TypeDefn
│ ┌────────────────────────────────────────────────
│ │ type : TypeId
│ │ causality : Direction
│ │ ────────────────────────────────────────────────
│ │ kind = Flow
│ └────────────────────────────────────────────────
└──────────────────────────────────────────────────
```

The causality of the flow is indicated by its direction. An object may be either producing the flow of data (initiating) or consuming it (responding).

$Direction_{7.1.9} ::= Initiating \mid Responding$

2.1.3. Signal type description

A basic unit of interaction defined in RM-ODP is a signal. It consists of a single, atomic action (message) between a basic computational object and a binding object. Signals are named, contain parameters (number, names and types) and have causality (direction).

```
┌─ SignalSignature 7.1.9 ─────────────────────────────
│ TypeDefn
│ ┌────────────────────────────────────────────────
│ │ sigName : Name
│ │ nrArguments : N
│ │ arguments : Name ↦ TypeId
│ │ causality : Direction
│ │ ────────────────────────────────────────────────
│ │ kind = Signal
│ │ #(dom arguments) = nrArguments
│ └────────────────────────────────────────────────
└──────────────────────────────────────────────────
```

2.1.4. Interface type description

There are three kinds of interface types. *Operational interfaces* consist of a set of operations (interrogations or announcements). Thus, operational interface types are defined as a set of operation signatures. The second kind of interface is a *stream interface*, which consists of a set of information flows. The final kind of interface is a *signal interface* which contains a set of signal signatures.

```
┌─ ComputationalInterfaceSignature 7.1.10 , 7.1.17 , 7.1.31 ──────
│ TypeDefn
│ ┌────────────────────────────────────────────────
│ │ defn : OperationalInterface⟪F OperationSignature⟫
│ │      | StreamInterface⟪F FlowSignature⟫
│ │      | SignalInterface⟪F SignalSignature⟫
│ │ ────────────────────────────────────────────────
│ │ kind = Interface
│ └────────────────────────────────────────────────
└──────────────────────────────────────────────────
```

Including a description of some interface semantics as part of an interface description is for further study, and not presented in this specification.

2.1.5. Relationship type description

ODP-based systems should support system evolution to reflect the changing nature of an open system. Thus an ODP-based Type Manager should allow new relationships to be introduced at run-time by a user and for these relationships to be later used for type matching and type checking. One approach to facilitate this is to store a meta-level description of each relationship and to allow new relationship descriptions (type descriptions) to be added at run-time.

Relationship types consist of two parts: syntactic and semantic. The syntactic elements of a relationship are captured by a set of role types. The following definition of relationship types is an extension of the ISO General Relationship Model [12].

```
__ RoleSignature _____
  rolename : Name
  roletype : TypeId

  required_cardinality : N
  permitted_cardinality : N
```

Each role models one participant in the relationship, and has a name, type, and cardinalities associated with it. Required cardinality corresponds to the minimum number of participants in that role, and permitted cardinality defines the maximum number. Each role may have additional described behaviour associated with it, however that is not modelled in this specification.

The semantics of a relationship are captured in two parts: a declaration of the characteristics of the relationship, and a definition rule which may be used to determine membership in a given relationship. The characteristics applicable for all relationships include the method of handling deletion of one of the roles. When a specified role is deleted, the possible options for dealing with the relationship are: (a) delete other roles as well; (b) release other roles (*e.g.* set them to be NULL); (c) prevent the deletion from occurring unless other roles are empty (NULL).

```
__ RelationshipSignature _____
  TypeDefn
 _____
  roles : F RoleSignature
  delete_all_in_roles : RoleSignature ↦ F RoleSignature
  release_all_in_roles : RoleSignature ↦ F RoleSignature
  only_if_none_in_roles : RoleSignature ↦ F RoleSignature
 _____
  kind = Relationship
  #roles > 0
  dom delete_all_in_roles ∈ roles ∧ ran delete_all_in_roles ⊆ roles
  dom release_all_in_roles ∈ roles ∧ ran release_all_in_roles ⊆ roles
  dom only_if_none_in_roles ∈ roles ∧ ran only_if_none_in_roles ⊆ roles
```

Homogeneous binary relationships (having exactly two roles, each of the same type) are a special case. For homogeneous binary relationships, additional information can be maintained about the characteristics of the relationship, such as whether it is reflexive, irreflexive, symmetric, *etc.* The definition may also record whether the relationship is intended for supporting type matching. The schema on the left shows the definition of a homogeneous binary relationship, while the definition on the right shows the possible characteristics.

```
┌─ BinaryRelationshipSignature ──────
│ RelationshipSignature
├──────────────────────────────────────
│ R : Reflexivity          ::=  Reflexive | Irreflexive | Antireflexive
│ S : Symmetry             ::=  Symmetric | Asymmetric | Antisymmetric
│ T : Transitivity         ::=  Transitive | Intransitive | Antitransitive
│ type_matching : B
├──────────────────────────────────────
│ #roles = 2
│ ∃ r₁, r₂ ∈ roles •
│   r₁ ≠ r₂ ∧ r₁.roletype = r₂.roletype
└──────────────────────────────────────
```

The semantics of constraints imposed by the binary relationship characteristics is captured by the relationship repository operations. When an instance of a relationship is added, it is checked to ensure that it does not violate the characteristics of the relationship description. For example, the RM-ODP subtyping relationship is reflexive, antisymmetric and transitive. Therefore when adding a new instance of this relationship, appropriate checks should be made to ensure that these constraints are not violated. As mentioned earlier, each relationship should have a definition rule associated with it. This rule can be used to enforce checking whether types can legally participate in the relationship (as explained below). For binary relationships it is modelled as a Z relation between two type identifiers. If two types may legally participate in the type relationship, they are members of the Z relation (the definition rule).

The set of characteristics includes an indication of whether automatic type matching is possible (*e.g.* if matching is based on syntax only). This knowledge, together with the definition rule provides the mechanism for allowing the Type Manager to automatically determine membership in a relationship, as can be done for structural subtyping (as just one example). If full automation of matching is not possible, the definition rule can be used by a tool supporting semantic type matching (*e.g.* a browsing tool for a user). Note that defining the semantics of each relationship allows users to define their own compatibility relationships, and allows multiple definitions of subtyping relationships to co-exist in the repository and be used for different matching purposes when necessary. This approach facilitates federation of different distributed environments which can use different compatibility relationships. The proposed Type Manager provides a tool for mapping between these relationships. It does not, however, prescribe a set of compatibility relationships, as this would contradict the goal of openness.

As an example, the RM-ODP structural subtyping relationship (as applied to operational interfaces) is shown by the following definition. Note that the definition is recursive: subtyping on operational interfaces will depend on subtyping of operations, which in turn will depend on subtyping of arguments and results. The definition of subtyping of operations, arguments and results is omitted from the following definition rule. These subtyping rules are not always straightforward to define (*e.g.* when recursive types are introduced). A description of RM-ODP subtyping rules for operational interfaces, including recursive types, can be found in Annex A of RM-ODP Part 3 [14].

The following rule first retrieves the two type descriptions to be compared ($t1\,defn$ and $t2\,defn$). If both type descriptions are for interfaces, then $t1$ is a subtype of $t2$ if for every operation in the supertype $t2\,defn$ (called x) there exists a corresponding operation (with the same name) in the subtype $t1\,defn$ (called y) such that operation y is a subtype of operation x.

$$
\begin{array}{l}
\_\,IsRMODPOperationalInterfaceSubtypeOf\,\_ : TypeId \leftrightarrow TypeId \\
t1\,defn, t2\,defn : \downarrow TypeDefn \\
\hline
\exists\,t1, t2 : TypeId\ \bullet \\
\quad t1\,defn = typedb(t1) \\
\quad t2\,defn = typedb(t2) \\[4pt]
\quad t1\ \underline{IsRMODPOperationalInterfaceSubtypeOf}\ t2 \Leftrightarrow \\
\quad\quad ((t1\,defn.kind = Interface \wedge t2\,defn.kind = Interface) \Rightarrow \\
\quad\quad\quad (\forall\,x \in \mathrm{ran}\ t2\,defn.defn\ \bullet \\
\quad\quad\quad\quad \exists\,y \in \mathrm{ran}\ t1\,defn.defn\ \bullet \\
\quad\quad\quad\quad\quad y.invocation.opName = x.invocation.opName \\
\quad\quad\quad\quad\quad y\ \underline{IsRMODPOperationSubtypeOf}\ x \\
\quad)) \\
\end{array}
$$

2.2. The type database

Types known to the Type Manager are stored in some form of specialised database. This is modelled as a partial function which maps type identifiers (*TypeId*) into type descriptions (*TypeDefn*). Initially the database is empty.

Only the operation to add a type to the database is shown here. Specification of other operations (*e.g.* deleting a type, performing a lookup of a type description) may be found in [5].

$$
\begin{array}{l}
\underline{TDB}\ \rule{6cm}{0.4pt} \\
\quad typedb : TypeId \nrightarrow \downarrow TypeDefn \\
\hline
\quad \underline{INIT}\ \rule{6cm}{0.4pt} \\
\quad typedb = \varnothing \\
\hline
\quad \underline{Add}\ \rule{6cm}{0.4pt} \\
\quad \Delta(typedb) \\
\quad tid? : TypeId \\
\quad defn? : \downarrow TypeDefn \\
\hline
\quad tid? \notin \mathrm{dom}\ typedb \\
\quad typedb' = typedb \cup \{tid? \mapsto defn?\} \\
\end{array}
$$

3. THE RELATIONSHIP REPOSITORY

There are two aspects to consider about relationships between types: the definition of the relationship and the instances of the relationship. The definition of a relationship includes the roles of the relationship, the predicate defining a membership rule as well as a description of the properties of the relationship (*e.g.* reflexivity, symmetry). This information is maintained in the form of a 'relationship type', and is stored in the type description repository as described in the previous section. The Type Manager also stores the set of instances for relationships between types. RM-ODP defines a Relationship Repository function, however it differs in scope from the following description in that RM-ODP relationships are defined between objects and interfaces, whereas our approach is for relationships between types, as required for type matching (*e.g.* subtyping relationships).

The following section discusses relationship instances, focusing on homogeneous binary relationships only. Binary relationships such as subtyping, inheritance and compatibility are the most interesting from the type matching point of view. These concepts are then used to specify the database of relationships maintained by the Type Manager.

3.1. Homogeneous binary relationship instances

Homogeneous binary relationship instances are building blocks of type hierarchies and can be stored in a directed graph structure. The following discussion is based around the assumption of storing binary relationship instances as a general directed graph (not necessarily fully connected), that is specialised for recording type relationships.

Using the previous definition of relationship types, we assume that the two roles of a binary relationship are called '*from*' and '*to*'. We use the Z notation $\succ$ as the name of the reflexive transitive closure of the relationship, and use an infix notation (*i.e. from $\succ$ to*).

Integrity checks are performed when adding a new instance of the relationship. Namely, the types being added to the graph are verified to ensure they are defined in the type database, and the pair of types being added is verified to ensure that it does not violate the characteristics of the relationship. For instance, the graph must be acyclic for a reflexive, antisymmetric and transtive relationship (*e.g.* subtyping). The reflexive transitive closure ($\succ$) is checked to ensure that this condition is not violated.

The relationship characteristics can be used for two purposes. The first is to ensure that adding a given pair of types to the relationship will not violate the relationship characteristics (*e.g.* adding $\langle a, a \rangle$ to an antireflexive relationship).

$$defn?.R = Antireflexive \wedge defn?.T = Antitransitive \Rightarrow from? \neq to?$$
$$defn?.R = Antireflexive \wedge defn?.T \neq Antitransitive \Rightarrow from? \nsucc to?$$
$$defn?.S = Antisymmetric \Rightarrow to? \nsucc from?$$
$$defn?.T = Antitransitive \Rightarrow from? \nsucc to?$$

The second use of the characteristics is to ensure duplicate information is not stored in the graph. A new instance is added if it cannot be deduced by reflexivity, symmetry or transitivity.

$$((defn?.R \neq Reflexive \vee from? \neq to?) \wedge$$
$$(defn?.S \neq Symmetric \vee to? \nsucc from?) \wedge$$
$$(defn?.T \neq Transitive \vee from? \nsucc to?))$$
$$\Rightarrow \succ' = \succ \cup \{(from?, to?)\}$$

3.2. The relationship database

Relationship instances are stored in a database, *RelnDB*, defined in this section. Initially the relationship instance database is empty.

```
┌─ RDB ──────────────────────────────────────────────────────
│
│  ┌──────────────────────────────────────────────────────
│  │ relndbinst : TypeId ↦ RelnInst
│  │ typedb : TypeDB
│  ├──────────────────────────────────────────────────────
│  │ ∀ r : dom relndbinst •
│  │         (r ∈ dom typedb.typedb ∧
│  │          typedb.typedb(r).kind = Relationship)
│  │
│  ├─ INIT ────────────────────────────────────────────────
│  │ relndbinst = ∅
│  │ typedb.INIT
│  └──────────────────────────────────────────────────────
```

One constraint that must always be true on the database is that for every relationship instance known in the instance database, the relationship definition for that relationship must also be known. Operations are provided to add an instance to the graph, to delete an instance, to find all types that are related to a given type and to find all types to which a given type is related. Specification of these operations, and their integrity checks may be found in [5].

4. THE TYPE MANAGER — A GLOBAL VIEW

The type description repository consists of the underlying type data base represented by the *typedb* function, and all the operations to manipulate/query the database. Similarly the relationship repository consists of the relationship database (*relndb*) and the operations upon it. A complete specification of these operations is given in [5].

```
┌─ TypeDB ──────────────────        ┌─ RelnDB ──────────────────
│ TDB                                │ RDB
│ TDB_Add                            │ RDB_AddInst
│ TDB_Delete                         │ RDB_DelInst
│ TDB_Lookup                         │ RDB_AreRelated
│ TDB_Exists                         │ RDB_RelatedTo
└───────────────────────────        │ RDB_RelatedFrom
                                     └───────────────────────────
```

The Type Manager consists of the type description repository (an instance of the class *TypeDB*) and the relationship repository (an instance of the class *RelnDB*). These two databases are separate, but related (*e.g.* relationship instances in RelnDB must be for a relationship whose type is defined in TypeDB). A summary of the functionality provided by the ODP-based type manager specified in this paper is given by the protocol diagram in Figure 1, showing the operations defined and their input and output parameters. Using these simple operations, type descriptions can be added, deleted and queried, and relationships can be stored and retrieved. More complex query operations can be built using this basic set of operations.

5. CONCLUSION

This paper presents an overview of the Object-Z specification of both basic types supporting interoperability of ODP-based systems and a Type Manager providing a persistent repository of

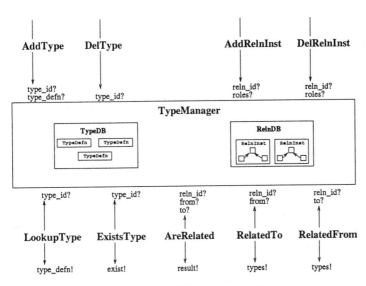

Figure 1: Type Manager Protocol

types. Users and ODP infrastructure components are able to access type information to provide type-safe dynamic selection of services, dynamic binding and dynamic operation invocation, while supporting system evolution.

The specification satisfies and extends the functionality required by the RM-ODP model. A specification of the two major components of an ODP-based Type Manager is given. The type description repository showed a concrete specification of some of the more important types referred to in the RM-ODP documents. In addition, it introduced the concept of relationship descriptions as types themselves since they can be expressed by a predicate. The relationship repository described the specification of a general directed graph of type identifiers. It introduced a specialised relationship database which associates a relationship name with a graph of instances and presented a specification of necessary integrity constraints.

The complete specification was used as the basis of our prototype Type Manager which enhances service selection and interoperability in the DCE platform for distributed computing [4]. Further work is being carried out on including a description of some semantics in type descriptions. The major goal of this work is to add elements of semantics specification which can be automatically or semi-automatically matched, improving type-safety of interoperability. Including elements of semantics in type descriptions can also support a user in deciding which types are semantically related.

REFERENCES

[1] A. Albano, G. Ghelli, and R. Orsini. "A Relationship Mechanism for a Strongly Typed Object-Oriented Database Programming Language". Technical Report FIDE/91/17, FIDE Project, 1991.

[2] APM Ltd, Cambridge UK. *ANSAware 4.1 Application Programmer's Manual*, Mar. 1992. Document RM.102.00.

[3] R. Balter. "Construction and Management of Distributed Office Systems Achievements and Future Trends". In *ESPRIT '89*, Proceedings of the 6th Annual ESPRIT Conference, pages 47–58. Brussels, November 27–December 1, 1989.

[4] C. J. Biggs, W. Brookes, and J. Indulska. "Enhancing Interoperability of DCE Applications: a Type Management Approach". In *Proceedings of the First International Workshop on Services in Distributed and Network Environments, SDNE'94*, 1994.

[5] W. Brookes and J. Indulska. "A Formal Specification of an ODP-based Type Manager". Technical Report 286, The University of Queensland, Department of Computer Science, Jan. 1994.

[6] L. Cardelli and P. Wegner. "On Understanding Types, Data Abstraction, and Polymorphism". *Computing Surveys*, 17(4):471–522, Dec. 1985.

[7] E. Cusack, S. Rudkin, and C. Smith. "An Object Oriented Interpretation of LOTOS". In *Proceedings of the Second International Conference on Formal Description Techniques (FORTE'89)*, pages 265–284. Vancouver, B.C., 5–8 December 1989.

[8] J. de Meer and R. Roth. "Introduction to algebraic specifications based on the language ACT ONE". *Computer Networks and ISDN Systems*, 23, 1992.

[9] J. S. Dong and R. Duke. "An Object-Oriented Approach to the Formal Specification of ODP Trader". *Proceedings of the International Conference on Open Distributed Processing, ICODP'93*. Berlin, Germany, 14–17 September, 1993.

[10] R. Duke, G. Rose, and A. Lee. "Object-oriented protocol specification". In *Proceedings of the Tenth International IFIP WG 6.1 Symposium on Protocol Specification, Testing and Verification*, pages 323–339. Ottawa, Ont., 12–15 June 1990.

[11] ISO IS 8807. LOTOS, A Formal Description Technique based on the Temporal Ordering of Observation Behaviour, 1988.

[12] ISO/IEC CD 10165-7. Information Technology—Open Systems Interconnection—Structure of Management Information. Part 7: General Relationship Model, 1993.

[13] ISO/IEC DIS 10746-2. Draft Recommendation X.902: Basic Reference Model of Open Distributed Processing — Part 2: Descriptive Model, Apr. 1994. Output of Geneva editing meeting 14–25 February 1994.

[14] ISO/IEC DIS 10746-3. Draft Recommendation X.903: Basic Reference Model of Open Distributed Processing — Part 3: Prescriptive Model, Apr. 1994. Output of Geneva editing meeting 14–25 February 1994.

[15] ISO/IEC JTC1/SC21. Working Document on Architectural Semantics, Specification Techniques and Formalisms, N4887, 1990.

[16] ISO/IEC JTC1/SC21/WG7 N743. "ANNEX G: Z Specification of the Trader", Nov. 1992.

[17] E. Najm and J.-B. Stefani. "A formal semantics for the ODP computational model". *Computer Networks and ISDN Systems*, to appear.

[18] Object Management Group. *Object Management Architecture Guide*, second edition, Sept. 1992. Revision 2.0, OMG TC Document 92.11.1.

[19] OMG and X/Open. *The Common Object Request Broker: Architecture and Specification*, 1992.

[20] W. Rosenberg and D. Kenney. *Understanding DCE*. Open System Foundation, 1992.

[21] J. M. Spivey. *The Z Notation: A Reference Manual*. Intl. Series in Computer Science. Prentice-Hall, 1989.

[22] S. Stepney, R. Barden, and D. Cooper, editors. *Object orientation in Z*. Workshops in Computing. Springer-Verlag, Published in collaboration with the British Computer Society, 1992.

[23] P. Stocks, K. Raymond, D. Carrington, and A. Lister. "Modelling open distributed systems in Z". *Computer Communications*, 15(2):103–113, Mar. 1992.

SESSION ON

Integrating Databases in Distributed Systems

34

Multiware Database: A Distributed Object Database System for Multimedia Support

C. M. Tobar[a] and I. L. M. Ricarte[b]

[a]Institute of Informatics, Pontifical Catholic University of Campinas (PUCCAMP), Brazil. e-mail: tobar@dca.fee.unicamp.br

[b]Dept. of Computer Engineering and Industrial Automation, State University of Campinas (UNICAMP), Brazil. e-mail: ricarte@dca.fee.unicamp.br

This paper describes the Multiware Database, a database capable to store and manage complex multimedia documents using the features of both object database management systems and open distributed systems. This multimedia database is a component of the Multiware project[1], a platform supporting distributed multimedia cooperative applications. Representation of multimedia documents in this system is normalized, in the sense that distinct aspects of documents, such as structural and exhibition aspects, are stored as distinct objects in the database. This normalization enables concise representations and integration of distinct multimedia standards within the same framework.

Keyword Codes: H.5.1; H.2.8; C.2.4
Keywords: Multimedia Information Systems; Database Applications; Distributed Systems.

1. INTRODUCTION

Due to recent technological developments, the concept of a "computer-based document" has evolved sensibly since the days dominated by word processing. Computer documents incorporating sounds and images (still or animated) are already a reality, and uncountable applications can benefit with the utilization of this type of multimedia data. The manipulation of multimedia documents and their related activities, such as authoring and exhibition, involves large amounts of data, which should be persistent and efficiently retrieved. In order to support these requirements, some type of storage server should be provided. Ideally, a multimedia database management system would provide these capabilities, along with the standard features of database systems, such as data independency (thus supporting data sharing) and consistency control. However, there is no satisfactory implementation of such a system. Furthermore, access and manipulation requirements are even more severe when the application is intrinsically distributed.

This paper describes one approach to the implementation of such a distributed multimedia database, the Multiware Database, on top of an object-oriented database management system (ODBMS for short). One of the key points in the design of Multiware Database is that it should seamlessly be integrated with an open distributed platform, and

---

[1]Multiware is supported by Grant No. 93/2617-0 from the State of São Paulo Foundation for Research Support (FAPESP), Brazil.

document representations should comply to existing or proposed multimedia standards.

The development of Multiware Database was motivated by the Multiware project [10]. Multiware is a platform to support multimedia cooperative applications incorporating ideas from the Reference Model for Open Distributed Processing (RM-ODP) and from existing industrial specifications such as the Common Object Request Broker Architecture (CORBA). Therefore, one of the major concerns on the design of this multimedia database system is to incorporate (not reimplement) services already provided by Multiware, whenever the platform service can improve efficiency of the final applications.

The Multiware platform project aims the development of a middleware level of software and hardware to allow the manipulation of multimedia information. Multiware's goal is to support the new generation of applications, which requires distributed heterogeneous computational environments, such as cooperative work (CSCW), decentralized artificial intelligence (DAI), and decision making support systems.

Multiware provides three functional layers, namely Basic Software and Hardware, Middleware, and Groupware (Figure 1). The strategy behind Multiware is to use, as much as possible, related available standards, specifications, and products, favoring the adoption of object-oriented technology. From a functional point of view, the Basic Software and Hardware layer provides the functions supported by operating systems and communication protocols, among others. Distributed processing facilities are provided by the Middleware layer, which aggregates ODP functionality to existing commercial distributed systems and supports functions specific to real-time multimedia processing, in special the multimedia database management system. The Groupware layer provides framework and components that support group work based on multimedia.

Implementation of multimedia information systems, and more specifically of multimedia databases, is a relatively new field. One important aspect is the integration of standards related to the many forms of representation of generic and multimedia data, since standards will be responsible to a more widespread use of multimedia support systems and object database systems.

Previous work on this field points toward the use of an ODBMS as the best way to support the basic requirements of storage of multimedia data and to describe and represent multimedia information [12]. Other related topics of interest to multimedia databases are user interaction, retrieval strategies, and standards, which are addressed in the following reviewed work.

Grosky [4] provides an overview on Multimedia Information Systems. He proposes a "generic architecture" for such systems, which would be logically composed by three interrelated repositories of data, namely a standard database (with non-multimedia data), a multimedia database (with uninterpreted multimedia objects), and a feature database (with information that could potentially help the retrieval of multimedia data). Although it is not required that these components are implemented as distinct databases, this architecture shows concern with the fact that current systems hardly can provide efficient manipulation of all types of data. However, the proposed architecture does not address aspects of distribution of such systems.

Bulterman [1] analyses multimedia document behavior during user interaction on a distributed multimedia application, and presents an infrastructure to support requests of real-time dynamic presentations of multimedia adaptable documents; that is, the appli-

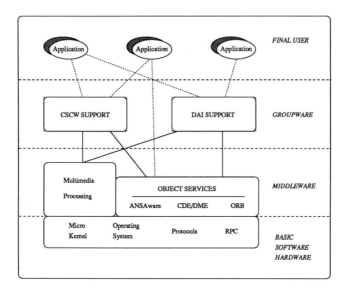

Figure 1: Multiware Platform.

cation can adapt to the available resources. Data transfer problems are not addressed by this infrastructure. The aspect of multimedia adaptation results in the reduction of cost related to authoring multimedia applications. The Multiware Database proposal presents a similar adaptation mechanism.

Koegel *et al.* [8] developed a distributed multimedia information system using a standard. This system constitutes one of the first efforts to validate the operational model of a multimedia standard, using an ODBMS to store document descriptions. Its three-layered architecture presents information of hypermedia content and structure, object instances, and application objects and their attributes. The authors intend to develop generic multimedia presentation services, avoiding reimplementation of the application layer for each new document type definition, as proposed for the Multiware Database.

Klas, Löhr, and Neuhold [7] present a tutorial on multimedia and its application requirements. Some multimedia standards are considered and integrated to a distributed and open extensible ODBMS system. They address some important features required by multimedia applications and not supported by object-oriented systems, such as time-dependent data support, user interaction mechanisms, and content-based query and retrieval techniques. Their multimedia database includes a synchronization manager, an interaction manager, and a continuous object manager. It is interesting to notice that in the Multiware Database design these functions are provided by isolated external servers.

The remaining of this paper presents the Multiware Database proposal, whose main goal is to provide multimedia storage and (subsequent) retrieval support for Multiware distributed applications. Synchronization and user interaction (related to information exhibition), information preparation (for transport and exhibition), and information transport are necessary multimedia services, not directly supported by Multiware Database but for which there must be some correspondent data inside the multimedia database, in order to facilitate all related multimedia activities, from capture through management. Multiware Database provides this support in a distributed heterogeneous environment.

The organization of the paper follows. Section 2 describes the Multiware Database, addressing its constitution in terms of manipulation and storage units. Section 3 discusses object services supported (or expected to be supported) by open distributed systems, and introduces a set of multimedia services which are provided by the Multiware Database. Section 4 presents the technological structure of Multiware Database and discusses multimedia services localization. Section 5 presents the functional structure of Multiware Database, along with its implementation based on the object-oriented approach. Finally, Section 6 presents conclusions and future work.

2. THE MULTIWARE DATABASE COMPOSITION

The Multiware Database provides multimedia information storage and retrieval services to the applications, considering their need of timeliness and feasible manipulation mechanisms. The basic structures of the Multiware Database are based on the object-oriented approach, and they aim the achievement of this goal. The description of these structures is the subject of this Section.

2.1. Multiware Manipulation Unit

From the user or application point of view, the Multiware Database stores and retrieves multimedia documents (here denoted MMDoc), considered to be the unit of manipulation for multimedia information.

In the Multiware context, MMDocs can present different classifications according to their *structure* (simple or complex), *organization* (hypermedia, simple-media, or multiple-media), *behavior* (static or dynamic — interactive, when exhibited), *continuity* (with or without temporal synchronization requirements), or *influence* (with or without modification of information, respectively active or inactive, during or after exhibition). All these perspectives are orthogonal and are considered in the Multiware Database design. Thus, it is possible to store a multiple-media complex MMDoc, which during its exhibition will be dynamic and active and will need synchronization. User interaction can exist during active and inactive MMDoc exhibitions, requiring information updating in the database.

From the point of view of object representation, an MMDoc is composed by two parts, one *multimedia object* and its (possibly many) *presentation forms* (Figure 2).

A multimedia object contains all the user information, which is to be stored, retrieved and exhibited. This information can be represented by several data pieces of different media types, and requires structural and, optionally, continuity (temporal) meta-data in order to relate the user data pieces.

A presentation form describes exhibition features, which are related to available hardware, security, and user selectivity possibilities, described in details in Subsection 2.2.

The representation structure of an MMDoc reflects the fact that it has information content and description, which are independent of presentation features. On the other

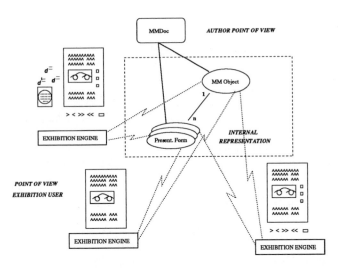

Figure 2: MMDoc Composition: it reflects the distinct multimedia capabilities of presentation terminals; in this example, interactive audio and video (top left), still images only (bottom left), and interactive video only (bottom right).

hand, presentation relates to exhibition aspects, which depend on available resources.

2.2. Multiware Storage Entities

The parts of an MMDoc described above yield the two storage entities of the Multiware Database, which are the multimedia object (denoted MMObj) and the presentation form.

An MMObj stores the (exhibition-independent) description of an MMDoc. It is composed by single media objects (such as an image, a text, or a sound), by hypermedia objects (such as hypertext, text linked with images or sound), and/or other composite multimedia objects. It also stores meta-information related to the multimedia document, such as author and creation date.

An MMObj should be able to represent the composition of single media objects and hypermedia objects. Thus, it can be recursively defined as follows:

> **a single media object** is an MMObj;
> **an hypermedia object or nested MMObj** is an MMObj; and
> **a composition of MMObj's** along with information describing the composition and relating the MMObj's, is also an MMObj.

A single MMObj is a piece of information created through one type of media representation and encoding format. For example, it could represent either an ASCII text or an MPEG item of video and audio.

One nested MMObj represents a kind of "hyper-MMObj", through which it is possible to navigate via links and to reach other simple or nested MMObj.

Each MMObj, besides the fact of being a component of other MMObj, has associated information, other than its structural information, related with intra-object synchronization properties, exhibition aspects of temporal and spatial synchronization, behavior, influence, identification and control. With this information it is possible to process the multimedia data, by an exhibition engine external to the Multiware Database, in order to *(i)* obtain proper synchronization between components, *(ii)* allow user interaction, and *(iii)* modify the information for later exhibitions.

The recursive nature of an MMObj, along with the disjoint information of its structure, provides two interesting forms of reuse for authoring purposes: reuse of description definition and reuse of multimedia pieces.

The information of one specific MMDoc is associated to a given description (that is, to one MMObj) and can be exhibited in multiple different forms. The reason for these different forms might be distinct hardware features, security-related aspects, or user selectivity. The characteristics of each possible exhibition format are stored and represented through a multimedia presentation form.

Presentation forms represent the information required to allow preparation, communication, and exhibition of an MMDoc on different distributed exhibition nodes. They include information about available hardware, location, quality (color, exhibition speed, resolution), completeness (partial or complete exhibition of each component of an MMDoc), fault resolution (delay, loss, acceleration), compression format, user selective choices, and so on. With this information and the description of the MMObj, it is possible to derive terminal-specific presentations of MMDocs, as further described in Section 5.

3. MULTIMEDIA DATABASE-RELATED SERVICES

Distributed processing comprises the class of information processing activities in which discrete components may be located at more than one location, and consequently requires explicit communication among components [6]. Heterogeneous distributed systems impose requirements of interoperability between components of different vendors, and if they present scalability characteristics, they are considered open systems.

One option to obtain distribution openess is through the adoption of a client-server-like architecture, where clients communicates to servers in order to obtain *services*.

The Object Management Group (OMG) effort has produced a terminology, used throughout this document, for object-oriented distributed services. The OMG aims the specification of standards to achieve interoperability, through a set of object services which provides database-like facilities for distributed objects. Database facilities usually provided by relational or object-oriented database systems have similar counterparts among OMG services, which cover most of the necessary services for multimedia manipulation.

Among OMG services there is a *brokering service* which is the key mechanism for the integration of other services and platforms, regardless the heterogeneities introduced by open distributed system architectures. The brokering server, known as Object Request Broker (ORB), is the main component of the OMG proposed architecture, and provides behavior invocation through a registration mechanism for location and naming.

This Section presents a brief description of services required by multimedia applications and not covered by current OMG services. The relationship between the complete set of services and the activities related to multimedia applications is then described.

3.1. Multimedia New Necessary Services

Multimedia manipulation requires specialized services not covered by the current OMG proposal, such as a specialized *retrieval service*, although the OMG query and archive services provide some retrieval functionality. Therefore, a specific retrieval service is considered from now on. *Wrap-up, transport, presentation*, and *performance monitoring* are four other necessary services for multimedia manipulation, not present in OMG specifications. The motivation for these multimedia services is described next.

The *Retrieval Service* is related (as far as possible) with content (semantic), structural, behavioral, and influence retrieval, along with whole or partial MMDoc description and content reuse.

The *Wrap-up Service* is responsible by the creation of multimedia objects in the necessary format for exhibition. Its main concerns are to generate adequate information for distributed transport and to filter unnecessary information related to presentation details, as discussed in Section 2.

The *Transport Service* provides efficient and reliable (as necessary) moving of multimedia objects. These objects, once in an exhibition node, are processed by presentation engines, which could demand further multimedia data transport, related to specific media types which composes the MMDoc to be exhibited.

The *Presentation Service* relates to aspects of processing of transported information in order to produce planned stimuli (upon creation) to a human being, *i.e.*, an exhibition. Synchronization, interaction, influence, security, and direct access are some of the issues related to this service.

The *Performance Monitoring Service* controls (fine tunes) and manages multimedia related aspects in the distributed environment through the generation of a log of performance events, such transmission initial time, final time, and interactive rate.

3.2. Multimedia Activities and Services

Among the activities related to the manipulation of multimedia information there are *capture, authoring, storage, communication, exhibition*, and *management*. Each of these activities requires support by a distinct set of services, as follows:

Capture: data interchange (conversion, compression), lifecycle (object instances), and archive.

Authoring: lifecycle, security, relationships, transactions, concurrency control, query, retrieval, archive and change management.

Storage: archive, lifecycle, security, relationships, transactions, interaction, concurrency control, change management, and replication.

Communication: retrieval, security, wrap-up (formatting, compression, etc.), and transport.

Exhibition: presentation (synchronization, interaction), security, and externalization.

Management: operational control, performance monitoring, startup, backup and restore, query, licensing, and security.

Trading, event notification, persistence, installation and activation, threads, time, and naming services belong to a set of basic support services necessary to all other services.

4. MULTIWARE DATABASE TECHNOLOGICAL STRUCTURE

The Multiware Database proposal, described in Section 2, presents an alternative architecture for multimedia database systems which does address distribution aspects. Therefore, it is important to analyze which services are provided by this platform in order to integrate seamlessly the database to Multiware. Several of the necessary services for multimedia manipulation are available in a ODBMS or can constitute autonomous specialized servers, available in an open architecture and accessible via an ORB.

In contrast to the ODBMS architecture, the ORB architecture enables each multimedia service to provide a separate piece of database technology. This allows the possibility of customized and expandable systems, with their components supplied by different vendors. A multimedia database system can adopt one or a combination of both architectures, if the combined approach is chosen, it will present several autonomous servers, interconnected through ORBs, some of which are distributed nodes of an ODBMS.

The Multiware Database has as framework one distributed ODBMS, several specialized media repository servers, and several autonomous multimedia object-related servers.

The ODBMS component of Multiware Database acts as a gateway to multimedia information, actually stored in specialized repository servers. The adoption of a distributed architectural organization for the Multiware Database permits the expansion of potential applications for multimedia services [9]. Two distinctive architectures are appropriate for this scenario, namely the master-slave and the federated organizations.

The ODBMS supporting the Multiware Database is a federated database server, that communicates via the brokering service (ORB). Each federated node presents a master-slave architecture between its object-oriented database manager and its related specialized media servers, as presented in Section 5.

In order to obtain a flexible and efficient access mechanism to MMDocs there is an Object Database Adaptor (ODA), through which the ODBMS is integrated to the ORB architecture [2]. ODA provides facilities to register with the ORB a set of stored objects identifiers and to allow access (direct access if desirable) to stored objects, as if they had been individually registered.

Depending upon the available functionalities of the integrated ODBMS, it is possible to locate each multimedia service either to the database system or to the platform specialized servers. There are three aspects to be considered in order to determine the location of each specific multimedia service:

> **broadness:** does the multimedia service apply only to the ODBMS or to several different clients?
> **actuation domain:** does the multimedia service act only upon objects stored in the ODBMS or upon other external objects?
> **efficiency:** which service does present better performance (CPU, I/O, and memory), the one in the ODBMS or the autonomous one?

Table 1 presents a cross-reference between the several multimedia services (previously referenced and identified in Subsection 3.1) and multimedia manipulation activities (introduced in Subsection 3.2) in order to present an analysis of the ideal service location, with the following notation: (A) capture; (B) authoring; (C) storage; (D) communication;

(E) exhibition; and (F) management. The *Relevance* column classifies each service taking into account the proposed aspects for allocating the service, and it uses the following notation: (1) server to multiple clients; (2) exclusive ODBMS server; (3) applicable to general objects; (4) applicable only to ODBMS objects; (5) critical efficiency; and (6) non critical efficiency.

OMG Service	(A)	(B)	(C)	(D)	(E)	(F)	Relevance
archive	X	X	X				(1) (3) (6)
backup/restore						X	(2) (4) (6)
change management		X	X				(2) (4) (6)
concurrency control		X	X				(2) (4) (6)
data interchange	X						(1) (3) (5)
externalization					X		(1) (3) (5)
licensing						X	(1) (3) (6)
lifecycle	X	X	X				(1) (3) (5)
operational control						X	(1) (3) (5)
performance monitoring						X	(1) (3) (5)
presentation			X		X		(1) (4) (5)
query		X				X	(2) (4) (5)
relationships		X	X				(2) (4) (5)
replication		X					(2) (4) (5)
retrieval		X		X			(2) (4) (5)
security		X	X	X	X	X	(1) (3) (6)
startup						X	(1) (3) (6)
transactions		X	X				(1) (3) (5)
transport			X				(1) (4) (5)
wrap-up			X				(1) (4) (5)

Table 1:   Cross-reference between multimedia services and activities.

The results of the location analysis can be summarized as follows: on one hand, the services of archive, lifecycle, licensing, operational control, data interchange, externalization, security, startup and transaction services are well suited to be provided by specialized autonomous servers, without the interference of any ODBMS. On the other hand, backup and restore, change management, query, concurrency control, retrieval, relationships and replication services are better supported by the ODBMS.

Specialized servers can offer wrap-up, presentation, and transport services, though applied specifically to multimedia objects, because there is no restriction imposing that all multimedia objects should be stored in the ODBMS.

5. MULTIWARE DATABASE FUNCTIONAL STRUCTURE

Two distinct types of objects represent an MMDoc inside the Multiware Database, namely *information objects* and *presentation objects*, which are related respectively to the multimedia object and the presentation form introduced in Subsection 2.1.

A third type, an *exhibition object*, created dynamically during the retrieval of one MMDoc for exhibition, is the combination of data from one information object and one presentation object. It is the multimedia communication unit.

One information object can be associated to many presentation objects. It contains, besides content information, all the information concerning structure, synchronization, behavior, influence, identification, and control.

The structure of an information object can contain pointers to other information objects, simple media objects or hypermedia objects, allowing the definition of recursive multimedia data structures. Except by information objects, these objects are stored in distributed and specialized media repository servers, outside the scope of the ODBMS.

The Multiware Database provides basic object types to represent simple media objects and basic hypermedia objects, along with their associated basic operations. Through the ODBMS constructs, the user is able to define new types.

The way an MMDoc is represented produces a layered structural view of the Multiware Database, presented in Figure 3.

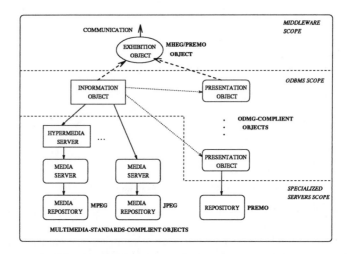

Figure 3: Multiware Database layered structure. The exhibition object is generated dynamically, not being permanently stored in the database scope. The distinct media repositories are also outside the scope of the object database.

One important aspect of multimedia information storage is that it should not be restricted to a central repository, since each type of media can be better supported by specialized servers. Thereafter, retrieval of a multimedia document involves retrieval operations from many individual distributed sources.

The open distributed architecture of Multiware presents potential heterogeneity problems between servers, platforms, and data formats. The adoption of standards mitigates the last problem, and also eases the integration with hardware and software products in order to support multimedia activities.

Some specialized media repository servers store hypermedia objects, which in turn have links to other simple or hypermedia objects. Each specialized media repository can be implemented according to a specific related media representation encoding format (standard or standard to be), such as MPEG, JPEG, G711, and MIDI.

One exhibition object is composed by all the information necessary for its exhibition, *i.e.*, components localization, synchronization, and specific presentation information. It is self-contained, being able to be processed by a PREMO engine, which in turn could emulate an MHEG engine.

Document-oriented standards of interest are MHEG and HyTime, both related to multimedia and hypermedia documents, and PREMO, related to presentation of MMDocs.

MHEG (Multimedia and Hypermedia information coding Expert Group) [3] defines an interchange format for multimedia and hypermedia objects as well as a specification of how to process the data being interchanged. HyTime (Hypermedia/Time-Based Structuring Language) [11], on the other hand, is strong in mechanisms for the synchronization of media data, specifying the structure of an MMDoc through an extension of the Standard Generalized Markup Language (SGML). PREMO [5] is a standard concerned with presentation aspects of a multimedia document, specifying mechanisms to represent and control input and output interactions.

The adoption of object models by these standards eases the integration of the particular aspects addressed by each standard, and provides a natural framework for their representation in a multimedia database system.

6. CONCLUSIONS

In this work, the architecture of Multiware Database is described. Multiware Database is the multimedia storage server system of Multiware distributed processing platform. The motivation for this work is to provide a multimedia database to support the storage of multimedia information, which should become available to applications using Multiware.

Since Multiware adheres to RM-ODP and CORBA, questions were raised regarding the ideal allocation of services related to multimedia activities; some of these services could be supported by both a specialized component of the platform and by an object database system. A framework to define allocation strategies was presented in Section 4.

Besides the service allocation discussion, another important aspect of Multiware Database is the role of multimedia-related standards, mainly PREMO and MHEG. Internally, objects are structured following the OMG object database standard, ODMG-93 [2].

The current version of Multiware Database focuses on one node of the federated ODBMS, storage and communication activities aspects, inactive MMDoc's, presentation and exhibition objects adherents to PREMO and MHEG standards (to be), information objects adherents to ODMG-93 standard.

Future versions will integrate the distributed servers and the manipulation of active MMDoc's according to the following strategy. During the second phase, support for the remaining multimedia manipulation activities (capture, exhibition, management, and authoring) will be added. There are many expected design dificulties to afford the support of the authoring activity, such as long transactions, incomplete objects, and so on. The third phase will support the federated ODBMS and implement basic interaction functions for dynamic MMDoc's. The final and fourth phase will incorporate the full features and facilities described in this document.

The authors are thankful to Manuel J. Mendes and Heinz D. Nevermman for their valuable comments.

References

[1] Dick C. A. Bulterman. Specification and support of adaptable networked multimedia. Published in Multimedia Systems, 1993. ftp: ftp.cwi.nl.

[2] R. G. G. Cattell, editor. *The Object Database Standard: ODMG-93.* Morgan Kauffman, 1994.

[3] Françoise Colaitis. Opening up multimedia object exchange with MHEG. *IEEE Multimedia*, 1(2):80–84, Summer 1994.

[4] William I. Grosky. Multimedia information systems. *IEEE Multimedia*, 1(1):12–24, Spring 1994.

[5] I. Herman, S. Carson, J. Davy, P. J. W. ten Hagen, D. A. Duce, W. T. Hewitt, K. Kansy, B. J. Lurvey, R. Puk, G. J. Reynolds, and H. Stenzel. PREMO: An ISO standard for a presentation environment for multimedia objects, October 1994. Submmited to ACM Multimedia'94 Conference.

[6] ISO/IEC JTC 1/SC 21. Basic reference model of ODP - part 2: Descriptive model. Working Draft N 7524, April 1993. 18 pages.

[7] Wolfgang Klas, Michael Löhr, and Erich J. Neuhold. Multimedia database systems. In *Proceedings of the Workshop on New Database Research Challenges*, pages 93–125, Rio de Janeiro, Brazil, September 1994. PUC-RJ/VLDB.

[8] J. Koegel, L. Rutledge, J. Rutledge, and C. Keskin. HyOctane: A HyTime engine for an MMIS. In *Proceedings ACM Multimedia*, August 1993.

[9] T. D. C. Little and A. Ghafoor. Spatio-temporal composition of distributed multimedia objects for value-added networks. *IEEE Computer*, pages 42–50, October 1991.

[10] W. P. D. C. Loyolla, E. R. M. Madeira, M. J. Mendes, E. Cardozo, and M. F. Magalhães. Multiware platform: an open distributed environment for multimedia cooperative applications. In *IEEE COMPSAC*, 1994.

[11] S. R. Newcomb, N. A. Kipp, and V. T. Newcomb. The HyTime, hypermedia/time-based document structuring language. *Communications of the ACM*, 34(11):67–83, November 1991.

[12] Darrell Woelk, Won Kim, and Willis Luther. An object-oriented approach to multimedia databases. In *Proceedings of the ACM SIGMOD International Conference on the Management of Data*, pages 311–325, 1986.

35

ObjectMap: Integrating high performance resources into a distributed object-oriented environment

Mike Sharrott, Stuart Hungerford and John Lilleyman,

Co-operative Research Centre for Advanced Computational Systems (ACSys)
CSIRO Division of Information Technology,
GPO Box 664, Canberra, ACT 2601, Australia.

ABSTRACT

As modern computing environments mature towards a distributed object-oriented architecture, the ability to effectively integrate a high performance computation resource into the environment becomes more difficult. This is due to the restricted operating system that such resources typically provide with respect to object-oriented programming techniques and distributed access. For similar reasons, the capability of the resource to access modern, persistent object-oriented databases is also limited.

This paper describes the ObjectMap software framework which is currently being developed to address these problems and to more efficiently integrate computation resources into main stream computing. The work focuses on data handling between persistent databases and computation resources across a network.

ObjectMap has adopted the emerging Common Object Request Broker Architecture (CORBA) standard from the Object Management Group (OMG) for its base architecture with the implementation being realised using the Orbix product from Iona Technologies. The paper describes the benefits obtained through using the CORBA and also the limitations that it has introduced.

The research is being performed by the CSIRO Division of Information Technology as a project within the Co-operative Research Centre (CRC) for Advanced Computational Systems (ACSys) in collaboration with the Australian National University's department of Computer Science. It is also being sponsored by Digital Equipment Corporation (Australia) Pty. Ltd.

KEYWORDS

CSIRO, DIT, ObjectMap, OMG, CORBA, Orbix, OODB, ObjectStore

1.   INTRODUCTION

Over the past few years, research has been undertaken at the CSIRO's Division of Information Technology (DIT) to allow distributed applications to efficiently access the Division's DECmpp 12000 (MasPar MP-1) massively parallel computer [1, 2, 3]. This research has highlighted the need for a software framework that can transparently handle data originating from a modern persistent object store and make it available for processing on a computation resource. Any such framework should encompass the benefits from an object oriented approach and a distributed architecture.

DIT is developing the software framework, through the ObjectMap CRC project, to be a suitable architecture to address these problems and to provide a test bed on which to experiment with some different data handling, mapping and caching policies.

The project will deliver a demonstration system, shown in Figure 1, which will consist of a client application accessing image data from an Object-Oriented DataBase (OODB) and performing image processing operations on a Massively Parallel Processor (MPP). The client will run on a PC under Windows NT (and possibly Chicago). ObjectStore is being adopted for the OODB which will be hosted by a Sun Sparc workstation. The MPP is a DECmpp 12000 hosted by a Digital Alpha.

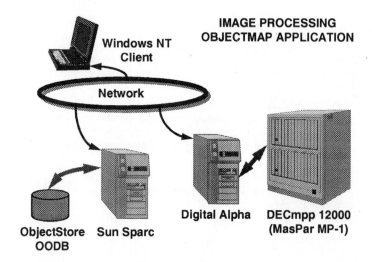

Figure 1 ObjectMap Demonstration System

This paper starts with a discussion of the main requirements that a suitable software framework must meet in order to achieve high efficiency and performance. This is followed by a description of the approach used by ObjectMap and its design components. A summary of the limitations and experiences encountered during the project is also given.

2. SOFTWARE FRAMEWORK REQUIREMENTS

One of the key requirements for a software framework to effectively integrate a computation resource into a distributed object-oriented environment is the need for efficient data handling.

A typical application session involves the client locating the data that it requires (from one of many stores available on the network) then making it available to a computation server where the required operation will be performed. This process possibly involves several data movements and is made worse by the typical requirement of a computation resource to load the data onto its custom processors before the operation may be performed.

These data movements must be kept to a minimum in order to keep the system efficiency, and hence performance, high. The policies adopted for this are store and resource dependent, hence the software framework should provide a solid test-bed on which different policies may be investigated.

The framework should provide access to different kinds of data from a wide variety of stores. These include files from a standard filesystem, C++ or CORBA objects from an Object Oriented DataBase (OODB) and records from a Relational DataBase (RDB). Features should be incorporated to enable the user to browse the data independently of the associated store.

A further requirement of the framework is that it supports multiple users and multiple user access to the computation resource even when the resource itself may not. Either the framework or the resource should have a mechanism for swapping out jobs or segmenting memory to enable multi-user access.

As well as the specific requirements, there are a number of more general features that any such distributed framework should fulfil. The framework should be robust and incorporate exception handling such that it can handle network nodes becoming unavailable during a session. The system should be modular to allow new applications to easily be incorporated. The Graphical User Interface (GUI) of the application client should be able to remain sensitive whilst framework operations are in progress. This allows users, if required, to interrupt the currently active operation or to start setting up the next operation request. These general requirements may be met by adopting a standard core distributed processing package.

3. THE OBJECTMAP APPROACH

This section describes the general approach that has been adopted by the ObjectMap framework in order to meet the requirements outlined in Section 2.

3.1. Object-Oriented (OO)

A high level abstraction of the requirements for the ObjectMap framework identifies four main components to the system; data servers, data objects, computation servers and application clients. A data server accesses and manages data from a store, making it available as data objects to the other framework components. A computation server performs operations on the data objects using its associated high performance resource. Finally, an application client can browse and select data and operations from the data and computation servers respectively.

By adopting an OO approach, each of these components map to an object. This enables the complexity of the framework to be hidden through inheritance from core objects.

3.2. Distributed Technology and CORBA

Typically, on a network, the host machines for the data store and for the computation resource are not the same. This means that the ObjectMap framework requires to have distributed technology incorporated in order to allow the components to talk to each other.

Currently the technology of distributed object computing is expanding rapidly due to the release of the Common Object Request Broker Architecture (CORBA) standard [4] from the Object Management Group (OMG). This architecture is designed to allow objects to communicate across heterogeneous platforms whilst still separating the object's interface from its implementation. The CORBA is the communication core of the OMG's larger Object Management Architecture (OMA) [5] which defines a standard set of object services and common facilities that should be provided in a computing environment.

By adopting the CORBA as the underlying communication mechanism, the ObjectMap framework can take advantage of the features that the technology offers. These include a dynamic binding interface which enables client objects to make requests on server objects with no prior compile-time knowledge of the server object's interface or location. This information may be obtained dynamically through repositories on the network. A further feature of the dynamic interface is the ability to make asynchronous function calls between objects.

A disadvantage to the CORBA is that it only deals with CORBA objects which must be defined using the Interface Definition Language (IDL). ObjectMap requires to access other kinds of data from stores, such as C++ objects and file data. Hence the ObjectMap design needs to provide a method for mapping these objects into CORBA data objects whilst still retaining efficient data handling.

There are several implementations of the CORBA available (or soon to be released). These include Orbix (Iona Technologies), ObjectBroker (Digital), Distributed Objects Everywhere (SunSoft) and the System Object Model (IBM). For the first prototype of ObjectMap, it was decided to use Orbix [6,7,8] as it was the most established and robust CORBA implementation available. Orbix also provides a C++ binding (that has been submitted to the OMG as the standard C++ binding [9]) which simplifies application development and maps well to the ObjectMap requirement to access C++ object data from persistent stores.

3.3. Data Objects and Factories

Typically a data server will access many data components from its store and make them available as data objects. There are several options for managing these data objects ranging from each data object having its own process to all the objects sharing the same process. The optimal approach is application dependent.

The approach used by the ObjectMap framework is to provide a factory object which manages the creation and deletion of the data objects. By this mechanism different options may be tried in order to find the optimal solution for the application.

Many high performance computation resources require that the source data be loaded onto their custom processors before an operation may be performed. Similarly to make the resultant data available at the network level, it must be transferred back from the processors. Typically the resources provide some high bandwidth mechanism for this data movement between the host and the processors. By adopting the factory object to manage the data objects, the ObjectMap framework is able to take advantage of these communication channels.

4. OBJECTMAP COMPONENTS

The ObjectMap architecture, shown in Figure 2, is split into a number of key components.

- **ObjectServer**
- **ObjectFactory**
- **ComputeServer**
- **DemoClient**

Each of these components are separate CORBA processes and hence may be distributed across a heterogeneous network with CORBA handling the communication. Each component is a single CORBA object with the exception of the ObjectFactory, which consists of multiple CORBA objects: a Factory and a number of DataObjects.

4.1. ObjectServer

The Object Server provides the public interface to an associated persistent store of data, which may be a filesystem, OODB or RDB.

By making calls to ObjectFactory objects, the ObjectServer makes data available from the store as CORBA objects called DataObjects (shown in Figure 2 as Data 1 and Data 2). These DataObjects may then be accessed by other CORBA clients. By the same process, the ObjectServer allows new or modified data objects to be incorporated back into the store.

The ObjectServer allows efficient data handling and caching policies to be included between the store and the DataObjects.

4.2. ObjectFactory

The ObjectFactory is split into a Factory object and a number of DataObjects (shown in Figure 2 as Data 1 and Data 2). The Factory is responsible for creating and maintaining the DataObjects in response to requests by an ObjectServer. The DataObject is the network representation of a data component from a persistent store. The DataObject communicates with its associated ObjectServer in order to access its data from the store which may involve data copying and caching policies.

4.3. ComputeServer

A ComputeServer performs computation on DataObjects in response to a request from a client. The request specifies the required computation parameters including object references to source and destination DataObjects.

Typically the ComputeServer will perform the computations on an associated high performance resource such as a Massively Parallel Processor (MPP), e.g. DECmpp 12000 (MasPar MP-1). This may require the ComputeServer to move or copy data from the DataObjects to the custom processors of the resource before the computation may be performed. This will take advantage of any high performance IO gateways into the resource.

4.4. DemoClient

The DemoClient is the application and ObjectMap client and typically provides an event-driven GUI. The DemoClient binds to ObjectServers in order to retrieve data from a persistent store or to create new data which maps to the store. The DemoClient does not see the ObjectFactory but deals directly with references to DataObjects. The DemoClient may also bind to ComputeServers to perform computations on the data. At any stage the DemoClient may be attached to multiple ObjectServers and ComputeServers.

The ObjectMap architecture allows the DemoClient to remain sensitive whilst any ObjectServer or ComputeServer operations are in progress.

5.   OBJECTMAP ARCHITECTURE

This section provides further details on the design of the individual components of the ObjectMap architecture.

5.1.   ObjectServer

5.1.1.   Architecture

The design of an ObjectServer, shown in Figure 3, has adopted a layered architecture for both the object interface and its internal structure.

The object interface is split into two parts:

- Core Functions
- Data-Specific Functions

The core functions must be implemented by all ObjectServers and must provide the base interface necessary to identify and provide general access to the ObjectServer and the data that it serves. The data-specific interface functions provide more efficient access to the data via individual functions for each served data type.

The internal structure of the ObjectServer is layered to adopt maximum portability and reusability across ObjectServers. For example all store specific operations are held within a StoreManager object. This allows the ObjectServer to be mapped from one store to another (e.g. Filesystem to OODB based) simply by rewriting the StoreManager object and not the whole ObjectServer.

5.1.2.   Data Handling

When a client requests data, the ObjectServer calls a function in the ObjectFactory to create a DataObject of the required type. However, at this point, the data will not be loaded from the store into the DataObject. The ObjectServer will simply maintain a record consisting of a handle to the data in the store and an object reference to its associated DataObject.

When the client (or a different client) binds to the DataObject and calls a member function, the DataObject will recognise that it has not yet been initialised and will bind to the ObjectServer in order to do so. At this point the ObjectServer will access the required data from the store and will transfer it to the DataObject.

Using this handling mechanism, a data caching policy may be inserted between the store and the DataObject, such that the DataObject only contains a working subset of the data that it requires. Using the underlying core functionality, the ObjectMap framework provides a test-bed on which to experiment with different data caching and handling policies.

5.1.3. Object Pointers

If the store associated with an ObjectServer is an OODB, then it is likely that the store contains trees of objects where one object may point to several others. This creates a problem when one such object is activated in how to handle the child objects. There are several approaches possible which range from making the requested object and all of its dependants active to making only the requested object active. ObjectMap has adopted the latter approach.

The reason for ObjectMap's "*make active only on demand*" approach is efficiency and the granularity of the objects proposed to be used in the framework. Because each data component is likely to be large (satellite images or 3D medical data sets) then making the object, and all of its dependant objects active, is likely to use up all the system's resources.

When an object, which has child objects, is accessed from a store, then the associated DataObject must mark its child pointers to signify that they do not yet have their own DataObject. Hence if a client tries to access a child pointer, the DataObject will first create a new DataObject for the child before processing the request. This is achieved through calls to the Factory back to the ObjectServer.

5.1.4. Persistence

The OMG is currently defining a standard for CORBA object persistence through the definition of the standard object services. This will enable a CORBA object to be mapped between being active and being swapped to disk whilst still retaining all state information. This occurs invisibly to the client.

The ObjectServer is providing a similar persistence service for non-CORBA objects. For example, an ObjectServer may handle C++ objects from an OODB or objects stored in a custom binary format in a filesystem.

5.2. ObjectFactory

5.2.1. Architecture

The ObjectFactory consists of a single Factory object and a number of DataObjects. The Factory is responsible for the creation and deletion of the DataObjects.

When a DataObject is created, it is assigned a unique identifier (UID) which is used for all communications with the ObjectServer in order to uniquely identify the object's associated data in the store. The data handling between a DataObject and its entry in a store is described in Sections 5.1.2 and 5.1.3 above.

5.2.2. DataObjects and Processes

In a network environment it may be required to split the DataObjects over several nodes for efficiency purposes. For example, by mapping the DataObject to the host of a computation resource, the resource may be able to take advantage of a high bandwidth path to load the data onto its custom processors. This is described further below in Section 5.3.2. To support this, each DataObject requires to be mapped to its own process so they may be split over the network. The ObjectMap framework provides policies in the Factory object to control the DataObject to process mapping.

5.3.   ComputeServer

5.3.1.   Architecture

The ComputeServer interface is split into two main components:

- Core Functions
- Computation-Specific Functions

The core functions must be implemented by all ComputeServers and will be used by clients whilst browsing several ComputeServers to locate the operations that they require. The core functions identify the ComputeServer and list the operations that it provides. The computation-specific functions provide access to the actual computation operations. The function arguments specify all the required operation parameters including object references to the source and destination DataObjects.

5.3.2.   Data Mapping

A high performance computation resource typically loads all computation data onto its custom processors before the operation may be performed. Similarly the resultant data is transferred back off the custom processors to the host before it can be accessed at the network level. These data transfers require to be optimised in order to ensure that the performance of the resource still remains high. Most computation resources provide a high bandwidth IO connection between the host and the custom processors. For example, the DECmpp 12000 (MasPar MP-1) [10] provides a RAM based file system (accessible to the network through NFS) which has a connection directly to the parallel processors. By configuring the DataObjects to actually store their data in this filesystem, the full bandwidth connection may be utilised.

For some applications, the system performance may be further improved by configuring the ComputeServer to cache as much computation data as possible on the custom processors pending possible further requests

The ObjectMap framework provides the core functionality such that it may be used as a test-bed for research to be performed into more efficient mappings of DataObjects between the network and the computation resources.

5.3.3.   Parallel Computation Service (PCS)

High performance computation resources are generally single user machines where one user grabs the whole system for the duration of their work. Other users must queue to access the system.

Research has been carried out at DIT to look at how a system may be made multi-user by providing a job manager. This manager schedules jobs and controls all job swapping as well as providing mechanisms for optimal data handling on the custom processors This research resulted in the Parallel Computation Service (PCS) [1] being developed. A second design of PCS is now underway which is based on the CORBA rather than UNIX socket and pipe connections.

The PCS integrates with the ObjectMap framework seamlessly by configuring each ComputeServer to be a PCS client. Further information on the PCS may be found in [1].

5.4. DemoClient

5.4.1. ObjectServer and ComputeServer Browsing

Being the user interface to an application, the DemoClient requires to be able to browse ObjectServers and ComputeServers in order to locate the data and services that it requires. ObjectMap provides this functionality through a combination of the core interfaces of the servers and the standard CORBA object location and binding services.

5.4.2. Sensitive User Interface

Typically the computations being performed by a ComputeServer will not be real time. Hence it is important that the user interface of the DemoClient remains sensitive to user requests whilst operations are in progress. This allows the user to start building the next request or to interrupt the currently active operation. This is achieved in the ObjectMap framework by implementing the function calls asynchronously using the CORBA dynamic binding interface.

In order to notify the DemoClient that a computation has completed, the framework provides two mechanisms of call-back. If the DemoClient is itself a CORBA object, then the ComputeServer can be configured to call a function within the DemoClient on completion. Alternatively the DemoClients may poll a computation status flag within the ComputeServer in order to wait for completion. This polling may be integrated with the event loop of the DemoClient. Further mechanisms for notification will be investigated during the course of the project.

6. OBJECTMAP LIMITATIONS

A number of limitations have been identified with the ObjectMap framework. These can be attributed to both the design, the probable applications and to the CORBA. The following sections summarise these findings.

6.1. Interface Definition Language (IDL)

The IDL language does not yet provide some of the constructs that would be very useful. In particular, there is no one-to-one mapping between C++ and IDL, hence trying to represent C++ objects as CORBA objects creates several problems. This is not something that the CORBA was designed for as the CORBA assumes that the starting point was an IDL interface.

With respect to C++, the following constructs are missing:

- No constructors or destructors
- No function overloading
- No concept of "by pointer" or "by value"
- No operator functions
- No templates

These limitations may be addressed in future versions of IDL as the technology becomes more established.

6.2.   Object Granularity

If an application consists of hundreds of objects of different types, then the overhead costs involved in integrating it with the ObjectMap framework may be quite high. This is made worse if the objects are highly interconnected through pointers and object references.

The ObjectMap framework is biased towards applications consisting of fairly large granularity objects (e.g. satellite images or 3D medical data sets) where typically less than twenty were active at any one time.

One specific application that did not map well to the ObjectMap framework required to access and process many small and highly interconnected C++ objects from an OODB. In this case a specific C++ distributed system was more applicable. For example, the ObjectStore OODB provides an automatic method of persistence for object hierarchies, with the kernel handling all data mappings [11]. A further possible approach is to use a distributed C++ class packages such as NetClasses [12].

6.3.   Data Handling

Care must be taken when handling pointers in the Orbix implementation of CORBA. For example, if an object member function takes a pointer argument, then Orbix will automatically copy the whole of the "pointed to" data to the address space of the destination object as part of the function call. A similar copy process occurs when a function returns a pointer but with the added limitation that Orbix automatically deletes the function pointer and what it points to after the return has completed. Hence if the function copy is required to be kept then it must be copied prior to the function return.

Obviously at some stage the data requires to be copied across the network between the source and destination object's address spaces, but these movements should be minimised in order to prevent thrashing.

6.4.   Deadlock

If a CORBA object is processing a request from one client and then it receives another request from a different client, then the second request will be blocked pending completion of the first. Hence if any circular dependencies are present, a deadlock will occur. This may be avoided to a limited extent by making use of threads and time-outs, but this is a general limitation which is present in most distributed or parallel systems.

7.   CONCLUSIONS

The ObjectMap software framework addresses the main requirements that are necessary for the integration of a high performance computation resource into a distributed object-oriented environment. It also provides a sound test-bed on which different data caching and handling policies may be tested. By taking the object-oriented approach, the ObjectMap framework will hide much of the complexity through inheritance.

The CORBA has provided a sound model on which to develop the framework. In particular the distributed and dynamic binding features are being extensively used. The Orbix implementation has provided a robust development platform and its C++ binding has greatly simplified the development effort.

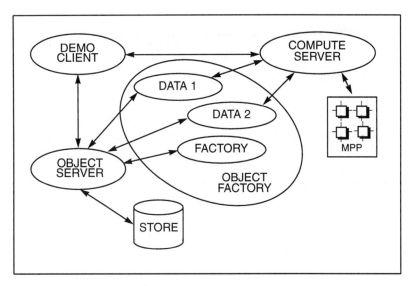

Figure 2 High Level Architecture

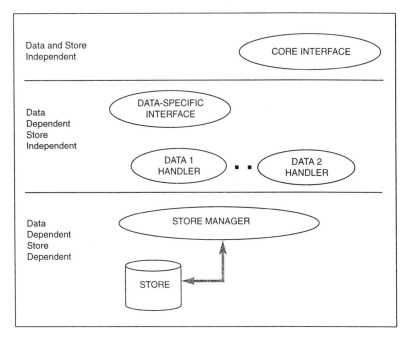

Figure 3 ObjectServer Architecture

ACKNOWLEDGEMENTS

The ObjectMap research project is being sponsored by Digital Equipment Corporation (Australia) Pty. Ltd. under the Co-operative Research Centre (CRC) for Advanced Computational Systems (ACSys). This is being performed in collaboration with the Australian National University's department of Computer Science.

REFERENCES

1. M. R. Sharrott, and J. M. Lilleyman, The Parallel Computation Service (PCS), Proc. DICTA-93, Digital Image Computing: Techniques and Applications, Macquarie University, Australia, December 1993, vol. 1, pp.390-397.

2. D. Keightley, K. Tsui, J.M. Lilleyman, and K. Moore, A software framework for an applications-driven parallel image processing and display system (PIPADS), Proc. SPIE Conference on Recent Advances in Sensors, Radiometric Calibration and Processing of Remotely Sensed Data, vol. 1938, Orlando, Florida, April 1993, pp. 368-379.

3. K. Tsui, P. A. Fletcher and M. A. Hutchins, PISTON - A Scaleable Software Platform for Implementing Parallel Visualisation Algorithms, Proc. Computer Graphics International '94, Melbourne, Australia, June 1994.

4. Object Management Group, The Common Object Request Broker: Architecture and Specification, Revision 1.1, OMG Document Number 91.12.1, December 1992.

5. Object Management Group, Object Management Architecture Guide, Revision 2.0, OMG Document Number 92.11.1, September 1992.

6. Iona Technologies, Orbix - Getting Started, Version 1.2, February 1994

7. Iona Technologies, Orbix - Programmer's Guide, Version 1.2, February 1994.

8. Iona Technologies, Orbix - Advanced Programmer's Guide, Version 1.2, February 1994.

9. Iona Technologies, Orbix - IDL to C++ Mapping, Version 1.2, February 1994.

10. T. Blank, The MasPar MP-1 Architecture, Proc. CompCon '90, 1990, pp. 20-24.

11. Object Design Inc., ObjectStore User Guide: Library Interface, Release 3 for UNIX Systems, December 1993.

12. PostModern Computing Technologies Inc., NetClasses - An Object-Oriented Communications Toolkit, Technology Overview.

SESSION ON

DCE Experiences

36

Experiences With the OSF Distributed Computing Environment

John Dilley, Hewlett-Packard Laboratories

This paper describes the Open Software Foundation's Distributed Computing Environment (OSF DCE) in the context of the ISO Reference Model of Open Distributed Processing (RM-ODP). It presents a critical assessment of the DCE technology, and suggests future work for both DCE and ODP. We suggest how DCE can support Open Distributed Processing, summarize the experiences we had building a DCE-based prototype, and discuss the lessons we learned from this activity. As a result of the prototype effort we built a set of DCE services to improve the usability and utility of DCE, including an object-oriented programming system, an event notification service, and a name space browser/editor for the DCE directory service. Current research efforts are underway to study fault tolerance and to provide generic data replication.

Keyword Codes: C.2.4; C.2.2; D.1.5
Keywords: Distributed Computing; Network Protocols; Object-Oriented Programming

1. BACKGROUND

In 1991 Hewlett-Packard's Networked Systems Architecture group began to develop prototype software to learn more about the Open Software Foundation's Distributed Computing Environment (OSF DCE). The prototypes were done in the context of HP's Cooperative Computing Environment (CCE [1]) Architecture Framework. Many of the goals of the CCE framework were similar to those of the ISO ODP effort, described in [2][3][4][5], and summarized in a very readable way in [6]. The structure of the CCE framework was based upon the ISO ODP viewpoints, applied to a constrained set of objectives that focused on HP's business interests. Both efforts defined a set of crucial services and mechanisms required in a distributed environment and a vocabulary for describing distributed components, applications, and systems.

At that time the ODP viewpoints did not consider legacy environments, which made it difficult to support HP's business. The ODP viewpoints were also not focused on any business areas. In trying to apply the descriptions of the Enterprise and Information viewpoints to HP's CCE we concluded that the Enterprise viewpoint described a business model and would be hard to generalize and still be a useful model for application design. For our CCE prototype, the Enterprise viewpoint became a set of requirements to support business processes and assumptions.

The Information model seemed to overlap with the Enterprise and Computational viewpoints, therefore we developed our own model for information. The model centered on a particular information system and its application: information was described in four aspects, Input, Output, Presentation and Storage. Within each of these aspects there were attributes such as Security and Consistency that would need to be supported at the Engineering level.

The Computational viewpoint was well defined but based around a linguistic framework (ANSA DPL). Although most agreed that this would be the ideal way to model the Computational viewpoint, it did not reflect reality—since programmers did not want to learn another language. HP had an installed base that could not be moved to a new linguistic framework and therefore needed a more evolutionary approach to the computational model.

The CCE computational viewpoint described the interaction model and development model for distributed computing systems and identified development and runtime services that would be required to support the viewpoint and integrate the installed base of applications and system level services that had been written in a multitude of languages and would not be re-written.

2.   OSF DCE OVERVIEW

The Open Software Foundation's Distributed Computing Environment (OSF DCE) is a software infrastructure for development of distributed computing applications. DCE facilities are accessed through programmatic and RPC interfaces, both defined in the DCE Application Environment Specification (AES [7]); some of the services are also accessible through user-level commands.

The OSF took a different approach than the ODP group at defining an open distributed computing environment. The DCE provides a set of distributed computing technologies, and uses those technologies to define its distributed environment. On the one hand this does not provide the high-level aspects provided by ODP, such as a common vocabulary for specifying viewpoints for distributed computing. On the other hand, DCE provides specific software technology, with a reference (or sample) implementation, assisting application portability and assuring interoperability. Since ODP does not currently specify a technology viewpoint (i.e., it does not provide strict specifications for APIs or wire protocols), it is likely that early "ODP-conformant" applications will not interoperate.

The components of DCE are summarized in relation to ODP in the following sections.

2.1.   Communications

The Remote Procedure Call (RPC) component allows client processes to communicate with servers across a local or wide-area network. DCE RPC uses the procedure call/return paradigm for remote communication—a remote procedure call looks to the developer like a local procedure call, however there are some significant differences in call latency and fault semantics which developers need to be aware of when writing RPC applications.

RPC development starts with the definition of data types and remotely-callable procedures associated with an application. These types and operations are specified using the DCE Interface Definition Language (IDL). A compiler provided with DCE converts IDL specifications into RPC communication stubs. DCE RPC stubs provide for data marshaling, which prevents the application developer from having to explicitly handle data conversion between different systems. In RPC, parameters in remote operations, which can be complex data types (including pointers), are automatically traversed and converted into and out of a network transmissible form. RPC provides access transparency, as defined by ODP in [3][6].

2.2.   Directory Services

Each remotely accessible object in a distributed system has a distinct network location which can vary over time. To access an object (or service) requires first determining its network location using some type of location or directory service. The benefit a directory service provides is that service locations do not need to be known at compile time (and statically bound), or even at program load time—but rather can be dynamically located, and even relocated at run time.

The DCE includes facilities called the Cell Directory Service (CDS)[1] and Global Directory Service (GDS). The CDS is a cell-local directory service, whose main purpose is to provide a name-to-address lookup to allow clients to locate servers in the cell. DCE RPC includes a Name Service Interface (NSI), which clients use to access server location information in the CDS name space. NSI allows DCE clients to locate servers at run-time in a location-independent manner. In addition, the directory supports access to arbitrary information via the X.500 Directory Service (XDS [8]) interface. GDS allows multiple DCE cells to be connected to allow name space searches to traverse cell boundaries.

The directory service and the RPC runtime library together provide location transparency. An additional aspect of object location that ODP specifies is trading [9]. The DCE CDS does not provide a full constraint-based trading capability, however the NSI search facility effectively provides trading based upon the interface (type), object identity, and desired network transport protocol the client can use. We have found in most cases this is sufficient, though a full Trader in DCE would provide additional value.

2.3. Time

The DCE Distributed Time Service (DTS) maintains a single, global virtual clock, which is then used to keep machine clocks synchronized across the network. A consistent global time is required by the DCE Security Service authentication mechanism and the Distributed File System. DTS assures that each DCE cell will maintain a consistent, monotonically increasing time. Global DTS servers must be connected with an external time provider that provides the actual time signal. The DTS protocol then keeps host clocks consistent with Coordinated Universal Time (UTC). RM-ODP does not define an equivalent service [10].

2.4. POSIX Threads

The DCE Pthreads package defines a multi-tasking capability for distributed applications. DCE uses Pthreads to allow a single process to have many logical threads of control; these threads can initiate or serve RPCs concurrently. The Pthreads package provides synchronization primitives, including a mutual exclusion locking facility (mutex) which allows multiple threads to wait for a signal of a particular condition becoming true (condition variables). RM-ODP does not currently define a threading or concurrency model.

2.5. Security

Having an integrated security service is a key benefit of the DCE environment. Providing secure communications in systems that cannot trust all components of the network is quite difficult to achieve. The DCE Security Service, based upon MIT's Kerberos implementation, is fully integrated with the RPC mechanism. DCE Security provides mutual authentication of requests between client and server, so the client and server can be assured that each other's identity is valid. Based upon the trusted identity of a client, DCE applications can provide authorization of client requests to control access to resources using an Access Control List (ACL). The Security Service can utilize data encryption for RPC calls to prevent listeners on the network

---

1. A cell is the DCE unit of administrative division.

from being able to intercept or tamper with RPC requests and responses. The DCE security system performs some of its security checks in the RPC run-time; certain authentication and authorization checks occur before user code is even invoked. Only if the caller is using a proper level of security does the call proceed into user code.

The DCE Security Service provides several of the security functions called for in ODP.

- Access Control Function is provided through DCE's Access Control List (ACL) facility.
- Security Audit Function is provided in the 1.1 release of DCE.
- Authentication Function is provided using the security service as the trusted third party, with principals stored in the DCE registry.
- Integrity Function is provided via authenticated RPC using encryption and checksums of data in RPC communications.
- Confidentiality Function is provided through the user data privacy option that encrypts data in RPC calls.
- Non-repudiation Function is not currently provided by DCE; there is no way to authoritatively convince a third party that a particular object was involved in an RPC. (The client and server are of course convinced.)
- Key Management Function is provided by DCE, in a way that relies upon the host file system security.

2.6.   Distributed File System

DCE includes a Distributed File System (DFS), which provides a shared global file system visible to hosts in all connected cells. The DFS provides all users with the same view of the global file system; other distributed file systems present a different view of files depending upon whether one is on a file server or a client system. DFS uses file caching so that once a remote file is accessed, a copy is cached locally to improve performance. RM-ODP does not specify a distributed file system.

3.   THE DCE PROTOTYPE

In 1991-1992, our group built a multimedia mail application to experiment with the fledgling Distributed Computing Environment [11][12]. Our intention was not to create a production-quality mail system, but rather to learn more about DCE. The goals of the DCE prototype effort were as follows.

- Validate the Cooperative Computing Environment framework through actual experience.
- Demonstrate the advantages of DCE-based client/server distributed computing over more traditional forms of distributed computing.
- Determine the aspects of distributed computing for which DCE was well suited. No distributed computing platform will provide an exact match for the requirements of every application; we wanted to figure out the tasks for which DCE was well suited, and those for which DCE did not do so well.
- Characterize DCE performance. One key aspect of a distributed environment is its implications on system and application performance. A key part of the prototype effort involved understanding the performance of DCE RPC the DCE directory service [13].

- Learn about management of distributed systems and applications. Operation and management of distributed applications is inherently more complex than for non-distributed applications.

- Identify areas where DCE should be enhanced. The first release of any technology is bound to be limited—we wanted to determine those areas where additional DCE tools and services were required to support distributed application development.

- Propose enhancements to improve the quality of DCE. Since HP is a DCE vendor, we wanted to make sure our efforts could be used to improve the product offering from HP as well as OSF's DCE in general.

3.1. The MultiMedia Mail System

The application we chose to exercise DCE was a multimedia mail system that could compose and send messages containing text, graphics, and voice annotations. We felt this system would require enough of the facilities of the infrastructure to provide valuable insight into DCE, distributed application design and development, and distributed application management.

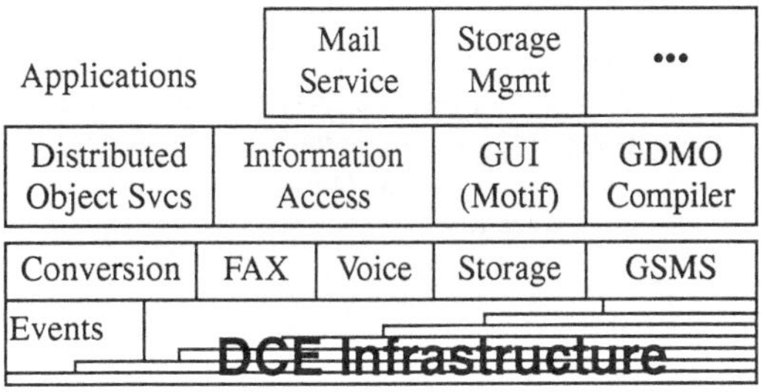

Applications	Mail Service	Storage Mgmt	...
Distributed Object Svcs	Information Access	GUI (Motif)	GDMO Compiler
Conversion	FAX	Voice	Storage	GSMS
Events	DCE Infrastructure			

FIGURE 1. MultiMedia System Components

The distributed system we built consisted of several major components. Each of these components used DCE to provide distribution of service to clients across the local area network. The components we built were as follows:

- The mail service implemented a user mailbox and used store-and-forward networking to transmit mail messages between users. The mailbox and mail submission protocols were defined in IDL and implemented using RPC.

  There were two mail service user interfaces, one based upon Andrew Mail and the other a custom DCE-based application. Through this we learned how to wrap existing applications that were not written with DCE in mind.

- A Fax service provided a transport for mail messages and a way to send graphics or text images to users who did not have an email box. Fax images could be separately stored and used as components in mail messages.

- A voice service provided recording and playback of voice messages. Voice messages could be used as annotations in mail messages.

- A distributed information storage system provided storage and retrieval of bulk data from servers on the network, including text, graphics, and voice and fax annotations. Multimedia messages were often quite large, so custom file servers with large disk arrays were helpful for bulk information storage.

- A conversion service converted data among the formats required by the various interfaces. For example, the conversion server could convert from the Fax format to the format used in the mail GUI. This type of service is often embedded in, for example, a mail service; we wanted to make the conversion a generic service for use by mail, fax, or other system components, to allow openness and the ability to dynamically add new system services.

- The Event Service provided asynchronous notification of faults and status information for the services in the environment.

- A generic service management interface provided access to the state of the servers and services in the network, and allowed system managers to start, stop, and restart services. The Event Service was used to convey state change notifications from application components to the management service. An RPC to a per-node management application was used to start server instances.

- A configuration service allowed services to store configuration data for use at start-up and during execution. The configuration state was globally available; this facility provided a standard configuration interface, instead of requiring each service to have a custom interface to its configuration data.

3.2. DCE Lessons Learned

Some of the initial challenges of using DCE were a steep learning curve, difficulty of application development, and lack of debugging tools for developers. Furthermore, initial installation and configuration of the DCE environment was time consuming, and determining the health of DCE at that time was non-trivial. Finally, early DCE was not integrated with the existing operating system and distributed computing software we were using.

All of the interfaces in our system were defined using DCE IDL and accessed via DCE RPC. The RPC mechanism provides a powerful client/server communication model; RPC keeps the programmer from having to deal with location or distribution issues in the system. By defining formal interfaces, we found that applications were easier to enhance and evolve. Provided that the interface does not change, we were able to evolve applications from providing quickly-developed but incomplete capabilities, to providing more sophisticated implementations as time permits. On the down side, the RPC interface was not easy to learn and use.

The Cell Directory Service provides the host independence part of the location transparency available to RPC applications. The CDS allows easy reallocation of services to systems in the network: the service simply registers its location upon start-up after which new clients will be able to locate it.[2] CDS does not provide the ability to trade for services based on an arbitrary set of constraints. Having a trader available in DCE would be beneficial. It also did not at that time provide tools to assist with name space maintenance. There was a graphical browser and a command-line interface to the directory, but the former only displayed directory contents; the latter allowed modification, but not in the context of the browser, so it was more difficult to use.

One of the key challenges of writing DCE-based applications turned out to be dealing with their multi-threaded nature. The difficulty was not so much in writing thread-safe code as in using existing system software that was not written to be thread-safe. Until such a time when system and third party software is thread-safe, DCE developers must use caution when using non-DCE libraries, and may have to protect unsafe calls using mutual exclusion (mutex) locks.

---

2. This is not service migration; migration typically involves relocation without the client faulting or having to restart.

4. DCE ENHANCEMENTS

Based upon the areas of improvement we identified building prototypes, we proposed a set of contributions we could make to enhance the DCE offering. HP has developed many of these facilities and delivered them as part of subsequent DCE releases. As and when the ODP offering includes a technology platform, these types of enhanced services will likely be required.

4.1. Distributed Debugging Environment

DDE provides a graphical-based user interface for DCE application debugging. It is a thread-aware debugger, meaning you can manipulate the state of the threads in your application. DDE lets you view the threads in your process, enable or disable threads for execution, cancel threads, as well as do a full set of regular debug activities. While DDE is not a truly distributed debugger (in the sense that you cannot single step across process boundaries and debug the server executing a client request), it goes a long ways towards making DCE applications easier to debug.

4.2. Trace/Log Facility

The Trace/Log Facility allows developers to instrument their code with trace statements at "interesting" points. When the trace statement is executed, it emits a line of output according to a predefined format, which includes the time of the trace (to millisecond precision), the thread which executed the statement, a trace identifier (each trace object can have a separate identifier in its output), and a user-defined message. Global time (UTC) is used in the trace output to allow traces from multiple processes in different time zones to be sorted into a single trace. This is necessary to see the order of certain operations. Each trace statement can identify a selector-level, which allows run-time control over which messages are emitted.

4.3. Instrumented IDL Compiler

The DCE IDL compiler generates client and server communication stubs that provide distribution transparency (data marshaling, sending network packets). Sometimes when debugging distributed applications it is nice to determine where in the RPC process the program was when it aborted (distributed programs sometimes terminate none too gracefully). The Instrumented IDL Compiler is a version of the DCE IDL compiler that adds tracing statements (using the facility described above) in key points in the stubs. When the program runs, the IDL compiler's trace selectors can be turned up such that they will emit trace messages when they enter and exit the communication stubs, when they begin and finish marshaling data, and when they begin and finish sending the request. Elapsed times are included in the trace output to provide an indication of how long each step in the process took.

4.4. Sample Applications

The sample applications are a set of programs that demonstrate the use of DCE by application developers. They attempt to show the various facilities of DCE, from a very simple "get-started" application, to applications focused on each of the DCE facilities, to a more representative application that uses RPC, naming, threads, and security together. These applications are extensively commented with a discussion of each new DCE API routine being used, the parameters and their types, their return value, and a description of the expected operation. The sample applications can be read as tutorial information, or leveraged to create new applications.

4.5.   Object-Oriented DCE Framework

Two limitations of the early DCE were difficulty of use for developers, and lack of support for C++. We felt that by encapsulating the facilities of DCE in a C++ class library, we could provide greater ease of use than regular DCE. While this could have been done by adding a C library with wrappers for the cumbersome data structures and API routines, this would not have provided access for C++ developers, and could not have provided the same value. One reason is that DCE, as originally designed, included an object-oriented model for accessing distributed objects. Also, object-oriented analysis and design techniques are well suited to the requirements of distributed computing and distributed application development.

HP OODCE provides a C++ framework for DCE application developers [14]. OODCE consists of an enhanced IDL compiler, which converts interface definitions into C++ classes, and a class library that encapsulates much of the complexity of the DCE programming environment. The classes generated by the enhanced IDL compiler inherit from framework classes provided in the OODCE library. These framework classes define the structure of object-oriented DCE applications, and provide a significant amount of code to interact with the DCE run-time environment, for example to register location information and to interact with the security subsystem. In addition to the framework classes, a set of C++ utility classes are provided to make it more convenient to deal with regular DCE data types.

An earlier ICODP paper reported on the use of DCE and C++ as a programming environment for ODP [15]. The focus of that work was on a direct mapping from ODIN to DCE and C++. This work by contrast uses standard DCE IDL for interface definition, and is focused on making the existing DCE easy to use.

4.6.   On-Line Help Facility

The help files for the sample applications are delivered as a set of README files, but in addition these files have been integrated with the Visual User Environment (VUE) hypertext help facility. This allows users to search for topics by keyword, and locate individual lessons in the various sample applications.

4.7.   Enhanced CDS Browser/Editor

Being able to graphically view and manipulate the contents of the directory service is an essential capability in any distributed environment. The CDS browser provides an X11-based interface to the CDS name space, displaying it as a tree of objects. Through the browser the user can see the various RPC server entries, groups, and profiles in the directory name space. The DCE includes a basic CDS browser with viewing capabilities. The enhanced browser/editor we provided added the ability to interact with CDS to add, remove, or modify the entries in the name space, and to view and change any ACLs associated with an object.

4.8.   System Administration

A set of tools have been deployed to allow more powerful administration of the distributed environment, including central point software installation, cell configuration management built into the standard System Administration Manager (SAM) tool, a cell validation test suite, and a snapshot facility to help troubleshoot cell problems. The validation suite and snapshot facility help administrators determine if the cell is operational, and to assess the state of DCE servers.

The DCE configurator provides a single point of configuration for a DCE cell; it can discover the current cell configuration as a starting point for reconfiguration. The configurator also provides a cell planning facility, allowing administrators to plan and pre-test a cell configuration before committing the changes.

4.9. Legacy System Integration

The first action a user takes on a system is to log in—to specify and validate their identity by giving a user name (login name or principal in DCE terms) and a secret password. The original release of DCE required users to validate themselves to the system and to DCE separately; but the current release of DCE provides for an integrated login, allowing the user to establish their identity only once, and have DCE credentials automatically refreshed whenever the user gives their password to the system (for example to unlock their display).

The DCE Distributed File System provides an advanced distributed file system; however, the ONC Network File System (NFS) is currently the most widely deployed distributed file system. In order to allow use of both file systems, HP has supplied an NFS-DFS gateway, which allows NFS clients to access a DFS file system. This gateway allows gradual migration to DCE: servers upgraded to DCE and DFS can run the NFS-DFS gateway to allow existing clients the same access to their data as before. New clients can use DFS directly to gain the addition benefits DCE/DFS provides.

4.10. Event Service

DCE does not have a standard asynchronous notification (event) service at the time of this writing. As part of the prototype we developed and used one to support distributed systems management. The model we used was subscribe/publish, where each agent interested in receiving event notifications sent a subscription request to an event service indicating the event type, severity, and application generating the event. The event service receives subscription requests and maintains a database of interested subscribers; it also receives asynchronous events and redistributes them to event clients based upon the subscription database.

Using this service, event producers do not have to be aware of the entities interested in receiving events, nor with the details of distribution of events to them (and having to deal with the fact that some of them may be unreachable), nor do they have to be aware of whether anyone is in fact interested in an event. Instead, producers send all events to the Event Service and forget about them. The Event Service is then responsible for sending the event to all interested event consumers. Furthermore, event consumers do not need to be aware of specifically where events are generated. Instead, they receive any event that is of the type and severity they requested a subscription to.

4.11. Systems Management

While polling the state of the system was an important part of our prototype, we noted that a more efficient approach was to manage by exception. In other words, instead of relying upon an occasional poll to tell us that a component had gone down or come back up, we used the Event Service for this purpose. This way there was minimal traffic on the network in the steady state, but when some potentially interesting event occurred, a notification was sent; management agents could register their interest in that event type and then receive the notifications when they occurred.

While this is far from a full system management solution, it did enable us to determine at a crude level whether the system was functional, and if not, where the fault lay.

4.12. Data Replication Architecture

In a distributed environment, data that needs to be highly available is typically replicated in several locations, to protect against a single failure causing data or services to be unavailable. Our goal in the data replication area is to make it possible to develop application logic independent of the availability strategy. We believe that the availability requirements not only depend on the nature of the application but also on the operating environment in which it is deployed. For example, the desired failure recovery strategy of the CDS depends on what kind of applications are depending on its availability. OLTP developers have told us that absolute correctness and consistency between CDS database replicas is a requirement for their applications. In other words, all replicas must be synchronized so that no name space operations are lost if a master crashes before propagating updates to a read-only replica. By contrast, certain other mission critical applications would rather have automatic fail-over of a read-only slave into a writable master even if some entries are temporarily lost so that new services could continue to export their bindings and continue running while CDS recovers.

In this CDS example, it would be nice to be able to write the CDS namespace database logic independently from the mechanism which will handle replication, fault detection, and recovery. For that reason, we have drafted a data replication architecture to describe the components of a replication system in a general way. We designed a general-purpose interface to which applications can be coded. The value is that replication systems with different configurations and policies may be implemented behind the standard interface. That frees the application developer from having to care about replication and failure recovery. Choosing the appropriate replication mechanism can be done at link time. This approach also provides a consistent way for an application to access data replication.

5.   FUTURE DIRECTIONS

There is still much work to be done in the area of enhancing DCE in the areas of system and application management, fault tolerance and data replication, and asynchronous notification. In addition, the DCE effort would benefit from the vocabulary defined by RM-ODP for definition of Enterprise, Information, and Computational viewpoints (DCE already provides equivalents for the Engineering and Technology viewpoints).

In addition, the ODP effort may be able to benefit from industry experiences such as ours with distributed computing and DCE. Some specific issues and areas which merit further study are listed below.

- Should DCE make use of the Enterprise, Information, and Computational viewpoints from RM-ODP? Can DCE be described in terms of Engineering and Technology viewpoints?

- Can ODP make use of DCE technologies to provide interoperation between ODP-compliant applications? Could ODP use the DCE Engineering and Technology viewpoints, if defined? Without a technology mapping, interoperation is difficult to achieve. This has been illustrated by the CORBA effort, which initially specified only an architecture, but is now working to address interoperability between implementations [16].

- Distributed systems and application management is an area of continued investigation. End users want to be able to determine how to get their task done—they should not be expected to be network management and protocol experts.

6. ACKNOWLEDGMENTS

We would like to thank Deborah Caswell for providing the information on fault tolerance, recovery, and data replication; and Jeff Morgan for providing the historical view of the CCE. Thanks also to the following for their review comments on this work: Lee Boswell, Bob Fraley, Richard Friedrich, Mickey Gittler, Marta Kosarchyn, Joe Martinka, Bob Price, Walt Tuvell. Thanks also to the pioneering members of the original "Exploitation of DCE" prototype team, and especially to the team that made HP's DCE into a viable distributed computing platform.

REFERENCES

[1] NewWave Computing Framework, Vol 1-3, Version 1.1. Hewlett-Packard, Information Architecture Group, October 1990.

[2] ISO/IEC JTC1/SC21/WG7 N838: Basic Reference Model of Open Distributed Processing—Part 1: Overview and Guide to Use. July, 1993.

[3] ISO/IEC CD 10746-2.3: Basic Reference Model of Open Distributed Processing—Part 2: Descriptive Model. July, 1993.

[4] ISO/IEC CD 10746-3.2: Basic Reference Model of Open Distributed Processing—Part 3: Prescriptive Model. July, 1993.

[5] ISO/IEC JTC1/SC21/WG7 N839: Basic Reference Model of Open Distributed Processing—Part 4: Architectural Semantics. July, 1993.

[6] Reference Model of Open Distributed Processing: A Tutorial. K. Raymond. CRC for Distributed Systems Technology, CITR, University of Queensland. Proceedings of 2nd ICODP, Berlin, 1993.

[7] OSF DCE Application Environment Specification/Distributed Computing RPC Volume. ISBN 0-13-043688-7.

[8] X/Open Directory Service API Specification. X.400 API Association, X/Open Company Limited, 1990.

[9] ODP-Trader. M. Bearman. CRC for Distributed Systems Technology. Proceedings of 2nd International Conference on Open Distributed Processing, Berlin, 1993.

[10] Is DCE a Support Model for ODP? A. Beitz, P. King, K. Raymond. CRC for Distributed Systems Technology. Proceedings of 2nd International Conference on Open Distributed Processing, Berlin, 1993.

[11] Experiences with Building a Multimedia Mail System Prototype on the OSF DCE Platform. Information Architecture Group, Hewlett-Packard Document NSA-92-012. June, 1992.

[12] The DCE Experience. Information Architecture Group, Hewlett-Packard Document NSA-92-014. July, 1992.

[13] A Performance Study of the DCE 1.0.1 Cell Directory Service and Implications for Application and Tool Programmers, J. Martinka, et al. Lecture Notes in Computer Science 731, Springer-Verlag, 1993

[14] OODCE: A C++ Framework for the OSF Distributed Computing Environment. J. Dilley, Hewlett-Packard. To be published at Usenix 1995.

[15] A Distributed Object-Oriented Platform Based on DCE and C++. P. Bosco, G. Martini, C. Moiso, CSELT. Proceedings of 2nd International Conference on Open Distributed Processing, Berlin, 1993.

[16] Universal Networked Objects. J. Nichol, D. Curtis, D. Vines, N. Holt, O. Hurley, G. Lewis. OMG Document 94-9-32; Object Management Group, Inc. September, 1994.

37

The TRADEr: Integrating Trading Into DCE

K. Müller–Jones, M. Merz, W. Lamersdorf

University of Hamburg, Department of Computer Science
Vogt–Kölln–Str. 30, D–22527 Hamburg, Germany
(kmueller,merz,lamersd)@dbis1.informatik.uni-hamburg.de

Abstract

Client support for locating, accessing, and using arbitrary services in open system environments emerges as one of the most interesting, complex, and practically relevant tasks of realising realistic open distributed systems applications. In the context of *Open Distributed Processing* (ODP), current standardisation efforts for a *trading* function play an increasingly important role for open system integration.

In addition to ongoing ODP standardisation activities official and de–facto standard system environments such as, e.g., the OSF *Distributed Computing Environment* (DCE), support efficient development and portability of distributed system applications. Therefore, time seems now ready to analyse and evaluate the use of such platforms also for developing efficient implementations of higher–level system support services, e.g., an ODP trading function.

This paper first elaborates on the specific potential of OSF DCE for supporting *implementations* of ODP trader functions. It then presents an architecture and reports on experiences with such an implementation in the context of system support for general service access, management and coordination in open distributed environments within the *TRADE* project. Finally, the paper draws the attention to still existing *limitations* and deficiencies of OSF DCE for realising ODP trader functions and proposes respective extensions to OSF DCE both at a conceptual and a concrete systems implementation level. According to such prototype implementation experiments, ODP trading functions can be integrated smoothly into a uniform standard system support platform (like DCE) and can be implemented efficiently by extending DCE by additional trading functions which specifically support service management, mediation and access for open distributed applications.

Keyword Codes: C.2.4 D.2.6 H.4.3
Keywords: Distributed Systems, Programming Environments, Communications Applications

1 Introduction

Based on rapid recent developments of telecommunication and networking technologies, users of distributed systems are now increasingly confronted with multitudes and varieties of service offerings in an, in principle, world–wide *open market of services* [MML94]. Faced with the complexity of such open distributed environments, one of the main tasks is to support users and application programs to locate and utilise such services in effective and efficient manners. In this context, one of the most promising efforts is to extend open (operating) system platforms by unified service *trading* or *broking* components as an important structuring technique for efficient design of open distributed systems. Accordingly, the specification of a trading component is currently — among other issues — subject of the *Open Distributed Processing (ODP)* international standardisation activities as the so–called *ODP trading function* [ISO94].

Beyond ongoing standardisation efforts for a unified trader component specification, it is now increasingly important to also examine possible *implementations* of such trading concepts. Common foundations for various distributed application implementations are existing and evolving distributed systems architectures, as, e.g., proposed and developed by various vendors and consortia like OSF's DCE [Fou92], and OMG's CORBA [OMG91]. In particular, the integration of a trading component into these architectures has to be evaluated. A good starting point for such an evaluation is the *Distributed Computing Environment (DCE)* from the Open Software Foundation (OSF) [Fou92] which has gained wide commercial acceptance and is one of the important system platforms for distributed application development in the future. In particular, this paper focuses on

- an analysis of DCE mechanisms to support service management and service access,

- the limitations of DCE for supporting service mediation,

- a proposal for an architecture of the *TRADEr*, a DCE trading component, which has been developed at the University of Hamburg within the TRADE (*Tr*ading and *Coor*dination *E*nvironment) project and its co–project COSM (*C*ommon *O*pen *S*ervice *M*arket),

- some details of a smooth integration of the TRADEr implementation into DCE, especially into existing DCE concepts for service management and access,

- extensions of the *service type* abstraction as proposed by ODP which will also be integrated into the TRADEr, and

- current and future work within the TRADE and COSM projects to support client/server mediation in a distributed open systems environment.

2   The ODP Trading Function

The main task of a trader function is the *mediation* and *management* of services in open distributed systems. For this purpose, the trader first offers mechanisms for arranging and categorising various service types, and then supports potential service clients with specific service selection strategies. Thus, the functionality of a trader component can be compared to, e.g., a *yellow pages* service which categorise service kinds and provides service selection support based on different service properties. The most important formal concept underlying such a trading function is the notion of a *service type* [ISO94]. A service type may contain *interface types* which specify the operational service interfaces in terms of operation signatures as well as *service property types* which add additional semantic details to the service type description. A second important mechanism for service structuring in open distributed systems is based on service *contexts* in which service offers can be grouped and located (e.g. in a hierarchical organised name space).

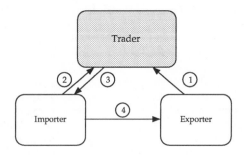

Figure 1: The ODP Trader and its Users

Possible interactions of clients and servers with a trader component in an open distributed environment are shown in Figure 1.

According to the ODP trader function, service provider, a service *exporter*, first registers its service by supplying the service type, the current service property values, and a context in which the service shall be exported (step 1). Then a service client, the *importer*, may ask for servers offering a specific service (step 2). Such an inquiry contains — among other things — the desired service type, the desired service properties, and a search context. Based on this information, the trader then determines appropriate service providers, and selects — if necessary — the best matching service offer. Subsequently, the necessary binding information is returned to the client (step 3), and the client is finally able to execute remote operations offered by the service provider directly (step 4).

3   DCE Service Concept and Support Infrastructure

The OSF Distributed Computing Environment (DCE) provides an integrated set of support services and application programming interfaces (so–called *middleware* [Ber93]) for the development of distributed applications in open heterogeneous environments. The main goal of DCE is to provide users and applications of distributed computing environments with a *homogeneous* view which hides much of the complexity of the underlying hardware and system software components. Therefore, DCE provides ways to develop platform–independent distributed applications, based on de–facto standardised application programming interfaces. Basic DCE services are the *Remote Procedure Call (DCE RPC)*, the *Thread Service*, the *Cell Directory Service (CDS)*, the X.500 compliant *Global Directory Service (GDS)*, the *Security Service*, and the *Distributed Time Service (DTS)*. Additional services are the *Distributed File Service (DFS)* and the *Diskless Client Support*. Based on standard protocol implementations, each of these services provides a single unified application programming interface (with the exception of the Cell Directory Service which offers two such interfaces).

3.1   DCE Support for Service Management and Access

The basic unit of service management and structuring in DCE is the so–called *cell*. Each cell includes its own Security Service, a Cell Directory Service, and a Distributed Time Service. The Cell Directory Service plays a central role for storing all information concerning actual services of the DCE cell (e.g. configuration and binding information). Access to information of foreign cells is provided by an X.500 compliant Global Directory Service which offers a world–wide available name space connecting several different organisation domains.

DCE service structuring is conceptually based on service *interfaces*, i.e. sets of service operations. In order to offer a service interface in a distributed environment, a service provider has to first describe its interface in terms of an abstract *interface definition language (IDL)*. An IDL service description includes information about the offered operation types as well as data types for parameters and results. *Universal Unique Identifiers (UUIDs)* are used for unique identification of interfaces and have to be included in the interface description. They basically provide a very simple type system based on interface names[1] as a basis for managing service providers in the DCE Cell Directory Service. In addition to the interface, a UUID *version number* can be used in order to describe relationships between different versions of an interface (with a given UUID). In this way, support for *inclusion polymorphism* [CW85] can be provided, for example for expressing service evolution (e.g. by step–wise addition of new operations to an interface). Since there is no central type management component included in DCE, however, the DCE application programmer stays solely and fully responsible for the correct definition and use of such interface relationships and compatibilities.

In the following, we give a concrete example of how to establish a binding between a client and a service provider using the DCE Cell Directory Service. Figure 2 shows the involved DCE system components and denotes the necessary execution steps for establishing such a binding between a DCE client and a remote DCE server. In a DCE based open systems implementation environment, a new service provider has to execute the following steps in order to register a new interface[2]:

1. The first step is to inform the RPC runtime system about the offered interface type using an *interface handle*. In addition, a *type UUID* for local type identification, an *object UUID* for unique interface access, and an *Entry Point Vector (EPV)* which serves as a local pointer to the server operations, have to be provided.

2. Subsequently, *binding vectors* are generated which contain binding information (e.g supported communication protocols and dynamic communication endpoints). These along with the interface type and the object identification are then registered at the local host's *Endpoint Mapper*. It manages the mappings from interface and object identifiers to communication endpoints of current running service providers at the local host.

---

[1]In contrast, ODP defines the notion of an *interface type* for this purpose.
[2]In principle, a server can provide several interfaces at a time.

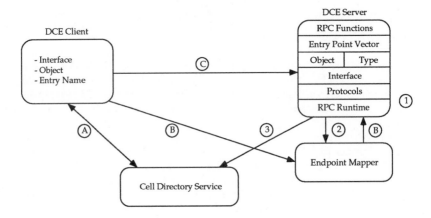

Figure 2: CDS Based Binding

3. In the last step, the same information (except for the dynamic communication end-points) are registered in the DCE Cell Directory Service under a distinct entry name.

In order to obtain the necessary binding information for a server offering a desired interface, a client has to execute the following steps:

A. First, the DCE Cell Directory Service must be called in order to obtain server binding handles of a distinct server instance. For this service instance, the distinct server entry name, the desired interface identifier, and the object identification have to be supplied.

B. Using these binding handles, the client can then bind to the server and start to execute its remote procedure call. Since the DCE Cell Directory Service manages only incomplete bindings without dynamic communication endpoints, the first call is directed (by the RPC runtime system) to the Endpoint Mapper located at the server's host. There, the dynamic communication endpoints are added to the binding handle, and the call is forwarded to the actual service provider.

C. Subsequent remote procedure calls can then be transferred directly to the corresponding service provider.

3.2   DCE Support for Service Mediation

Although DCE — as briefly reviewed above — provides several useful prerequisites for vendor–independent implementations of distributed open system applications, it still lacks some important features, especially in order to support a *trading function* as, e.g. specified by ODP. Currently, there is no support for service mediation in DCE, i.e. the clients (resp. application programmers) are entirely responsible for the selection of appropriate service offerings.

3.2.1 DCE Cell Directory Service

A simple basis for such support can, as described above, be realised by the DCE Cell Directory Service. The corresponding *NSI application programming interface*, specifically designed for registration of RPC interfaces with the Cell Directory Service, provides simple access and management functions for storing service offers in a hierarchical organised name space. Since there is no direct support for search functionalities the usage of the NSI interface is mainly restricted to simple name–based lookup operations[3].

3.2.2 DCE Global Directory Service

In addition to the Cell Directory Service, the programmer could also use the more powerful functions of the X.500 compliant DCE Global Directory Service. In contrast to the Cell Directory Service, however, there is no direct support for RPC interface registration. Therefore, application programmers have to write their own registration functions. One of the main advantages of the DCE Global Directory Service is the ability for attribute–based searching in the name space. It also provides a world–wide accessible name space which facilitates interworking capabilities, e.g. within trader federations. Section 4.2 explains the use of the DCE Global Directory Service for implementing the TRADEr's service offer management capabilities.

4 The TRADEr: Service Mediation on Top of DCE

This section describes a prototype implementation of a trading component based on DCE, called the *TRADEr* (i. e. TRADE trader), developed recently in the *TRADE* project at the University of Hamburg, Germany. In the TRADE project, prototype trader functions are implemented and step–wise extended as part of a general system support environment for service access, management and coordination in open distributed systems [MJM94, MJML95]. The corresponding *COSM/TRADE* prototype system is currently realised on a heterogeneous cluster of interconnected Sun SPARC and IBM RS/6000 workstations; the *TRADEr* implementation has been developed on IBM RS/6000 workstations with AIX and DCE.

Figure 3 gives an architectural overview of the main TRADEr components. As shown, the TRADEr is structured into several sub–modules which — in turn — realise the main sub–tasks of a composite trading function. According to the trader's role in service management, selection, and access in open distributed systems, the TRADEr's core components are the *service offer manager*, the *service selection manager*, the *trader interworking manager*, and the *type manager*. Additionally, *access control management* is provided as an extended option. As an orthogonal extension of OSF DCE functions, the TRADEr is basically realised as a DCE RPC server using authenticated *Remote Procedure Calls* as interface both for service exporters and importers as well as trading administrators. In addition, the TRADEr prototype implementation uses DCE *Threads* for efficient execution of the different TRADEr functions. DCE *Cell Directory Service* and *Global Directory*

---

[3]CDS groups and profiles provide simple additional name space organisation, they can be used for user–provided search routines (see section 4.2).

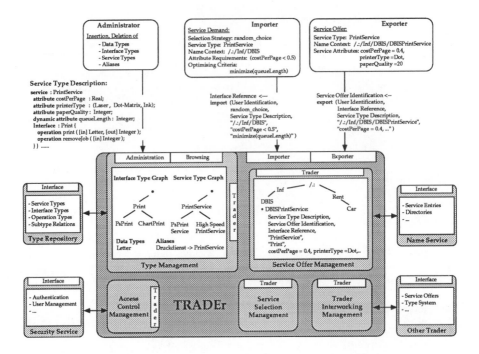

Figure 3: TRADEr Components and Implementation Architecture

Service together provide the basis for realising the service offer management as one of the main functional units of the TRADEr. Figure 3 also gives some implementation details of, e.g., the TRADEr's *export* and *import* functions used by service offering and accessing nodes respectively. (The example specifications given as part of the exporting and importing functions in Figure 3 refer to a basic *Printing Service*).

The following subsection concentrates specifically on the trading functions of *type* and *service offer management* in order to explain some details of design and implementation decisions of the current TRADEr's prototype implementation as realised on top of OSF DCE.

4.1   Extending Type Management within the TRADEr

One of the most important components of a composite trader function is the *type manager*. It provides the basis for a common understanding and for comparison of services types as a main structuring technique for service requests and offerings in open distributed environments. As shown above, different notions of *service types* serve in this context as formal abstractions of *service characteristics*, i. e. common properties of classes of service instances of a distinct service type. Because of the significance of the service type concept, standard typing mechanisms, as known from modern programming languages, and further extensions to such service type concepts have to be evaluated for the trader's type manager component. Therefore, we briefly mention in the following some alternative forms of well understood type mechanisms which are capable to fulfill the requirements for service typing and trading in general, and address extensions to such basic type management techniques. Such type extensions are main candidates for being integrated into an extended type manager component of a trader and, accordingly, into future releases of the TRADEr prototype implementation.

In general, several levels of type management can be distinguished within a trader's type manager (based on very simple up to very rich and complex service type descriptions). Some possible steps in such a continuum from simple name–based service (type) descriptions up to (ideal) full formal semantic specifications of a service type are listed below. They are used as a basis for step–wise extension of type description and managenement functions as already available in, e.g., DCE, up to what is really needed for a future trader's type manager components.

1. In its simplest form, classification ("typing") of services is based on type *names*. Limited type flexibility can be achieved using *subtype polymorphism* as one kind of type polymorphism as, e.g., explained in [CW85]. Subtype polymorphism means that a type A which is a subtype of type B can also be used in a type safe way when B is required. For example, a service S offering one additional operation at its interface as compared to a service K can then also be used by a client instead of K. Since this kind of subtyping is based on names only, subtype relationships have to be defined *explicitly*, i.e. whenever new interface types are inserted, their respective sub- and supertypes must be listed explicitly by, e.g., the trader administrator. As explained in the previous section, DCE provides only very little support for subtype polymorphism (based on interface type name, and version numbers for interfaces) but no support for checking the defined relationship between interfaces automatically.

2. Adding explicit *attributes* to the interface type is one way to enrich service descriptions which are to be matched in an open distributed trading environment. Service attributes play an important role to extend the specification of semantics of a given type by including some selected, predefined properties into service (type) descriptions. This allows further discrimination between service types offering similar operations. Subtype polymorphism for attributes can then be extended using *semantic substitutability* as explained in, e.g., [IBR93]. Currently, however, there is no support for service attributes in existing releases of OSF DCE. Service attributes can only to be *simulated* using explicit "get" and "set" operations, similar to the operations defined in the CORBA standard [OMG91] for attribute access.

3. As a subsequent step in generalising and enriching service type descriptions, simple name based subtyping mechanism can be replaced by a *structural* subtyping mechanism. Here, decisions of type relationships can be made *automatically* by the system based on a structural analysis of the corresponding service type descriptions. This concept is known as *type conformance* [BHJ+87] and enforces an *implicit* (or

automatic) style of subtype checking which frees the application programmer from this failure prone standard programming tasks.

4. Experiences with the development of system support for open distributed applications at our group have shown that additional service description techniques are required in order to express semantical aspects of services which go beyond standard (i.e. programming language) type concepts. As one such extension, for example, *finite state machines*, have been introduced into COSM service (type) description as a first step into protocol specifications and are used to support users of a so–called *Generic Client* in accessing "unclassified" services which are not known in advance [ML93, MML94]. They provide a way to formally express how to use a service, e .g. which sequence of operations may be executed by a remote user, e. g. via a Generic Client component, as part of its service description.

5. A final extension to service (type) descriptions considered in the TRADE and COSM projects so far is concerned with coordination of complex distributed services which are comprised of a set of more basic ones. It uses *Coloured Petri Nets* as a formal description technique of such coordination problems and is currently evaluated for supporting workflow modeling and execution within complex open distributed client/server environments [MMML94, MJML95]. Here, Coloured Petri Nets are used for describing service coordination in order to support the execution of workflows in open service environments in an adequate manner.

4.1.1 A Name Based Type Manager with Explicit Subtyping

The current TRADEr prototype provides dynamic type management based on explicit subtyping as mentioned in items 1 and 2 above. Internally, all type information is managed within the type manager component as two directed acyclic type graphs which represent the type relationships between interface and service types. At the current prototype state the administrator still has to explicitly list all supertypes when inserting new interface and service types. Simple type checking is supported by an internal interface comparing service type descriptions as offered by exporters with existing previously defined service types. The trader's external *administration interface* and *browsing interface* provide functions for insertion, deletion and browsing of service and interface types at runtime.

Currently the TRADEr's type manager is extended to support implicit type checking as mentioned in item 3 above as well. In this context, the CORBA type model (respectively the CORBA IDL) is evaluated for its capability to describe services more adequately. Further extensions (see item 5) shall be integrated in the future, if possible.

Future versions of the TRADEr prototype implementation will also comprise an *interface repository* which is developed currently. It will be used to store all kinds of service type descriptions. In the mean time, the TRADEr prototype still stores type management information as local data in volatile memory.

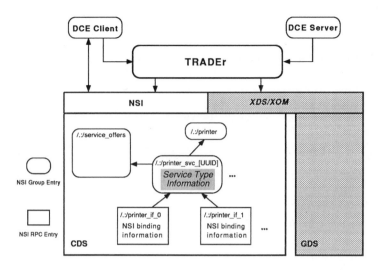

Figure 4: Use of DCE CDS and GDS for Service Offer Management

4.2 Service Offer Management

Another important core component of the TRADEr is the *service offer management*. In the current TRADEr prototype, service offer management is entirely based on the DCE Cell Directory and Global Directory services. The respective interface contains operations for a variety of service offer management functions, e.g. inserting, deleting, reading, and modifying of service offers. Also attribute–based search operations are provided which facilitates the implementation of the TRADEr's service selection strategies. For a service provider to advertise a service offer in the TRADEr, it is possible either to use CDS entry names, such as /.:/services/Laserprinter_1, or to use X.500 entry names, such as /.../C=DE/O=Hamburg University/OU=services/CN=Laserprinter_1. Service offers are stored in a special format representing the offered service type and the corresponding interface types, the current values of the static service attributes, the interface reference for server binding, and, optionally, a service type description. This service type description can be used by programmers to develop corresponding DCE client applications and to generate the RPC stubs necessary for communicating with the DCE server.

The current TRADEr prototype implementation of the service offer management operations is based on the two DCE name service application programming interfaces, namely the *Name Service Interface (NSI)* and the *X/Open Directory Service and X/Open OSI–Abstract–Data Manipulation (XDS/XOM)* interface. Using both of these programming interfaces allows a smooth integration of the TRADEr's trading functions into the existing DCE RPC/CDS interface registration concept. Figure 4 shows the usage of the two programming interfaces in some more detail.

In order to enable former Cell Directory Service based clients to access servers advertising their service offers in the TRADEr, service offers exported into the CDS name space are at first generated using the NSI interface. This is necessary because the NSI interface expects a special internal format for name service entries. Subsequently, the service attributes used by the TRADEr are added to the CDS entry via the XDS/XOM interface. Therefore, former CDS based servers are able to use the capabilities of the more powerful TRADEr functions without affecting existing distributed applications.

In addition to standard name service operations, the XDS/XOM interface provides powerful functions for searching the X.500 name space. The corresponding search operation, namely the ds_search function, is extensively used for the realisation of the service selection strategies offered by the TRADEr. In the current DCE release, however, the ds_search function is restricted to the X.500 name space and can not be used for searching the CDS name space as well. Because of this limitation, the TRADEr simulates searching the CDS name space by using the CDS group concept by collecting all service offers in a well-known group entry for each CDS directory.

5   Conclusions and Outlook

Work reported in this article is based on the importance of early evaluations of prototype implementations of trading concepts in the context of existing distributed systems platforms as, for example, the OSF Distributed Computing Environment (DCE). A concrete implementation goal is an orthogonal and smooth integration of basic trader functions into, in particular, already available service registration and management mechanisms.

Related experiences from the TRADEr prototype implementation as available so far are twofold: On the one hand, DCE already provides powerful functions for distributed application programming, many of which could also be used beneficially for the TRADEr's trading function implementations. On the other hand, however, DCE concepts already available for service (type) description are still unsatisfactory and corresponding DCE functions still lack important (especially: type management) mechanisms which are necessary for service management and mediation in open distributed environments. Therefore, extended type management functions have to be developed separately, based on modern programming language (polymorphic type) concepts and on specific service description extensions (as, e. g., for protocol and workflow management specifications) as needed in open distributed environments. In summary, experiences with the TRADEr prototype implementation have shown that trading can be smoothly integrated into DCE, but DCE functions have to be extended substantially in order to support and use, in particular, realistic service descriptions based on modern type management functions.

In the TRADEr prototype implementation, such extensions currently concentrate on more elaborate *type management* functions (as described above), support for distributed *trader interworking*, and support for inclusion of *dynamic* (i. e. time varying) *attributes* into extended service descriptions. In particular, techniques to extend service descriptions with additional semantic information are still evaluated and will be integrated step–wise into the COSM/TRADE prototype implementation. In this context, early experiences with service type extensions like finite state machine description of service protocols [ML93] have motivated recent work on using of more powerful description methods also for coordinating *composite* services (aspects of *workflow management*) based on Petri nets

[MMML94, MJML95] as mentioned in section 4.1 of this paper.

References

[Ber93]   P. A. Bernstein. Middleware – an architecture for distributed system services. Technical Report CRL 93/6, Digital Equipment Corporation, Cambridge Research Lab, March 1993.

[BHJ+87]   A. Black, N. Hutchinson, E. Jul, H. Levy, and L. Carter. Distribution and abstract types in Emerald. *IEEE Transactions on Software Engineering*, 13(1):65–76, 1987.

[CW85]   L. Cardelli and P. Wegner. On understanding types, data abstraction, and polymorphism. *ACM Computing Surveys*, 17(4):471–522, December 1985.

[Fou92]   Open Software Foundation. *Introduction to OSF DCE*. Prentice–Hall, Englewood Cliffs, New Jersey, 1992.

[IBR93]   J. Indulska, M. Bearman, and K. Raymond. A type management system for an ODP Trader. In *Proceedings of the IFIP TC6/WG6.1 International Conference on Open Distributed Processing*. North–Holland, Elsevier Science Publishers B.V., September 1993.

[ISO94]   ISO/IEC JTC 1/SC 21. Recommendation X.9tr/Draft ODP Trading Function ISO/IEC 13235: 1994/Draft ODP Trading Function, July 1994. ISO/IEC JTC 1/SC 21 N9122.

[MJM94]   K. Müller, K. Jones, and M. Merz. Service mediation and management in open distributed systems. In B. Wolfinger, editor, *Innovationen bei Rechen– und Kommunikationssystemen — Eine Herausforderung für die Informatik*, Informatik Aktuell, pages 219–226. Springer–Verlag, Berlin,Heidelberg, August 1994. in German.

[MJML95]   K. Müller-Jones, M. Merz, and W. Lamersdorf. Cooperative applications: Integrated application process control and service mediation in open distributed systems. University of Hamburg, Germany, submitted for publication, in German, 1995.

[ML93]   M. Merz and W. Lamersdorf. Cooperation support for an open service market. In *Proceedings of the IFIP TC6/WG6.1 International Conference on Open Distributed Processing*, pages 329–340. North–Holland, Elsevier Science Publishers B.V., 1993.

[MML94]   M. Merz, K. Müller, and W. Lamersdorf. Service trading and mediation in distributed computing environments. In *Proceedings of the 14th International Conference on Distributed Computing Systems (ICDCS '94)*, pages 450–457. IEEE Computer Society Press, 1994.

[MMML94]   M. Merz, D. Moldt, K. Müller, and W. Lamersdorf. Workflow modeling and execution with coloured petri nets in COSM. University of Hamburg, Germany, submitted for publication, 1994.

[OMG91]   The Common Object Request Broker: Architecture and Specification. Digital Equipment Corporation, Hewlett–Packard Company, HyperDesk Corporation, NCR Corporation, Object Design Incorporated, SunSoft Incorporated, 1991.

38

Experiences Using DCE and CORBA to Build Tools for Creating Highly-Available Distributed Systems

E. N. Elnozahy[a,†], Vivek Ratan[b], and Mark E. Segal[c]

a School of Computer Science, Carnegie Mellon University, Pittsburgh, PA 15213, USA, mootaz@cs.cmu.edu

b Dept. of CSE, FR-35, University of Washington, Seattle, WA 98195, USA, vivek@cs.washington.edu

c Bellcore, 445 South Street, Morristown, NJ 07960, USA, ms@thumper.bellcore.com (contact author)

Open Distributed Processing (ODP) systems simplify the task of building portable distributed applications that can interoperate even when running on heterogeneous platforms. In this paper, we report on our experience in augmenting an ODP system with tools that allow developers to build highly available distributed objects with little or no additional programming effort. Our tools are implemented within the context of the DCE and CORBA standards for distributed computing. We describe the system that we built and how the combination of DCE and CORBA often helped our efforts and sometimes impeded them. Based on our laboratory experiences, we conclude that these standards generally have a good potential for developing tools for high availability that are portable and applicable to a variety of applications in a distributed computing environment. This potential, however, is hampered by several shortcomings and problems in the specifications of the standards. Such problems could impede other developers and researchers who plan to use these standards. We discuss these problems and suggest solutions for them.

1. INTRODUCTION

Open Distributed Processing (ODP) systems reduce the complexity of designing and implementing applications in distributed computing environments. Applications that run on an ODP system follow standards that allow them to be portable across heterogeneous platforms, and also allow them to interoperate with other distributed applications that follow the same standards. This paper describes our experiences augmenting an ODP system with a toolset that can automatically add high availability to distributed objects. These objects adhere to the Common Object Request Broker Architecture standard (CORBA) [OMG91, Vinoski93]. A highly available object continues to run in the presence of hardware or software faults, as well as planned maintenance activities such as hardware and software upgrades. Applications where high availability is important include financial transaction-processing systems, telecommunications, medical systems, and real-time process control.

The toolset includes a number of software-based techniques for providing high availability with little or no intervention from the programmer. Application developers implement their objects following the CORBA standard and link them with our toolset to provide the required high

---

† This author was supported in part by the National Science Foundation under grant Number CCR-9410116.

availability. The implementation uses a locally-developed CORBA library [Diener94] which runs over OSF's Distributed Computing Environment (DCE) [Millikin94, OSF91] on SparcStations running SunOS 4.1 and DEC Alphas running OSF/1.

We chose a CORBA-compliant platform to implement our toolset because we believe that many future distributed applications will adopt the CORBA standard. Thus, our toolset could be ported to other platforms and application domains. We decided also to rely on DCE to provide the networking support. There are several alternatives to this decision, each typically consisting of a CORBA package that implements its own name service, and interacts with the network directly through the socket layer in Unix® or using another RPC system. We decided against using these CORBA packages because their reliance on non-standard naming, and lack of security facilities. DCE does not have these problems and complements the CORBA library that we had with its naming, security, and RPC services. Our choice also offers a potential for interoperability with other "pure" (i.e. non-CORBA) DCE applications and tools.

The implementation of our toolset was able to benefit from many of the facilities that CORBA and DCE provide, such as the name server, the uniform Interface Definition Language, and RPC groups, among many others. These contributed to the simplification of the implementation effort and we were able to verify the benefits that both standards offer for the development of distributed applications. Unfortunately, our laboratory experiences revealed a number of problems with both CORBA and DCE. Though some problems were specific to our platform, others were resulting from the definitions of both standards and could impede other researchers and developers who would be involved in projects using CORBA and/or DCE. We discuss these problems and we suggest solutions to them.

The primary focus of this paper is our experiences using CORBA and DCE to build a high-availability toolset. A detailed description of the implementation of the toolset itself and a performance evaluation can be found elsewhere [Elnozahy95]. Section 2 includes an overview of the high availability toolset to provide the necessary background and context for describing our experience with CORBA and DCE, which is detailed in Section 3. We present a summary and our conclusions in Section 4.

2. THE DESIGN OF A HIGH AVAILABILITY TOOLSET

2.1. Design Goals and Overview

We have constructed a toolset that provides high availability to distributed applications with little or no additional support by the application programmer. This approach relieves the programmer from the mundane and often error-prone tasks of handling failures and recoveries at the application layer. Also, the approach has the potential of improving the availability of existing applications that were written without consideration for high availability.

The system uses a local implementation of the CORBA standard, called Touring Distributed Objects (TDO) [Diener94]. In this system, application programs consist of distributed server and client objects that communicate by remote method invocation according to the CORBA standard [OMG91]. Server objects follow a multithreaded programming model, where a thread is automatically started to execute the method invoked by a remote client object. It is assumed that the execution of a method invocation is short and the corresponding thread lives only during the course of serving the invoked method. This model is consistent with the familiar remote procedure call paradigm for client/server applications.

® Unix is a registered trademark of Novell, Inc.

TDO objects are written in C++ and use CORBA's Interface Definition Language (IDL) to define the exported methods. Figure 1 illustrates how the components that make up a TDO CORBA/DCE application fit together. Figures 2a and 2b show the C++ source code for a simple TDO CORBA/DCE application. The TDO runtime system provides support for exporting the server methods, implemented by server implementation (SI) objects, by assigning a unique name in a hierarchical name space for each instance of an exported class. The TDO compiler automatically generates two proxy classes, one for the client side and the other for the server. These classes act as stubs of remote communication for each IDL interface. The server proxy class (SP) handles all the details of translating the remote procedure calls into C++ method invocations, and of exporting the name of the class into the global name space. Client objects that wish to interact with a certain server must be linked with the corresponding client proxy of the server. The client can then access the server through normal invocations of C++ methods. The client proxy locates the required server and handles the details of translating the remote invocations into remote procedure calls.

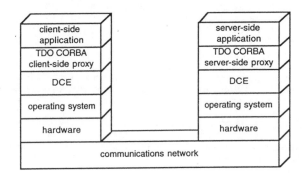

Figure 1 — The Structure of a TDO CORBA/DCE Application

The system uses DCE to provide the underlying support in the form of remote procedure calls, multithreading, naming, RPC groups, and security. Most notably, the proxy class uses DCE RPC to implement remote method invocation. The proxies also use the CDS name space of DCE to implement the global name space. Additionally, the implementation relies on DCE's multithreading support for implementing lightweight threads. The application, however, does not need to call the DCE layer directly, since TDO abstracts the support provided by DCE. This approach simplifies the task of the programmer.

2.2. Mechanisms for High Availability

The toolset provides a repertoire of mechanisms that can be used to add high availability to objects. These mechanisms differ according to their costs and reliability guarantees, with the cost increasing with the degree of reliability desired. High availability is added by linking the server proxy of a highly-available object with a library that includes the necessary support. Application developers can select the appropriate availability mechanism for their applications by linking with an appropriate library. The system is designed to make this choice as transparent as possible to the application. A detailed description of the availability mechanisms used by the system appears elsewhere [Abbott90, Birman93, Elnozahy93, Gray91, Huang93, Segal93]. We focus here on describing the implementation of one technique, namely the primary/backup scheme and how it made use of CORBA and DCE support. In this technique, several copies of

```
// simple TDO server
main(int argc, char *argv[])
{

  // create a proxy and an implementation
  simple proxy ;
  Simple impl ;
  // plug the implementation into the proxy
  proxy.represent(&impl);
  // "advertise" the proxy in the CDS so that
  // clients can bind to it...
  proxy.export("/.:/tm/simple.server");
  // fork a thread which listens for incoming messages
  CORBA::BOA::impl_is_ready();

  // wait for the thread to terminate (i.e. wait forever...
  // use ctrl C to kill off the server)
  CORBA::BOA::wait() ;
  exit(0) ;
}
```

Figure 2a — C++ Code for a Simple TDO CORBA Server

```
// simple TDO client
main()
{
  // create a proxy
  simple clientProxy;
  // "bind" it to its implementation using the
  // represent member function.
  clientProxy.represent("/.:/tm/simple.server");

  // send it a ping message
  clientProxy.ping() ;
  exit(0) ;
}
```

Figure 2b — C++ Code for a Simple TDO CORBA Client

a server object exist on different machines. One copy is designated the primary while the others are backups. The primary is the contact point for the client proxies, which are unaware of the presence of the backups. Each backup has its own private copy of each proxy class that is exported by the server. Only the primary's proxy, however, exports the corresponding class instance into the global name space.

During normal operation, the primary receives invocation requests from client proxies for the methods that the server object exports. The primary executes the appropriate method and sends to the backup objects the corresponding updates to the object's state due to the method invocation. After the backups acknowledge the receipt of these updates, the primary replies to the client object with the result. This sequence of steps is shown in Figure 3.

Support for high availability is added through a high-availability layer (HAL) that is linked with the proxy object at the primary and each of the backup (see Figure 4). The HAL intercepts all remote procedure calls intended for the object as well as those originating from the object. The HAL encapsulates all functions related to providing high-availability and is completely transparent to the application.

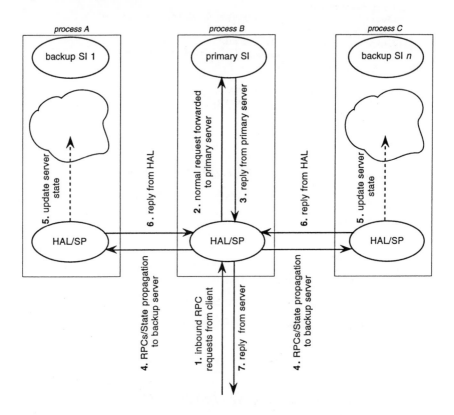

Figure 3 — Communication Between HALs, Primary and Backup Servers

The SP for the primary server object exports its interface to the global name space, enabling client proxies to bind to the replicated object. The HAL at the primary object also creates an RPC group in a special subtree of the CDS space which contains the names of the primary server and the backups. This naming scheme allows the primary and backup proxies to communicate among themselves at the HAL level to support high availability, but is completely transparent to client objects. Figure 5 shows how a highly-available server, my_app, would be registered in the CDS namespace.

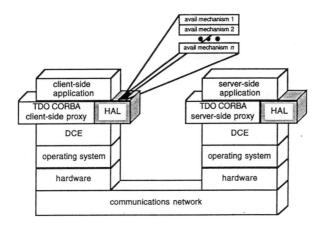

Figure 4 — The Architecture of a Highly Available Application

The HAL implements several functions related to high availability, such as failure detection, the election of a new primary after a failure, and the communication of state updates. These functions rely on the existence of the RPC group that was created by the primary. The HAL communicates using remote method invocation following the CORBA standard.

3. EVALUATION

In this section we describe our experiences using CORBA and DCE to build our high-availability toolset. Overall, the facilities provided by CORBA and DCE greatly simplified our implementation efforts. The HALs at the primary and backups use CORBA remote method invocation for communication. We found this layering to be very convenient and simpler than the alternative of performing the HAL-level communications directly through DCE. We initially considered using DCE to get better control of the underlying communications protocol and better performance. We eventually decided against it, however, because the low-level computation and communication abstractions offered by DCE would make our implementation efforts more difficult than working at the object-oriented CORBA layer. Thus, our code was operating at a high level which was independent of the hardware or the network protocol. We believe that this choice led to the simplicity of the design and implementation. The naming services made available by DCE, especially RPC group naming, were also very valuable in simplifying the implementation and reducing its complexity. We were also able to use vendor-supplied tools to monitor the CDS name space while our programs were executing, which was extremely valuable for debugging the HAL. The alternative of implementing a name service on top of a lower layer such as Unix sockets or some vendor-supplied name server would perhaps have required more coding and would not have been useful during debugging. DCE's excellent integration of RPC, threads, and exception handling provided a good foundation for building the CORBA layer as well as the HAL.

Besides these positive facts, our experience also outlined several problems with the CORBA and DCE standards. Some of these problems are specific to the nature of providing high availability, while others are more fundamental. We discuss these problems in the remainder of

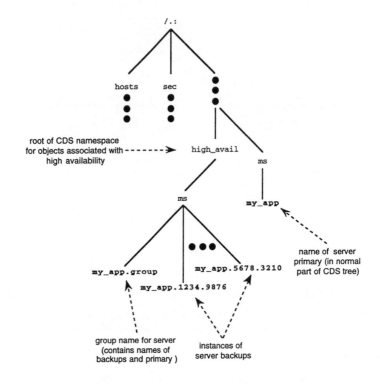

Figure 5 — A Highly-Available Application in the CDS Namespace

this section. We have been careful to isolate the problems that were merely bugs in our implementation or related to our environment. We do not discuss these problems here.

3.1. CORBA Standards Issues

The TDO CORBA platform we used was implemented locally and provided a subset of the CORBA standard and the likely C++ language mapping as they existed when TDO was written (early 1994). The first problem that we found with the CORBA standard is that the Interface Definition Language (IDL) restricts arrays to be of fixed size. This is fundamentally restrictive for general programming problems and ignores a wealth of experience in distributed computing which shows that communicating variable-sized data is a very common operation (e.g., the implementation of TCP/IP [Postel81]). In our case, this restriction proved costly when the primary object was transmitting state updates to the backups. Since it is impossible to predict at compile time the amount of state changes that would result from a remote method invocation, we had three alternatives. The first is to pick a large array size, which would force the transmission of useless data over the network for small updates, causing a lot of overhead. The second is to pick a small array size and perform several method invocations until the state update transfer is complete. This alternative also causes a lot of overhead in the form of repeated interrupts, and complicates the code because of the need to handle partial updates in the presence of failures. The third alternative was to export several methods for state transfer, each having a different array size and use the method with the most effective size during run time. This alternative reduces the overhead, but it complicates the implementation both at the sender and receiver. None of these alternatives is satisfactory (we currently use the second one). It may be argued that this problem can be addressed by using CORBA's sequence, a data structure type that does not have limitations on the size of the data to be transmitted. This data structure, however, is designed to support complex data structures such as trees and lists, and a typical implementation will have to perform marshalling and unmarshalling on transmission and receipt, respectively. Sending a variable-sized array as a sequence will thus incur unnecessary performance overhead. We believe that variable-sized arrays are essential and should be added to the standard.

The second problem that we found with the CORBA standard is the lack of specifications concerning threads and exception handling. CORBA-compliant objects are multi-threaded by default, yet the CORBA standard has intentionally omitted any specification related to the interactions between threads, exceptions, and the programming language. These problems are likely to produce CORBA implementations that differ in the way they handle threads and will create problems in porting applications among heterogeneous platforms, which is against the spirit of ODP. Perhaps a better way to solve this problem is to follow the approach used by DCE, which has integrated communication, multi-threading, and exception handling to some degree. We believe similar integration strategies should be added to the CORBA standard.

The third problem that we encountered was relatively minor. CORBA IDL does not have a specification for 64-bit data types. On a machine like the DEC Alpha, CORBA's long data type is actually mapped to a 32-bit quantity which is different from the machine's concept of a long variable, which is a 64-bit quantity. It is not clear how this problem could be solved, though. Supporting 64-bit types would make interoperability between 64-bit and 32-bit machines difficult. On the other hand, by only supporting 32-bit quantities, application programmers writing code on 64-bit machines should be aware of the difference, which may require additional marshalling and unmarshalling, and exposes the network issues at the application level. DCE, which already provides a 64-bit integer type (hyper), faces the same problem. On a 32-bit machine, hypers are converted to structures which makes even simple arithmetic operations (e.g., addition) difficult to implement.

3.2. RPC Groups in CDS

Our toolset uses CDS RPC groups to implement a scheme for keeping track of the backups and primary of a replicated object. The CDS RPC group mechanism allows a set of servers registered with CDS to be associated with a single group name. The documentation states that updates to a group are to be propagated consistently to each node that caches a copy of the group's list of names. However, the documentation is vague about how soon updates to a group should be propagated to all cached copies in the network, and what happens if nodes crash while a group update is being propagated. We found this vagueness to be a problem for three reasons.

First, each copy of the replicated object depends on the CDS to know the identities and locations of the other currently running copies of the object. This information is vital, especially when a primary process dies and the replicas have to decide who will become the new primary. The replicas must ensure that only one primary is chosen. As replicated objects die or are newly created, the RPC group in CDS containing information on the replicated copies changes. Since group lookups at each node go to the local CDS cache at that node, replicated objects can have an inconsistent view of the currently active set of objects due to DCE's loose cache coherency semantics.

Second, implementations of DCE adopt lax interpretations of these specifications to improve performance. For example, deleting a name from a group takes from an hour to a day to be propagated to all cached copies in one implementation, while it is propagated every 12 hours in another. Such implementations assume that changes to the name space are not frequent, and therefore they are not suitable for supporting the type of mechanisms that we implemented.

Third, if a node that is running the primary CDS or a cached backup copy of CDS crashes during a group update propagation, some CDS caches could be properly updated, while some may be incorrect. Even worse, we don't know when (or if) the CDS caches will be made consistent. Since the semantics of the group naming during failures is not specified, we cannot rely on this technique as a basis for our toolset in the long run.

We were able to work temporarily around the cache coherence problem by patching the DCE object code to force lookups of group members to go directly to the central CDS database, thereby bypassing the local cache. A permanent solution to all of these problems is to build a separate name server that is optimized to maintain a more consistent view of group membership. This name server could be integrated into CDS via the CDS junction mechanism, which would allow adding high availability to CDS without modifying CDS itself.

3.3. Proxies and Multicast

The primary/backup scheme discussed in Section 2.2 as well as other software-based mechanisms for high availability could take advantage of a multicast capability for propagating updates and synchronizing replicas. We are only interested in a multicast that provides FIFO, reliable message delivery in the presence of communication failures. We don't believe though that a multicast protocol with a sophisticated ordering semantics would be useful for our purposes. Because our TDO CORBA/DCE programs are multi-threaded and therefore may execute in a different order on different machines, the receivers will have to implement a mechanism to ensure that the different interleaving of thread executions would not violate consistency among the replicas. The end-to-end argument in system design then suggests that a sophisticated multicast ordering would not be very useful for our purposes. Currently, the closest thing to a multicast that DCE provides is unreliable broadcast RPC execution semantics to processes on the same local-area network, which is not useful for our purposes because we need reliable delivery to a specific group of processes.

We implemented our multicast primitives by explicitly coding the multicasting in the HAL. While this approach is inefficient, it hides complexity from the application programmer and allows us to experiment with multicasting in our toolset. If future versions of DCE or a CORBA-compliant platform provide multicast in a manner that can take advantage of multicasting network hardware, the efficiency of our toolset will increase dramatically.

3.4. General Observations on DCE

Perhaps the biggest criticism we have for DCE is its enormous complexity. To write even a small DCE program requires initially climbing a steep learning curve. If DCE is going to be used to build large systems, tools must be provided to abstract the enormous amount of details involved in order to make it easier to use. Based on our TDO experiences, we can say that a well-designed layer over DCE can hide many of its complexities and make it more manageable for new application developers.

Another significant problem we encountered with DCE is that its implementations are not robust. While constructing our toolset, the DCE implementations we used crashed frequently. These crashes made building our toolset more difficult, and would also increase the difficulty of building highly-available applications. Because the crashes occurred in different DCE implementations, we attribute these crashes to the complexity of the DCE standard, which makes the development of a robust DCE implementation difficult.

We have also observed that different DCE implementations behave differently. A stated advantage of ODP in general, and DCE in particular, is application portability across heterogeneous computing platforms. Common behavior across platforms is critically important for building a portable high availability toolset. We attribute these problems to the imprecision of the specification of many aspects of DCE, especially when it comes to defining behavior during failures. For example, server-side exceptions are supposed to be caught and possibly propagated to the client. The integration of communication and exception handling is an important feature of DCE, but these capabilities behave differently on different platforms. This is disappointing given the elaborate specification of the standard.

4. CONCLUSIONS AND FUTURE WORK

We have described our experiences in building a toolset to add high availability to distributed objects on a CORBA-compliant platform which, in turn, runs over DCE. The toolset allows application developers to create highly-available applications without having to worry about the underlying technical details of failure handling and recovery. The toolset provides a variety of mechanisms for increasing program availability that can be selectively incorporated into a program, depending on its needs.

We built the toolset over CORBA and DCE because we believe these technologies will be used to create many kinds of distributed applications in the future. Both CORBA and DCE hide many low-level inter-process communication details and provide facilities to make writing distributed programs easier. By using these technologies, we were able to build on their strengths and add capabilities to provide high availability to those applications that require it. Specifically, we found that building the HAL using CORBA was very convenient, and the high-level view of the network as a set of distributed objects has greatly simplified our implementation effort. We also found the naming and RPC group services of DCE to be very valuable.

Although CORBA and DCE offered many benefits to our project, we also discovered a number of problems with both standards that introduced additional complexity in our implementation. The CORBA standard for instance does not allow the transmission of variable-sized arrays over the network, something that we believe very essential for distributed applications. We also struggled because of the lack of specification of the interactions of exception handling and threads, especially during failures. These led to many ambiguities that are likely to be handled differently by different implementations of the standard. The CORBA standard should precisely define the behavior of CORBA-compliant objects to resolve these ambiguities, while maintaining the overall simplicity of the current standard and avoiding the complexity and overloading that characterize the DCE standard.

For DCE, we were disappointed that the specifications were imprecise in many situations, especially concerning the behavior during failures. The specifications were also lacking in defining what happens to CDS during failure. This imprecision led to implementations of DCE that behave differently, which is against the spirit of open distributed processing. We also believe that the lack of multicast support and the complexity of the standard are going to be serious problems that researchers and developers will have to face when building distributed applications. We also observed that even though DCE implementations are sold as finished products, they still have many bugs in them, which we attribute to the complexity of the standard that makes it very difficult to build robust implementations.

In spite of these criticisms, we believe that CORBA and DCE can form a platform that offers a good potential for building many kinds of distributed systems. Over time, we hope the problems we described will be resolved by the suppliers of CORBA and DCE products, as well as by the appropriate standards organizations. As people attempt to use these systems to build tools and applications of significant size, more problems will be discovered. As long as there are ways for people working "in the trenches" to affect the design of CORBA and DCE, these systems will continue to evolve into better tools for building large applications.

5. ACKNOWLEDGMENTS

Glen Diener implemented TDO and provided helpful suggestions on its use. Peter Bates, Gita Gopal, R. C. Sekar, and Marc Shapiro gave us many helpful comments on earlier versions of this work. Finally, the authors would like to thank the anonymous reviewers for their comments, which helped clarify a number of technical details and dramatically improved the readability of the paper.

REFERENCES

[Abbott90]   Abbott, R., "Resourceful Systems for Fault Tolerance, Reliability, and Safety", *ACM Computing Surveys*, 22(1), March, 1990.

[Birman93]   Birman, K., "The Process Group Approach to Reliable Distributed Computing", *Communications of the ACM*, 36(12), December, 1993.

[Diener94]   Diener, G., "Touring Distributed Objects", Internal Bellcore Memorandum, June, 1994.

[Elnozahy93]   Elnozahy, E. N., *Manetho: Fault Tolerance in Distributed Systems Using Rollback-Recovery and Process Replication*, Ph.D. Dissertation, Rice University, October, 1993. Also available as Technical Report 93-212, Department of Computer Science, Rice University.

[Elnozahy95]   Elnozahy, E. N., V. Ratan, and M. Segal, "A Toolset for Building Highly-Available Software Systems Using CORBA and DCE", in preparation.

[Gray91]   Gray, J., and D. Siewiorek, "High-Availability Computer Systems", *IEEE Computer*, 24(9), September, 1991.

[Huang93]   Huang, Y., and K. Chandra, "Software Implemented Fault Tolerance: Technologies and Experiences", 23rd Symposium on Fault Tolerant Computing, Toulouse, France, pp. 2–9, June, 1993.

[Millikin94]   Millikin, M., "DCE: Building the Distributed Future", *Byte*, 19(6), pp. 125–134, June, 1994.

[OMG91]   —, "The Common Object Request Broker: Architecture and Specification", Object Management Group, TC Document Number 91.12.1, Revision 1.1, December, 1991.

[OSF91]   —, "Introduction to OSF DCE", Open Software Foundation, 1991.

[Postel81]   Postel, J., "Internet Protocol", Internet Request for Comments RFC 791, September, 1981.

[Schneider90]   Schneider, F., "Implementing Fault-Tolerant Services Using the State Machine Approach", *ACM Computing Surveys*, 22(4), December, 1990.

[Segal93]   Segal, M., and O. Frieder, "On-the-Fly Program Modification: Systems for Dynamic Updating", IEEE Software, pp. 53–65, March, 1993.

[Vinoski93]   Vinoski, S., "Distributed Object Computing with CORBA", *C++ Report*, pp. 34–38, July–August, 1993.

PART THREE

Position Statements and Panel Reports

39

The Open Systems Industry and the Lack of Open Distributed Management

Lance Travis

Open Software Foundation,11 Cambridge Center, Cambridge, MA 02142, USA

cmt@osf.org

Abstract

This paper discusses reasons why the Open Systems industry has failed in its attempts to deliver interoperable vendor-independent distributed systems and network management products.

Keyword Codes: K.6.3; K.6.4

Keywords: Software Management, System Management

1. INTRODUCTION

The Open Systems industry has not provided a strong response to user needs for open distributed management. End users are quite clear in their requirements. They are asking for management systems that allow them to manage their entire enterprise. They have systems from multiple vendors running many different operating systems, networking protocols, and applications. They want to be able to manage the different components with one set of tools. In addition, they want to be able to integrate and manage new technologies without abandoning their legacy systems.

Currently, they have to buy management applications tailored to each individual platform. These applications cannot share data with other similar applications for other platforms and the end result is that end-users have to buy, learn, and use multiple dissimilar applications in order to manage their enterprise. Additionally, they are not able to manage parts of their enterprise because management applications don't exist.

The end-users believe that the Open Systems industry should provide the focal point for solving these problems. By having systems vendors and software vendors agree on standards for programming interfaces and management protocols, end-users hope to be able to buy management products that can be used to manage their entire enterprise and that can be used with other vendor's management products to solve more complex management problems. For example, in the ideal world of open systems management, an end-user should be able to buy a software distribution application that can be used to install applications from any other vendor and can be used to install applications on any target system. The software distribution application should be able to share data with configuration and licensing applications so that the user can have a basis for sophisticated asset management. Unfortunately, current products do not support this ideal. Instead, end-users can buy software distribution applications from a multitude of vendors that can be used on a limited set of platforms (usually only those of the vendor) and can only be used to install a limited set of applications (usually those of the vendors).

The remainder of this paper examines why standards efforts and the multiple vendor consortium investigating open management have failed to deliver much substance. The final section provides criteria necessary for a successful delivery of open systems distributed management.

2. STANDARDS

Lack of standards has not been the only cause of this problem. In some management areas, more than enough standards exist. In the software distribution example, PASC (POSIX) standards exist for packaging software and for the interfaces for a software distribution applications and standards exist for network connectivity, naming, and security.

This suite of standards should enable the ideal open systems software distribution application to exist, but the application does not exist. Obviously more than just standards need to exist.

Standards do not completely solve the problem for multiple reasons. Too many uncoordinated groups are working on standards in the management area. The Desktop Management Taskforce (DMTF), X/Open, Network Management Forum (NMF), Management

Interface Consortium (MIC), OSF, OMG, ISO, and IETF are all working on various sets of management related standards. Little formal or informal coordination exists between these organizations, which results in multiple models for defining management data (MIBs, MIFs, GDMO, OMG IDL) and mismatched management models. Vendors are forced to track and participate in multiple forums and end-users are confused as to which standards are important. Some vendors seem to be more interested in issuing press releases that extol their commitment to open systems by pointing to the large number of standards activities in which they participate than they are in actually implementing technology based on the standards.

SNMP has succeeded as the starting point for many management applications and the industry has been able to quickly develop a set of interoperable SNMP-based technologies. The problems associated with poorly coordinated efforts have begun to infiltrate the SNMP community. Many groups have begun to define private SNMP-based MIBS for their devices or applications. Unfortunately, these extensions to the official standard MIBs create problems for the application vendors since the applications must be aware of the extensions in order to make use of them. The groups defining different MIBs align their efforts with a small number of application vendors (or are themselves the application vendors). Unfortunately, other applications that can process SNMP data are not aware of the extensions and users now must by multiple SNMP-based management applications in order to manage the MIB extensions.

In other areas, standards exist, but do not cover the entire problem space. For example, for network management SNMP and CMIP and their related management object models define how to define and move management data. XMP provides a standard API for accessing this data. Many management platforms support these standards, but unfortunately, each vendor adds their own user interface development tools (e.g., OpenView Windows), event processing, name space management, and their means managing the management data databases and eliminate the portability and interoperability that that standards begin to provide. Also, standards exist for managing software license databases, but no standards exist for an API allowing applications to access a license database or for the format of the database.

Standards by themselves are not sufficient to guarantee interoperability. No matter how precise the definition of a protocol, two different implementors will create implementations that will not interoperate unless they cooperate during the implementation. Different interpretations of written text and inevitable holes in the specification will result in non-interoperable implementations. At a minimum, conformance and interoperability tests are required to ensure that implementations derived from a standard will interoperate. Unfortunately, the cost of developing such tests is high and the organization that creates the standard usually does not have the resources or charter to do such tests and they never get implemented. Some organizations such as COS, rely on donated tests for their interoperability certification. Since vendors have little incentive to incur the cost of developing this tests alone, few tests get donated.

3. VENDOR CONSORTIUM

Some vendors have recognized the problem of relying solely on standards to create open interoperable solutions and have formed consortium and joint development efforts in order to develop shared technologies as the basis of interoperable solutions. A common starting point for vendor products can greatly accelerate the development of interoperable products. For example, the release on TCP/IP source code as part of the BSD 4.2 release of Unix greatly expanded the use and deployment of TCP/IP. Vendors had a way to quickly implement the technology and were able to check interoperability against other implementations based on the same starting point. The original authors of the SNMP specification each implemented a version of the technology, resolved interoperability problems among their separate implementations and then licensed this technology to the world. Shortly after this, two free versions of SNMP became available on the Internet. Within two years, SNMP became the dominant network management protocol and interoperability problems among different vendor's products were virtually nonexistent. (The proliferation of non-standard and private MIBs has created problems within the SNMP community.) CMIP or CMOT for which no reference implementation was ever produced is virtually non-existent in LAN management.

Within the management area, several consortium have failed to deliver technology that can be used as the basis for a set of management products. Unix International ceased to exist long before the UI Atlas management platform became available. The companies working under the COSE process on system management never delivered any of the system management technology they announced two years ago and they have stopped trying. The OSF DME program was greatly scaled back and its system management and network management technology is being used by only a few of the OSF sponsor companies that funded its development. The DMTF was beginning to make progress but Microsoft has reduced their commitment to the consortium. For a consortium effort to be successful, vendors need to learn how to cooperate with their traditional rivals. Self interest can easily derail these efforts.

The cause of these failures seems to be the result of each vendor's concerns over protecting their installed base and maintaining account control. Vendors have invested greatly in developing their own management products. Replacing these products with products based on a common reference technology becomes very difficult unless tools and transition aids are provided that will allow the vendors existing installed based to interoperate with the new technologies. Since each vendors installed based widely diverges from the other vendors, efforts to agree on common technologies quickly become bogged down with endless discussions about whose programming interfaces and management model to preserve.

4. NECESSARY CONDITIONS

More than these numerous ad hoc standards and consortium efforts are needed. All of the following four items are needed to help drive the open systems industry towards complete interoperable management solutions.

- A strong coordinated effort for developing a complete suite of specifications. Individual management specification efforts (e.g., X/Open, OMG, MIC and others) should continue to develop the management models and specifications for their particular niches and areas of interest, but they should work to ensure that their work fits into a large enterprise mode for management. The Network Management Forum's (NMF) OMNIPoint specification is a good starting point for a complete reference model for management.

- Reference implementations based on the emerging specifications that can be used by vendors as the starting point for their products. This reference implementation needs to be easily accessible and should not reflect any specific vendor bias. The OSF pre-structured technology (PST) process is designed to allow vendors to work cooperatively on reference software. Teaching vendors how to work together cooperatively on joint projects is a very difficult task and is a focal point of the PSTG process. The reference implementations will have to provide the flexibility in order to allow for the integration of existing management applications.

- Formal specifications, conformance testing, and certification programs based on the reference implementations. X/Open has started this work for some of their technologies. A valid economic model for the creation of tests needs to be found.

- End user purchasing profiles to force vendors to make the products. End-users need to back their purchasing profiles and purchase specifications with their money. Vendors are continuing to make money selling their existing management products and promoting their proprietary management interfaces and their non-interoperable solutions. The vendors have little financial incentive to change unless the end-users force the issue by not buying the existing products. This may be the most important factor of all. If users specify open distributed management products but continue to purchase vendor proprietary products, the vendors will never build the open management products.

If any of the above four are missing (which is today's situation) the open systems industry will continue to fail to deliver open distributed management solutions.

40

Murky Transparencies: Clarity using Performance Engineering

Joseph Martinka, Richard Friedrich, Tracy Sienknecht

Hewlett-Packard Company - HP Laboratories, Palo Alto, California U.S.A.
{martinka, richf, tracy}@hpl.hp.com

This position paper highlights a daunting challenge facing the deployment of open distributed applications: the performance management component of the transparency functions. Applications operating in an ODP environment require distribution transparencies possessing comprehensive performance management capabilities including monitoring and modeling. The transparency functions are controlled by adaptive management agents that react dynamically to meet client QoS requirements given a current set of server and channel QoS capabilities. This technical challenge must work in a open environment with multiple autonomous administrative domains. For this goal to be realized, the ODP architecture must be enhanced. Distributed performance management of "operational" communications has been neglected in favor of the trendy multi-media "streams" communication in spite of the dominance of the former in current and future applications.
Keyword Codes: C.4, I.6.3, C.2.4
Keywords: Performance of systems, Simulation, Distributed systems

1   ODP'S CHALLENGE TO APPLICATION PERFORMANCE

The ultimate goal for applications in Open Distributed Processing (ODP) is to insulate the application design and programming from the effects of distribution [1]. Such a goal is at once noble and daunting. The design space for even simple distributed applications using current API's such as the Distributed Computing Environment (DCE), Corba, Banyan VINES, or other systems, discourages the average application designer. The ODP goal to insulate the application from its distributional complexities decreases the time, risk and expertise needed to design workable ODP applications. However, the inevitable consequence of making the application transparent to distribution is that the infrastructure must assume the role of providing the resulting transparency mechanisms. As Hamlet reflects: *ay, there's the rub*. The architectural specification for the mechanisms that support the application in meeting its Quality of Service (QoS) and functional goals while providing transparent distribution missing from the Reference Model for ODP (RM-ODP).

1.1 The changing application environment
Meeting this requirement for distributional transparency will be difficult enough even if simply limited to a single distributed systems infrastructure (e.g., DCE) which permits heterogeneous hardware. The future ODP infrastructure is heterogenous in software and hardware. This fact compounds the performance management challenges. Systems will be composed of multiple vendor operating environments running various distribution software (e.g., DCE, CORBA, SNMP, TINA-based DPE) that must interoperate. The communication channels will deliver concurrent asynchronous packets for operational processes and isosynchronous stream traffic with radically different control and QoS criteria.

Since the ODP architecture is the only unifying core to unite this Babel of systems, it is insufficient. Few programs will be written without substantial assistance and reliance on middleware services and libraries. The use of middleware will be composed of many "black box" functions that are harder to characterize and tune since they are not directly accessible by the same tools developers use for their own code. The application itself becomes more obtuse. Having parts of a client /server application written by the same developers is likely to become the exception than the norm. Applications will increase their performance dependence on services not under the control of the application designer. Implications of the functional choices available to the developer are often hidden and may not match the assumptions of the middleware design. In some cases, these middleware services will be based on legacy application 'back-ends' expected to exist in an ODP environment for many years. As the distribution infrastructures mature, some applications will outlive the infrastructure on which they were built. Optimizations appropriate for one environment can be sub-optimal for its replacement. Design trade-offs for future infrastructures are impossible to predict or anticipate. Some means to cope with this change must be provided during the application's lifetime.

Networking channels introduce variable and uncertain latency factors into an application's performance not faced in conventional monolithic applications. Network distance between a client and server is not controllable at design time, and may range from a co-located process in the same node to a node on another continent. The result is wide deviations of average latencies of communication, compounded by variability due to network congestion effects. These latencies are troublesome even as bandwidth improves markedly in the future.

The performance challenge in ODP should be met by applying both existing performance engineering techniques and those requiring breakthrough technologies (e.g., automated performance management control). The breakthroughs need an aggressive research and development effort and architectural support from the RM-ODP if the goals of ODP are to be realized.

1.2 Performance Issues with Transparencies and Domains

Most ODP transparencies have significant performance components, particularly *location, migration* and *replication*. Each needs to provide performance management functions that will maximize the performance behavior of the environment.

In *location* transparency, a server object instance is found for the client object. The selection of this server among several choices should be made based upon performance factors, including which server is best-suited to meet the client's QoS needs. These factors should be based in part on the channel latencies and contention between the client and a proposed server object, current loading of participating nodes, etc. We believe that to satisfy this transparency, automatic agents must manage the location transparency (trading and binding) mechanisms.

In *migration* transparency, an object is moved from one node to another in order to maximize the ability of the server object to meet a larger fraction of the client binding requests with satisfactory QoS. This transparency enhances the system's ability to balance the load, reduce latency, move object servers to follow periodic or unexpected workload patterns, and to unload hardware nodes for administrative or maintenance work.

In *replication* or group transparency, the use of mutually behaviorally compatible objects act together to support an interface and enhance *performability*: the performance and availability of a service. Replication carries a performance cost. As it is utilized, the overhead to maintain consistency between servers increases the resource consumption among replicates. Performance management agents must ensure that the number and geographical placement of replicas achieve the most efficient gain in performability.

An additional requirement for performance management is that automated performance management agents interact in a federation or domain. These agents, for the sake of scalability and complexity, must negotiate and trade information among themselves so that transparencies can function efficiently on behalf of clients. Agent management operation will minimize human interaction to the extent possible to allow greater administrative spans of control, and handle interactive loads several orders of magnitude larger than current day systems. Negotiations with agents outside that federation (e.g., services provided to or for another commercial entity) must be in terms of common QoS agreements, the methods and terms of which are little understood today.

If automation is to be used, only systems that are sufficiently and thoroughly understood can have automatic transparencies. Precise definitions of systems boundaries and adequate abstraction of the world outside these boundaries are areas where RM-ODP must assist.

2 PERFORMANCE MANAGEMENT: CAN IT ASSIST?

The dynamic nature of shared network channels and binding implications on QoS require an automatic element in ODP performance. Decisions to select efficient bindings, migrate objects to improve efficiency and replicate objects for load-balancing require a performance decision-making apparatus which is dynamic and mostly autonomous. Predictable system behavior results only if we architect appropriate performance management functionality, a common performance management control language, and develop objects with expectations to cooperate with the environment.

2.1 Instrumentation and monitoring

Pervasive, heterogeneous, distributed application instrumentation is essential for the performance management of the ODP environment. The management difficulty is increased by using many disparate technologies, distribution protocols and operating environments. We developed consistent performance metric requirements for this instrumentation [2] and developed a specification for distributed monitoring in DCE [3]. These efforts provide insight into the performance management needs of ODP applications such as providing users with a single coherent view of application behavior regardless of application object location. Distributed performance management needs extensions to RM-ODP in two areas: standardized performance metrics and standardized access and control mechanisms.

Pervasive, standard metrics provide the crucial foundation for distributed application performance management. These metrics support evaluating computational, engineering and technology viewpoint behavior in relation to the enterprise and information viewpoint requirements. We define implementation instances of the metrics as sensors. Sensors must be pervasive in object, cluster and capsule software services to realize end-to-end QoS goals. Thus standard metrics allow consistent interpretation of data collected anywhere in the domain.

However, standard metrics are necessary but insufficient. Standard sensor access and control mechanisms are required. These mechanisms should be implemented as a Performance Measurement Interface exported by all objects. An additional mechanism is necessary to support efficient and scalable performance measurement. Sensors must support management by exception semantics through a threshold evaluation technique. Configuring a value and percentile results in sensors that report data only when the threshold conditions are met. There is also a critical need for performance tracing capabilities in ODP systems so as to understand behavior of a workload's transactions. The trace provides logical topology information needed to model the resource use of a transaction as it demands various application services.

A scalable measurement infrastructure must be defined to collect and transport performance data efficiently without noticeable resource utilization. The measurement infrastructure must also support correlating metrics from objects that reside on different computing nodes.

2.2 Performance modeling

The requirements for distributed modeling extends the traditional modeling boundaries concentrating on classic computer node problems. The system boundaries now include the network as a vital component to application throughput and response times. Transitions between abstraction levels are usually painful because current modeling techniques and assumptions for capacity planning are ill-suited for this increased scope. The unit of work, the user transaction, which normalizes the resource demands of the system must be distinguishable and traceable through the layers of middleware as well as opaque services/objects.

The interactions of diverse resource behaviors makes modeling more challenging. We found that our models' outputs depend less on the detailed specification of node's resource use, and more on the understanding of the application's use of communication channels. Model details become entwined with the location and frequency of access to various server objects separated by uncertain network latencies. Consequently this complexity will prompt the model to be delivered with the object. The ODP application modelling will become an operational capability for critical ODP transparencies beyond its traditional role in capacity planning and performance tuning. It is also needed in ODP application design to extend or extrapolate prototype or benchmark measurements to larger target systems.

3   CONCLUSION

Instrumentation and monitoring must be integrated with modelling so that automatic parameterization of models can support the decisions of location, migration and replication transparency agents. Models can be used to assist the synthesis of end-to-end QoS expectations where comprehensive measurement is impossible or burdensome. Proxy instrumentation can occasionally or constantly validate the models as conditions change operationally or as new applications are fitted into the environment. There is substantial interest and research in the issues of multi-media streams QoS. However, the prosaic operational channels also need additional ODP architectural support to ensure that the promise of distributional transparency for performance can be built into supporting agents and managers. Specifications in QoS for operational performance must be enhanced and combined with robust performance models providing "real-time" results. This synergy must be developed to support distribution transparencies. Monitoring, collection, modeling and controlling of the system will need largely be automated to realize performance critical transparencies. These technologies are not yet understood nor pervasive in today's distributed environments. ODP requires them tomorrow.

1   ISO/IEC JTC1/SC21/WG7 N885, *Reference Model for Open Distributed Processing Part 1*, November 1993.
2   Richard Friedrich, *The Requirements for the Performance Instrumentation of the DCE RPC and CDS Services*, OSF DCE RFC 32.0, June 1993.
3   Richard Friedrich, Steve Saunders and Dave Bachmann, *Standardized Performance Instrumentation and Interface Specification for Monitoring DCE Based Applications*, OSF DCE RFC 33.0, November 1994.

41

Quality of Service Workshop

Jan de Meer[a]   jdm@fokus.berlin.gmd.d400.de
Andreas Vogel[b]   andreas@dstc.edu.au

[a] GMD-FOKUS
Hardenbergplatz 2
10623 Berlin, Germany

[b] CRC for Distributed Systems Technology
DSTC, Level 7, Gehrmann Laboratories
University of Queensland, 4072, Australia

The 3rd Workshop on Quality of Service (QoS) was held in February 1995 in Brisbane, Australia, in conjunction with the International Conference on Open Distributed Processing. The first workshop took place in May 1994 at the Université de Montreal, Canada, and the second one during the European RACE Conference on Integrated Services and Networks (IS&N) in September 1994 in Aachen, Germany. The first two workshops were supported by the European RACE project on QoS TOPIC, and the recent workshop in Brisbane qualified as an IFIP workshop as well. Because of the interest in and the feedback to this series of workshops we are looking forward to another ICODP-workshop on QoS, possibly early in 1996.

The Brisbane workshop addressed the problem of introducing QoS concepts and mechanisms on the different levels of abstraction in a system. The invited talk by GMD-FOKUS Berlin opened the workshop by highlighting the paradigm change which occurs, when moving from message-based systems to continuously transmitting ones. Whereas the former can be checked by a "probe-and-observe" technique, the latter cannot be. There are two major differences. One is the continuous flow of information, which requires continuous observations, and the second is the distribution of applied quality control algorithms. In a client server system, you have admission control procedures applied to the server and deficiency control procedures applied at the client. These QoS control elements cooperate to provide the required QoS level. In order to be confident of the distributed QoS control procedure one has to observe the end-to-end behaviour instead of partial behaviours. Message-based systems are separated into protocol entities, which are checked independently because of their point-to-point communication nature.

In a later session, there was a presentation and a discussion of how to classify service qualities for high performance storage systems. An approach to QoS negotiation for multimedia on-demand systems was presented by the Universite de Montreal. HP California looked at how the performance of distributed processing could be improved by the integration of continuously measurement and decision models into the system design and maintainance. Despite success in improving system efficiency, many problems are still waiting for a solution. It was recognised that standardised interfaces and the object-based approach of ODP is crucial for the integration of QoS. Thus middleware could aid the decomposition of sophisticated application objects into more measurable ones. The University of Dresden, Germany, presented a tool-based approach which allows the specification, application and monitoring of user-defined QoS parameters for high-speed transport protocols. There is a need for the integration of QoS aspects into the services of distributed platforms. Reconfiguration, an important concept for QoS customisation, addressed by the Norwegian Computing Centre, which participated in an European RACE project on Service Creation. The approach taken is based on the Formal Description Technique SDL and comprises the demands for QoS management of reconfiguration in accordance with invariances of service qualities. A flexible approach to dynamic routing was jointly presented by the Dutch PTT and University of Enschede. Heuristics support the search for an optimal allocation of communication and computation resources when a reconfiguration occurs.

The presentation of papers was followed by a panel which addressed the role of middleware for QoS issues. The panel consisted of Richard Friedrich (HP California), Jerry Rolia (Carleton University Ottawa), Jan de Meer (GMD-FOKUS Berlin), Jacob Slonim (Centre of Advanced Studies IBM Toronto) and Andreas Vogel (DSTC). It was said that cost and security must play an important role in the evaluation of service qualities. The acceptance of the needs of specialised groups of users is not yet sufficiently studied. Despite the technical qualities, user groups expect sophistication of security and privacy policies at reasonable cost. The more technical memebers of the audience thought that economic factors such as cost lead to a discussion outside the ODP framework. It was recognised that economic aspects are of great importance for the discussion on QoS. However, a relation was tackled to trading based on service qualities. Formal description techniques did not play an important role during discussion.

We would like to thank all panelists and others who contributed to the success to this ICODP workshop. We hope that the issues on QoS addressed in this workshop become more aligned to the user's and supplier's needs.

INDEX OF CONTRIBUTORS

Arsenis, S.   285

Bauer, M.   34
Bearman, M.   133, 185
Beitz, A.   185
Benford, S.   197
Berry, A.   55, 79
Bond, A.   67
Bowman, H.   399, 413
Brookes, W.   67, 425
Burger, C.   208

Colban, E.   105

de Meer, J.   512
de Paula Lima Jr., L.A.   173
Derrick, J.   399, 413
Dilley, J.   465
Dupuy, F.   105

Eckardt, T.   93
Elnozahy, E.N.   488

Franken, L.J.N.   384
Friedrich, R.   347, 508

Geihs, K.   157
Goodchild, A.   145
Gudermann, F.   157

Hafid, A.   335
Haverkort, B.R.H.M.   384
Heineken, M.   309

Horstmann, T.   297
Hungerford, S.   451
Hutschenreuther, T.   359

Indulska, J.   67, 425

Janssen, P.   384

Kinane, B.   117
Kitson, B.   233
Koch, T.   259
Kong, Q.   79
Krämer, B.   259

Lamersdorf, W.   476
Lee, O.-K.   197
Lilleyman, J.   451
Linington, P.   15
Louis, S.   323

Madeira, E.R.M.   173
Magedanz, T.   93
Markwitz, S.   157
Martinka, J.   347, 508
Mazaher, S.   372
Merz, M.   476
Meyer, B.   271, 309
Mittasch, C.   359
Müller-Jones, K.   476

Popien, C.   271, 309
Puder, A.   157

Raeder, G.   372
Ratan, V.   488
Raymond, K.   3, 55
Ricarte, I.L.M.   439

Saunders, S.   347
Schill, A.   359
Segal, M.E.   488
Sharrott, M.   451
Sienknecht, T.   347, 508
Simoni, N.   285
Slonim, J.   34
Steen, M.   399, 413

Teaff, D.   323
Tobar, C.M.   439
Travis, L.   503

v. Bochman, G.   335
v. Liempd, G.   384
Virieux, P.   285
Vogel, A.   185, 512

Warner, M.   221
Wasserschaff, M.   297
Waugh, A.   133
Wildenhain, F   359
Wittmann, A.   93

Yang, Z.   6

Zhou, W.   245

KEYWORD INDEX

Accounting   221
Activities   93
AI general   384
ANSAware   259
Applicative (functional) programming   117

Charging   221
Client/server model   245
Communications
  applications   208, 476
  management   173, 309, 384
Computational objects   93
Computer
  communication networks   3, 359
  systems organization, general   208
Conceptual graphs   157
Concurrent programming   425
Configuration languages   372
Control methods and search   384
CORBA   297, 451
CSCW   297
CSIRO   451

Database applications   439
Design   55, 105, 399
Distribution
  configuration   372
  and maintenance   105
Distributed
  applications   34, 285
  computing   465
  systems   3, 15, 55, 67, 79, 117, 133, 145,
    173, 185, 197, 208, 233, 245, 309, 347,
    359, 413, 425, 439, 476, 508
  systems management   259, 271
  targeting   372
DIT   451

Fault-tolerant computing   245
Federated trading   197

General Model   79

Groupware   297

Information
  infrastructure   34
  search and retrieval   145
  storage and retrieval   323
  systems   79
    application   185
    general   208

Life-long learning   34

Management
  functions   93
  services   93
Modelling   285
Multimedia
  applications   335
  information systems   439

Network
  operations   173, 359
  protocols   465

Object
  orientation   285
  -oriented programming   117, 465
ObjectMap   451
ObjectStore   451
ODP   93
OMG   451
Online information services   221
OODB   451
Open distributed processing   245, 372
Open systems   34
Optimization   384
Orbix   451

Performance of systems   309, 323, 347, 384,
  508
Policy enforcement   259
Problem solving   384

Production rules   259
Programming environments   67, 476

Quality management   335
Quality of service   335, 372, 384

Reliability   384
Remote procedure calls   245
Requirements/specifications   399, 425
Resource discovery   197

Simulation   347, 508
Smalltalk   297
Software

engineering   105
  management   503
Specification   413
System management   233, 503
Systems and software   323

TMN   93
Tools and techniques   105, 413
Trading   157
Type specification   157

User interfaces   145

Viewpoint specifications   221